Lecture Notes in Computer Science 5867

Commenced Publication in 1973
Founding and Former Series Editors:
Gerhard Goos, Juris Hartmanis, and Jan van Leeuwen

Editorial Board

David Hutchison
Lancaster University, UK
Takeo Kanade
Carnegie Mellon University, Pittsburgh, PA, USA
Josef Kittler
University of Surrey, Guildford, UK
Jon M. Kleinberg
Cornell University, Ithaca, NY, USA
Alfred Kobsa
University of California, Irvine, CA, USA
Friedemann Mattern
ETH Zurich, Switzerland
John C. Mitchell
Stanford University, CA, USA
Moni Naor
Weizmann Institute of Science, Rehovot, Israel
Oscar Nierstrasz
University of Bern, Switzerland
C. Pandu Rangan
Indian Institute of Technology, Madras, India
Bernhard Steffen
University of Dortmund, Germany
Madhu Sudan
Microsoft Research, Cambridge, MA, USA
Demetri Terzopoulos
University of California, Los Angeles, CA, USA
Doug Tygar
University of California, Berkeley, CA, USA
Gerhard Weikum
Max-Planck Institute of Computer Science, Saarbruecken, Germany

Michael J. Jacobson Jr. Vincent Rijmen
Reihaneh Safavi-Naini (Eds.)

Selected Areas in Cryptography

16th Annual International Workshop, SAC 2009
Calgary, Alberta, Canada, August 13-14, 2009
Revised Selected Papers

 Springer

Volume Editors

Michael J. Jacobson Jr.
University of Calgary, Department of Computer Science
2500 University Drive NW, Calgary, Alberta, T2N 1N4, Canada
E-mail: jacobs@cpsc.ucalgary.ca

Vincent Rijmen
K.U. Leuven, ESAT/COSIC
Kasteelpark Arenberg 10, 3001 Leuven-Heverlee, Belgium
E-mail: vincent.rijmen@esat.kuleuven.be

Reihaneh Safavi-Naini
University of Calgary, Department of Computer Science
2500 University Drive NW, Calgary, Alberta, T2N 1N4, Canada
E-mail: rei@ucalgary.ca

Library of Congress Control Number: 2009937877

CR Subject Classification (1998): E.3, K.6.5, D.4.6, E.2, K.4.4, I.1

LNCS Sublibrary: SL 4 – Security and Cryptology

ISSN   0302-9743
ISBN-10 3-642-05443-9 Springer Berlin Heidelberg New York
ISBN-13 978-3-642-05443-3 Springer Berlin Heidelberg New York

springer.com

© Springer-Verlag Berlin Heidelberg 2009
Printed in Germany

Typesetting: Camera-ready by author, data conversion by Scientific Publishing Services, Chennai, India
Printed on acid-free paper   SPIN: 12788668   06/3180   5 4 3 2 1 0

Preface

The 16th Workshop on Selected Areas in Cryptography (SAC 2009) was held at the University of Calgary, in Calgary, Alberta, Canada, during August 13-14, 2009. There were 74 participants from 19 countries. Previous workshops in this series were held at Queens University in Kingston (1994, 1996, 1998, 1999, and 2005), Carleton University in Ottawa (1995, 1997, and 2003), University of Waterloo (2000 and 2004), Fields Institute in Toronto (2001), Memorial University of Newfoundland in St. Johns (2002), Concordia University in Montreal (2006), University of Ottawa (2007), and Mount Allison University in Sackville (2008).

The themes for SAC 2009 were:

1. Design and analysis of symmetric key primitives and cryptosystems, including block and stream ciphers, hash functions, and MAC algorithms
2. Efficient implementations of symmetric and public key algorithms
3. Mathematical and algorithmic aspects of applied cryptology
4. Privacy enhancing cryptographic systems

This included the traditional themes (the first three) together with a special theme for 2009 workshop (fourth theme).

We received 86 submissions, of which one was withdrawn. The review was double-blinded. Each paper was reviewed by three members of the Program Committee and submissions that were co-authored by a member of Program Committee received two additional reviews. No member of Program Committee reviewed their own submission. The average quality of submissions was high and this made final selection of the papers a challenging task. We accepted 28 papers with 10 papers in the area of hash functions. The high number of papers in this area could be partially attributed to the interest generated in this area by the NIST competition. The remaining 18 papers were on block and stream ciphers, public key schemes, implementation, and privacy-enhancing cryptographic systems.

In addition, the program included two invited talks:

– Jan Camenisch — Privacy-Enhancing Cryptography: Theory and Practice
– Andreas Enge — Elliptic Complex Multiplication in Cryptography

We would like to thank the Program Committee for their hard work and careful reviews. We also benefited from the expertise of many external reviewers who helped the Program Committee with high-quality reviews. A list of all external referees appears here.

We also would like to thank Coral Burns and Elmar Tischhauser for technical support, and Hadi Ahmadi, Mina Askari, Martin Gagné, Kris Narayan, Arthur Schmidt, Michal Sramka, and Mohammed Tuhin, whose effort ensured smooth running of the workshop.

Finally, we gratefully acknowledge the generous support of the Faculty of Science and Department of Computer Science of the University of Calgary, the University of Calgary University Research Grants Committee, the informatics Circle of Research Excellence (iCORE), the Pacific Institute for the Mathematical Sciences (PIMS), and Microsoft Research for their generous financial support.

September 2009   Michael J. Jacobson, Jr.
                                                              Vincent Rijmen
                                                         Reihaneh Safavi-Naini

16th Annual Workshop on Selected Areas in Cryptography

August 13–14, 2007, Calgary, Alberta, Canada

in cooperation with the
International Association for Cryptologic Research (IACR)

Conference Co-chairs

Michael J. Jacobson, Jr.	University of Calgary, Canada
Vincent Rijmen	Katholieke Universiteit Leuven, Belgium and Graz University of Technology, Austria
Reihaneh Safavi-Naini	University of Calgary, Canada

Program Committee

Masayuki Abe	NTT, Japan
Mikhail J. Atallah	Purdue University, USA
Roberto Avanzi	Ruhr University Bochum, Germany
Feng Bao	Institute for Infocomm Research, Singapore
Paulo Barreto	University of São Paulo, Brazil
Jan Camenisch	IBM Research, Switzerland
Vassil Dimitrov	University of Calgary, Canada
Christophe Doche	Macquarie University, Australia
Orr Dunkelman	Ecole Normale Supérieure, France
Helena Handschuh	Katholieke Universiteit Leuven, Belgium
Thomas Johansson	Lund University, Sweden
Mike Just	University of Edinburgh, UK
Charanjit Jutla	IBM Research, USA
Liam Keliher	Mount Allison University, Canada
Xuejia Lai	Shanghai Jiao Tong University, PR China
Pil Jong Lee	Pohang University of Science and Technology, Korea
Mitsuru Matsui	Mitsubishi Electric Corporation, Japan
Shiho Moriai	Sony Corporation, Japan
Eiji Okamoto	University of Tsukuba, Japan
Josef Pieprzyk	Macquarie University, Australia
Bart Preneel	Katholieke Universiteit Leuven, Belgium
Matt Robshaw	Orange Labs, France
Francesco Sica	
Doug Stinson	University of Waterloo, Canada
Edlyn Teske	University of Waterloo, Canada

Nicolas Thériault   Universidad de Talca, Chile
Adam L. Young   MITRE, USA
Amr Youssef   Concordia University, Canada
Michael Wiener   Cryptographic Clarity, Canada

External Reviewers

Martin Ågren   Michael Naehrig
Toru Akishita   Anderson Clayton Nascimento
Elena Andreeva   Maria Naya-Plasencia
Kazumaro Aoki   Mehrdad Nojoumian
Adem Atalay   Raphael C.-W. Phan
Dan Bernstein   Daniel Rasmussen
Marina Blanton   Thomas Ristenpart
Charles Bouillaguet   Andy Rupp
Suresh Chari   Yu Sasaki
Joo Yeon Cho   Michael Scott
Ming Duan   Yannick Seurin
Sung Wook Eom   Igor Shparlinski
Keith Frikken   Paul Stankovski
Philippe Gaborit   Ron Steinfeld
Willi Geiselmann   Jiayuan Sui
Darrel Hankerson   Xiaorui Sun
Nadia Heninger   Daisuke Suzuki
Florian Hess   Koutarou Suzuki
Howard Heys   Elmar Tischhauser
Seok Hee Hong   Jalaj Upadhyay
Marko Hölbl   Berkant Ustaoglu
Sebastiaan Indesteege   Salil Vadhan
Kimmo Jarvinen   Vesselin Velichkov
Marcos A. Simplício Jr.   Frederik Vercauteren
Anindya Patthak   Huaxiong Wang
Nathan Keller   Yongtao Wang
Sun Young Kim   Ruizhong Wei
Kazukuni Kobara   Jian Weng
Dae Sung Kwon   Hongjun Wu
Tanja Lange   Jiang Wu
Gaëtan Leurent   Zhongming Wu
Ji Li   Liangyu Xu
Wei Li   Kan Yasuda
Julio Lopez   Muhammad Reza Z'Aba
Stefan Lucks   Greg Zaverucha
Yiyuan Luo   Erik Zenner
Alex May   Bo Zhu
Nicky Mouha

Sponsoring Institutions

The Faculty of Science and Department of Computer Science of the University of Calgary
The University of Calgary University Research Grants Committee
The informatics Circle of Research Excellence (iCORE)
The Pacific Institute for the Mathematical Sciences (PIMS)
Microsoft Research

Table of Contents

Block Ciphers

Modes of Operation

Implementation of Public Key Cryptography

Hash Functions and Stream Ciphers

Practical Collisions for SHAMATA-256

Sebastiaan Indesteege[1,2,*], Florian Mendel[3], Bart Preneel[1,2],
and Martin Schläffer[3]

[1] Department of Electrical Engineering ESAT/COSIC, Katholieke Universiteit
Leuven. Kasteelpark Arenberg 10, B–3001 Heverlee, Belgium
[2] Interdisciplinary Institute for BroadBand Technology (IBBT), Belgium
[3] Institute for Applied Information Processing and Communications
Inffeldgasse 16a, A–8010 Graz, Austria

Abstract. In this paper, we present a collision attack on the SHA-3
submission SHAMATA. SHAMATA is a stream cipher-like hash function
design with components of the AES, and it is one of the fastest submitted hash functions. In our attack, we show weaknesses in the message
injection and state update of SHAMATA. It is possible to find certain
message differences that do not get changed by the message expansion
and non-linear part of the state update function. This allows us to find
a differential path with a complexity of about 2^{96} for SHAMATA-256
and about 2^{110} for SHAMATA-512, using a linear low-weight codeword
search. Using an efficient guess-and-determine technique we can significantly improve the complexity of this differential path for SHAMATA-
256. With a complexity of about 2^{40} we are even able to construct
practical collisions for the full hash function SHAMATA-256.

Keywords: SHAMATA, SHA-3 candidate, hash function, collision
attack.

1 Introduction

A cryptographic hash function H maps a message M of arbitrary length to a
fixed-length hash value h. Informally, a cryptographic hash function has to fulfil
the following security requirements:

- *Collision resistance:* it is infeasible to find two messages M and M^*, with
  $M^* \neq M$, such that $H(M) = H(M^*)$.
- *Second preimage resistance:* for a given message M, it is infeasible to find a
  second message $M^* \neq M$ such that $H(M) = H(M^*)$.
- *Preimage resistance:* for a given hash value h, it is infeasible to find a message
  M such that $H(M) = h$.

The resistance of a hash function to collision and (second) preimage attacks
depends in the first place on the length n of the hash value. Regardless of how a
hash function is designed, an adversary will always be able to find preimages or

---

[*] F.W.O. Research Assistant, Fund for Scientific Research — Flanders (Belgium).

M.J. Jacobson Jr., V. Rijmen, and R. Safavi-Naini (Eds.): SAC 2009, LNCS 5867, pp. 1–15, 2009.
© Springer-Verlag Berlin Heidelberg 2009

second preimages after trying out about 2^n different messages. Finding collisions requires a much smaller number of trials. Due to the birthday paradox, collisions can be found in a generic way with an effort of only about $2^{n/2}$. A hash function is said to achieve *ideal security* if these bounds are guaranteed.

In the last few years, the cryptanalysis of hash functions has become an important topic within the cryptographic community. Especially the collision attacks on the MD4 family of hash functions (MD5, SHA-1) have diminished the confidence in the security of these commonly used hash functions. Therefore, NIST has started the SHA-3 competition [7] to find a successor for the SHA-1 and SHA-2 hash functions. The goal is to find a hash function which is fast and still secure within the next few decades.

Many new and interesting hash functions have been proposed. One of them is SHAMATA [1]. Out of the 51 first round candidates, SHAMATA is one of the fastest submissions having a speed of 8–11 cycles/byte on 64-bit and 15–22 cycles/byte on 32-bit platforms [1]. It is a register based design, similar to the hash function PANAMA [5] and also bears resemblance to the sponge construction [2].

In this work, we analyse the security of the hash function SHAMATA. After a description of SHAMATA in Sect. 2, we analyse some basic differential properties of the message injection and state update function in Sect. 3. We show how to efficiently linearise SHAMATA by considering special XOR differences with an equal difference in all bytes. In Sect. 4, we construct a good differential path for the linearised variant of SHAMATA using a low-weight codeword search. Section 5 explains how basic message modification techniques allows us to construct a collision attack with a complexity of 2^{96} for SHAMATA-256 and 2^{110} for SHAMATA-512, based on this differential path. For SHAMATA-256, the attack is improved further to a complexity of only 2^{40} SHAMATA rounds using a complex guess-and-determine strategy. This attack is practical, and we show a collision example in App. A. We conclude our analysis of the hash function SHAMATA in Sect. 6.

2   Description of SHAMATA

In this section, we give a brief description of the hash function SHAMATA. SHAMATA is a register based hash function design that operates on an internal state of 2048 bits and produces a hash value of 224, 256, 384 or 512 bits. The internal state consists of two parts: the main mixing register $B_3, \ldots, B_0$ and the second mixing register $K_{11}, \ldots, K_0$. Internally, SHAMATA uses rounds of the AES block cipher [6] as building blocks.

First, the message is padded to an integer number of 128-bit blocks using classical Merkle-Damgård strengthening, like in the MD4 family. The registers comprising the internal state of SHAMATA are set to their initial values, which depend on the digest length used. Then, each 128-bit message block is used once to update the internal state as described below. Finally, the finalisation phase of SHAMATA generates the output digest from the internal state. For a detailed description of the initialisation and finalisation phases of SHAMATA, we refer to [1], as these details are not relevant to our analysis.

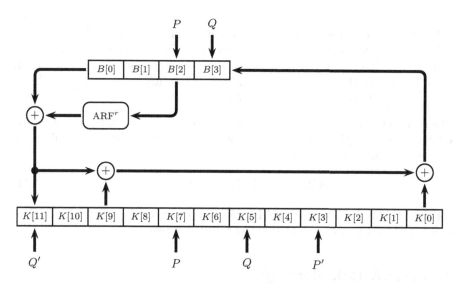

Fig. 1. The state update function of SHAMATA

2.1   The Message Injection

The message injection of SHAMATA updates the internal state using a 128-bit message block. The message block M is first expanded as follows:

$$P = MC\left(M^{\mathrm{T}}\right) \ , \qquad Q = MC\left(M\right) \ ,$$
$$P' = P(1) \,\|\, Q(0) \ , \qquad Q' = Q(1) \,\|\, P(0) \ . \tag{1}$$

Here, MC is the MixColumns operation from the AES block cipher [6] and M^{T} is the transpose of M, where M is viewed as a 4×4 matrix of bytes. The notation $P(i)$ denotes the i-th most significant 64-bit half of the 128-bit word P. Thus, P' and Q' are simply recombinations of the columns of P and Q. These expanded message words and a block counter *blockno* are then added to six words of the internal state using XOR:

$$B_2 \leftarrow B_2 \oplus P \oplus blockno \ , \qquad B_3 \leftarrow B_3 \oplus Q \oplus blockno \ ,$$
$$K_3 \leftarrow K_3 \oplus P' \ , \qquad\qquad K_5 \leftarrow K_5 \oplus Q \ , \tag{2}$$
$$K_7 \leftarrow K_7 \oplus P \ , \qquad\qquad K_{11} \leftarrow K_{11} \oplus Q' \ .$$

2.2   The State Update Function

After the expanded message words have been added, the state update function updates the internal state by clocking the registers of the internal state twice, as is shown in Fig. 1. Formally, these two clockings can be written as

$$
\begin{aligned}
feedK_1 &= ARF^r\,(B_2) \oplus B_0 \ , & feedB_1 &= feedK_1 \oplus K_9 \oplus K_0 \ , \\
feedK_2 &= ARF^r\,(B_3) \oplus B_1 \ , & feedB_2 &= feedK_2 \oplus K_{10} \oplus K_1 \ , \\
B_i\ \ \ \ &\leftarrow B_{i+2}\ \ \text{for}\ \ i=0,1 \ , & K_i\ \ \ \ &\leftarrow K_{i+2}\ \ \text{for}\ \ i=0,\dots,9 \ , \quad (3) \\
B_2\ \ \ \ &\leftarrow feedB_1 \ , & K_{10}\ \ \ \ &\leftarrow feedK_1 \ , \\
B_3\ \ \ \ &\leftarrow feedB_2 \ , & K_{11}\ \ \ \ &\leftarrow feedK_2 \ .
\end{aligned}
$$

The function ARF^r consists of r rounds of the AES block cipher [6], omitting subkey additions. Thus, the ARF function consists of the SubBytes, ShiftRows and MixColumns operations:

$$
ARF(X) = MC\,(SR\,(SB\,(X)))\ . \tag{4}
$$

For SHAMATA-224 and SHAMATA-256, the number of rounds r is equal to one. For SHAMATA-384 and SHAMATA-512, r is two.

3 Basic Attack Strategy

In this section, we describe the basic attack strategy to construct collisions for SHAMATA. The attack is similar to the attack on PANAMA [4,10], since we construct a collision in the internal state during the message injection phase. In this phase, the message input can be used to control the differences in the internal state. However, since the expanded message block is inserted several times into the internal state, finding a differential trail seems to be difficult at first. However, by exploiting some differential properties of the state update, we can find a differential trail for SHAMATA which results in a collision with a good probability.

3.1 Overview of the Attack

The main idea of the attack on SHAMATA is to insert special message differences Δ, which do not get changed by the message expansion and the non-linear function ARF^r. Then, the same difference Δ will be added to six positions of the internal state by the message injection. By imposing conditions on the input of ARF^r, we can ensure that the difference Δ does not get changed by this non-linear function. Hence, all parts of the state update are linear regarding the XOR difference Δ and we can search for a differential path using basic linear algebra.

3.2 Choosing the Message Difference

In the message expansion of SHAMATA, the 128-bit message word M is first arranged in a 4×4 array of bytes. Then, the MixColumns transformation is applied to both M and M^T and some columns are rearranged to get the expanded message blocks P, P', Q and Q'. All transformations are applied on the byte level and we can make the following observation.

Observation 1. A message difference Δ with equal differences in all 16 bytes, results in the same difference Δ in each of the expanded message words P, P', Q and Q'.

Transposition and rearranging columns does not change the value of byte differences. MixColumns applies the following linear transformation over $GF(2^8)$ to each column [6]:

$$
\begin{aligned}
b_0 &= 2 \bullet a_0 \oplus 3 \bullet a_1 \oplus 1 \bullet a_2 \oplus 1 \bullet a_3 \\
b_1 &= 1 \bullet a_0 \oplus 2 \bullet a_1 \oplus 3 \bullet a_2 \oplus 1 \bullet a_3 \\
b_2 &= 1 \bullet a_0 \oplus 1 \bullet a_1 \oplus 2 \bullet a_2 \oplus 3 \bullet a_3 \\
b_3 &= 3 \bullet a_0 \oplus 1 \bullet a_1 \oplus 1 \bullet a_2 \oplus 2 \bullet a_3
\end{aligned}
\tag{5}
$$

If all input values are equal to some value a, we get with $2 \bullet a \oplus 3 \bullet a = 1 \bullet a$:

$$
b_i = 2 \bullet a \oplus 3 \bullet a \oplus 1 \bullet a \oplus 1 \bullet a = 1 \bullet a = a .
\tag{6}
$$

and all output values are equal. Hence, for any message difference Δ with equal values in all bytes, the same difference Δ will be injected into the 6 state words B_3, B_2, K_{11}, K_7, K_5 and K_3.

3.3 Linearising ARF^r

The only non-linear part in SHAMATA is the modified AES-round ARF^r. The function ARF^r behaves linearly if a given input difference Δ results in the same output difference Δ. This is again possible for certain differences, by additionally imposing conditions on the input values of ARF^r:

Observation 2. There are input differences Δ of ARF^r with equal differences in all 16 bytes, which result in the same output difference Δ for certain conditions on the input values of ARF^r.

For example, in the case of ARF^1 (SHAMATA-256), the input difference $\Delta = $ 0xff,0xff,... results in the same output difference $\Delta = $ 0xff,0xff,... if all input byte values are equal to either 0x7e or 0x81. A more careful choice of the difference in the input bytes can improve the probability that the differential through ARF^r is followed.

For ARF^1 a careful examination of the difference distribution table (DDT) of the AES S-box reveals that the best choice is a difference of 0xc5 in each byte. Indeed, this difference passes through the S-box unchanged for input values {0x00, 0x1d, 0xc5, 0xd8} and hence, with an optimal probability of 2^{-6}. Using this difference, there are 4^{16} values for the input to ARF^1 which exhibit the desired differential behaviour, corresponding to a differential probability of 2^{-96}.

In the case of ARF^2 (SHAMATA-512), we can no longer view each S-box independently. Eliminating linear steps at the in- and output, ARF^2 reduces to SubBytes, followed by MixColumns and another SubBytes operation. Thus, each column is still independent here. We have performed an exhaustive search to find the best difference consisting of 16 equal bytes that passes through ARF^2 unchanged. The best choice is a difference of 0x18 in each byte, which passes through ARF^2 for $(22)^4$ values, corresponding to a differential probability of $2^{-110.16}$.

3.4   Basic Message Modification

In this section, we analyse the possibilities to fulfil the conditions on the input of ARF^r. For each active ARF^r function, the input value has to be such that the difference is passed unchanged. The probability of this event was optimised in the previous section. Note however that in each round, the expanded message word P is XORed directly to B_2. Hence, if the ARF^r function in the first clocking is active, we can simply choose M such that the input to ARF^r is X, which is fixed to one of the "good" values ensuring that the active ARF^r has the required differential behaviour:

$$M = (MC^{-1}(P))^T = (MC^{-1}(B_2 \oplus X))^T. \tag{7}$$

If the ARF^r function in the second clocking of a round is active, a similar approach can be used, as the message is also XORed to B_3 via Q, which forms the input to ARF^r in the second clocking:

$$M = MC^{-1}(Q) = MC^{-1}(B_3 \oplus X). \tag{8}$$

These basic message modification techniques do not work anymore as soon as two consecutive ARF^r functions of a single round are active. If we get a difference Δ in both B_2 and B_3 after the message injection, we can adjust only one input of the following two ARF^r functions. The main problem here is that we do not have enough freedom to fulfil the conditions on the message input imposed by both active ARF^r functions. Hence, in this case, one of them has to be satisfied probabilistically. The best probability is 2^{-96} for ARF^1 and $2^{-110.16}$ for ARF^2, as was shown in Sect. 3.3.

Hence, we will aim for a differential path with a low number of consecutive active ARF^r functions (see Sect. 4). Unfortunately, in any differential path, we always get a difference in both, B_2 and B_3 after the first message injection. However, in Sect. 5.2, we show how we can still fulfil both conditions for SHAMATA-256 with much less effort, such that the attack becomes practical.

4   Finding a Good Differential Path

In this section, we first show how to find an efficient collision path for SHAMATA. Recall from Sect. 3.4 that the new message freedom in each round of SHAMATA allows an adversary to linearise the ARF^r function in one of the two clockings in a round. Thus, we aim to find a collision differential path that activates the ARF^r function in at most one clocking of each round as well. However, it was already pointed out in Sect. 3.4 that it is impossible to avoid this in the round where the first difference is introduced, but we can aim to avoid this in all the other rounds. We describe two methods to achieve this. The first method is based on searching low-weight codewords of a linear code and the second method is a simple exhaustive search. The former is more general and can also be used to find differential paths spanning a long message. The latter is only feasible for short messages, but it is simpler. In the case of SHAMATA, either of the methods can be used to achieve the same result.

4.1   Low-Weight Codewords

For a fixed number of message blocks, all differential paths under consideration can be seen as the codewords of a linear code. We show that searching for low-weight codewords in this code is a useful tool to construct good differential paths for SHAMATA. The use of low-weight codeword search techniques to construct differential paths was proposed by Rijmen and Oswald [9] and extended by Pramstaller *et al.* in [8].

A codeword of the code under consideration contains, for each round, the message difference and the differences in the internal state registers immediately after the new message block was added. As we consider only Δ differences, each of these differences is represented by a single bit. Let $\Delta m^{(i)}$, $\Delta b_3^{(i)}, \ldots, \Delta b_0^{(i)}$ and $\Delta k_{11}^{(i)}, \ldots, \Delta k_0^{(i)}$ denote these bits for round i. With N the fixed number of message blocks used, a codeword of the code is then given by

$$\left[\Delta m^{(1)} \cdots \Delta m^{(N)} \parallel \Delta b_3^{(1)} \cdots \Delta k_0^{(1)} \parallel \cdots \parallel \Delta b_3^{(N)} \cdots \Delta k_0^{(N)} \right] . \tag{9}$$

We now construct the generator matrix $\mathbf{G}$ of this code. The differences in a SHAMATA state immediately after the message addition in round i can be represented by an 1×16 binary vector $\Delta s^{(i)}$,

$$\Delta s^{(i)} = \left[\Delta b_3^{(i)} \cdots b_0^{(i)} \; k_{11}^{(i)} \cdots k_0^{(i)} \right] . \tag{10}$$

As the ARF^r function is assumed to behave linearly with respect to the Δ difference, the state difference vector in round i, $\Delta s^{(i)}$, can be written in function of the state differences vector in round $i - 1$, $\Delta s^{(i-1)}$, as follows

$$\Delta s^{(i)} = \Delta s^{(i-1)} \cdot \mathbf{A} \oplus \Delta m^{(i)} \cdot w . \tag{11}$$

Here, w is a 1×16 vector indicating to which positions of the internal state a new message block is added. It is easy to see that

$$w = \left[1\,1\,0\,0\,1\,0\,0\,0\,1\,0\,1\,0\,1\,0\,0\,0 \right] . \tag{12}$$

The 16×16 matrix $\mathbf{A}$ is a transition matrix corresponding to the two clockings in the round. It is given by

$$\mathbf{A} = \begin{bmatrix} 0 & 1 & & 0 & & & & \\ 1 & 1 & & 1 & & & & \\ 0 & & 1 & 0 & & & & \\ 1 & & & 1 & & & & \\ 0 & & & 0 & 1 & & & \\ 0 & & & 0 & & 1 & & \\ 1 & & & 0 & & & 1 & \\ 0 & & & 0 & & & & 1 \\ \vdots & & & \vdots & & & & \ddots \\ 0 & & & 0 & & & & 1 \\ 1 & & & 0 & & & & & 1 \end{bmatrix}^2 . \tag{13}$$

Now, consider the $N \times 17N$ generator matrix $\mathbf{G}_{\mathrm{all}}$ given by

$$\mathbf{G}_{\mathrm{all}} = \left[\begin{array}{c|ccccc} & w & w\mathbf{A} & w\mathbf{A}^2 & \cdots & w\mathbf{A}^{N-1} \\ & & w & w\mathbf{A} & \cdots & w\mathbf{A}^{N-2} \\ \mathbf{I}_{N\times N} & & & w & & \vdots \\ & & & & \ddots & w\mathbf{A} \\ & & & & & w \end{array}\right]. \quad (14)$$

This is the generator matrix of a linear code that contains all length N differential paths of the type we consider. As we are only interested in collision differentials, it is required that the last internal state has no difference. This can be achieved by using Gaussian elimination to force zeroes in the last 16 columns of $\mathbf{G}_{\mathrm{all}}$. This gives the generator matrix $\mathbf{G}$, which generates a linear code containing all differential paths that result in a collision.

Due to the possibility of message modification in either of the clockings in a SHAMATA round, but not both (see Sect. 3.4), a good differential path for SHAMATA activates the ARF^r function in at most one clocking per round. As was already noted, it is impossible to avoid activating ARF^r in both clockings of the round where a difference is first introduced. But we aim to avoid this in the remainder of the differential path.

Intuitively, a codeword with a low weight in Δb_2 and Δb_3, which are the input differences to ARF^r, is more likely to satisfy this property than a random codeword. Thus, we look for low-weight codewords in this code, considering only the weight of these bits, using an algorithm similar to that of Canteaut and Chabaud [3]. For each codeword below a certain threshold weight, we check if it satisfies the condition mentioned above. If it does, a suitable collision differential path has been found. If not, the search is simply continued. Note that this search method can find collision differential paths shorter than N rounds. Indeed, nothing prevents the search from padding a shorter differential path to N rounds by adding rounds without a difference, as we indeed observed. The shortest collision differential path we found is shown in Table 1. It consists of 25 rounds and, except for the first round, only activates ARF^r in at most one of the clockings of a round.

4.2   An Alternative Approach

Note that, for a given length of N rounds, there are only 2^N possible differential paths of the type we consider. Indeed, as each message block can only have a Δ difference or no difference at all, there are only 2^N possible message differences. Given the message difference, exactly one differential path follows. Hence, when N is not too large, a simple brute force search can also be a viable approach.

As the more general approach given above resulted in a differential path of only 25 rounds, a brute force approach is indeed practically feasible. We have exhaustively searched all differential paths of length up to 25 rounds. As expected, this search also found the differential path given in Table 1. Moreover, there is only one differential path of 25 rounds, and no shorter differential paths of this type exist. Hence, the differential path in Table 1 is optimal.

Table 1. The differential path for 25 rounds of SHAMATA with differences after each clocking. For differences at the input of ARF^r (word B_1, grey column), the differential probabilities of each round are given in the last two columns for SHAMATA-256 (ARF^1) and SHAMATA-512 (ARF^2).

round	M	B_3	B_2	B_1	B_0	K_{11}	K_{10}	K_9	K_8	K_7	K_6	K_5	K_4	K_3	K_2	K_1	K_0	ARF^1	ARF^2
1	Δ	Δ	Δ	Δ		Δ	Δ				Δ		Δ		Δ				
		Δ	Δ	Δ	Δ	Δ	Δ	Δ				Δ		Δ		Δ		2^{-192}	$2^{-220.32}$
2	Δ			Δ	Δ	Δ		Δ	Δ		Δ						Δ		
			Δ			Δ	Δ		Δ	Δ		Δ							
3	Δ	Δ		Δ		Δ		Δ	Δ						Δ				
		Δ	Δ		Δ		Δ		Δ		Δ					Δ		2^{-96}	$2^{-110.16}$
4	Δ	Δ				Δ	Δ	Δ		Δ	Δ	Δ	Δ		Δ		Δ		
			Δ					Δ	Δ		Δ	Δ	Δ	Δ		Δ			
5			Δ		Δ		Δ	Δ	Δ		Δ	Δ	Δ	Δ		Δ			
					Δ	Δ		Δ	Δ	Δ		Δ	Δ	Δ	Δ			2^{-96}	$2^{-110.16}$
6		Δ				Δ		Δ	Δ	Δ		Δ	Δ	Δ	Δ				
			Δ			Δ		Δ	Δ	Δ		Δ	Δ	Δ					
7	Δ	Δ	Δ			Δ	Δ		Δ	Δ	Δ		Δ	Δ	Δ				
		Δ	Δ	Δ	Δ	Δ		Δ	Δ		Δ	Δ		Δ	Δ	Δ		2^{-96}	$2^{-110.16}$
8	Δ			Δ			Δ	Δ	Δ	Δ		Δ	Δ	Δ	Δ				
					Δ		Δ	Δ	Δ	Δ		Δ	Δ	Δ	Δ				
9		Δ				Δ			Δ	Δ	Δ	Δ		Δ	Δ				
		Δ	Δ				Δ		Δ	Δ	Δ	Δ		Δ					
10	Δ					Δ		Δ		Δ			Δ		Δ				
						Δ		Δ		Δ		Δ		Δ	Δ				
11							Δ		Δ		Δ		Δ						
								Δ		Δ		Δ		Δ		Δ			
12		Δ								Δ		Δ		Δ					
			Δ						Δ		Δ		Δ						
13	Δ	Δ				Δ			Δ							Δ			
			Δ		Δ	Δ		Δ			Δ							2^{-96}	$2^{-110.16}$
14		Δ			Δ	Δ	Δ		Δ			Δ							
		Δ	Δ			Δ	Δ		Δ	Δ		Δ							
15	Δ	Δ							Δ			Δ							
			Δ					Δ			Δ								
16		Δ		Δ		Δ					Δ			Δ					
			Δ		Δ		Δ				Δ			Δ				2^{-96}	$2^{-110.16}$
17	Δ	Δ	Δ			Δ	Δ	Δ			Δ			Δ		Δ			
		Δ	Δ	Δ		Δ	Δ	Δ	Δ		Δ				Δ			2^{-96}	$2^{-110.16}$
18	Δ	Δ			Δ		Δ	Δ	Δ	Δ				Δ		Δ			
		Δ	Δ			Δ		Δ	Δ	Δ	Δ				Δ				
19	Δ								Δ		Δ			Δ		Δ			
		Δ							Δ		Δ	Δ		Δ					
20			Δ								Δ		Δ	Δ			Δ		
				Δ	Δ	Δ						Δ		Δ	Δ		Δ	2^{-96}	$2^{-110.16}$
21				Δ	Δ	Δ							Δ		Δ		Δ		
		Δ				Δ		Δ							Δ		Δ		
22			Δ				Δ		Δ						Δ				
		Δ		Δ		Δ		Δ	Δ								Δ	2^{-96}	$2^{-110.16}$
23			Δ		Δ	Δ		Δ		Δ									
				Δ			Δ		Δ		Δ							2^{-96}	$2^{-110.16}$
24		Δ			Δ	Δ			Δ		Δ		Δ						
		Δ	Δ			Δ				Δ		Δ		Δ					
25	Δ																		

5 Collision Attack on SHAMATA

In this section, we put together the various pieces that were introduced, and present our collision attack on SHAMATA. We search for a message pair which follows the differential path in Table 1.

5.1 Collisions for SHAMATA-256 and SHAMATA-512

In rounds where none of the ARF^r functions is active, the differential path is always followed, regardless of the message block. Hence, in those rounds, we make an arbitrary choice for the message block. In rounds with exactly one active ARF^r function, the message modification technique presented in Sect. 3.4 is used to deterministically construct a message block that ensures that the differential path is followed. This takes only negligible time, *i.e.*, no more than computing a single round of SHAMATA.

However, in the first round where a difference is introduced, the ARF^r function is active in both clockings. The message modification technique of Sect. 3.4 can only deterministically satisfy the conditions for one of them. As discussed in Sect. 3.4, the probability that the path is still followed is 2^{-96} for ARF^1 (SHAMATA-256) and $2^{-110.16}$ for ARF^2 (SHAMATA-512). A prefix with no difference is used to provide the required message freedom.

Thus, a conforming pair for the first round of the differential path can be found by performing about 2^{96} trials for SHAMATA-256 and about 2^{110} trials for SHAMATA-512. Once such a pair has been found, a colliding message pair can be constructed with negligible additional effort. Thus, the overall complexity of our attack is about 2^{96} SHAMATA rounds for SHAMATA-256, and about 2^{110} SHAMATA rounds for SHAMATA-512. The attack requires only negligible memory and is easily parallelisable. Hence, for both variants of SHAMATA, the attack is significantly faster than a brute force attack. Note that the attack also applies to SHAMATA-224 and SHAMATA-384.

5.2 Practical Collisions for SHAMATA-256

In the case of SHAMATA-256, a more efficient approach exists to control the values which are input to the ARF^r function in both clockings of a round. This approach exploits the fact that in SHAMATA-256 only a single AES round is used, *i.e.*, $r = 1$. Hence, this method can not be applied to SHAMATA-512, where $r = 2$.

Assume we aim to fix the inputs to the ARF^1 function in both clockings of round i to X_1 and X_2, respectively. Let $B^{(i)}$ denote the B-register at the beginning of round i. Then, this requirement can be written as

$$\begin{cases} B_2^{(i)} \oplus P^{(i)} \oplus i = X_1 \\ B_3^{(i)} \oplus Q^{(i)} \oplus i = X_2 \end{cases}. \tag{15}$$

Using the definition of the state update function of SHAMATA in (1)–(3), this can be rewritten in a function of the internal state at the beginning of round $i-1$ and the message blocks M_{i-1} and M_i, yielding the following

$$\begin{cases} M_{i-1} = MC^{-1}\left(D_1 \oplus SB^{-1}\left(C_1 \oplus SR^{-1}\left(M_i\right)\right)\right) \\ M_{i-1}{}^{\mathrm{T}} = MC^{-1}\left(D_2 \oplus SB^{-1}\left(C_2 \oplus SR^{-1}\left(M_i{}^{\mathrm{T}}\right)\right)\right) \end{cases}, \qquad (16)$$

where C_1, C_2, D_1 and D_2 are constants defined by

$$\begin{aligned} C_1 &= SR^{-1}\left(MC^{-1}\left(B_0^{(i-1)} \oplus K_9^{(i-1)} \oplus K_0^{(i-1)} \oplus i \oplus X_1\right)\right), \\ C_2 &= SR^{-1}\left(MC^{-1}\left(B_1^{(i-1)} \oplus K_{10}^{(i-1)} \oplus K_1^{(i-1)} \oplus i \oplus X_2\right)\right), \\ D_1 &= B_2^{(i-1)} \oplus (i-1), \\ D_2 &= B_3^{(i-1)} \oplus (i-1). \end{aligned} \qquad (17)$$

These constants only depend on the internal state of SHAMATA-256 at the beginning of round $i-1$, and are thus known. Now, we search for message blocks M_{i-1} and M_i such that the conditions of (16) are satisfied.

A straightforward approach to find the message blocks M_{i-1} and M_i would be to guess one of them, compute the other using the first equation of (16) and then, check if the second equation of (16) holds as well. This procedure is expected to find a solution after about 2^{128} trials. We propose a guess-and-determine approach which performs significantly better. Our approach is as follows

1. Assume we know the four bytes of M_i indicated in the pattern in Fig. 2 (a). Note that this pattern is symmetric, *i.e.*, it is invariant under matrix transposition. This implies that also the same pattern of bytes of $M_i{}^{\mathrm{T}}$ is known. Note that in (16), M_i and $M_i{}^{\mathrm{T}}$ are input to the inverse ShiftRows operation or SR^{-1}. This operation performs a circular right shift of the rows of the state over 0, 1, 2 or 3 bytes for the first, second, third and fourth row, respectively. Hence, the bytes of M_i indicated in Fig 2 (a) form the first column of $SR^{-1}(M_i)$. Similarly, the first column of $SR^{-1}\left(M_i{}^{\mathrm{T}}\right)$ is known. All other operations in (16) treat the four columns independently, so knowledge of the first columns of $SR^{-1}(M_i)$ and $SR^{-1}\left(M_i{}^{\mathrm{T}}\right)$ suffices to compute the first columns of M_{i-1} and $M_{i-1}{}^{\mathrm{T}}$. The latter is equal to the first row of M_{i-1}, which overlaps with the first column of M_{i-1} in exactly one byte.

   Thus, we investigate all 2^{32} guesses for four bytes of M_i as indicated in Fig. 2 (a). For each guess, we compute the first column and the first row of M_{i-1} using (16). Then, we verify if the overlapping byte matches, and if so, we save the candidate in a list L_1. As this imposes an 8-bit condition, about 2^{24} candidates are expected to remain.

2. The same procedure is repeated with the patterns in Fig. 2 (b), Fig. 2 (c) and Fig. 2 (d). Each pattern is invariant under matrix transposition, and results in one column after applying the SR^{-1} operation. This results in four lists, L_1, L_2, L_3 and L_4 of about 2^{24} elements each.

3. An element of the list L_1 contains candidate values of the first row and column of M_{i-1}. Similarly, an element of the list L_2 contains the second row and column of M_{i-1}. Note that these overlap in two byte positions. Thus, we can merge both lists and store all matching combinations in a new list, L_A. The expected number of entries in the new list L_A is $2^{24} \times 2^{24} \times 2^{-16} = 2^{32}$.

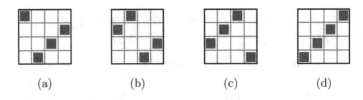

(a)   (b)   (c)   (d)

Fig. 2. Patterns used in the guess-and-determine phase

If the lists L_1 and L_2 are sorted according to the overlapping bytes, this merge operation can be performed very efficiently.

4. The same procedure is used to merge the lists L_3 and L_4, resulting in a new list L_B which is also expected to contain about 2^{32} entries.

5. Finally, the lists L_A and L_B are merged. The entries in these lists overlap in eight byte positions, which corresponds to a 64-bit condition. Again, if both lists are sorted according to these bytes, merging them can be done efficiently. The number of expected matches is $2^{32} \times 2^{32} \times 2^{-64} = 1$.

It is easy to verify that each final match will satisfy (16), and also that every solution to (16) will be found by this procedure. The time complexity of this algorithm is dominated by the merging of lists L_A and L_B, which takes 2^{32} operations. Using hash tables as the data structure to store the lists, an explicit sorting step can be avoided. The memory complexity is determined by one of the lists L_A or L_B, as only one of them really needs to be stored in memory, while the elements of the other can be computed on-the-fly. This corresponds to a memory requirement of about 2^{32} AES states.

For a practical implementation, it is better to reduce the memory requirements of the algorithm, at the expense of an increase in its time complexity. This can be done by, for instance, fixing the byte in the first row and last column of M_{i-1} a priori. Then, the lists L_1 and L_4 are only expected to contain 2^{16} elements each, and the lists L_A and L_B are reduced to about 2^{24} elements. Thus, the total memory complexity is reduced to about 2^{24} AES states, or 256 MB. However, as one byte was fixed a priori, the entire procedure has to be repeated 2^8 times, increasing the time complexity to 2^{40} operations. We have implemented our attack. The guess-and-determine phase was run on a cluster using 256 jobs with a running time of about 5 minutes each. The rest of the attack takes only negligible time using message modification, as explained in Sect. 3.4. A collision example for SHAMATA-256 is given in App. A.

6   Conclusion

In this paper, we have presented a practical collision attack on the SHA-3 submission SHAMATA. Due to weaknesses in the message injection and state update function of SHAMATA it is possible to find certain message differences, that do not get changed by the message expansion or the non-linear part of the state update function. These symmetric XOR differences need to be equal in each byte of the 128-bit words. Using these differences, the non-linear ARF^r function behaves linearly and we can search for a differential path using a linearised variant of SHAMATA. Moreover, since we use the same difference in every 128-bit word, we can represent each word of the internal state by a single bit.

The main weakness in SHAMATA is the relatively light message injection followed by a low number of register clockings. The message injection allows us to efficiently fulfil many conditions using basic message modification. This results in an attack complexity of about 2^{96} for SHAMATA-256 and 2^{110} for SHAMATA-512. Using an efficient guess-and-determine technique we are able to improve the complexity of the attack on SHAMATA-256 to about 2^{40} round computations and present a practical collision for SHAMATA-256. Possible improvements for SHAMATA include increasing the number of times the internal registers are clocked and the use of constants to avoid the use of symmetric differences.

Acknowledgements

This work was supported in part by the IAP Programme P6/26 BCRYPT of the Belgian State (Belgian Science Policy), and in part by the European Commission through the ICT programme under contract ICT-2007-216676 ECRYPT II. The collision example for SHAMATA-256 was obtained utilizing high performance computational resources provided by the University of Leuven, http://ludit.kuleuven.be/hpc.

References

1. Atalay, A., Kara, O., Karakoç, F., Manap, C.: SHAMATA Hash Function Algorithm Specifications. Submission to NIST (2008),
   http://www.uekae.tubitak.gov.tr/uekae_content_files/crypto/
   SHAMATASpecification.pdf,
   http://www.uekae.tubitak.gov.tr/home.do?ot=1&sid=601&pid=547
2. Bertoni, G., Daemen, J., Peeters, M., Assche, G.V.: Sponge functions. In: ECRYPT Hash Workshop, Barcelona, Spain, May 24-25 (2007),
   http://sponge.noekeon.org/SpongeFunctions.pdf
3. Canteaut, A., Chabaud, F.: A New Algorithm for Finding Minimum-Weight Words in a Linear Code: Application to McEliece's Cryptosystem and to Narrow-Sense BCH Codes of Length 511. IEEE Transactions on Information Theory 44(1), 367–378 (1998)
4. Daemen, J., Assche, G.V.: Producing Collisions for Panama, Instantaneously. In: Biryukov, A. (ed.) FSE 2007. LNCS, vol. 4593, pp. 1–18. Springer, Heidelberg (2007)
5. Daemen, J., Clapp, C.S.K.: Fast Hashing and Stream Encryption with PANAMA. In: Vaudenay, S. (ed.) FSE 1998. LNCS, vol. 1372, pp. 60–74. Springer, Heidelberg (1998)
6. Daemen, J., Rijmen, V.: The Design of Rijndael: AES — The Advanced Encryption Standard. Springer, Heidelberg (2002)

7. National Institute of Standards and Technology: Announcing Request for Candidate Algorithm Nominations for a New Cryptographic Hash Algorithm (SHA-3) Family. Federal Register 27(212), 62212–62220 (November 2007), http://csrc.nist.gov/groups/ST/hash/documents/FR_Notice_Nov07.pdf
8. Pramstaller, N., Rechberger, C., Rijmen, V.: Exploiting Coding Theory for Collision Attacks on SHA-1. In: Smart, N.P. (ed.) Cryptography and Coding 2005. LNCS, vol. 3796, pp. 78–95. Springer, Heidelberg (2005)
9. Rijmen, V., Oswald, E.: Update on SHA-1. In: Menezes, A. (ed.) CT-RSA 2005. LNCS, vol. 3376, pp. 58–71. Springer, Heidelberg (2005)
10. Rijmen, V., Van Rompay, B., Preneel, B., Vandewalle, J.: Producing Collisions for PANAMA. In: Matsui, M. (ed.) FSE 2001. LNCS, vol. 2355, pp. 37–51. Springer, Heidelberg (2002)

A   Colliding Message Pair for SHAMATA-256

```
m1 =
00000000: 10 37 fd e7 65 30 1c c0 e3 61 6e 41 24 6f cb b9  |.7..e0...anA$o..|
00000010: 7f 28 81 17 81 4a d1 3f bf 4e ca da 92 f5 35 d0  |.(...J.?.N....5.|
00000020: f0 f0 dc 19 73 d5 a7 07 8c 0b bc 3d b6 85 46 57  |....s......=..FW|
00000030: 02 92 d1 24 00 df 40 67 ca 2c fa 5b 9d 70 2c ce  |...$..@g.,.[.p,.|
00000040: de 38 51 f5 01 3c 3b aa d8 ba 38 0e a1 40 b1 91  |.8Q..<;..8..@..|
00000050: 7b 18 18 24 cc d9 76 c0 f7 4a 61 28 86 06 30 8e  |{..$..v..Ja(..0.|
00000060: 30 8d ab a3 62 52 aa ee 5d 66 2b 13 ec 71 6b ca  |0...bR..]f+..qk.|
00000070: e3 29 f2 2c b3 ed 3d 7e f7 f2 fd 0b 1e c7 d6 e5  |.).,.,.="........|
00000080: aa bc bf ab f9 fb 56 d1 b5 8e df 57 ce 90 e8 fe  |......V....W....|
00000090: 1e 93 a2 80 e6 4c 6f 43 b3 9a 57 9f 0c c2 69 b6  |.....LoC..W...i.|
000000a0: 7e 29 61 77 24 b7 48 d9 45 27 30 13 b8 19 12 d6  |~)aw$.H.E'0.....|
000000b0: ac b4 56 92 00 c5 d6 b3 60 2d 52 6c ef bc 22 6d  |..V.....'-Rl.."m|
000000c0: e5 83 e5 09 3b 2d e2 80 55 13 94 0d 2c a6 e3 d8  |....;-..U...,...|
000000d0: 53 e9 01 66 72 ae 8d cf 68 25 8a b6 ae 64 e7 c1  |S..fr...h%...d..|
000000e0: 5a 39 6b 5a ff 41 0e 5f 6e 60 cb 5d 1c ed ca 01  |Z9kZ.A._n'.]....|
000000f0: 70 af 0a ab dd ed 2c 32 00 c0 3f 2c 66 22 04 c0  |p.....,2.?,f"..|
00000100: 3b 97 65 9d 01 64 98 7b e6 63 d4 d6 4b 77 00 bb  |;.e..d.{.c..Kw..|
00000110: bb ac 35 e3 27 66 55 34 0c 0f db d7 2f 16 19 ae  |..5.'fU4..../...|
00000120: 5b 6f 1a 5a b0 28 b9 1e 89 84 7b a5 71 46 a7 e2  |[o.Z.(....{.qF..|
00000130: f5 b1 8d d2 9e b9 04 9e 79 43 ca df 65 cf 9f c1  |........yC..e...|
00000140: bb f6 43 f9 cd 88 af 13 ea 2f 93 e8 cd 39 8c a0  |..C....../...9..|
00000150: 3e ba 1b ef e2 d5 0d 6b 59 89 11 cb cf b8 ad c4  |>.....kY.......|
00000160: 1a 3f 2f 9d a3 1d 82 3c e0 75 9d 83 b2 ac 3c bf  |.?/....<.u....<.|
00000170: e0 27 0c c5 af b0 be a9 94 1e de 9d 50 69 10 cb  |.'..........Pi..|
00000180: 69 3a 97 08 f4 9b a6 6d df 71 4d 44 40 ec 05 7e  |i:.....m.qMD@..~|
00000190: a6 21 6d 89 f6 7b f4 4f 04 05 1a d3 bd c7 97 27  |.!m..{.O.......'|

SHAMATA-256(m1) =
00000000: 6e a3 b1 a1 29 75 8d 3f f5 60 f8 1b 6b 11 02 9a  |n...)u.?.'..k...|
00000010: 14 b9 b2 d9 b3 2a b6 02 2a f5 83 ab e3 4c 1a 2a  |.....*..*....L.*|


m2 =
00000000: 10 37 fd e7 65 30 1c c0 e3 61 6e 41 24 6f cb b9  |.7..e0...anA$o..|
00000010: 80 d7 7e e8 7e b5 2e c0 40 b1 35 25 6d 0a ca 2f  |..~.~...@.5%m../|
00000020: 0f 0f 23 e6 8c 2a 58 f8 73 f4 43 c2 49 7a b9 a8  |..#..*X.s.C.Iz..|
00000030: fd 6d 2e db ff 20 bf 98 35 d3 05 a4 62 8f d3 31  |.m... .5...b..1|
00000040: 21 c7 ae 0a fe c3 c4 55 27 45 c7 f1 5e bf 4e 6e  |!......U'E..^.Nn|
00000050: 7b 18 18 24 cc d9 76 c0 f7 4a 61 28 86 06 30 8e  |{..$..v..Ja(..0.|
00000060: 30 8d ab a3 62 52 aa ee 5d 66 2b 13 ec 71 6b ca  |0...bR..]f+..qk.|
00000070: 1c d6 0d d3 4c 12 c2 81 08 0d 02 f4 e1 38 29 1a  |....L........8).|
00000080: 55 43 40 54 06 04 a9 2e 4a 71 20 a8 31 6f 17 01  |UC@T....Jq .1o..|
00000090: 1e 93 a2 80 e6 4c 6f 43 b3 9a 57 9f 0c c2 69 b6  |.....LoC..W...i.|
000000a0: 81 d6 9e 88 db 48 b7 26 ba d8 cf ec 47 e6 ed 29  |.....H.&....G..)|
000000b0: ac b4 56 92 00 c5 d6 b3 60 2d 52 6c ef bc 22 6d  |..V.....'-Rl.."m|
000000c0: e5 83 e5 09 3b 2d e2 80 55 13 94 0d 2c a6 e3 d8  |....;-..U...,...|
000000d0: ac 16 fe 99 8d 51 72 30 97 da 75 49 51 9b 18 3e  |.....Qr0..uIQ..>|
000000e0: 5a 39 6b 5a ff 41 0e 5f 6e 60 cb 5d 1c ed ca 01  |Z9kZ.A._n'.]....|
000000f0: 8f 50 f5 54 22 12 d3 cd ff 3f c0 d3 99 dd fb 3f  |.P.T"....?.....?|
00000100: 3b 97 65 9d 01 64 98 7b e6 63 d4 d6 4b 77 00 bb  |;.e..d.{.c..Kw..|
00000110: 44 53 ca 1c d8 99 aa cb f3 f0 24 28 d0 e9 e6 51  |DS.......$(..Q|
00000120: a4 90 e5 a5 4f d7 46 e1 76 7b 84 5a 8e b9 58 1d  |....O.F.v{.Z..X.|
00000130: 0a 4e 72 2d 61 46 fb 61 86 bc 35 20 9a 30 60 3e  |.Nr-aF.a..5 .0'>|
00000140: bb f6 43 f9 cd 88 af 13 ea 2f 93 e8 cd 39 8c a0  |..C....../...9..|
00000150: 3e ba 1b ef e2 d5 0d 6b 59 89 11 cb cf b8 ad c4  |>.....kY.......|
00000160: 1a 3f 2f 9d a3 1d 82 3c e0 75 9d 83 b2 ac 3c bf  |.?/....<.u....<.|
00000170: e0 27 0c c5 af b0 be a9 94 1e de 9d 50 69 10 cb  |.'..........Pi..|
00000180: 69 3a 97 08 f4 9b a6 6d df 71 4d 44 40 ec 05 7e  |i:.....m.qMD@..~|
00000190: 59 de 92 76 09 84 0b b0 fb fa e5 2c 42 38 68 d8  |Y..v.......,B8h.|

SHAMATA-256(m2) =
00000000: 6e a3 b1 a1 29 75 8d 3f f5 60 f8 1b 6b 11 02 9a  |n...)u.?.'..k...|
00000010: 14 b9 b2 d9 b3 2a b6 02 2a f5 83 ab e3 4c 1a 2a  |.....*..*....L.*|
```

Improved Cryptanalysis of the Reduced Grøstl Compression Function, ECHO Permutation and AES Block Cipher

Florian Mendel[1], Thomas Peyrin[2], Christian Rechberger[1],
and Martin Schläffer[1]

[1] IAIK, Graz University of Technology, Austria
[2] Ingenico, France
thomas.peyrin@gmail.com, martin.schlaeffer@iaik.tugraz.at

Abstract. In this paper, we propose two new ways to mount attacks on the SHA-3 candidates Grøstl, and ECHO, and apply these attacks also to the AES. Our results improve upon and extend the rebound attack. Using the new techniques, we are able to extend the number of rounds in which available degrees of freedom can be used. As a result, we present the first attack on 7 rounds for the Grøstl-256 output transformation[1] and improve the semi-free-start collision attack on 6 rounds. Further, we present an improved known-key distinguisher for 7 rounds of the AES block cipher and the internal permutation used in ECHO.

Keywords: hash function, block cipher, cryptanalysis, semi-free-start collision, known-key distinguisher.

1 Introduction

Recently, a new wave of hash function proposals appeared, following a call for submissions to the SHA-3 contest organized by NIST [26]. In order to analyze these proposals, the toolbox which is at the cryptanalysts' disposal needs to be extended. Meet-in-the-middle and differential attacks are commonly used. A recent extension of differential cryptanalysis to hash functions is the rebound attack [22] originally applied to reduced (7.5 rounds) Whirlpool (standardized since 2000 by ISO/IEC 10118-3:2004) and a reduced version (6 rounds) of the SHA-3 candidate Grøstl-256 [14], which both have 10 rounds in total.

Many hash functions [1, 2, 6, 12, 14, 16, 17] use concepts or parts of the block cipher AES [25] as basic primitives, and research on AES-related hash functions is ongoing [15, 22, 27]. In this direction, a new attack model has been recently introduced for block ciphers [18]. In this model, the secret key is known to the adversary and the goal is to distinguish the block cipher from a random permutation. In particular, reduced versions of the AES have been studied in this setting [18,24] and recently, an attack on full AES-256 has been published [5].

---

[1] Note that the 7-round semi-free-start collision attack on Grøstl-256 in the preproceedings version of this paper does not have enough freedom to succeed, see Sect. 6.1.

M.J. Jacobson Jr., V. Rijmen, and R. Safavi-Naini (Eds.): SAC 2009, LNCS 5867, pp. 16–35, 2009.
© Springer-Verlag Berlin Heidelberg 2009

Table 1. Summary of results for Grøstl, ECHO and AES

target	rounds	computational complexity	memory requirements	type	section
	6	2^{112}	2^{64}	semi-free-start collision	see [22]
Grøstl-256	6	2^{64}	2^{64}	semi-free-start collision	Sect. 6.1
	7	2^{55}	-	permutation distinguisher	Sect. 6.1
	7	2^{56}	-	output transf. distinguisher	Sect. 6.1
ECHO	7	2^{896}	-	permutation distinguisher	see [2]
	7	2^{384}	2^{64}	permutation distinguisher	Sect. 6.3
AES	7	2^{56}	-	known-key distinguisher	see [18]
	7	2^{24}	2^{16}	known-key distinguisher	Sect. 6.2

In the rebound attack [22], two rounds of the state update transformations are bypassed by a match-in-the-middle technique using the available degrees of freedom in the state. The characteristic used in the attack is then constructed by moving the "most expensive" parts into these two rounds. The "cheaper" parts are then covered in an inside-out manner, called outbound phase. Other work in parallel to this explores the application of the rebound idea to other SHA-3 candidates [21, 28]. Recently, the rebound attack has been extended to attack the full compression function of Whirlpool [19] and LANE [20] by using the additional freedom of the key schedule (Whirlpool) or other parts of the state (LANE).

In this work, we present improved techniques to use the freedom available in only a single state. The effect of both techniques we present are an extension of the number of rounds in which degrees of freedom can be used to improve the work from the two to four rounds. As a result, we present the best known attacks on the reduced Grøstl-256 permutation and output transformation (up to 7 out of 10 rounds), and also significantly improve upon the first known-key distinguisher [18] for 7-round AES and 7 rounds of the internal permutation used in ECHO. A summary of our results is given in Table 1.

2   Description of AES-Based Primitives

In this paper, we show improved attack strategies for AES based cryptographic primitives. We apply the ideas to the Grøstl and ECHO hash function, and to the block cipher AES. In the following, we describe these functions in more detail.

2.1   AES

The block cipher Rijndael was designed by Daemen and Rijmen and standardized by NIST in 2000 as the Advanced Encryption Standard (AES) [25]. The AES follows the wide-trail design strategy [7, 8] and consists of a key schedule and state update transformation. Since we do not use the key schedule in our attack, we just describe the state update here.

In the AES, a 4×4 state of 16 bytes is updated using the following 4 round transformations, with 10 rounds for AES-128 and 14 rounds for AES-256:

- The non-linear layer SubBytes (SB) applies a S-Box to each byte of the state independently
- The cyclical permutation ShiftRows (SR) rotates the bytes of row j left by j positions
- The linear diffusion layer MixColumns (MC) multiplies each column of the state by a constant MDS matrix
- AddRoundKey (AK) adds the 128-bit round key K_i to the state

Note that a round key is added prior to the first round and the MixColumns transformation is omitted in the last round of AES. For a detailed description of the AES we refer to [25].

2.2 The Grøstl Hash Function

Grøstl was proposed by Gauravaram *et al.* as a candidate for the SHA-3 competition [14]. It is an iterated hash function with a compression function built from two distinct permutations P and Q. Grøstl is a wide-pipe design with proofs for the collision and preimage resistance of the compression function [13]. A t-block message M (after padding) is hashed using the compression function $f(H_{i-1}, M_i)$ and output transformation $g(H_t)$ as follows:

$$H_0 = IV$$
$$H_i = f(H_{i-1}, M_i) = H_{i-1} \oplus P(H_{i-1} \oplus M_i) \oplus Q(M_i) \quad \text{for } 1 \leq i \leq t$$
$$h = g(H_t) = trunc(H_t \oplus P(H_t)),$$

The two permutations P and Q are constructed using the wide trail design strategy. The design of the two permutations is very similar to AES, instantiated with a fixed key input. Both permutations of Grøstl-256 update an 8×8 state of 64 bytes in 10 rounds each. The round transformations of Grøstl-256 are briefly described here:

- AddRoundConstant (AC) adds different one-byte round constants to the 8×8 states of P and Q
- the non-linear layer SubBytes (SB) applies the AES S-Box to each byte of the state independently
- the cyclical permutation ShiftBytes (ShB) rotates the bytes of row j left by j positions
- the linear diffusion layer MixBytes (MB) multiplies each column of the state by a constant MDS matrix

2.3 The ECHO Hash Function

The ECHO hash function is a SHA-3 proposal submitted by Benadjila *et al.* [2]. It is also a wide-pipe iterated hash function and uses the HAIFA [3] domain extension algorithm. Its compression function uses an internal 2048-bit permutation that can be seen as a big AES manipulating 128-bit words instead of bytes. More precisely, an appropriately padded t-block message M and a salt s are hashed using the compression function $f(H_{i-1}, M_i, c_i, s)$ where c_i is a bit counter:

$$H_0 = IV$$
$$H_i = f(H_{i-1}, M_i, c_i, s) \quad \text{for } 1 \le i \le t$$
$$h = trunc(H_t),$$

The compression function of ECHO is built upon a 2048-bit permutation F, composed of 8 rounds (resp. 10 rounds) in the case of a 256-bit output (resp. 512-bit output). Its internal state can be modeled as a 4×4 matrix of 128-bit words. The concatenation of the input chaining variable and the incoming message block are the plaintext input of the permutation F which is tweaked by the input counter c_i and the salt s. A feed-forward of the plaintext is then applied to the internal state V:

$$V = F_{c_i, s}(H_{i-1} || M_i) \oplus (H_{i-1} || M_i)$$

and the 512-bit output chaining variable H_i is the xor of all the 128-bit word columns of V for a 256-bit hash digest size. In the case of a 512-bit hash value, the 1024-bit output chaining variable H_i is the xor of the two left and the two right 128-bit word columns of V.

A permutation round is very similar to an AES round, except that 128-bit words are manipulated. One round is the composition of three sub-functions $BigMC \circ BigShR \circ BigSW$:

- The non-linear layer BigSubWords (BigSW) applies two AES rounds to each of the 16 128-bit words of the internal state. The round keys, always different, are composed of the salt value and a counter value initialized by c_i.
- The cyclical permutation BigShiftRows (BigShR) rotates the location in the matrix of the 128-word of row j left by j positions
- The linear diffusion layer BigMixColumns (BigMC) multiplies each column of the state by a constant MDS matrix

3   Basic Attack Properties

Before describing attacks for Grøstl, ECHO and AES in detail, we give an overview of the round transformation properties used by the attacks. Since we mostly use Grøstl to describe the attacks and the properties of MixColumns and MixBytes are rather similar, we use MixBytes to describe their common properties in the following. We will use differential properties of the SubBytes and MixBytes transformation and exploit the diffusion property of both, ShiftBytes (ShiftRows) and MixBytes. Together, this leads to an efficient guess-and-determine strategy for both, differences and values at the input and output of SubBytes and MixBytes.

Since we exploit the differential properties of SubBytes and MixBytes, we define the notation and state variables according to these two transformations. We denote the SubBytes layer of round i by SB_i, its input state by SB_i^{in} and the output state by SB_i^{out}. An equivalent notation is used for the MixBytes transformation. The MixBytes transformation of round i is denoted by MB_i, its input state by MB_i^{in} and the output state by MB_i^{out}. We will use MC_i for the MixColumns transformation of ECHO and AES. Further, counting from 0, we denote the byte in row r and column c by $[r, c]$, i.e. $SB_i^{in}[r, c]$ for the input of the S-box in round i.

3.1   Improving on the Rebound Attack

The main idea of the rebound attack [22] is to start close to the middle of a (truncated) differential path, connect using the available degrees of freedom in the middle and finally propagate outwards. Our attack works rather similar for Grøstl-256, ECHO and AES, and in the following we use Grøstl-256 to describe the attacks and discuss then, the application to ECHO and AES. Similar to the rebound attack, we start with a truncated differential path with a full active state in the middle of the trail. Fig. 1 shows the truncated differential path used in both permutations P and Q of our improved attack on Grøstl-256. In the rebound attack, the middle part of the differential trail is solved first for both differences and values by exploiting the available degrees of freedom (inbound phase). Then, differences and values are propagated outwards probabilistically (outbound phase) to find semi-free-start collisions, free-start collisions, or non-random properties of the permutations or compression function.

In this work, we improve on the middle part of the attack where we exploit the available degrees of freedom of the state values and differences. The idea is to first find differences and values for the middle (4-round) part of the differential trail, with the following number of active bytes at SubBytes:

$$1 \xrightarrow{r_1} 8 \xrightarrow{r_2} 64 \xrightarrow{r_3} 8 \xrightarrow{r_4} 1$$

3.2   Exploiting Properties of the Round Transformations

In this section, we briefly describe which properties of the round transformations are used for the attacks in the following sections. Note that some used properties, especially those of MixBytes, are specific to a truncated differential path with a minimum number of active S-boxes such as the one given in Fig. 1.

SubBytes. In our attacks, we use some differential properties of the AES S-box. Most of these properties can simply be verified by computing the differential distribution tables (DDT) [4] of the S-box (or inverse S-box).

- For a given input (or output) difference of the AES S-box, the number of possible output (or input) differences is restricted to 127.

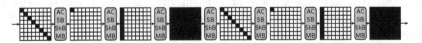

Fig. 1. The position of active bytes of the 7 round differential path for both permutations P and Q

- For a given input and output difference, the number of possible input values is limited to either 2 or 4 values.
- For a given input and output difference, the AES S-box behaves linear due to the fact that there are only 2 or 4 solutions per S-box (see Sect. 4.2 for more details).

In the following sections, we use some differential S-box tables to efficiently carry out the attacks. We call S_F^δ the table that contains all input byte pairs (a, b), such that we get the difference δ at the output of the AES S-box, i.e. such that $Sbox(a) \oplus Sbox(b) = \delta$. Each table S_F has 256 entries with 127 possible input differences $a \oplus b$ of the S-box. More precisely, for any difference $\delta \neq 0$ on the output of the S-box, 129 input differences are not possible, 126 differences have two candidates and 1 difference has 4 candidates. The table S_B^δ contains all the output byte pairs (a, b), such that we get the difference δ at the input of the S-box, i.e. after the application of the inverse AES S-box. For a fast implementation of the attacks, these tables are precomputed and sorted accordingly.

ShiftBytes. The ShiftBytes (or ShiftRows) transformation moves bytes and thus, differences to different positions but does not change their value. The good diffusion property of ShiftRows allows us to choose certain differences and values of the subsequent MixBytes layer independently. Assume we have one active column in MixBytes. Then, we get after the adjacent ShiftRows application one active byte in each new column. Hence, we can determine these single active bytes by the subsequent MixBytes transformation independently.

MixBytes. In the case of MixBytes (or MixColumns), we use the property of an $n \times n$ MDS matrix that, given any n bytes of input and output, the other n bytes can be uniquely determined. Since MixBytes is linear, this also holds for differences. In the following attacks, we use differential paths with a minimum number of active S-boxes. Hence, also the number of differences in the MixBytes transformation is minimal and every active MixBytes operation contains zero differences in exactly 7 (3 for MixColumns) input/output bytes. It follows, that choosing a single byte difference uniquely determines all other 8 (4 for MixBytes) differences.

Hence, for a fixed position of active bytes, we get 255 possibilities for the difference propagation of MixBytes (bundles in [9]). These cases can be precomputed and stored in tables. We call M_F^i the table that contains all possible input differences of MixBytes (or MixColumns), such that we get only one non-zero byte at row i in the output. We call M_B^i the same table but for the inverse of the MixBytes (or MixColumns) transformation. Since the MixBytes transformation is linear, the same tables can be used for values and differences.

3.3   Known-Key Distinguishers

In the following, we will describe known-key distinguisher attacks against AES and the internal permutations used in Grøstl and ECHO. We refer to [18] for the details of this setting. However, in this paper, our distinguishers will consist in finding a pair of plaintext for the keyed permutation (when the key is randomly chosen but known by the attacker) such that some plaintext and ciphertext words contain no difference. For the distinguishing attack to be valid, the complexity should be lower than expected in the case of a random permutation. Assume an n-bit permutation with differences in i bytes of the plaintext and in i bytes of the ciphertext. Then, assuming that the positions of the byte-differences are fixed, the complexity of the generic attack is greater or equal (depending on the values of i and n) to $2^{(n-8\cdot i)/2}$.

4   A Linearized Match-in-the-Middle Attack

In this section, we present a method which allows us to find a state pair with differences according to the truncated differential path of Fig. 1 with a complexity of about 2^{48}. The main idea is to first search for differences according to the 4-round middle part $(1 \rightarrow 8 \rightarrow 64 \rightarrow 8 \rightarrow 1)$ of the path. We can find such differences with a complexity of about 1 by guess and determine (see Sect. 4.1). In the second phase, we try to solve for the corresponding values of the state. The main idea is that we can do this *linearly*. Since the differential of each S-box is fixed we get either 2 or 4 possible values for the AES S-box (see Sect. 3.2). In these cases, the S-box behaves linearly and hence, we can find the correct values by solving a linear system of equations (see Sect. 4.2). Note that we need to repeat the solving phase with new differences if no solution was found.

4.1   Filtering for Differential Paths

In this section, we filter for candidate differences which follow the middle part $(1 \rightarrow 8 \rightarrow 64 \rightarrow 8 \rightarrow 1)$ of the differential path of Fig. 1 with a high probability. Fig. 2 shows the corresponding round transformations and the differential path in detail. In the attack, we use properties of the SubBytes (SB) and MixBytes (MB) transformations to filter for differential paths. Hence, we are interested in the input and output of these transformations. The first and second column show differences at the input and output of the S-boxes (SB_i^{in} and SB_i^{out}), and column three and four show differences at the input and output of the MixBytes transformations (MB_i^{in} and MB_i^{out}). To determine possible input and output differences of these two transformations, we use their corresponding lookup tables M_F^j, M_B^j, S_F^δ and S_B^δ (see Sect. 3.2).

Column 1. We start with the differences of the first column (marked by "1" in state MB_2^{in} and MB_2^{out}) of the MixBytes operation of round 2 (MB_2). Since 7 input byte differences are required to be zero, choosing one of the remaining 9 non-zero differences, uniquely determines all other differences of MB_2. Since the ShiftBytes and AddRoundConstant operations are linear, we get the same differences for the bytes marked by "1" in states SB_3^{out} and SB_3^{in}. It follows that we can choose from 255 non-zero differences for the first byte of SB_3^{in}, and this choice determines all differences marked by "1" between state SB_2^{out} and SB_3^{in}.

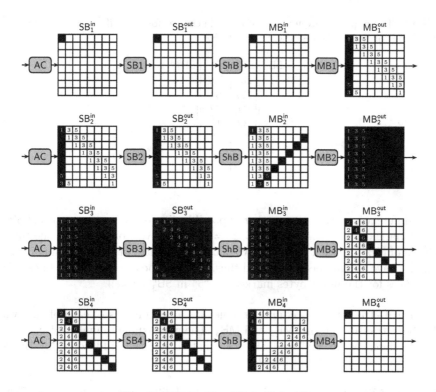

Fig. 2. Filtering for differential paths

Column 2. Next, we continue with the differences of the first column of MB_3 (marked by "2" in states MB_3^{in} and MB_3^{out}). Again, 7 differences of MB_3 are zero and choosing one byte determines all differences of the first column of MB_3. Note that the input of the first column of SB_3 and thus, the difference of $SB_3^{in}[0,0]$, has already been fixed in the previous step. Due to the differential behavior of the AES S-box (see Sect. 3.2), we can choose from only 127 differences for the corresponding output byte of SB_3 ($SB_3^{out}[0,0]$). Choosing one of these possible 127 differences uniquely determines all differences marked by "2" between states SB_3^{out} and SB_4^{in}.

Column 3. Then, we continue with the second column of MB_2 (marked by "3" in states MB_2^{in} and MB_3^{out}). Again, 7 bytes of the input differences are required to be zero. Additionally, one output difference of SB_3 ($SB_3^{out}[1,1]$) has already been fixed due to **Column 2**. Again, we can only choose from 127 possible input differences for SB_3 ($SB_3^{in}[1,1]$) and get 127 possible differences for the bytes marked by "3" between SB_3^{in} and SB_2^{out}.

Column 4-5. We proceed with the second column of MB_3, marked by "4" in states MB_3^{in} and MB_3^{out}. Note that the input bytes of two S-boxes ($SB_3^{in}[0,1]$ and $SB_3^{in}[7,0]$) have already been fixed due to **Column 1** and **Column 3**. These two

Table 2. The approximate number of possible choices for the differences at the input and output of the 3 S-boxes SB_2, SB_3 and SB_4

SB_2^{in}	SB_3^{in}	SB_3^{out}	SB_4^{out}
1	255	127	1
	127	64	
	64	32	
	32	16	
	16	8	
	8	4	
	4	2	
	2	1	

input differences restrict the number of possible differences for the output of SB_3 (bytes marked by "4") to about $256/2^2 = 64$ values. We continue with the third column of MB_2 (marked by "5"). Two output differences of the corresponding S-box SB_3 have already been fixed and thus, we can choose from about 64 possible differences for the input bytes marked by "5" in SB_3^{in} as well.

Column 6-16. This procedure continues for all 8 columns of each of the two MixBytes transformations MB_2 and MB_3. The approximate number of possible S-box differences for SB_3^{in} and SB_3^{out} are halved for each additional MixBytes column and are shown in Table 2.

MB_1 and MB_4. Until now, we have determined differences for the states SB_2^{out}, SB_3^{in}, SB_3^{out} and SB_4^{in}. Since all differences in SB_2^{out} and SB_4^{in} have already been determined, we have only about $255/2^8 \sim 1$ difference left for SB_2^{in} and SB_4^{out}. Note that choosing the difference for one byte determines all other differences as well due to the restrictions by MixBytes.

Note that we can find one possible differential characteristic with a complexity of about one, since we filter though each MixBytes and S-box transformation only once. The total number of possible differential paths can be determined by considering the number of choices we have at the input and output of S-box SB_3, the input of S-box SB_2 and the output SB_4. The approximate number of choices are listed in Table 2 and by multiplying these numbers we can get up to $\sim 2^{64}$ possible differential paths or starting points for the next phase.

4.2   Solving for Conforming State Pairs

After we have found a differential path we need to search for a valid pair of the state. Since the differential of each active S-box is fixed there are only either 2 or 4 input pairs possible. In both cases, the S-box behaves linearly [10] and hence, we can easily solve the resulting linear system of equations. In the following description we assume that we have only 2 possible input pairs for each active S-box (note that in this case, all S-boxes behave linearly).

Consider the diagonal of SB_3^{out} respectively the first column of MB_3^{in} (denoted by "2" in Fig. 2). For each S-box we have 2 possible inputs k_i and k_i' for $0 \le i < 8$

such that the differential path holds. In other words, we have 2^8 possible inputs for the diagonal of SB_3^{out}. Let $x \in \{0, 1\}^8$ then the possible values for the diagonal of SB_3^{out} are given by:

$$k \oplus x \cdot (k \oplus k')$$

where $k = [k_0, \ldots, k_7]$ and $k' = [k'_0, \ldots, k'_7]$.

Next, we compute the first byte of SB_4^{in} by going forward ShiftBytes, MixBytes and AddRoundConstant.

$$SB_4^{in}[0, 0] = (k \oplus x \cdot (k \oplus k')) \cdot L$$

where L denotes composition of ShiftBytes, MixBytes and AddRoundConstant. Since these transformations are all linear L is a linear transformation as well.

Since we have 2 possible values a and a' for $SB_4^{in}[0, 0]$ such that the differential trail holds, the following equation has to be fulfilled.

$$(k \oplus x \cdot (k \oplus k')) \cdot L = a \oplus y \cdot (a \oplus a')$$

where $y \in \{0, 1\}$.

By doing the same for the other diagonals (corresponding to columns 2-8 of MB_3^{in}) we get a system of 64 equations in 64+8=72 variables which has to be fulfilled to guarantee that the differential trail holds in the forward direction. In a similar way we also get a system of 64 linear equations in 72 variables by going backward from SB_3^{in} to SB_2^{out}. However, since the values of SB_3^{in} and SB_3^{out} are related, we get in total a system of 128 equations in 80 variables when we combine them. In other words, to find a valid pair, we have to backtrack and try about 2^{48} differential paths and thus, solve the linear system of equations 2^{48} times. Since we can start with up to 2^{64} differential paths, we can only find about $2^{64-48} = 2^{16}$ pairs after the linear solving step.

Note that the attack works similar if one has 4 possible input pairs for the S-box. By choosing the differences in the previous step (Sect. 4.1) in a way, to maximize the number of differentials with 4 possible pairs for the S-box, the overall complexity can be reduced slightly (by about 2^2 to 2^5). The total complexity of the attack is given by the number of times we need to solve the resulting linear system of equations (we assume here that this corresponds to about one compression function call).

4.3 Application to AES

The same technique applies to the block cipher AES as well. In this case, we start with a differential path with the following sequence of active S-boxes:

$$1 \rightarrow 4 \rightarrow 16 \rightarrow 4 \rightarrow 1$$

Hence, we get 64 conditions (equations) for the S-box layers of round 2, 3 and 4. Since we have 64 equations in 24 variables, we need to repeat the attack $2^{64-24} = 2^{40}$ times to find a valid pair. Note that in the case of AES, we get a better complexity if we first fix the differential path for rounds 1-3 $(1 \rightarrow 4 \rightarrow$

16 → 4) and then, solve for the pair. In this case, we get only 32 conditions and the complexity to solve for a pair is about 2^{12}. Since we need to repeat the attack only 2^{24} times to fulfill the last MixColumns operation we get a total complexity of only 2^{36} in this case.

5 A Start-from-the-Middle Technique

In this section, we describe another attack that uses the available freedom degrees in the middle. The truncated differential path considered here will be the same than in the previous section or in the rebound attack [22]: in the case of Grøstl, we use the one from Fig. 1. More precisely, the attack will first focus on a 3-round part of the middle of the path, the following sequence of active bytes:

$$1 \xrightarrow{r_1} 8 \xrightarrow{r_2} 64 \xrightarrow{r_3} 8$$

We can find a conforming state pair according to this path with only a few operations by choosing "good" differences in advance and exploiting the available degrees of freedom. We start at the last MixBytes transformation of the 3-round trail (MB_3 in Fig. 2) and compute backwards. The attack can be divided into three main phases:

1. In Phase 1, we start with 1-byte differences at the output of each MixBytes column MB_3 (MB_3^{out}) and compute backwards to the input of SB_3 (SB_3^{in}). Each column of MixBytes MB_3 can be computed independently. Then, we maintain as much freedom as possible in the input difference of SB_3 (SB_3^{in}) by using the precomputed differential tables of the S-box.
2. In Phase 2, we have enough degrees of freedom to choose the differences for SB_3^{in} such that each of the eight MB_2 MixBytes transitions from 8 to 1 active byte in backward direction is satisfied.
3. In Phase 3, we get more degrees of freedom since both (a, b) and (b, a) are valid solutions for each byte of SB_2^{in}. Hence, we can randomize each active byte of SB_2^{in} and get enough pairs such that the last single MixBytes transformation MB_1 can be fulfilled as well.

At this point, all available degrees of freedom have been used and we rely on a probabilistic behavior for the remaining transitions in backward and forward direction.

5.1 Application to Grøstl-256

Phase 1. We randomly select non-zero differences for the eight active bytes of SB_4^{in}, i.e. for $SB_4^{in}[i, i]$ with i ranging from 0 to 7. Those differences will remain unchanged when computing backward to MB_3^{out}. Since the MixBytes transformation is linear, we apply its inverse (Phase 1.A in Fig. 3) to MB_3^{out} and deterministically get 64 byte differences for MB_3^{in} and thus, for SB_3^{out}. We denote by $\delta[i, j]$ the byte difference of $SB_3^{out}[i, j]$. For each output difference $\delta[i, j]$ in SB_3, we compute all valid byte pairs $SB_3^{in}[i, j]$ such that the S-box differential holds

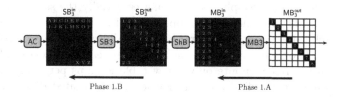

Fig. 3. Phase 1 of the attack

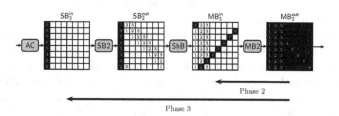

Fig. 4. Phase 2 and Phase 3 of the attack

(Phase 1.B in Fig. 3). As discussed in Sect. 3.2 we can choose from 127 possible input differences for $SB_3^{in}[i,j]$ using the S-box differential table. For each of these XOR difference, we get two possible pairs (a,b) and (b,a). Hence, for each byte of SB_3^{in}, we get a list (denoted by capital letters in Fig. 3) of 254 valid candidate pairs which are sorted by input difference and stored in table $S_F^{\delta[i,j]}$. Note that *any* choice of these pairs will conform to the expected differential path from SB_3^{in} up to SB_4^{in}.

Phase 2. We now take care of the differential path from SB_3^{in} to SB_2^{in}. Since we can choose a candidate pair for each byte SB_3^{in} independently, we will process independently for each column of MB_2 as well. More precisely, for each column j of SB_3^{in} (or MB_2^{out}), we will use the inverse MixBytes table M_B^j to choose each byte difference of SB_3^{in}, such that they result in only one active byte at the input of MB_2 (MB_2^{in}). For each of the 255 differences of M_B^j, we check if some candidate pairs of SB_3^{in} (computed during Phase 1 and stored in $S_F^{\delta[i,j]}$) can fit the 8-byte difference of MB_2^{out} (see Fig. 4). Since for each byte of SB_3^{in} we can choose from 127 possible output differences of the S-box, the probability of success is $127/255 \simeq 1/2$.

Thus, for an entire column of MB_2 we get a probability of $(127/255)^8 \simeq 2^{-8}$ such that one valid candidate pair can be found. Since we can start with 255 input differences for each column of MB_2, we can find one solution for a column with probability $1 - (1 - (127/255)^8)^{255} \simeq 0,62$. We continue for all eight columns of SB_3^{in}. The probability of success is about $(0,62)^8 \simeq 2^{-5,5}$ and we have to restart at Phase 1 about $2^{5,5} = 46$ times to find a solution. At the end of Phase 2, we get a set of byte pairs for SB_3^{in}, which conforms to the differential path from SB_2^{in} up to SB_4^{in}.

Note that these two first phases are doing essentially the same work as the rebound attack [22], but need fewer operations to complete (on average the rebound attack takes about one operations per valid candidate, but this whole step required 2^{64} operations). Here, we need to repeat the process $2^{5.5}$ times to find a solution, but compute only a few table lookups per iteration. Thus, we consider that we can find one solution for the truncated differential path $1 \rightarrow 8 \rightarrow 64 \rightarrow 8$ with about one computation of Grøstl-256 on average.

Phase 3. It seems that at this phase, the differences in $\mathsf{SB}_2^{\mathsf{in}}$ and $\mathsf{SB}_4^{\mathsf{out}}$ can not be chosen anymore. However, an observation allows us to actually get some control over the differences in $\mathsf{SB}_2^{\mathsf{in}}$. We denote by S a 64-byte solution of $\mathsf{SB}_3^{\mathsf{in}}$ (found at the end of Phase 2) and by $(a, b)^{[i,j]}$ the byte pair of row i and column j in S. In Fig. 4, we can see that the active bytes of $\mathsf{SB}_2^{\mathsf{in}}$ are located in the first column. By looking at this figure, it is easy to check that the differences of the active bytes located at row j of $\mathsf{SB}_2^{\mathsf{in}}$ depend only on the byte pairs of the j-th column of $\mathsf{MB}_2^{\mathsf{out}}$ (or $\mathsf{SB}_3^{\mathsf{in}}$). We know that $(a, b)^{[0,j]}, ..., (a, b)^{[7,j]}$ are valid solutions for this column, and switching a and b in any of the pairs actually maintains the validity of those candidates (the differences values of each byte will remain the same in MB_2 and MB_3).

Thus, one solution for each column of $\mathsf{SB}_3^{\mathsf{in}}$ provides us in fact 2^8 valid candidates[2]. Each of these solutions will lead to a random difference on the corresponding active byte $\mathsf{SB}_2^{\mathsf{in}}[j, 0]$, independently of all other differences in SB_2. Now, if we can hit any of the elements of M_B^0 for MB_1 from the newly available differences in $\mathsf{SB}_2^{\mathsf{in}}$, we directly get a solution for the differential path from $\mathsf{SB}_1^{\mathsf{out}}$ to $\mathsf{SB}_4^{\mathsf{out}}$. Since we have 255 elements in M_B^0, we expect about 2^7 solutions on average (2^8 solutions, but half of them may be repeating, see footnote 2).

We did not succeed to control the differences in $\mathsf{SB}_4^{\mathsf{out}}$ as well. Thus, if the differences are uniformly distributed, the success probability for the 8 to 1 active byte transition from the MixBytes layer MB_4 is equal to $2^{-8 \times 7} = 2^{-56}$.

5.2   Application to AES

Again, also this technique can be applied to the AES block cipher. We use the same differential path as in Fig. 1, except that we manipulate a 4×4 state and that no MixColumns transformation is applied in the last round:

$$4 \rightarrow 1 \rightarrow 4 \rightarrow 16 \rightarrow 4 \rightarrow 1 \rightarrow 4 \rightarrow 4$$

Phase 1. This step is analog to the Grøstl-256 case.

---

[2] We have 2^8 different combinations by switching a and b for each column. However, we must take in account that some repeating combinations are counted here (given a combination, inverting everything will obviously lead to exactly the same behavior in the differential path). Thus, instead of having 64 bits degrees of freedom left (8 for each column) we intrinsically loose one of them and get 63 degrees of freedom.

Phase 2. This step is similar to the case of Grøstl-256. However, the probability computation changes when looking for a match between MB_2^{out} and SB_3^{in}. For each column, we now get a probability of $(127/255)^4 \simeq 2^{-4}$ such that at least one valid candidate pair can be found. Since we have 255 differences in M_B^i, we will immediately find one solution for each starting difference SB_4^{in} of the attack. In fact, we expect up to about 2^4 solutions for each column.

Phase 3. Again, we try to control the differences in SB_2^{in}. We use the same technique as for Grøstl-256: for each active byte at row i in SB_2^{in}, we can randomize its difference by randomizing the solutions on the column i in SB_3^{in}. By switching a and b, we directly get 2^4 solutions per column. Moreover, we also have to consider the fact that for the AES case, we already had 2^4 solutions per column. Thus, we get in total about 2^8 solutions per column (see footnote 2). Each of those solutions will lead to a random difference on the corresponding active byte of SB_2^{in}, independently of the other active bytes of SB_2^{in}. Now, if we can hit any of the elements of M_B^0 using the available differences in SB_2^{in}, we get a solution for the differential path between SB_1^{out} and SB_4^{out}. Since we have 255 elements in M_B^0, the whole attack will find about 2^7 solutions on average (2^8 solutions, but half of them may be fully repeating ones, see footnote 2).

Extending the Path. Propagating from SB_4^{in} to SB_5^{in} according to the truncated differential path has a success probability of $2^{-3 \cdot 8} = 2^{-24}$. Thus, we can find a pair corresponding to the path from SB_1^{in} to SB_5^{in} with about 2^{24} round computations on average.

6   Results

In the previous two sections, we have proposed two new techniques to find differences and values for a 4-round truncated differential path with $1 \rightarrow 8 \rightarrow 64 \rightarrow 8 \rightarrow 1$ active bytes for Grøstl-256. In the following, we apply these results to the permutation, compression function and output transformation of Grøstl-256, the AES block cipher and the ECHO permutation.

6.1   Grøstl-256

Both proposed techniques can be used to improve the complexity of the 6-round semi-free-start collision attack of [22]. However, due to the limited degrees of freedom, a semi-free-start collision attack on 7 rounds of the Grøstl-256 compression function is not possible.

6 Rounds. Both proposed techniques (described in Sect. 4 and Sect. 5) can be used to find a valid pair for the 6 round trail of P and Q, given in [22]:

$$8 \xrightarrow{r_1} 1 \xrightarrow{r_2} 8 \xrightarrow{r_3} 64 \xrightarrow{r_4} 8 \xrightarrow{r_5} 8 \xrightarrow{r_6} 64$$

Using the linearized match-in-the-middle attack, we can omit the conditions on SB_4. Hence, the number of equations is reduced to 64 and we expect to find a solution (in fact 2^8 solutions) for already the first differential path. The complexity to find a match for the 8 active bytes (64 bits) at the input, and at the output prior to the (linear) MixBytes transformation is 2^{32} each. Hence, the total complexity to find a semi-free-start collision for 6 rounds of Grøstl-256 is about 2^{64} in time and memory.

Using the start-from-the-middle technique, we can construct a differential path with active bytes $8 \rightarrow 1 \rightarrow 8 \rightarrow 64 \rightarrow 8 \rightarrow 8 \rightarrow 64$) with only a few operations. As a proof of concept, we give in Appendix A a valid input pair for the permutations P and Q on 6-rounds of Grøstl-256 which conforms to this truncated differential path. We get a final complexity of 2^{64} operations and memory for a semi-free-start collision on Grøstl-256 reduced to 6 rounds.

7 Rounds. Again, both techniques can be used to find a valid pair conforming to the 4-round part in the middle of Fig. 1 ($1 \rightarrow 8 \rightarrow 64 \rightarrow 8 \rightarrow 1$) with a relatively low complexity (2^{48} and 2^{56}). This path can be extended by one round in backward and two rounds in forward direction to give a differential path of the form:

$$8 \rightarrow 1 \rightarrow 8 \rightarrow 64 \rightarrow 8 \rightarrow 1 \rightarrow 8 \rightarrow 64,$$

However, using both techniques we can only find 2^{16} pairs conforming to this truncated differential path and one can convince himself with a counting argument: In the middle of the differential path where all bytes of the state are active, one can start with approximatively $2^{512} \cdot 2^{512} = 2^{1024}$ different pairs. However, only a portion 2^{-56} will follow a MixBytes transition $8 \rightarrow 1$, and only a portion $2^{-56 \cdot 8} = 2^{-448}$ will follow a MixBytes transition $64 \rightarrow 8$ (because we have a probability of 2^{-56} for each column). Since we have two $64 \rightarrow 8$ and two $8 \rightarrow 1$ transitions and consider them to be independent, only $2^{1024-448 \cdot 2 - 56 \cdot 2} = 2^{16}$ valid pairs will remain for the 4-round path in the middle ($1 \rightarrow 8 \rightarrow 64 \rightarrow 8 \rightarrow 1$) and thus, also for the 7-round path.

Note that due to this lack of freedom a semi-free-start collision using this truncated differential path is not possible. For a collision at the end of 7 rounds, we need about 2^{64} pairs for each, P and Q. Otherwise, a birthday attack on 128 bits (8 active bytes at the input, 8 active bytes prior to MixBytes at the output) is not feasible. By using different positions of active bytes in round 2 and 6, but the same column for P and Q, we can construct about $2^6 \cdot 2^6 \cdot 2^3 \cdot 2^3 = 2^{18}$ different truncated paths. By far not enough for a collision attack. However, one could think of a free-start near-collision attack on 7 rounds of Grøstl-256 but this property gets destroyed by the output transformation.

Therefore, we can only get a distinguisher for the permutation or output transformation of Grøstl-256 reduced to 6.5 rounds (without the final MixBytes transformation). The complexity is 2^{48} instead of 2^{224} for a random 512-bit permutation or 2^{112} for a random 256-bit function. We can get a distinguisher for the full 7 rounds by applying the subspace distinguisher proposed in [19]. Note that the input and the output differences of the 6.5 round attack form a vector space of dimension 64 at the input and output. Since the Mixbytes transformation is linear also the output differences after 7 rounds form a vector space of dimension 64. Hence, we can apply the subspace distinguisher with parameters $N = 512$, $n = 64$, $t = 128$ (generic complexity: $2^{115.4}$) to distinguish 7 rounds of the permutation P and Q. To construct a vector space of size $t = 128$, we need to repeat our attack on the comression function 2^7 times. Hence, the total complexity for the subspace distinguisher of the permutation is about 2^{55} permutation calls with negligible memory.

Similarily, we can use the subspace distinguisher to distinguish the output transformation of Grøstl-256 as well. Note that the 8 active bytes of the input are added to the output by the feed-forward. However, due to the truncation at the end the output differences will still form a vector space of dimension 64. Since Grøstl-256 truncates columns and MixBytes works on columns, we keep only half of the vector space. Hence, we can apply a subspace distinguisher with parameters $N = 256$, $n = 64$, $t = 256$ (generic complexity: $2^{75.9}$) and need to repeat our attack 2^8 times to get a vector space of size $t = 256$. Hence, the total complexity for the subspace distinguisher on 7 rounds of the Grøstl-256 output transformation is about 2^{56} output transformation calls and negligible memory.

6.2 AES Block Cipher

Both proposed techniques apply to the block cipher AES in the known-key distinguisher setting as well. The resulting 7-round differential path for AES is computed by simply extending the 4-round path in both forward and backward direction to give the following sequence of active bytes:

$$4 \to 1 \to 4 \to 16 \to 4 \to 1 \to 4 \to 4$$

Note that the last MixColumns operation is omitted in the AES. Since we aim for 4 active bytes in both, plaintext and ciphertext, we would expect to find such a pair with about 2^{48} computations for a random permutation. Note that an equivalent generic attack needs to find a pair with only 4 active bytes at the input and output as well. Hence, the best generic method is to start with 4 active bytes at the input and search for a near-collision on 12 non-active bytes at the output with complexity $2^{(12 \cdot 8)/2} = 2^{48}$.

Using the linearized match-in-the-middle attack, we get a known-key distinguisher for 7-rounds of AES with a complexity of about 2^{36} and negligible memory. However, the start-from-the-middle technique allows us to further improve the complexity for the known-key distinguisher to about 2^{24} in time and negligible memory for 7-rounds of AES. Additionally, one may think of other applications of these attack such as near-collisions on a compression function built upon the 7-round reduced AES in Davies-Meyer mode [6, 23], or a collision attack on the compression function for 5 rounds.

6.3 Internal Permutation of ECHO

It is possible to apply the start-from-the-middle technique to other AES-based hash functions, such as ECHO [2] whose internal 2048-bit permutation can been seen as a big AES processing 128-bit words instead of bytes. This directly gives us an improved distinguisher on 7 rounds whose complexity is $2^{3 \cdot 128} = 2^{384}$ operations (compared to the previous one with complexity 2^{896}) and memory requirements are 2^{256}. However, we can improve the memory requirements by storing a differential lookup table for the AES super box [9] with size $2^{32} \cdot 2^{32} = 2^{64}$, instead of a differential lookup table for two full AES rounds with size $2^{128} \cdot 2^{128} = 2^{256}$. This is possible due to the fact that one can combine the last MixColumns transformation of the AES with the subsequent BigMixColumns transformation of ECHO, since both transformations are linear. Note that this attack only allows to distinguish 7 rounds of the ECHO internal permutation from a random 2048-bit permutation, but does not apply to the compression function due to the word compression at its output.

7   Conclusion and Future Work

In this paper, we have proposed two new ways to mount attacks on the SHA-3 candidates Grøstl and ECHO, and the block cipher AES. Our results improve upon and extend the rebound attack. Both techniques are an extension of the number of rounds in which degrees of freedom can be used to improve from two to four rounds. As a result, we present the best known attacks on constructions where (reduced variants of) *permutations* are used. We improve on the attack on the reduced Grøstl-256 compression function (up to 6 out of 10 rounds), and present a distinguisher for 7-rounds of the Grøstl-256 permutation and output transformation. Further, we improve upon the distinguisher for 7-rounds of the internal permutation of ECHO and the known-key distinguisher for 7-rounds of the block cipher AES. Nevertheless, a comfortable security margin for these SHA-3 candidates remain. Not only because both proposals have a higher number of rounds, but also because in an attack on the hash function much less degrees of freedom are available (compared to an attack on the compression function or permutation).

On the other hand, the new techniques of this paper have been optimized for this setting and do not directly apply to other settings where more degrees of freedom are available. Sources for such degrees of freedom are salt, counter, or key inputs. While the analysis typically gets more complicated if more freedom is available, much better attacks can be expected. As an example we refer to a recent extension of the rebound attack on the full 10-round Whirlpool compression function [19]. Note that Whirlpool is a block cipher based construction which offers additional degrees of freedom through its conservative key schedule. Some SHA-3 candidates use block-cipher based compression functions with key-schedules less conservative than Whirlpool. Hence, more degrees of freedom are available to an attacker and better results may be expected along those lines.

In general, the rebound attack and its extensions as described in this paper, work with any differential or truncated differential. However, the diffusion properties of AES based hash functions allow a very simple construction of good truncated differential paths, which facilitates the analysis. Nevertheless, future

work will include the application of the rebound idea on other hash constructions, even though this may require sophisticated tools to obtain good results,
as was the case for *e. g.* SHA-1 [11].

Acknowledgments

We would like to thank Joan Daemen for the idea on the linearized match-in-the-
middle attack, Henri Gilbert, Mario Lamberger and the anonymous referees for
useful comments and discussions, and Vincent Rijmen for comments on the preproceedings version of this paper. The work in this paper has been supported in
part by the European Commission under contract ICT-2007-216646 (ECRYPT
II) and by the IAP Programme P6/26 BCRYPT of the Belgian State (Belgian
Science Policy).

References

1. Barreto, P.S.L.M., Rijmen, V.: The WHIRLPOOL Hashing Function. Submitted to
   NESSIE, revised May 2003 (September 2000),
   http://www.larc.usp.br/~pbarreto/WhirlpoolPage.html (2008/12/11)
2. Benadjila, R., Billet, O., Gilbert, H., Macario-Rat, G., Peyrin, T., Robshaw, M.,
   Seurin, Y.: SHA-3 Proposal: ECHO. Submission to NIST (2008),
   http://crypto.rd.francetelecom.com/echo/
3. Biham, E., Dunkelman, O.: A Framework for Iterative Hash Functions - HAIFA.
   Cryptology ePrint Archive, Report 2007/278 (2007), http://eprint.iacr.org
4. Biham, E., Shamir, A.: Differential Cryptanalysis of DES-like Cryptosystems. J.
   Cryptology 4(1), 3–72 (1991)
5. Biryukov, A., Khovratovich, D., Nikolic, I.: Distinguisher and Related-Key Attack
   on the Full AES-256. In: Halevi, S. (ed.) CRYPTO 2009. LNCS, pp. 231–249.
   Springer, Heidelberg (2009)
6. Cohen, B., Laurie, B.: AES-hash. Submission to NIST: Proposed Modes (2001),
   http://csrc.nist.gov/groups/ST/toolkit/BCM/documents/proposedmodes/
   aes-hash/aeshash.pdf
7. Daemen, J., Rijmen, V.: The Wide Trail Design Strategy. In: Honary, B. (ed.)
   Cryptography and Coding 2001. LNCS, vol. 2260, pp. 222–238. Springer, Heidelberg (2001)
8. Daemen, J., Rijmen, V.: The Design of Rijndael. Information Security and Cryptography. Springer, Heidelberg (2002)
9. Daemen, J., Rijmen, V.: Understanding Two-Round Differentials in AES. In: De
   Prisco, R., Yung, M. (eds.) SCN 2006. LNCS, vol. 4116, pp. 78–94. Springer, Heidelberg (2006)
10. Daemen, J., Rijmen, V.: Plateau characteristics. IET Information Security 1(1),
    11–17 (2007)
11. De Cannière, C., Rechberger, C.: Finding SHA-1 Characteristics: General Results
    and Applications. In: Lai, X., Chen, K. (eds.) ASIACRYPT 2006. LNCS, vol. 4284,
    pp. 1–20. Springer, Heidelberg (2006), http://dx.doi.org/10.1007/11935230_1
12. Fleischmann, E., Forler, C., Gorski, M.: The Twister Hash Function Family. Sub-
    mission to NIST (2008),
    http://ehash.iaik.tugraz.at/uploads/3/39/Twister.pdf

13. Fouque, P.A., Stern, J., Zimmer, S.: Cryptanalysis of Tweaked Versions of SMASH and Reparation. In: Avanzi, R., Keliher, L., Sica, F. (eds.) SAC 2008. LNCS, vol. 5381, pp. 136–150. Springer, Heidelberg (2009)
14. Gauravaram, P., Knudsen, L.R., Matusiewicz, K., Mendel, F., Rechberger, C., Schläffer, M., Thomsen, S.S.: Grøstl – a SHA-3 candidate. Submission to NIST (2008), http://www.groestl.info
15. Khovratovich, D.: Cryptanalysis of hash functions with structures. In: Jacobson, M.J., Rijmen, V., Safavi-Naini, R. (eds.) SAC 2009. LNCS, vol. 5867, pp. 108–125. Springer, Heidelberg (2009)
16. Khovratovich, D., Biryukov, A., Nikolic, I.: The Hash Function Cheetah: Specification and Supporting Documentation. Submission to NIST (2008), http://ehash.iaik.tugraz.at/uploads/c/ca/Cheetah.pdf
17. Knudsen, L.R., Rechberger, C., Thomsen, S.S.: The Grindahl Hash Functions. In: Biryukov, A. (ed.) FSE 2007. LNCS, vol. 4593, pp. 39–57. Springer, Heidelberg (2007)
18. Knudsen, L.R., Rijmen, V.: Known-Key Distinguishers for Some Block Ciphers. In: Kurosawa, K. (ed.) ASIACRYPT 2007. LNCS, vol. 4833, pp. 315–324. Springer, Heidelberg (2007)
19. Lamberger, M., Mendel, F., Rechberger, C., Rijmen, V., Schläffer, M.: Rebound Distinguishers: Results on the Full Whirlpool Compression Function. In: Matsui, M. (ed.) ASIACRYPT 2009. LNCS, vol. 5912. Springer, Heidelberg (to appear, 2009)
20. Matusiewicz, K., Naya-Plasencia, M., Nikolić, I., Sasaki, Y., Schläffer, M.: Rebound Attack on the Full LANE Compression Function. In: Matsui, M. (ed.) ASIACRYPT 2009. LNCS, vol. 5912. Springer, Heidelberg (to appear, 2009)
21. Mendel, F., Rechberger, C., Schläffer, M.: Cryptanalysis of Twister. In: Abdalla, M., Pointcheval, D., Fouque, P.A., Vergnaud, D. (eds.) ACNS 2009. LNCS, vol. 5536, pp. 342–353. Springer, Heidelberg (2009)
22. Mendel, F., Rechberger, C., Schläffer, M., Thomsen, S.S.: The Rebound Attack: Cryptanalysis of Reduced Whirlpool and Grøstl. In: Dunkelman, O. (ed.) FSE 2009. LNCS, vol. 5665, pp. 260–276. Springer, Heidelberg (2009)
23. Menezes, A.J., van Oorschot, P.C., Vanstone, S.A.: Handbook of Applied Cryptography. CRC Press, Boca Raton (1996), http://www.cacr.math.uwaterloo.ca/hac/
24. Minier, M., Phan, R.C.W., Pousse, B.: Distinguishers for Ciphers and Known Key Attack against Rijndael with Large Blocks. In: Preneel, B. (ed.) AFRICACRYPT 2009. LNCS, vol. 5580, pp. 60–76. Springer, Heidelberg (2009)
25. National Institute of Standards and Technology: FIPS PUB 197, Advanced Encryption Standard (AES). Federal Information Processing Standards Publication 197, U.S. Department of Commerce (November 2001)
26. National Institute of Standards and Technology: Announcing Request for Candidate Algorithm Nominations for a New Cryptographic Hash Algorithm (SHA-3) Family. Federal Register 27(212), 62212–62220 (November 2007), http://csrc.nist.gov/groups/ST/hash/documents/FR_Notice_Nov07.pdf (2008/10/17)
27. Peyrin, T.: Cryptanalysis of Grindahl. In: Kurosawa, K. (ed.) ASIACRYPT 2007. LNCS, vol. 4833, pp. 551–567. Springer, Heidelberg (2007)
28. Wu, S., Feng, D., Wu, W.: Cryptanalysis of the LANE Hash Function. In: Jacobson, M.J., Rijmen, V., Safavi-Naini, R. (eds.) SAC 2009. LNCS, vol. 5867, pp. 126–140. Springer, Heidelberg (2009)

A   Message and Chaining Variable Example for the 6-Round Differential Path of Grøstl-256

We give here in hexadecimal display a chaining variable and message pair example ($[H_1, M_1], [H_2, M_2]$) that verifies the 6-round differential path for Grøstl-256.

$$H_1 = \text{fdab6faf65da3531e5a7f611baba937d}$$
$$\text{b18648152738a5fe4bd38ca5a8b050e7}$$
$$\text{3d734623aed6f7a35e3fb3d72eba5e60}$$
$$\text{1712a3d23d76fe79ccbba10461dddee0}$$

$$M_1 = \text{66b16a712984a23ca99283090e5818c7}$$
$$\text{c7f46fcd74c54b7a9950a4bfcb2861b1}$$
$$\text{1f90846a04c92172af57a58ad9b747a3}$$
$$\text{a26dca926c18f410ad0f40f52800d27b}$$

$$H_2 = \text{21ab6faf65da3531e51bf611baba937d}$$
$$\text{b186c5152738a5fe4bd38c88a8b050e7}$$
$$\text{3d734623ecd6f7a35e3fb3d72e6c5e60}$$
$$\text{1712a3d23d767779ccbba10461ddde66}$$

$$M_2 = \text{f8b16a712984a23ca9ef83090e5818c7}$$
$$\text{c7f434cd74c54b7a9950a40fcb2861b1}$$
$$\text{1f90846a29c92172af57a58ad95547a3}$$
$$\text{a26dca926c18d710ad0f40f52800d27f}$$

Cryptanalyses of Narrow-Pipe Mode of Operation in AURORA-512 Hash Function

Yu Sasaki

NTT Information Sharing Platform Laboratories, NTT Corporation,
3-9-11 Midoricho, Musashino-shi, Tokyo, 180-8585 Japan
sasaki.yu@lab.ntt.co.jp
The University of Electro-Communications,
1-5-1 Choufugaoka, Choufu-shi, Tokyo, 182-8585 Japan

Abstract. We present cryptanalyses of the AURORA-512 hash function, which is a SHA-3 candidate. We first describe a collision attack on AURORA-512. We then show a second-preimage attack on AURORA-512/-384 and explain that the randomized hashing can also be attacked. We finally show a full key-recovery attack on HMAC-AURORA-512 and universal forgery on HMAC-AURORA-384. Our attack exploits weaknesses in a narrow-pipe mode of operation of AURORA-512 named "Double-Mix Merkle-Damgård (DMMD)," which produces 512-bit output by updating two 256-bit chaining variables in parallel. We do not look inside of the compression function. Hence, our attack can work even if the compression function is regarded as a random oracle. The time complexity of our collision attack is approximately 2^{236} AURORA-512 operations, and $2^{236} \times 512$ bits of memory is required. Our second-preimage attack works on any given message. The time complexity is approximately 2^{290} AURORA-512 operations, and $2^{288} \times 512$ bits of memory is required. Our key-recovery attack on HMAC-AURORA-512, which uses 512-bit secret keys, requires 2^{257} queries, 2^{259} off-line AURORA-512 operations, and a negligible amount of memory. The universal forgery on HMAC-AURORA-384 is also possible by combining the second-preimage and key-recovery attacks.

Keywords: AURORA, DMMD, collision, second preimage, HMAC.

1 Introduction

Hash functions are important cryptographic primitives used for various purposes. Currently, the National Institute of Standards and Technology (NIST) is conducting a SHA-3 competition for determining a new hash standard algorithm [1]. In the SHA-3 competition, 51 algorithms were accepted as candidates. One of the most important design aspects is the size of the internal state. To make hash algorithms efficient and compact, the internal state size should be as small as possible. If the internal state size is the same as the hash size, the structure is called a narrow-pipe mode [2]. On the other hand, to make hash algorithms

M.J. Jacobson Jr., V. Rijmen, and R. Safavi-Naini (Eds.): SAC 2009, LNCS 5867, pp. 36–52, 2009.
© Springer-Verlag Berlin Heidelberg 2009

secure, the internal state size should be larger than the hash size. Such a structure is called a wide-pipe mode [2]. Therefore, there is a trade-off of efficiency and security in the choice between the narrow-pipe or wide-pipe modes.

AURORA [3] is one of the hash algorithms submitted for SHA-3, which was designed by Iwata et al. AURORA mainly has four algorithms: AURORA-224, AURORA-256, AURORA-384, and AURORA-512. The output of AURORA-224 is obtained by truncating the last output value of AURORA-256. Similarly, the output of AURORA-384 is obtained by truncating the last output value of AURORA-512. Hence, two algorithms AURORA-256 and AURORA-512 are important to evaluate the security of AURORA. AURORA operates in narrow-pipe mode. In AURORA-256, a hash value is computed by iteratively applying a compression function that takes a 256-bit chaining variable and a 512-bit message block as input and a 256-bit chaining variable as output. AURORA-512 adopts a different mode of operation named Double-Mix Merkle-Damgård (DMMD), which produces 512-bit output by updating two 256-bit chaining variables in parallel. Update of 256-bit chaining variables is done by using almost the same compression functions as AURORA-256. A unique characteristic of the DMMD structure is computing 512-bit chaining variables by combining the component of AURORA-256. This gives a large advantage with respect to the size of the component to be implemented because components for AURORA-512 and AURORA-256 can be shared. Due to this structure, AURORA is efficient, especially in hardware. Hence, evaluating the security of AURORA-512 is useful for the cryptographic community to understand the tradeoff between efficiency and security in hash function design.

1.1 SHA-3 Requirements and Claimed Security of AURORA

NIST requires SHA-3 candidates to satisfy several security properties [1], e.g.,

- Preimage resistance of n bits,
- Second-preimage resistance of $n - k$ bits for any message shorter than 2^k blocks,
- Collision resistance of $n/2$ bits,
- Resistance on randomized hashing [4] of $n - k$ bits (See Section 2.2 for details.),
- $2^{n/2}$ queries and 2^n off-line computations against distinguishing attacks on HMAC [5].

According to Iwata et al. [3], DMMD has provable security for collision resistance and preimage resistance. It was proven that any adversary needs at least 2^{201} computations to find a collision of AURORA-512, and needs at least 2^{512} computations to find a preimage of AURORA-512. On the other hand, it was claimed that security of AURORA-512 is 256 bits for collision resistance, 512 bits for preimage resistance, and $(512 - k)$ bits for second preimage resistance of 2^k-block messages. It is also mentioned that AURORA can be securely used as randomized hashing and as a HMAC.

1.2   Our Contribution

We investigate the weaknesses of the DMMD mode of operation adopted in AURORA-512. We first show a collision attack on AURORA-512, where the time complexity is approximately 2^{236} AURORA-512 operations and requires $2^{236} \times 512$ bits of memory. We then show a second-preimage attack on AURORA-512 and -384. Our attack generates a second preimage of any given message. Generated messages are 8 blocks long. The time complexity is approximately 2^{290} AURORA-512 operations and requires $2^{288} \times 512$ bits of memory. We then explain that the randomized hashing can also be attacked. These attacks use the multi-collision attack on a Merkle-Damgård structure proposed by Joux [6]. However, direct application of [6] to AURORA-512 does not work regarding a collision attack and is not efficient regarding a second-preimage attack. This is due to the mixing function of AURORA-512, which is designed to prevent attacks using multi-collisions such as [6]. We show that AURORA-512 is vulnerable against multi-collision attacks even if the mixing function is adopted. Note that a similar approach was taken by Knudsen et al. [7] to attack MDC2 [8]. We finally show a full key-recovery attack on HMAC-AURORA-512 with 512-bit secret keys, which require 2^{257} queries, 2^{259} off-line AURORA-512 operations, and negligible amount of memory. The universal forgery on HMAC-AURORA-384 is also possible by combining the second-preimage and key-recovery attacks. Results of our attacks are summarized in Table. 1.

Outline. In Section 2, we describe the specifications of AURORA-512, randomized hashing, and HMAC. We then introduce Joux's multi-collision attack. In Section 3, we discuss a collision attack on AURORA-512. In Section 4, we discuss a second-preimage attack on AURORA-512 and -384. We then explain that randomized hashing can also be attacked. In Section 5, we present a key recovery attack on HMAC-AURORA-512 and a universal forgery attack on

Table 1. Summary of attacks on AURORA

Attack type	Hash size	Reference	Time	Memory	
Collision	512	[9][†]	$2^{234.4}$	$2^{229.6}$	
Collision	512	[9][†]	2^{249}	-	
Collision	512	Ours	2^{236}	2^{236}	
2nd-preimage	512/384	[9][†]	2^{291}	$2^{31.5}$	
2nd-preimage	512/384	Ours	2^{291}	2^{288}	
Randomized hash	512/384	Ours	2^{291}	2^{288}	
Attack type	Hash size	Reference	Time	Memory	Query
HMAC key recovery	512	Ours	2^{259}	–	2^{257}
HMAC universal forgery	384	Ours	2^{291}	2^{288}	2^{256}

† Ferguson and Lucks [9] also explained attacks on AURORA. Our work is independent of [9]. We describe the relationship between these two works in the Appendix.
‡ After our submission, Joux and Lucks showed improved analyses on AURORA [10].

HMAC-AURORA-384. In Section 6, we summarize what we can learn from these attacks and conclude this paper.

2 Related Works

2.1 Description of AURORA-512 and AURORA-384

We briefly describe the specifications of AURORA-512 and AURORA-384. Please refer to Ref. [3] for details.

An input message M is padded to be a multiple of 512 bits by the standard MD message padding, namely, a single bit '1', necessary numbers of '0's, and a 64-bit string representing the block length of M are appended to the end of M. Then, the padded message is divided into 512-bit message blocks $(M_0, M_1, \ldots, M_{N-1})$.

The computation for AURORA-384 is the same as AURORA-512 but for the initial value and truncating the last 512-bit value to 384 bits. Hence, we explain data processing in AURORA-512. AURORA-512 adopts a narrow-pipe mode of operation named DMMD, where two half-size (256-bit) chaining variables are updated independently by using the same message in each block. However, if all blocks are updated independently, the construction becomes vulnerable to Joux's multi-collision attack [6]. To prevent this attack, DMMD periodically computes the mixing function, which takes concatenation of two half-size chaining variables as input, to introduce the dependency of two chaining variables.

More strictly, in AURORA-512, compression functions $F_0, F_1, \ldots, F_7, G_0, G_1,$ $\ldots, G_7 : \{0,1\}^{256} \times \{0,1\}^{512} \rightarrow \{0,1\}^{256}$, two functions $MF, MFF : \{0,1\}^{512} \rightarrow \{0,1\}^{512}$, and two 256-bit initial values (IVs) H_0^U and H_0^D are defined. The algorithm to compute a hash value is as follows. This is also illustrated in Fig. 1. In the procedure below, we use k' to denote $k \bmod 8$.

1. for k=0 to $N - 1$ {
2. $H_{k+1}^U \leftarrow F_{k'}(H_k^U, M_k)$
3. $H_{k+1}^D \leftarrow G_{k'}(H_k^D, M_k)$
4.   if$(0 < k < N - 1) \wedge (k \bmod 8 = 7)$ {
5.   temp $\leftarrow H_{k+1}^U \| H_{k+1}^D$
6. $H_{k+1}^U \| H_{k+1}^D \leftarrow MF(\text{temp})$
7.   }
8. }
9. Output $MFF(H_N^U \| H_N^D)$

2.2 Description of Randomized Hashing

Randomized hashing [4] improves the security of the digital signature schemes from the collision attacks on the hash functions. It makes it difficult for the attacker to obtain a signature for one of the colliding messages from the signer and produce it as a signature for the other colliding message because the signer always uses a different random value in every signature generation process. One may note the discussion on its security by Gauravaram and Knudsen [11].

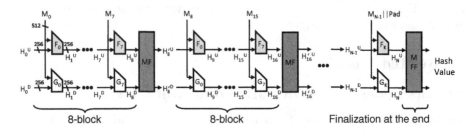

Fig. 1. Hash computation in AURORA-512

The algorithm of randomized hashing takes a message M and a key K as input and outputs a randomized message M^R. The procedure is as follows.

1. Process M to the padding procedure to make sure that the processed message M' is longer than K. Let M'_L and K_L be the length of M' and K, respectively.
2. Set $counter \longleftarrow \lfloor M'_L/K_L \rfloor$ and $remainder \longleftarrow M'_L \bmod K_L$.
3. Let R be concatenation of $counter$ copies of the K and the $remainder$ leftmost bits of the K.
4. Output $M^R \longleftarrow K \| M' \oplus R \| K_{L(2)}$, where $K_{L(2)}$ is a 16-bit binary representation of K_L.

According to NIST [1], SHA-3 candidates with n-bit output must provide $(n-k)$-bit security in the following attack: 1) The attacker chooses a message M_1, which is shorter than 2^k blocks. 2) Then, randomization value K_1 is chosen without control of the attacker. 3) The attacker finds M_2 and K_2 s.t. $(M_1, K_1) \neq (M_2, K_2)$, which yield the same randomized hash value.

2.3   Description of HMAC

HMAC [5] is an algorithm to compute a MAC when a key and a message are input. According to Krawczyk et al. [12], the minimal recommended length for the secret key is L, where L is the size of the hash function output. Therefore, it is reasonable to use 512-bit keys for HMAC with 512-bit output hash functions, and use 384-bit keys for 384-bit output hash functions. The HMAC algorithm to compute an output with a hash function H and an initial value H_0 when a key K and a message M are input is as follows.

$$K_0 \leftarrow Pad(K), \tag{1}$$

$$\text{temp} = H(H_0, (K_0 \oplus \texttt{ipad}) \| M), \tag{2}$$

$$\text{HMAC-}H(M) = H(H_0, (K_0 \oplus \texttt{opad}) \| \text{temp}), \tag{3}$$

where, `ipad` and `opad` are constant values defined in the specification of HMAC, and $Pad(\cdot)$ is a padding process of K. In $Pad(\cdot)$, if the size of K is shorter than the block length, zeros are appended to the end of K to make its length the same as the block length signified as K_0. If the size of K and the block length are identical, K is signified as K_0.

SHA-3 candidates with n-bit output are required to be secure against distinguishing attacks that require much fewer than $2^{n/2}$ queries and significantly less computation than a preimage attack.

2.4   Description of Joux's Multi-collision Attack

Let t-collision be t different messages that result in the same hash value. Joux showed that 2^k-collision of any n-bit iterated hash function can be found with a complexity of $k \cdot 2^{\frac{n}{2}}$ [6]. For chaining variables H_j and a compression function $CF(H_j, M_j) = H_{j+1}$, Joux's attack generates (M_j, M'_j) such that $CF(H_j, M_j) = CF(H_j, M'_j) = H_{j+1}$ for $j = 0, 1, \ldots, k - 1$. Any choice of (M_j, M'_j) for $j = 0, 1, \ldots, k - 1$ will result in the same H_k, hence a 2^k-collision is generated.

Joux applied this technique to a cascaded construction. Let $A(\cdot)$ be an n-bit iterated hash function and $B(\cdot)$ be an n-bit hash function. For an input message M, the cascaded construction outputs a $2n$-bit value $A(M)\|B(M)$. Intuitively, the cascaded construction has $2n$-bit security. However, with Joux's attack, collisions and preimages can be found with a complexity of $\frac{n}{2} \cdot 2^{\frac{n}{2}}$ and $n \cdot 2^{\frac{n}{2}} \cdot 2^n$, respectively. To find a collision, the attacker generates a $2^{\frac{n}{2}}$-collision of $A(\cdot)$. From these $2^{\frac{n}{2}}$ messages, two paired messages will also collide with each other by computing $B(\cdot)$. To find a preimage, the attacker generates a 2^n-collision of $A(\cdot)$. Then exhaustively searches for a message that connects the collision value to the n-bit of the given hash value for $A(\cdot)$. Since there are 2^n messages that match the n-bit of given hash value, one of the messages will also satisfy the n-bit of the given hash value for $B(\cdot)$.

Restriction of Joux's Multi-collision Attack. Joux's multi-collision attack is useful if a compression function includes two independent parts through several blocks like AURORA-512. In fact, if two independent parts in AURORA-512 continues for 256 blocks or 512 blocks, the Joux's attack can be applied to find collisions or second preimages. However, the mixing function inserted at every 8 blocks guarantees that the independent part continues for at most 8 blocks, and this prevents efficient application of Joux's attack.

3   Collision Attack on AURORA-512

Our attack finds collisions of 8-block messages with a complexity of 2^{236}.

Attack Procedure. The attack procedure is as follows. The attack is also illustrated in Fig. 2

1. Randomly choose $2^{224}(= 2^{256 \cdot \frac{7}{8}})$ M_0, and compute $H_1^U \leftarrow F_0(H_0^U, M_0)$ for each M_0. This yields an 8-collision ($=2^3$-collision) of H_1^U.
2. By applying Joux's attack [6] to M_1 through M_6, we obtain a 2^{21}-collision of H_7^U. Let these 7-block messages yielding the 2^{21}-collision be $M_{[06]}^{(i)}, 0 \leq i \leq 2^{21} - 1$.
3. Compute $H_{k+1}^D \leftarrow G_k(H_k^D, M_k^{(i)}), 0 \leq k \leq 6$ for all i. Let $H_7^{D(i)}$ be the corresponding 2^{21} H_7^Ds.

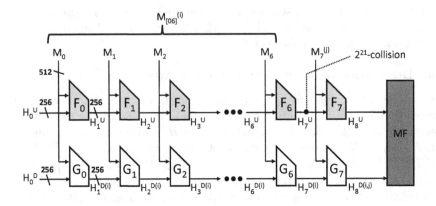

Fig. 2. Collision construction on AURORA-512

4. Set M_7 to be a randomly chosen value, and compute $H_8^{D(i)} = G_7(H_7^{D(i)}, M_7)$ for all i. Check whether or not a collision exists among 2^{21} $H_8^{D(i)}$.
5. If not, go back to Step 4 and try a different M_7. If a collision is found, let the corresponding 'i's be $i1$ and $i2$, and corresponding M_7 be $M_7^{(j)}$. Then, $M_{[06]}^{(i1)} \| M_7^{(j)}$ and $M_{[06]}^{(i2)} \| M_7^{(j)}$ are the colliding pair.

At Step 4, since there are 2^{21} $H_8^{D(i)}$, we can make roughly $2^{41}(= (2^{21})^2/2)$ pairs of $H_8^{D(i)}$. Therefore, the probability that a collision will be found is $2^{-215}(= 2^{-256} \cdot 2^{41})$. As a result, after 2^{215} iterations of Step 4, we expect to obtain a colliding pair.

Complexity Evaluation. Steps 1 and 2 cost $7 \cdot 2^{224}$ F_k-operations. Step 3 costs $7 \cdot 2^{21}$ G_k-operations. At Steps 4 and 5, the complexity of Step 4 for a chosen M_7 is 2^{21} G_k-operations. Therefore, 2^{215} iterations cost $2^{236}(= 2^{21} \cdot 2^{215})$ G_k-operations. Hence, the time complexity of this collision attack is $7 \cdot 2^{224} + 7 \cdot 2^{21} + 2^{236} \approx 2^{236}$ F_k or G_k operations. At Steps 1 and 2, we need to prepare $2^{236} \times 512$ bits of memory to find a 2^3-collision.

Remark on Success Probability of Generating Multi-collision. At Steps 1 and 2 of the attack procedure, the success probability of generating multi-collisions is much lower than $1/2$. Suzuki et al. [13] gives us the complexity for finding s-collisions of n-bit value with a probability of approximately $1/2$:

$$(s!)^{1/s} \times (2^{n \cdot \frac{s-1}{s}}) + s - 1. \tag{4}$$

The value of this equation is $2^{225.91} \approx 2^{226}$ when $n = 256$ and $s = 2^3$. However, considering that our attack generates 2^3-collisions 7 times at Steps 1 and 2, we need to dramatically increase the success probability. For this purpose, our attack computes 2^{230} different messages to find a 2^3-collision for each block. Since $2^{230-226} = 16$, the success probability for Steps 1 and 2 becomes $(1-(1/2)^{16})^7 \approx 1$.

Under this strategy, the attack complexity is $7 \cdot 2^{230} + 7 \cdot 2^{21} + 2^{236} = 2^{236.150}$ F_k or G_k operations, which is approximately 2^{236} AURORA-512 operations.

4   Second-Preimage Attack on AURORA-512 and -384

Our attack can generate second-preimages of any given message. Generated second-preimages are 8 blocks long. The time complexity of our attack is approximately 2^{290} AURORA-512 operations. Since the complexity is much lower than 2^{384}, the attack can also be applied to AURORA-384. Strictly speaking, the attack complexity depends on the output distribution of the compression function. We first assume that the output distribution is perfectly balanced, then discuss other cases later.

Attack Procedure. The attack procedure for some given message is as follows. The attack is also illustrated in Fig. 3.

1. Compute a hash value of the given message. Let T^U and T^D be the upper 256 bits and the lower 256 bits of the input values for the MFF function, respectively.
2. Choose an M_0 and compute $H_1^U \leftarrow F_0(H_0^U, M_0)$. Repeat this computation with changing M_0 until a 2^{32}-collision of H_1^U is obtained.
3. Following the first block, we apply Joux's attack [6] to M_1 through M_6. In total, we obtain a $2^{32 \times 7} = 2^{224}$-collision of H_7^U.
4. Compute $H_8^U \leftarrow F_7(H_7^U, M_7 \| \mathrm{Pad})$ for $2^{288} (= 2^{256} \cdot 2^{32})$ different M_7s, where Pad is the padding string for 8-block messages and the length of $M_7 \| \mathrm{Pad}$ must be 1 block. If the output distribution of F_7 is perfectly balanced with respect to $M_7 \| \mathrm{Pad}$, namely, the output distribution of $F_7(H_7^U, \cdot)$ is balanced,

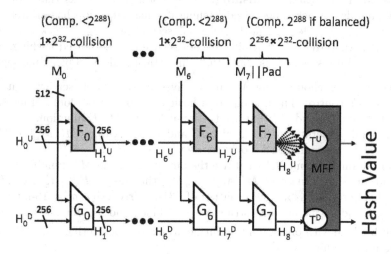

Fig. 3. Second-preimage construction for AURORA-512

we obtain 2^{32}-collisions for all possible values of H_8^U. Therefore, we obtain a 2^{32}-collision of $M_7\|\texttt{Pad}$ that maps H_7^U to T^U. Consequently, we obtain $2^{256}(=2^{224}\cdot 2^{32})$ messages $M_0\|M_1\|\cdots\|M_7\|\texttt{Pad}$ that produce T^U.

5. Compute $H_{k+1}^D \leftarrow G_k(H_k^D, M_k), 0 \leq k \leq 7$ for all $M_0\|M_1\|\cdots\|M_7\|\texttt{Pad}$ obtained at Step 4. Since we have 2^{256} different choices, we expect that one will match T^D. The matched message $M_0\|M_1\|\cdots\|M_7$ is a second preimage of the given message.

Complexity Evaluation. At Steps 2 and 3, if we try $2^{288}(=2^{256}\cdot 2^{32})$ different M_k for each block, we obtain a 2^{32}-collision due to the pigeonhole principle. The time complexity is at most $7 \cdot 2^{288}$ F_k operations, and the success probability is 1. Step 4 costs exactly 2^{288} F_7-operations if the output distribution of $F_7(H_7^U, \cdot)$ is perfectly balanced. Step 5 costs $8 \cdot 2^{256}$ G_k-operations. Therefore, the total time complexity of this attack is $7 \cdot 2^{288} + 2^{288} + 8 \cdot 2^{256} \approx 2^{291}$ F_k or G_k-operations, which is approximately 2^{290} AURORA-512 operations. At Steps 2 and 3, we need to prepare $2^{288} \times 512$ bits of memory.

Remark on Output Distribution. At Steps 2 and 3, we need only one 2^{32}-collision. Therefore, the attack complexity lessens if the distribution is not balanced. At Step 4, we need one 2^{32}-collision that produces T^U. If the distribution is not balanced and T^U is produced more frequently than other values, the complexity lessens. However, if T_U is not produced as much as other values, 2^{288} trials may not be enough to produce a desired 2^{32}-collision. In such a case, one solution is simply trying more messages until we obtain a 2^{32}-collision. Another solution is keeping other multi-collisions of H_7^U at Step 3, and start to compute F_7 by replacing the value of H_7^U.

Attack on Randomized Hashing. Second-preimage attacks that work for any IV can also attack randomized hashing if a hash function has an iterative structure, e.g., Merkle Damgård. Since our second-preimage attack can work for any IV, AURORA-512 and -384 are not secure in randomized hashing. The attack procedure is as follows. Note that this attack finds a 16-block message.

1. The attacker chooses any M and receives K that is chosen without the attacker's control. Then, compute a hash value of the randomized message and obtain T^U and T^D that are the input for the MFF function.
2. Randomly generate a 1-block value K' and a 7-block value $M_1'\|M_2'\|\cdots\|M_7'$.
3. Process the randomized 8-block message $K'\|K'\oplus M_1'\|K'\oplus M_2'\|\cdots\|K'\oplus M_7'$, and obtain $H_8'^U$ and $H_8'^D$ that are the output from the MF function.
4. Find an 8-block message $M_8'\|M_9'\|\cdots\|M_{15}'$ that maps $(H_8'^U\|H_8'^D)$ to $(T^U\|T^D)$, where M_{15}' is a concatenation of 431-bit free value m_{15}', 16-bit value $K'_{L(2)}$, and 65-bit padding string for a 15-block and 447-bit message. This can be done with our second-preimage attack by considering $(H_8'^U\|H_8'^D)$ as the initial value.
5. Output the key K' and the message $M_1'\|M_2'\|\cdots\|M_7'\|K'\oplus M_8'\|K'\oplus M_9'\|\cdots\|$ $\lceil K'\rceil^{431} \oplus m_{15}'$, where $\lceil K'\rceil^{431}$ represents the 431 left-most bits of K'.

The attack complexity is the same as that for the second-preimage attack. Note that at Step 1 of the procedure, the message M can be randomly given. Hence, this attack is stronger than breaking randomized hashing.

Remark on Iterated Compression Function Scenario. During the 8-step computation between two MF computations, AURORA uses 16 different functions $F_0, \ldots, F_7, G_0, \ldots, G_7$. It is interesting to observe the scenario where $F_0, \ldots, F_7$ are replaced with the same function F and $G_0, \ldots, G_7$ are replaced with G. In this scenario, the attack complexity can be reduced by generating multi-fixed-points.

In this attack, for a given H_0^U, the attacker generates 2^{32} messages denoted by $M^{(S)}$ that make $F(H_0^U, M^{(S)}) = H_0^U$. This requires the time complexity of approximately 2^{288} F computations. Then, self-concatenation of any choice of $M^{(S)}$ for 7 blocks guarantees that H_7^U is equal to H_0^U because any $M^{(S)}$ maps H_0^U to H_0^U during 7 blocks. This enables us to save the complexity of generating a multi-collision 6 times. The rest of the attack is exactly the same as the one for standard AURORA-512. Finally, the time complexity becomes $2^{289} (= 2 \times 2^{288})$, which is better than the attack on standard AURORA-512.

5   Key Recovery Attack on HMAC-AURORA

In this section, we present a full key recovery attack on HMAC-AURORA-512 when 512-bit secret keys are used and the MAC length is 512-bit long. Our attack requires 2^{257} queries and the off-line complexity is 2^{259} AURORA-512 operations. The attack can be carried out with a negligible amount of memory. This attack does not make any impact on security of AURORA as a SHA-3 candidate, however, the complexity is significantly less than that of the exhaustive search for a 512-bit key. Our attack can also recover the inner-key of HMAC-AURORA-384 with almost the same complexity as in HMAC-AURORA-512. This attack does not recover the outer-key of HMAC-AURORA-384, but universal forgery is possible by combining the inner-key recovery and second-preimage attacks. Different from collision and second-preimage attacks, this attack does not use multi-collisions. Hence, this attack reveals another security weakness of AURORA-512 and -384.

5.1   Full Key Recovery Attack on HMAC-AURORA-512

In this attack, we mainly ask 1-block messages (including padding bits) as queries. The structure to process a 1-block message in HMAC-AURORA-512 is illustrated in Fig. 4.

Attack Procedure

1. Prepare 2^{257} different messages that are the same length but shorter than 448 bits so that the length of padded messages does not exceed 1-block. Let M^i be prepared messages. Ask all M^i to the oracle, and obtain corresponding $\text{HMAC}_K(M^i)$.

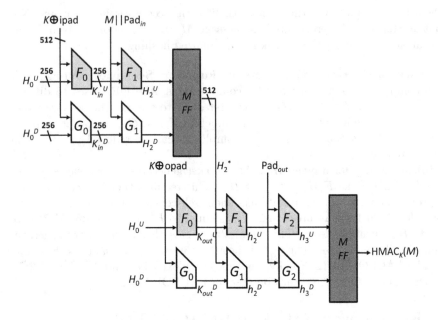

Fig. 4. Structure for processing a 1-block message in HMAC-AURORA-512

2. Find message pairs $(M^j, M^{j'})$ in which $\mathrm{HMAC}_K(M^j)$ and $\mathrm{HMAC}_K(M^{j'})$ are a collision. Due to the computation structure, a pair of messages has the following five possibilities to be a collision.
   Case 1: H_2^Us collide and H_2^Ds collide.
   Case 2: Case 1 does not occur and H_2^*s collide.
   Case 3: Case 1 and 2 do not occur and h_2^Us collide and h_2^Ds collide.
   Case 4: Case 1, 2, and 3 do not occur and h_3^Us collide and h_3^Ds collide.
   Case 5: Case 1, 2, 3, and 4 do not occur and HMAC values collide.
   Therefore, we expect to obtain several collisions in this Step.
3. To detect a Case-1 collision in Step 2, ask $M^j \| \mathrm{Pad}_{in} \| x$ and $M^{j'} \| \mathrm{Pad}_{in} \| x$ for any x to the oracle, and check whether $\mathrm{HMAC}_K(M^j \| \mathrm{Pad}_{in} \| x)$ and $\mathrm{HMAC}_K(M^{j'} \| \mathrm{Pad}_{in} \| x)$ are a collision or not. If they are a collision, $(M^j, M^{j'})$ is a desired pair with a negligible error probability.
4. Let $(M^{j1}, M^{j1'})$ be a colliding pair of Case 1 in Step 2. First, we exhaustively search for K_{in}^U by computing $F_1(K_{in}^U, M^{j1})$ and $F_1(K_{in}^U, M^{j1'})$ for all 2^{256} K_{in}^U and check whether the computed values are a collision or not. If they are a collision, the corresponding K_{in}^U is the correct value. Similarly, we detect K_{in}^D by computing $G_1(K_{in}^D, M^{j1})$ and $G_1(K_{in}^D, M^{j1'})$ for all 2^{256} K_{in}^D and check whether the computed values are a collision or not.
5. For all HMAC collision pairs $(M^j, M^{j'})$ obtained in Step 2, we compute the values of H_2^* and $H_2^{*'}$ with recovered K_{in}^U and K_{in}^D. If H_2^* and $H_2^{*'}$ are a collision, we discard that pair. Note, each of the remaining collision pairs are of Cases 3, 4 or 5 in Step 2.

6. Take a collision pair $(M^{j2}, M^{j2'})$ from all remaining collision pairs, and assume this pair is a collision of Case 3. We then recover K_{out}^U and K_{out}^D by the same method as Step 4. Namely, we exhaustively search for K_{out}^U such that $F_1(K_{out}^U, H_2^{*j2}) = F_1(K_{out}^U, H_2^{*j2'})$ and K_{out}^D such that $G_1(K_{out}^D, H_2^{*j2}) = G_1(K_{out}^D, H_2^{*j2'})$.

7. With recovered K_{in} and K_{out}, compute $\text{HMAC}_K(M)$ for any M that are already asked to the oracle, and check whether its HMAC value matches with the one obtained from the oracle. If matched, that K_{out} is the correct value. Otherwise, discard the pair $(M^{j2}, M^{j2'})$ and go back to Step 6. Repeat the attack by choosing a different collision pair until K_{out} is recovered.

Complexity and Success Probability. At Step 1, we ask 2^{257} queries to the oracle. At Step 2, the probability that the collision of each case is obtained can be considered as independent. According to [14, Theorem 3.2], the probability of obtaining a collision for a $\log_2 N$-bit output hash function, with trying $\theta \cdot N^{1/2}$ different messages is as follows.

$$1 - e^{-\frac{\theta^2}{2}} \tag{5}$$

Eq. 5 becomes approximately 0.86 when $\theta = 2$. Therefore, we expect to obtain a collision of each case with a probability of 0.86. To successfully recover K_{in} and K_{out}, we need to obtain a Case-1 and a Case-3 collision. By 2^{257} queries, the probability of obtaining these two collisions is $(0.86)^2 \approx 0.75$. This is higher than the probability of obtaining a single collision with 2^{256} queries, which is approximately 0.39. For simplicity, we assume that five collisions in total, a single collision in each case, are obtained. At Step 3, we need two queries for each collision. Hence, if we obtained five collisions, we need eight queries in the worst case, which is negligible compared to Step 1. At Step 4, we compute F_1 $2 \cdot 2^{256}$ times to recover K_{in}^U. For each guess of K_{in}^U, the probability that $F_1(K_{in}^U, M^{j1}) = F_1(K_{in}^U, M^{j1'})$ is expected to be 2^{-256}. Hence, we can expect that only one K_{in}^U is chosen as the correct guess. Similarly we compute G_1 $2 \cdot 2^{256}$ times to recover K_{in}^D. As a result, the time complexity for this Step is $2 \cdot 2^{256}$ F_1-operations $+ 2 \cdot 2^{256}$ G_1-operations $\approx 2^{257}$ AURORA-512 operations. Step 5 costs negligible time. In our assumption, three collisions, one for Cases 3, 4, and 5, will remain. Step 6 costs the same complexity as Step 4, which is 2^{257} AURORA-512 operations, and this is repeated three times in the worst case due to Step 7. Therefore, the time complexity for Steps 6 and 7 is $3 \cdot 2^{257}$ AURORA-512 operations. Finally, the total time complexity is 2^{257} AURORA-512 operations for Step 4 and $3 \cdot 2^{257}$ AURORA-512 operations for Step 6, which is 2^{259} AURORA-512 operations.

This attack can be easily carried out if we have a large amount of memory. Moreover, if we apply the memoryless collision search [15] for Step 2, all Steps can be carried out with a negligible amount of memory. To apply the memoryless collision search, we use the HMAC values obtained from the oracle as the next query. Therefore, Step 1 becomes adaptive. The memoryless collision search of our attack requires a message space of 512 bits[1]. Hence, we use 2-block messages as queries. Due to the increment of the message block, at Step 2, a message pair has six possibilities to be a collision. However, since this collision is filtered out at Step 3 with two additional queries, this does not impact the total attack complexity.

5.2   Universal Forgery on HMAC-AURORA-384

Inner Key Recovery Attack. AURORA-384 supports HMAC for a 384-bit MAC length. The structure for processing a 1-block message in HMAC-AURORA-384 is illustrated in Fig. 5.

The inner-key recovery procedure for HMAC-AURORA-384 is almost the same as that of HMAC-AURORA-512. For HMAC-AURORA-384, at Step 2 of the attack procedure, a pair of messages has the following six possibilities to be a collision.

Case 1: H_2^Us collide and H_2^Ds collide.
Case 2: Case 1 does not occur and H_2^*s collide.
Case 3: Case 1 and 2 do not occur and H_2^Ts collide.

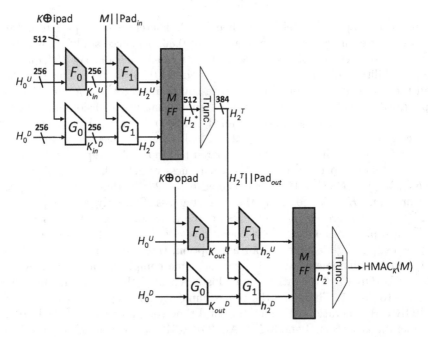

Fig. 5. Structure for processing a 1-block message in HMAC-AURORA-384

---

[1] If the message space is much smaller than 512 bits, for example 447 bits, the randomness for the memoryless collision search will collide after $2^{223.5}$ trials and we cannot make 2^{257} different queries.

Case 4: Case 1, 2, and 3 do not occur and h_2^Us collide and h_2^Ds collide.
Case 5: Case 1, 2, 3, and 4 do not occur and h_2^*s collide.
Case 6: Case 1, 2, 3, 4, and 5 do not occur and HMAC values collide.

Remember that H_2^T and HMAC values are 384 bits. By asking 2^{256} queries, we will obtain a single collision pair of Cases 1, 2, 4, and 5, and $2^{127}(=2^{256 \cdot 2-1-384})$ collision pairs of Cases 3 and 6; therefore, we expect to obtain $2^{128}+4$ collisions in total. To recover the inner-key, we need to detect the collision pair of Case 1. At Step 3 of the attack procedure, this can be achieved by asking two additional queries $M^j \| \mathrm{pad}_{in} \| x$ and $M^{j'} \| \mathrm{pad}_{in} \| x$ for each collision pair $(M^j, M^{j'})$. The inner-key recovery procedure at Step 4 is exactly the same, in which we need a time complexity of 2^{257} AURORA-384 operations.

Finally the inner-key is recovered with $2^{257}+2 \cdot (2^{128}+4) \approx 2^{257}$ queries and a time complexity of 2^{257} AURORA-384 operations. This attack can be performed with a negligible amount of memory.

Universal Forgery by Combining the Inner-Key Recovery and Second-Preimage Attacks. Although our attack cannot recover the outer-key, we can perform a universal forgery on HMAC-AURORA-384 by using the recovered inner-key and applying the second-preimage attack, which is explained in Section 4 or by Ferguson and Lucks [9].

In a universal forgery attack, the attacker has access to the oracle which returns $\mathrm{HMAC}_k(\cdot)$. For any given message M, our attack can find the value of $\mathrm{HMAC}_k(M)$ without asking M to the oracle. After revealing the inner-key, our attack requires one query and the same off-line complexity and memory as that of the second-preimage attack on AURORA-512, which are 2^{290} AURORA-512 operations and $2^{288} \times 512$ bits of memory in Section 4 of this paper and 2^{291} AURORA-512 operations and $2^{31.5}$ message blocks of memory in [9]. The attack procedure is as follows.

Target:
0. Receive M.
Preparation:
1. Recover the inner-key K_{in} with the attack explained in Section 5.2.
Universal forgery:
2. For the given M, find a second-preimage M' s.t. $\mathrm{AURORA}-384(K_{in}, M) = \mathrm{AURORA}-384(K_{in}, M')$ by using the second-preimage attack.
3. Ask M' to the oracle, and receive $\mathrm{HMAC}_k(M')$.
4. $\mathrm{HMAC}_k(M')$ is the HMAC value of M.

6   Discussion and Conclusions

The designers of AURORA showed proof for preimage resistance but did not show proof for second-preimage resistance. In fact, we showed that AURORA does not satisfy second-preimage resistance. Therefore, it would be useful to consider the differences of these two properties. We give some intuition by summarizing observations obtained from our attacks.

Assume that a hash function H is a sequence of several independent functions $H_1, H_2, \ldots, H_j$. The preimage resistance can be guaranteed if at least one of the functions is preimage resistant. However, this is not true for second-preimage resistance. To guarantee second-preimage resistance, all functions should be secure. The security bound of the second-preimage resistance is dependent on the weakest part of the hash function. AURORA can be regarded as consisting of two parts; the first 8-block H_1 and the MFF function H_2. Because the MFF function is secure, AURORA is secure on preimage resistance. However, because the first 8 blocks is not secure, AURORA does not satisfy second-preimage resistance.

From this observation, designing hash functions which are provably secure for second-preimage resistance seems harder than designing hash functions which are provably secure for preimage resistance.

7   Conclusion

We pointed out the weakness of the DMMD mode of operation. We first presented a collision attack on AURORA-512. We then presented a second-preimage attacks on AURORA-512 and -384, then explained that randomized hashing could also be attacked. Finally, we showed a full key-recovery attack on HMAC-AURORA-512 and a universal forgery attack on HMAC-AURORA-384.

Acknowledgements. I would like to thank Praveen Gauravaram (Department of Mathematics, DTU) for his comments on randomized hashing and thank Orr Dunkelman (Ecole normale superieure) for sharing his observations on the second-preimage attack in iterated compression function scenario. I also thank anonymous referees for helpful comments.

References

1. U.S. Department of Commerce, National Institute of Standards and Technology: Federal Register 72(212), Friday (November 2, 2007) Notices,
   http://csrc.nist.gov/groups/ST/hash/documents/FR_Notice_Nov07.pdf
2. Lucks, S.: A failure-friendly design principle for hash functions. In: Roy, B. (ed.) ASIACRYPT 2005. LNCS, vol. 3788, pp. 474–494. Springer, Heidelberg (2005)
3. Iwata, T., Shibutani, K., Shirai, T., Moriai, S., Akishita, T.: AURORA: A Cryptographic Hash Algorithm Family. Initial submission version (October 31, 2008), AURORA home page,
   http://www.sony.net/Products/cryptography/aurora/index.html,
   NIST home page:
   http://csrc.nist.gov/groups/ST/hash/sha-3/index.html
4. U.S. Department of Commerce, National Institute of Standards and Technology: Randomized Hashing for Digital Signatures (NIST Special Publication 800-106) (February 2009),
   http://csrc.nist.gov/publications/nistpubs/800-106/
   NIST-SP-800-106.pdf

5. U.S. Department of Commerce, National Institute of Standards and Technology: The Keyed-Hash Message Authentication Code (HMAC) (Federal Information Processing Standards Publication 198) (July 2008),
http://csrc.nist.gov/publications/fips/fips198-1/FIPS-198-1_final.pdf
6. Joux, A.: Multicollisions in iterated hash functions. Application to cascaded constructions. In: Franklin, M. (ed.) CRYPTO 2004. LNCS, vol. 3152, pp. 306–316. Springer, Heidelberg (2004)
7. Knudsen, L.R., Mendel, F., Rechberger, C., Thomsen, S.S.: Cryptanalysis on MDC-2. In: Joux, A. (ed.) EUROCRYPT 2009. LNCS, vol. 5479, pp. 106–120. Springer, Heidelberg (2009)
8. International Organization for Standardization: ISO/IEC 10118-2:1994, Information technology – Security techniques – Hash-functions – Part 2: Hash-functions using an n-bit block cipher algorithm (1994) (Revised in 2000)
9. Ferguson, N., Lucks, S.: Attacks on AURORA-512 and the Double-Mix Merkle-Damgaard transform. Cryptology ePrint Archive, Report 2009/113, Ver. 20090311:092718 (2009), http://eprint.iacr.org/2009/113
10. Joux, A., Lucks, S.: Improved generic algorithms for 3-collisions. Cryptology ePrint Archive, Report 2009/305 (2009), http://eprint.iacr.org/2009/305
11. Gauravaram, P., Knudsen, L.R.: On randomizing hash functions to strengthen the security of digital signatures. In: Joux, A. (ed.) EUROCRYPT 2009. LNCS, vol. 5479, pp. 88–105. Springer, Heidelberg (2009)
12. Krawczyk, H., Bellare, M., Canetti, R.: HMAC: Keyed-Hashing for Message Authentication. The Internet Engineering Task Force (1997),
http://www.ietf.org/rfc/rfc2104.txt
13. Suzuki, K., Tonien, D., Kurosawa, K., Toyota, K.: Birthday paradox for multicollisions. IEICE Transactions on Fundamentals of Electronics, Communications and Computer Sciences E91-A(1), 39–45 (2008)
14. Vaudenay, S.: A Classical Introduction to Cryptography: Applications for Communications Security. Springer, Heidelberg (2006)
15. Quisquater, J.J., Delescaille, J.P.: How easy is collision search. New results and applications to DES. In: Brassard, G. (ed.) CRYPTO 1989. LNCS, vol. 435, pp. 408–413. Springer, Heidelberg (1990)

A   Relationship between Our Work and [9]

This paper mainly presents three attacks on AURORA-512; a collision attack, a second-preimage attack and its application to randomized hashing, and a key-recovery attack on HMAC. Ferguson and Lucks independently found similar results on collision and second-preimage attacks [9].

Our work on collision and second-preimage attacks was motivated by the discussion by Ferguson, Lucks, and Iwata during a presentation on AURORA by Iwata at the first SHA-3 conference on 27th February 2009. We found our collision attack immediately after Iwata's presentation and informed it to the AURORA team that same day. On the other hand, Ferguson and Lucks mentioned that "at that point of time, (the concerns) had not been thought through" [9][Sec. 6]. Hence, we believe that we first found collision attack.

Regarding second-preimage resistance, we found the attack in a few days after the conference. Hence, the work is independent of Ferguson and Lucks. However, we heard they independently found the second-preimage attack during the SHA-3 conference before we found it.

From a technical viewpoint, the attack found by Ferguson and Lucks [9] and ours are in the same framework. However, we use 8-block multi-collisions, whereas [9] uses 9-block multi-collisions. Hence, the attack complexity of [9] is superior to ours in both collision and second-preimage attacks. The evaluation for the amount of memory is significantly different, in which our attack requires 2^{288}, whereas their attack [9] requires only $2^{31.5}$. This difference is based on the assumption on compression functions rather than attack techniques. Ferguson and Lucks assumes that compression functions are "balanced", whereas our attack also considers the case where the output distribution is very biased.

More on Key Wrapping

Rosario Gennaro and Shai Halevi

IBM T.J. Watson Research Center
Hawthorne, NY 10532, USA
rosario@us.ibm.com, shaih@alum.mit.edu

Abstract. We address the practice of key-wrapping, where one symmetric cryptographic key is used to encrypt another. This practice is used extensively in key-management architectures, often to create an "adapter layer" between incompatible legacy systems. Although in principle any secure encryption scheme can be used for key wrapping, practical constraints (which are commonplace when dealing with legacy systems) may severely limit the possible implementations, sometimes to the point of ruling out any "secure general-purpose encryption." It is therefore desirable to identify the security requirements that are "really needed" for the key-wrapping application, and have a large variety of implementations that satisfy these requirements.

This approach was developed in a work by Rogaway and Shrimpton at EUROCRYPT 2006. They focused on allowing deterministic encryption, and defined a notion of *deterministic authenticated encryption* (DAE), which roughly formalizes "the strongest security that one can get without randomness." Although DAE is weaker than full blown authenticated encryption, it seems to suffice for the case of key wrapping (since keys are random and therefore the encryption itself can be deterministic). Rogaway and Shrimpton also described a mode of operation for block ciphers (called SIV) that realizes this notion.

We continue in the direction initiated by Rogaway and Shirmpton. We first observe that the notion of DAE still rules out many practical and "seemingly secure" implementations. We thus look for even weaker notions of security that may still suffice. Specifically we consider notions that mirror the usual security requirements for symmetric encryption, except that the inputs to be encrypted are random rather than adversarially chosen. These notions are all strictly weaker than DAE, yet we argue that they suffice for most applications of key wrapping.

As for implementations, we consider the key-wrapping notion that mirrors authenticated encryption, and investigate a template of Hash-then-Encrypt (HtE), which seems practically appealing: In this method the key is first "hashed" into a short nonce, and then the nonce and key are encrypted using some standard encryption mode. We consider a wide array of "hash functions", ranging from a simple XOR to collision-resistant hashing, and examine what "hash function" can be used with what encryption mode.

Keywords: Deterministic Encryption, Key Wrapping, Modes of Operation, Symmetric Encryption.

M.J. Jacobson Jr., V. Rijmen, and R. Safavi-Naini (Eds.): SAC 2009, LNCS 5867, pp. 53–70, 2009.
© Springer-Verlag Berlin Heidelberg 2009

1   Introduction

Key-wrapping roughly refers to encrypting one cryptographic key with another. In this paper we focus on the use of this practice for symmetric encryption, where the main application is key-management. Key-management architectures often include a hierarchy (or tree) of keys, with a master key encrypting several lower keys, which in turn encrypt even lower keys, and with the leaf keys used to encrypt "the real data" (cf. [17, Chapter 6]). Another typical case where key-wrapping is used is retrofitting an encryption system to work with an incompatible key-management architecture, for example an AES encryption system with a 3DES key-management. In such cases, one can add a "glue layer" in between the encryption system and the key management architecture, that generates *data keys* as expected by the encryption system (e.g., AES keys) and uses the keys from the key-management architecture to wrap these data keys (e.g., using 3DES).

A similar situation arises when the encryption system must use its keys in a restricted manner, but the key-management architecture is not designed to keep track of these restrictions. For example, one system that we encountered was using the GCM encryption mode, and needed to comply with the following requirement from the NIST standard for GCM [9]:

> *The total number of invocations of the authenticated encryption function shall not exceed 2^{32}, including all IV lengths and all instances of the authenticated encryption function with the given key.*

However, that system was using a key-management architecture that did not keep track of the number of times that any single key is being served, and hence was not able to certify that the requirement from above is being met. Here too, the solution was to add a key-wrapping adapter layer that generates a new GCM key every time and wraps it with the given key from key-management.

1.1   What Is a Secure Key-Wrapping?

It is clear that any secure encryption scheme is in particular also a secure key-wrapping scheme. But using secure encryption may be an overkill for the application to key-wrapping. In particular, the usage of key-wrapping as an adapter between legacy systems sometimes imply severe practical limitations on its implementation, perhaps to the point of excluding general-purpose secure encryption. We therefore seek weaker notions of security that can be implemented even in cases where standard secure encryption is impossible, but are still strong enough for the purpose of key-wrapping.

This approach was taken by Rogaway and Shrimpton in [22], where they focused on allowing deterministic procedures. Specifically, they investigated *deterministic authenticated encryption* (DAE), which roughly formalizes "the strongest security that one can get from deterministic procedures." Although achieving less than standard authenticated encryption, DAE appears to be sufficient for key-wrapping: since the key itself is already random, it seems that randomness is not really needed in the encryption procedure itself. Indeed, Rogaway and Shrimpton included in the full version of their work an appendix in which they prove that DAE is good enough for applications that encrypt high entropy plaintext. (See more discussion in our Appendix A.)

However, even DAE may sometimes be too much to ask for. In this paper we show several examples of practical "seemingly secure" schemes that nevertheless fail to meet the notion of DAE. We thus aim lower, looking for even weaker notions that still suffice for key wrapping. Noting that the difference between key wrapping and general-purpose encryption is that the plaintext to be encrypted is a symmetric cryptographic key (and therefore is random), we consider notions of security that mirror the usual notions for symmetric encryption, except that the attack model postulates that the plaintext to be encrypted is random rather than adversarially controlled. Specifically, in Section 2 we present notions that mirror CPA-security, CCA-security, and integrity of ciphertext. We argue that these notions suffice for many application of key-wrapping. We also prove formally that they suffice for the typical application in which a master-key is used to encrypt data keys, which themselves are used to encrypt real data.

1.2 How to Achieve Secure Key-Wrapping?

Implementing secure key wrapping can be done using standard secure symmetric encryption, perhaps using generic composition techniques such as encrypt-then-authenticate [5, 6, 14] (which work for key-wrap just as well as for regular encryption). Another solution was given by Rogaway and Shrimpton [22], who designed a mode of operation called SIV that they prove to meet the stronger notion of DAE. However, applications of key wrapping sometimes place restrictions on the implementation. (For example, being deterministic, or using a specific encryption mode because that mode is already implemented in hardware, etc.) The thrust of this paper is therefore to examine many different plausible constructions, trying to separate secure constructions from insecure ones.

We focus specifically on an approach for achieving authenticated key-wrap that we call *Hash-then-Encrypt* (HtE). In this method, the key is first "hashed" into a short nonce, and then the nonce and key are encrypted using some standard encryption mode. There are several reasons to look at this approach: First, we may be able to get away with using a very simple "hash function" (maybe as simple as just XOR), which could be very efficient. Perhaps more importantly, this template could allow re-use of components that is already implemented in existing systems.

In this work we consider a wide array of "hash functions", and examine what "hash function" can be used with what encryption mode. We show that all the modes that we considered can be turned into a secure authenticated key-wrapping scheme by using a second-preimage-resistant function for the hash function, and many of them (except ECB and maybe CBC) can also use universal hashing. But resisting collisions is not really necessary in most cases. We show that for all modes except CTR, a simple fixed linear function is already enough to get some level of security (but this level of security deteriorates quickly with the

Table 1. Security of various Hash-then-Encrypt constructions

Encryption/Hash	XOR	Linear	2nd-preimage resistant	universal hashing
CTR	broken	broken	secure	secure
ECB	broken	somewhat*	secure	broken
CBC	broken	somewhat*	secure	?
masked ECB/CBC	somewhat*	somewhat*	secure	secure
XEX	secure	secure	secure	secure

* "somewhat" means concrete security that is worse than the birthday bound

length of the data key), and when using masked versions of ECB and CBC then even a simple XOR of the key blocks suffices to get the same level of security. Finally, when using a tweakable encryption mode [16] such as XEX [20], a simple XOR suffices to get security upto the birthday bound, regardless of the length of the data keys.[1] These results are summarized in Table 1.

1.3 Related Work

We already discussed the work of Rogaway and Shrimpton [22] on the key-wrap problem. A somewhat similar definition to DAE was later formulated by Amanatidis et al. in the context of searchable encryption [2]. An and Bellare studied authentication via redundancy in the context of standard symmetric encryption [3]. They argued that public redundancy function is not very useful for achieving authenticated encryption (although work by Gligor and Donescu [10], Jutla [12], and Rogaway [19] demonstrated that simple public redundancy is sufficient when using masked CBC or masked ECB for encryption). In our case we show that even very simple public redundancy is sufficient in most cases. Also Bellare and Namprempre [5] and Krawczyk [14] deal with generic composition techniques of encryption and authentication, and some further results were described by Canetti et al. [6].

Other related work was done in the area of KEM/DEM schemes for public key encryption, where different conditions on the KEM and/or DEM parts were investigated (e.g., [1, 7, 15]). Also, some recent work addressed public-key deterministic encryption [4]. Finally, we mention that encryption of "random messages" was also considered in the very different context of "Entropic security" by Russell and Wang [23] and Dodis and Smith [8]: in these works they attempted to provide statistical security for random messages using as little key material per message as possible.

Organization. Due to space limitations, some of the results and proofs were deferred to the final version.

---

[1] Observe that the masked modes and XEX are obtained by adding very simple masking to ECB and CBC modes, so it makes sense to talk about them here, even though our paper is focused on dealing with legacy systems.

2   Defining Security for Key Wrapping

Below we adapt the usual notions of security for symmetric encryption to the case of key-wrapping. The only difference between our notions and the standard ones is that the plaintext is chosen at random rather than being controlled by the attacker. We focus on the simplest case of fixed input length and no associated data, extensions and variations are discussed in the long version.

Syntactically, a key-wrapping scheme is identical to an encryption scheme. Namely, it includes a wrapping procedure Wrap that takes plaintext and wrapping-key and returns ciphertext, and an unwrapping procedure Unwrap that takes ciphertext and wrapping key and returns the plaintext (or an error symbol $\perp$). We have the usual validity condition, asserting that for any wrapping key K and plaintext D it holds that $\text{Unwrap}_K(\text{Wrap}_K(D)) = D$. Below we usually refer to the plaintext as a *data key*. We insist that wrapping keys (as well as data keys) are uniformly random bit strings of some given length. Hence key-generation is implicitly specified as choosing a random key of the appropriate length. We denote the length of the wrapping key by k, and the length of the plaintext/data-keys by ℓ. (One can think of k as the security parameter and ℓ is typically also a parameter.)

2.1   Security for Key-Wrap

Let $\mathcal{KW} = (\text{Wrap}, \text{Unwrap})$ be a key-wrapping scheme. All the security definitions are based on probabilistic games, involving an attacker A and the procedures Wrap and Unwrap. Our definitions use the "left-or-right" style. The basic game is the Random-Plaintext Attack (RPA), mirroring the usual chosen-plaintext attack: First a wrapping key W is chosen uniformly at random in $\{0,1\}^k$, together with random "challenge bit" b. Then the attacker interacts with the wrapping procedure as follows: whenever A invokes the wrapping procedure, two data keys D_0, D_1 are chosen uniformly at random in $\{0,1\}^\ell$, and A receives both D_0, D_1, and also the ciphertext $C = \text{Wrap}_W(D_b)$. The attacker A can keep making such queries, and eventually it halts and outputs a guess for the value of the challenge bit b. The RPA-*advantage* of A is defined as

$$\mathbf{Adv}_{\mathcal{KW}}^{\text{kw.rpa}}(A) \stackrel{\text{def}}{=} \Pr[A^{LR\$\text{Wrap}_W} \Rightarrow 1 | b = 1] - \Pr[A^{LR\$\text{Wrap}_W} \Rightarrow 1 | b = 0] \quad (1)$$

where $LR\$\text{Wrap}$ is the "left-or-right" procedure described above, $A \Rightarrow 1$ is the event where A outputs the bit '1', and the probability is taken over all the probabilistic choices in this game.

The *Chosen-Ciphertext Attack* (CCA) game is similar, except that the attacker is also given access to the unwrapping procedure that on query C returns $\text{Unwrap}_W(C)$, but the attacker is prevented from querying it on ciphertexts that were previously returned from the procedure Wrap. Then the *CCA-advantage* of A is defined as

$$\mathbf{Adv}_{\mathcal{KW}}^{\text{kw.cca}}(A) \stackrel{\text{def}}{=}$$
$$\Pr[A^{LR\$\text{Wrap}_W, \text{Unwrap}_W} \Rightarrow 1 | b = 1] - \Pr[A^{LR\$\text{Wrap}_W, \text{Unwrap}_W} \Rightarrow 1 | b = 0] \quad (2)$$

The *integrity of ciphertext* game (INT) is defined similarly to the RPA game, except that wrapping queries return only one random data key D and the corresponding ciphertext $C = \mathsf{Wrap_W}(D)$, and the attacker's goal is to output any ciphertext C^*, different than the ones that were returned from the Wrap procedure, that the unwrap procedure does not reject. Namely, the advantage of A is defined as

$$\mathbf{Adv}_{\mathcal{KW}}^{\mathsf{kw.int}}(A) \overset{\text{def}}{=} \Pr[A^{\$\mathsf{Wrap}} \Rightarrow C^* : C^* \text{ is "new" and } \mathsf{Unwrap}(C^*) \neq \bot] \quad (3)$$

where $Wrap is the procedure above that returns both a random data key and its encryption.

As usual, we extend the advantage notations to talk about the advantage of "any attacker within the given limited resources". For example, $\mathbf{Adv}(\mathsf{enc} = q)$ means any attacker that makes at most q queries to its encryption oracle. We will explicitly specify the relevant resources whenever we use this convention. We informally say that a scheme is "secure" when the advantage of an attacker is no more than the birthday bound (i.e., $O(q^2/2^n)$ where q is the resource bound and n is a relevant security parameter). We say that a scheme is "somewhat secure" where the advantage is exponentially small in n but larger than the birthday bound, and otherwise we say that the scheme is "broken." Clearly, RPA-security captures the notion of secrecy for the key against eavesdroppers, CCA-security ensures key secrecy also against active attackers, and the last notion adds explicit authentication.

Below we refer to a scheme which is both RPA-secure and has integrity of ciphertext as *authenticated key-wrap*. Just as for encryption [5, 13], an easy argument shows that RPA-security and integrity-of-ciphertext imply CCA-security. It is also easy to see that requiring both is strictly stronger than requiring CCA-security (e.g., a random permutation is CCA-secure but does not provide integrity of ciphertext).

In the long version we discuss some extensions of these definitions, e.g., to handle variable-input-length, associated data, etc.

2.2   Key-Wrapping Is Weaker Than DAE

We note that the security notions from above are all strictly weaker than the notion of deterministic authenticated encryption (DAE) of Rogaway and Shrimpton [22]: DAE requires that an attacker that interacts with the Wrap and Unwrap procedures (with some fixed random secret key W) cannot distinguish them from a dummy Wrap that returns only random bits, and a dummy Unwrap that rejects any "new" ciphertext (i.e., any ciphertext that was not returned by the Wrap procedure). Obviously, when interacting with the dummy procedures we have $\mathbf{Adv}^{\mathsf{kw.rpa}}(A) = \mathbf{Adv}^{\mathsf{kw.int}}(A) = 0$ (and therefore also $\mathbf{Adv}^{\mathsf{kw.cca}}(A) = 0$), so DAE implies all of our notions. In fact, Rogaway and Shrimpton proved in an appendix of the long version of [22] that DAE implies a similar (but stronger) notion of security, for a case where part of the plaintext is random and another part is chosen by the attacker. See discussion in our Appendix A.

On the other hand, in the DAE game the attacker can query the wrapping procedure on inputs of its choice, so it is easy to find examples of schemes that satisfy our notions but are not DAE. (In fact, some of our "provably secure" constructions from Section 3 below fail to meet the notion of DAE.) One easy example is a wrapping procedure based on a block cipher, that wraps a one-block data-key D using the wrapping key W by setting $C = \langle E_W(D), E_W(D+1) \rangle$. It is clear that in all the games as described above, an attacker making at most q queries to the wrapping procedure has advantage at most $O(q^2/2^n)$, since the inputs to the block ciphers will be all disjoint except with that probability. On the other hand, a DAE attacker that can specify the data keys only needs to encrypt two keys D_0, D_1 such that $D_1 = D_0 + 1$ and check that the first block in C_1 is the same as the second block in C_0.

2.3 Key-Wrapping Is Sufficient for Applications

Although weaker than DAE, we claim that our notions are sufficiently strong for most application of key wrapping. That is, for any application that uses a key-wrapping procedure to wrap random keys, it is sufficient for the key-wrapping to satisfy the notions that we defined above. In some sense this statement is true by definition: our notions ensure that an attacker cannot distinguish the "real keys" from random unrelated keys, which means that even after seeing (and perhaps even manipulating) the wrapped keys, they are still just random secret keys from the attacker's perspective. Below we demonstrate formally that secure key-wrapping is sufficient to get secure symmetric encryption.

Specifically, let $\mathcal{KW} = (\mathsf{Wrap}, \mathsf{Unwrap})$ be a key-wrapping scheme and let $\mathcal{SE} = (\mathsf{Enc}, \mathsf{Dec})$ be a symmetric encryption scheme. Consider the composite symmetric encryption scheme $\mathcal{C}$, whose key space is that of $\mathcal{KW}$ and whose message space is that of $\mathcal{SE}$. On a given key W and plaintext message M, the composite encryption chooses a new random data-key D from the key-space of $\mathcal{SE}$, wraps it using W to get $C_1 \leftarrow \mathsf{Wrap}_W(D)$, uses it to encrypt the message, getting $C_2 \leftarrow \mathsf{Enc}_D(M)$, and outputs the composite ciphertext $C = (C_1, C_2)$. The composite decryption first recovers $D \leftarrow \mathsf{Unwrap}_W(C_1)$ and if $D \neq \perp$ then computes and returns $M \leftarrow \mathsf{Dec}_D(C_2)$.

The following lemma asserts that if $\mathcal{KW}$ and $\mathcal{SE}$ are secure then so is $\mathcal{C}$. More specifically, if $\mathcal{KW}$ is RPA-secure and $\mathcal{SE}$ is CPA-secure then the composite is CPA-secure, if both are CCA-secure then so is the composite, and if both have integrity of ciphertext and the key-wrapping is RPA-secure then the composite also has integrity of ciphertext. One point to note is that since the application uses each data key only once, then the underlying encryption $\mathcal{SE}$ need only be secure under encryption of a single message.

Lemma 1. *Let q, q' be bounds on the number of encryption/wrapping queries and the number of decryption/unwrapping, respectively. Then*

$$\mathbf{Adv}_{\mathcal{C}}^{\mathsf{enc.cpa}}(enc = q) \leq \mathbf{Adv}_{\mathcal{KW}}^{\mathsf{kw.rpa}}(wrap = q) + q \cdot \mathbf{Adv}_{\mathcal{SE}}^{\mathsf{enc.cpa}}(enc = 1),$$

$$\mathbf{Adv}_{\mathcal{C}}^{\mathsf{enc.cca}}(enc = q, dec = q') \leq$$

$$\mathbf{Adv}^{\mathsf{kw.cca}}_{\mathcal{KW}}(wrap = q, unwrap = q') \; + \; q \cdot \mathbf{Adv}^{\mathsf{enc.cca}}_{\mathcal{SE}}(enc = 1, dec = q'),$$

$$\mathbf{Adv}^{\mathsf{enc.int}}_{\mathcal{C}}(enc = q) \leq$$
$$\mathbf{Adv}^{\mathsf{kw.rpa}}_{\mathcal{KW}}(wrap = q) + \mathbf{Adv}^{\mathsf{kw.int}}_{\mathcal{KW}}(wrap = q) \; + \; q \cdot \mathbf{Adv}^{\mathsf{enc.int}}_{\mathcal{SE}}(enc = 1).$$

The running-time bounds on the various attackers that are hidden in the expressions above are all about equal: they differ by at most the time that it takes to compute q encryptions/wrappings and q' decryptions/unwrappings.

3   Authenticated Key-Wrap

As we said before, in principle one can use any authenticated encryption scheme to achieve authenticated key-wrap. Another "clean solution" for obtaining authenticated key-wrap is *wrap-then-authenticate*, where one first employs any RPA-secure scheme for wrapping and then authenticates the ciphertext with any secure MAC. As for encryption [5, 6, 14], here too one gets RCCA-security from any MAC and authenticated key-wrap when the MAC is "strongly unforgeable".[2]

Yet another option is to use the carefully-engineered SIV mode of Rogaway and Shrimpton [22]: In SIV the data key D (and associated data A) are first fed into a pseudorandom function to get a nonce, $N = PRF_w(\mathsf{D}, \mathsf{A})$, and then the data key is encrypted with an IV-based encryption scheme (such as CTR mode or CBC mode, with a key which is independent of the PRF key). Shrimpton and Rogaway proved that SIV realizes their notion of DAE (and therefore is also an authenticated key-wrap) for any PRF and any "pseudorandom IV-based encryption."[3] They suggested implementing the pseudorandom function using a variant of CBC-MAC, and the encryption using CTR mode.

But as we argued in the introduction, there are still cases where one may want to use other implementation strategies. Below we analyze a wide range of solutions that may be appealing in practice, with a goal to determine what works and what doesn't.

3.1   Simplified SIV May Not Work

We remind the reader that the main difference between Rogaway and Shrimpton's notion of DAE and our notions of security is that the attacker in their model can choose the plaintext, whereas in our model the data-key is always chosen at random. One could therefore hope that we can get a secure scheme even if we weaken SIV by replacing the pseudorandom function with a "weak pseudorandom function." (Recall that a weak pseudorandom function [18] is a

---

[2] A MAC is "strongly unforgeable" if the attacker cannot even produce a new valid authentication tag for a previously-authenticated message.

[3] A "pseudorandom IV-based encryption" is one where the ciphertext in a chosen-plaintext attack is indistinguishable from random. Shrimpton and Rogaway called this notion "conventional IV-based encryption".

function F, such that no attacker can distinguish $F(x)$ from random as long as the points x themselves are chosen at random.)

Unfortunately, this intuition fails: For example, for an n-bit block cipher E and $2n$-bit data keys K_1, K_2, it is easy to see that the function $F_w(K_1, K_2) = E_w(K_1 \oplus K_2)$ is a weak pseudorandom function (upto the birthday bound). But implementing a key-wrap using this function and CTR mode is completely broken: an attacker that sees the ciphertext C_0, C_1, C_2 that corresponds to the key K_1, K_2 can trivially obtain a ciphertext for related keys simply by XOR-ing the same non-zero block Δ into both C_1 and C_2, which would be a valid ciphertext for the key $K_1 \oplus \Delta, K_2 \oplus \Delta$. Similarly, using this function F with CBC encryption is insecure, since if C_0, C_1, C_2, C_3 is a valid ciphertext for the key K_1, K_2, K_3, then C_0, C_2, C_1, C_3 is a valid ciphertext for the key $K_1, K_2 \oplus C_1 \oplus C_2, K_3 \oplus C_1 \oplus C_2$.

3.2   Hash-then-Encrypt

Next we examine the solution template of "Hash-then-Encrypt" (HtE). That is, the data-key K to be wrapped is first compressed into a one-block nonce using some "hash functions", $N = H(K)$, and then the nonce and key together are encrypted using a standard encryption mode. Here we consider using CTR mode, ECB, and CBC. (In the long version we also explore the masked variants ECB-X and CBC-X, and "narrow block tweakable modes" [16] such as Rogaway's XEX [20].) Hash-then-Encrypt is similar to SIV when one thinks of $E(H(K))$ as the pseudorandom function of SIV. However, below we also consider weak versions of H for which $E(H(K))$ is not a PRF.[4]

For any "hash function" H, given a wrapping key W (that includes a cipher key w) and a data key $K = \langle K[1], \dots, K[\ell] \rangle$ (where each $K[i]$ is an n-bit block), compute $N = H(K)$ and $C[0] = E_w(N)$ and then for $i = 1, \dots, \ell$ set:

HtCTR. $C[i] = K[i] \oplus E_w(C[0] + i - 1)$ where the addition is modulo 2^n (say).

HtECB. $C[i] = E_w(K[i])$.

HtCBC. $C[i] = E_w(C[i-1] \oplus K[i])$.

HtECB-X. $C[i] = E_w(X[i] \oplus K[i]) \oplus X[i]$, where the $X[i]$'s are "XOR universal" and derived from a different part of the wrapping key W.

HtCBC-X. $C[i] = E_w(C[i-1] \oplus K[i]) \oplus X[i]$, where the $X[i]$'s are "XOR universal" and derived from a different part of the wrapping key W.

HtXEX. $C[i] = E_w(X[i] \oplus K[i]) \oplus X[i]$, where the $X[i]$'s are computed as $X_i \leftarrow \alpha^{i-1} \cdot E_w(C[0])$, with α a primitive element of $GF(2^n)$.

The modes HtCTR, HtECB, and HtCBC are depicted in Figures 1, 2, and 3, respectively. For the "hashing" part we analyze several different functions, both keyed and un-keyed. For each encryption mode we seek sufficient and/or necessary conditions on the hash function to get authenticated key-wrap.

---

[4] Another technical difference is that in our case the key used by E in the "PRF part" is the same as the key used by E in the "encryption part."

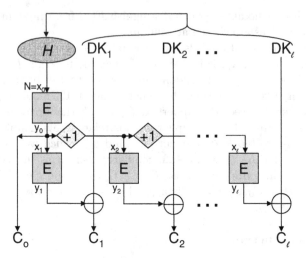

Fig. 1. HtCTR

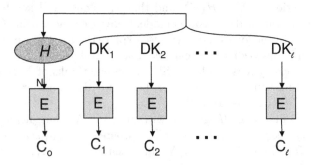

Fig. 2. HtECB

3.3 Hash-then-CTR

When using counter-mode encryption, it turns out that a necessary and sufficient condition on the hash function H is that it resists second-preimage collisions. However, we point out that in our case, the function H is allowed to have a secret key and moreover the attacker does not get to see the hash value, so it is easier to get second preimage resistance than in the usual settings where H is public. (In particular, any universal hash function is second preimage resistant in our setting.) The definition below formalizes the notion of second-preimage-resistance that we use:

Resisting second-preimage collisions. Let H be a function that can depend on a secret key and/or on public parameters. The definition below is formalized for the case of fixed input length, so we have a parameter ℓ that denotes the input length of H. We also have parameters n, k', k'' denoting the output length

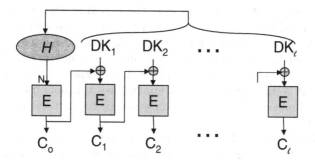

Fig. 3. HtCBC

and the lengths of the secret key and public parameters (if any). The attack scenario that we consider is where the secret key, public parameters, and the first preimage are chosen at random, $\mathsf{sk} \in_R \{0,1\}^{k'}$ $\mathsf{pp} \in_R \{0,1\}^{k''}$, $X \in_R \{0,1\}^{\ell}$, the attacker is given the public parameters and the first preimage, and its goal is to find a second preimage that collides with the first under H. We denote the second-preimage-advantage of an attacker A by

$$\mathbf{Adv}_H^{\mathsf{spr}}(A) \stackrel{\text{def}}{=}$$
$$\Pr_{\mathsf{sk,pp},X} [A(\mathsf{pp}, X) \Rightarrow X' \ : \ X' \neq X \text{ and } H(\mathsf{sk}, \mathsf{pp}, X) = H(\mathsf{sk}, \mathsf{pp}, X')] \quad (4)$$

The case where the secret key is empty corresponds to the usual notion of second preimage resistance for public hash functions (of function families). Below we also denote by α_H the probability that two random inputs collide under a random key, namely $\alpha_H \stackrel{\text{def}}{=} \Pr_{\mathsf{sk,pp},X,X'}[H(\mathsf{sk}, \mathsf{pp}, X) = H(\mathsf{sk}, \mathsf{pp}, X')]$. (Clearly $\alpha_H \leq \mathbf{Adv}_H^{\mathsf{spr}}$, but α_H could sometimes be much smaller.)

Next we show that the HtCTR construction is secure in the sense of authenticated key-wrap if and only if the hash function H is second-preimage resistant according to the notion above.

Lemma 2. *Let H be a (potentially keyed) hash function as above with input length ℓn and output length n, and consider the HtCTR construction using H for the hash function and with a truly random permutation for the block cipher. Then for any bound q on the number of wrapping queries, we have*

$$\mathbf{Adv}_{\mathrm{HtCTR}}^{\mathsf{kw.rpa}}(wrap = q) \leq 2\binom{q}{2}\alpha_H + O(q^2\ell/2^n),$$

$$\mathbf{Adv}_{\mathrm{HtCTR}}^{\mathsf{kw.int}}(wrap = q) \leq q\mathbf{Adv}_H^{\mathsf{spr}} + \binom{q}{2}\alpha_H + O(q^2\ell^2/2^n),$$

$$\mathbf{Adv}_{\mathrm{HtCTR}}^{\mathsf{kw.int}}(wrap = 1) \geq \mathbf{Adv}_H^{\mathsf{spr}}$$

The running-time bounds on the various attackers that are hidden in the expressions above are all about equal: they differ by at most the time that it takes to compute q wrappings.

Some constructions that are likely to meet the second-preimage resistant condition that is needed in Lemma 2 include most of the known cryptographic hash functions such as SHA1 or SHA256. Observe that when used with AES as the underlying cipher for encryption, we are limited to using only 128 bits of the output of the hash function. But second-preimage resistance is a very weak requirement, so it is likely that the SHA family meet this notion even if we only take 128 bits of output. Another solution would be to key these functions (e.g., using HMAC).

Another class of practical functions that meet this condition are universal hash functions (e.g., the polynomial-evaluation hash, or linear hash functions). These functions can be proven to meet the condition of second-preimage resistance as formulated in Eq. (4), but we stress that they only meet it if the hashing key is kept secret (as part of the key-wrapping key).

When the data-key K is a key for a block cipher E, it may even be plausible to use $N = E_K(\mathsf{const})$ as a checksum, where const is some public constant. For contemporary ciphers like AES, it may be reasonable to assume that the public function $H(K) = AES_K(\mathsf{const})$ is a second-preimage resistant function.

3.4   Hash-then-ECB and Hash-then-CBC

At first glance, one may suspect that the hash-then-encrypt method cannot be used with ECB encryption, since in ECB the hash value does not influence in any way the encryption of the data key itself. Below we show that this is not really the case, indeed ECB and CBC mode behave very similarly in our context (with one exception that is described below). For example, in Lemma 4 we prove that even a public linear function can result in an authenticated key-wrap when combined with ECB or CBC modes.

We begin with examining a composition of second-preimage resistant hashing with ECB and CBC. For both modes, we prove below that using a *public* second-preimage-resistant hashing is secure (under an additional mild structural condition). Perhaps surprisingly, however, it turns out that at least for ECB, when using a hash function that depends on a secret key, second-preimage resistance (or even universality) is not sufficient. (For CBC we still don't know if universal hashing suffices. We suspect that it is, but so far could not prove it.)

univHash-then-ECB may be insecure. We show a hash function with secret key (from $2n$ to n bits), which is second-preimage resistant and yet has the property that for any X, Y, Z, it holds that $X = H(Y, Z) \Leftrightarrow Y = H(X, Z)$, and we show how to use this property in an attack (since in ECB we use the same procedure to encrypt the nonce as we do the key blocks). Consider the following Hash-then-ECB scheme for wrapping a two-block data key: the hash function uses a block cipher E and depends on two secret cipher-keys, which we denote by h_1, h_2. Specifically, our hash function is defined as

$$H_{h_1,h_2}(Y, Z) = E_{h_1}^{-1}\left(E_{h_1}(Y) \oplus E_{h_2}(Z)\right)$$

It is not hard to see that this function H is second-preimage-resistant as per the definition from Eq. (4): if we replace the cipher with two random permutations then we have $\mathbf{Adv}_H^{\mathrm{spr}} = 2^{-n}$. (In fact, H is nearly a pairwise-independent hash function in this case.) On the other hand, if $X = H_{h_1,h_2}(Y,Z)$ then

$$E_{h_1}(H_{h_1,h_2}(X,Z)) =$$
$$= E_{h_1}(X) \oplus E_{h_2}(Z) = (E_{h_1}(Y) \oplus E_{h_2}(Z)) \oplus E_{h_2}(Z) = E_{h_1}(Y)$$

and therefore also $H_{h_1,h_2}(X,Z) = Y$. An attacker on HtECB, after seeing a ciphertext $C = \langle C_0, C_1, C_2 \rangle$ can therefore produce the valid forged ciphertext $C^* = \langle C_1, C_0, C_2 \rangle$.

publicSPR-then-ECB/CBC is secure. When H is a public function, on the other hand, we show that second-preimage-resistance is sufficient, under a mild structural condition. Specifically we need to assume that for a random input data-key K, the nonce $N = H(K)$ is also (close to being) a random n-bit block. Below we call a function with that property *well-spread*.

Lemma 3. *Let H be a public well-spread hash function with input length ℓn and output length n, and consider the construction HtECB, using H for the hash function and with a truly random permutation for the block cipher. Then for any bound q on the number of wrapping queries, we have*

$$\mathbf{Adv}_{\mathrm{HtECB}}^{\mathrm{kw.rpa}}(wrap = q), \ \mathbf{Adv}_{\mathrm{HtCBC}}^{\mathrm{kw.rpa}}(wrap = q) \le \binom{q}{2}(\alpha_H + \ell^2/2^n)$$

$$\mathbf{Adv}_{\mathrm{HtECB}}^{\mathrm{kw.int}}(wrap = q), \ \mathbf{Adv}_{\mathrm{HtCBC}}^{\mathrm{kw.int}}(wrap = q) \le O(q\ell) \cdot \mathbf{Adv}_H^{\mathrm{spr}}$$

The running-time bounds on the various attackers that are hidden in the expressions above are all about equal: they differ by at most the time that it takes to compute q wrappings.

Proof. (sketch): Below we only prove the bound on $\mathbf{Adv}_{\mathrm{HtECB}}^{\mathrm{kw.int}}$, the proof for $\mathbf{Adv}_{\mathrm{HtCBC}}^{\mathrm{kw.int}}$ is similar, and the RPA-bounds are straightforward. Denote the transcript of a q-query attack against HtECB by

$$\{(K_i, C_i) \ : \ K_i = \langle K_i[1], \ldots, K_i[\ell] \rangle, \ C_i = \langle C_i[0], C_i[1], \ldots, C_i[\ell] \rangle\}_{i \in [1,q]}$$

and let $C^* = \langle C^*[0], C^*[1], \ldots, C^*[\ell] \rangle$ be the attempted forged ciphertext. Also let $K^*[j] = E^{-1}(C^*[j])$ for $j = 0, 1, \cdots, \ell$ and $N^* = H(K^*[1], \ldots, K^*[\ell])$, so C^* is valid iff $N^* = H(K^*[0])$.

We have three types of ciphertext C^* to consider: either $C^*[0]$ is different from all the $C_i[j]$'s, or it is equal to one of the $C_i[0]$'s, or it is equal to one of the $C_i[j]$'s for $j > 0$. Denote the probability of C^* of the first type being valid by ε_1 and probability of C^* of the second type being valid by ε_2, and the probability of C^* of the third type being valid by ε_3. We show three collision-finders for H: one with success probability $\varepsilon_1 - O(q\ell/2^n)$, the second with with success probability ε_2/q, and the third with success probability $\varepsilon_3/q\ell$.

The first collision finder (that needs to work when $C^*[0] \ne C_i[j]$ for all i,j) gets a random input X and computes $N = H(X)$. (Recall that H is well spread and public, so N is nearly uniform and we can compute it.) Now the collision finder plays the integrity-of-ciphertext game with the attacker, choosing at random values for the K_i's and for the permutation E and its inverse E^{-1} as needed. When the attacker outputs C^*, the collision finder sets $E^{-1}(C^*[0]) = N$, which is a valid assignment with probability $1 - q\ell/2^n$, and returns K^* as the second preimage. Note that K^* is different from X with overwhelming probability, and the ciphertext C^* is valid iff indeed $H(K^*) = N$.

The second collision finder (that needs to work when $C^*[0] = C_i[0]$ for some i) also begins by getting some random input X and computing $N = H(X)$. Again, the collision finder plays the integrity-of-ciphertext game with the attacker, but now it chooses at random a query i and uses X as the data-key K_i. Clearly If C^* is a valid forgery and $C^*[0] = C_i[0]$ (which happens with probability ε_2/q) then K^* is a second preimage of N.

The third collision finder (that needs to work when $C^*[0] = C_i[j]$ for $j > 0$) also begins by getting some random input X and computing $N = H(X)$. Again, the collision finder plays the integrity-of-ciphertext game with the attacker, but now it chooses at random a query i and a block h, and uses N as the data-key block $K_i[j]$. Again, if C^* is a valid forgery and $C^*[0] = C_i[j]$ (which happens with probability $\varepsilon_3/q\ell$) then K^* is a second preimage of N. $\qquad\square$

XOR-then-ECB/CBC is not secure. It turns out that second preimage resistance is not a necessary condition when using ECB or CBC. Below we show that even a simple public linear function may be sufficient in this case. However, not every linear function works, and in particular just taking the XOR of the key blocks is *not* secure. Let $K[1], K[2]$ be a two-block data-key, and let $C[0], C[1], C[2]$ be the ciphertext corresponding to it using XOR-then-ECB key wrapping. Then one can check that the ciphertext $C[1], C[0], C[2]$ is a valid ciphertext, corresponding to the data key $K[1] \oplus K[2], K[2]$. The same attack works also for CBC.

Linear-then-ECB/CBC may be secure. Below we show, however, that the "permutation attack" from above is in some sense the only one that matters when using ECB or CBC with a public linear function. Specifically, we show that using a public linear function of the form $H(K[1], \ldots, K[\ell]) = \sum_j \alpha_j K[j]$ where the α_j's are linearly independent, is already enough to get some level of security. (For example, we can use $\alpha_j = \alpha^j$ where α is a primitive element in $GF(2^n)$.) However, the security level deteriorates quickly with ℓ: the advantage bound that we prove is only $(q(\ell+1))^{\ell+1}/2^n$. For the typical case $\ell = 2$ this means security level of $O(2^{n/3})$, which may be sufficient in many applications. But for longer keys this construction may not be secure enough to be used in practice.

Lemma 4. *Fix some $\ell < n$, and let $H(K[1], \ldots, K[\ell]) \stackrel{\text{def}}{=} \sum_{j=1}^{\ell} \alpha_j K[j]$, where the α_j's are linear operations over $\{0,1\}^n$ such that the set $\{1, \alpha_1, \ldots, \alpha_\ell\}$ is linearly independent. (That is, there is no nontrivial 0-1 combination of the α's and 1 that sums up to zero.)*

Consider the HtECB and HtCBC constructions using H for the hash function, and with a truly random permutation for the block cipher. Then for any bound q on the number of wrapping queries, we have

$$\mathbf{Adv}_{\mathrm{HtECB}}^{\mathrm{kw.rpa}}(\mathit{wrap} = q), \; \mathbf{Adv}_{\mathrm{HtCBC}}^{\mathrm{kw.rpa}}(\mathit{wrap} = q) \leq O(q^2 \ell^2 / 2^n)$$

$$\mathbf{Adv}_{\mathrm{HtECB}}^{\mathrm{kw.int}}(\mathit{wrap} = q), \; \mathbf{Adv}_{\mathrm{HtCBC}}^{\mathrm{kw.int}}(\mathit{wrap} = q) \leq O((q(\ell + 1))^{\ell+1} / 2^n)$$

In the long version we also include analysis for the constructions Hash-then-ECB-X/CBC-X and Hash-then-XEX. Some variations of our constructions for variable-input-length and associated-data are mentioned in Appendix B.

4 Conclusions

In this work we examined the practice of key-wrapping, and in particular the implementation template of Hash-then-Encrypt. We argued that this template may be attractive in practice, especially in cases where the key-wrapping is used to "glue" together existing incompatible systems. We considered a wide array of "hash functions" and encryption modes, showed how to break some combinations and proved security bounds for others. Although none of the combinations that we considered meets the notion of deterministic authenticated encryption due to Rogaway and Shrimpton [22], we argued that some of them are still secure enough for key-wrapping. To make this argument, we measured them against weaker notions of security, which are arguably sufficient for most applications of key-wrapping.

We would like to stress again that given the choice, one should prefer more robust implementations, such as standard authenticated encryption or the SIV mode of Rogaway and Shrimpton. But in cases where these options are not available, we believe that our results may provide guidance to what can or cannot be used safely.

References

1. Abe, M., Gennaro, R., Kurosawa, K., Shoup, V.: Tag-KEM/DEM: A new framework for hybrid encryption and a new analysis of Kurosawa-Desmedt KEM. In: Cramer, R. (ed.) EUROCRYPT 2005. LNCS, vol. 3494, pp. 128–146. Springer, Heidelberg (2005)
2. Amanatidis, G., Boldyreva, A., O'Neill, A.: Provably-secure schemes for basic query support in outsourced databases. In: Barker, S., Ahn, G.-J. (eds.) Data and Applications Security 2007. LNCS, vol. 4602, pp. 14–30. Springer, Heidelberg (2007)
3. An, J.H., Bellare, M.: Does encryption with redundancy provide authenticity? In: Pfitzmann, B. (ed.) EUROCRYPT 2001. LNCS, vol. 2045, pp. 512–528. Springer, Heidelberg (2001)
4. Bellare, M., Boldyreva, A., O'Neill, A.: Deterministic and efficiently searchable encryption. In: Menezes, A. (ed.) CRYPTO 2007. LNCS, vol. 4622, pp. 535–552. Springer, Heidelberg (2007)

5. Bellare, M., Namprempre, C.: Authenticated encryption: Relations among notions and analysis of the generic composition paradigm. In: Okamoto, T. (ed.) ASIACRYPT 2000. LNCS, vol. 1976, pp. 531–545. Springer, Heidelberg (2000)

6. Canetti, R., Krawczyk, H., Nielsen, J.B.: Relaxing chosen-ciphertext security. In: Boneh, D. (ed.) CRYPTO 2003. LNCS, vol. 2729, pp. 565–582. Springer, Heidelberg (2003)

7. Cramer, R., Shoup, V.: Design and analysis of practical public-key encryption schemes secure against adaptive chosen ciphertext attack. SIAM Journal on Computing 33(1), 167–226 (2003); Preliminary version in CRYPTO 1998

8. Dodis, Y., Smith, A.: Entropic security and the encryption of high entropy messages. In: Kilian, J. (ed.) TCC 2005. LNCS, vol. 3378, pp. 556–577. Springer, Heidelberg (2005)

9. Dworkin, M.: Recommendation for block cipher modes of operation: Galois/counter mode (GCM) and GMAC. NIST Special Publication 800-38D (2007)

10. Gligor, V.D., Donescu, P.: Fast encryption and authentication: XCBC encryption and XECB authentication modes. In: Matsui, M. (ed.) FSE 2001. LNCS, vol. 2355, pp. 92–108. Springer, Heidelberg (2002)

11. Halevi, S., Krawczyk, H.: Strengthening digital signatures via randomized hashing. In: Dwork, C. (ed.) CRYPTO 2006. LNCS, vol. 4117, pp. 41–59. Springer, Heidelberg (2006)

12. Jutla, C.S.: Encryption modes with almost free message integrity. In: Pfitzmann, B. (ed.) EUROCRYPT 2001. LNCS, vol. 2045, pp. 529–544. Springer, Heidelberg (2001)

13. Katz, J., Yung, M.: Characterization of security notions for probabilistic private-key encryption. Journal of Cryptology 19(1), 67–95 (2006); Earlier version in STOC 2000, pp. 245–254

14. Krawczyk, H.: The order of encryption and authentication for protecting communications (or: How secure is SSL?). In: Kilian, J. (ed.) CRYPTO 2001. LNCS, vol. 2139, pp. 310–331. Springer, Heidelberg (2001)

15. Kurosawa, K., Desmedt, Y.: A new paradigm of hybrid encryption scheme. In: Franklin, M. (ed.) CRYPTO 2004. LNCS, vol. 3152, pp. 426–442. Springer, Heidelberg (2004)

16. Liskov, M., Rivest, R.L., Wagner, D.: Tweakable block ciphers. In: Yung, M. (ed.) CRYPTO 2002. LNCS, vol. 2442, pp. 31–46. Springer, Heidelberg (2002)

17. Meyr, C.H., Matyas, S.M.: Cryptography: A New Dimension in Computer Data Security. John Wiley & Sons, Chichester (1982)

18. Naor, M., Pinkas, B., Reingold, O.: Distributed pseudo-random functions and KDCs. In: Stern, J. (ed.) EUROCRYPT 1999. LNCS, vol. 1592, pp. 327–346. Springer, Heidelberg (1999)

19. Rogaway, P.: Authenticated-encryption with associated-data. In: ACM Conference on Computer and Communications Security - ACM-CCS 2002, pp. 98–107. ACM, New York (2002)

20. Rogaway, P.: Efficient instantiations of tweakable blockciphers and refinements to modes OCB and PMAC. In: Lee, P.J. (ed.) ASIACRYPT 2004. LNCS, vol. 3329, pp. 16–31. Springer, Heidelberg (2004)

21. Rogaway, P.: Nonce-based symmetric encryption. In: Roy, B., Meier, W. (eds.) FSE 2004. LNCS, vol. 3017, pp. 348–359. Springer, Heidelberg (2004)

22. Rogaway, P., Shrimpton, T.: A provable-security treatment of the key-wrap problem. In: Vaudenay, S. (ed.) EUROCRYPT 2006. LNCS, vol. 4004, pp. 373–390. Springer, Heidelberg (2006)

23. Russell, A., Wang, H.: How to fool an unbounded adversary with a short key. In: Knudsen, L.R. (ed.) EUROCRYPT 2002. LNCS, vol. 2332, pp. 133–148. Springer, Heidelberg (2002)
24. Wagner, D.: A generalized birthday problem. In: Yung, M. (ed.) CRYPTO 2002. LNCS, vol. 2442, pp. 288–303. Springer, Heidelberg (2002)

A   The Rogaway-Shrimpton KIAE Notion

In an appendix to the long version of their paper [22], Rogaway and Shrimpton describe a notion of *key insertion authenticated encryption* that bares some similarities to our notion of authenticated key-wrapping: Specifically, they describe a setting where one applies a randomized encoding procedure to the adversarially-controlled message before encryption, and the attacker is given the corresponding ciphertext together with the randomness that was used in the encoding. The KIAE notion roughly requires that such an attacker cannot distinguish these (ciphertext, randomness) pairs from just random bits.

The crucial difference between our notions and KIAE is that we insist that the plaintext to be encrypted is completely random and outside of the attacker's control, whereas KIAE still allows the attacker to control parts of the plaintext.[5] Namely, our authenticated key-wrapping is a degenerate case of KIAE where the message and authenticated data are both empty. It thus follows that KIAE is still strictly stronger than all the security notions that we consider in this work.

In terms of usage, KIAE seems rather far removed from the way that key wrapping is typically used in practice: A typical applications would apply key-wrapping to a random data key, and then use that data key as a cryptographic key in some other scheme (say, to encrypt "real data" or a lower-level key in the hierarchy). The KIAE case seems to be targeted at applications where the data key (which is used as a cryptographic key elsewhere) is wrapped together with some "real data" in the same ciphertext. Hence in that notion the attacker gets to choose this "real data", while at the same time the data key is assumed to be random.

B   Variations and Extensions

Variable input length. All the constructions and proofs in this section extend also to the case of variable input-length. (Although for key-wrap this case may not be very interesting, since symmetric keys are typically all of the same length.) When using second-preimage resistant hash function, we need to assume that it is second-preimage resistant even for variable input-length. The proofs for the linear hash functions need to be extended by considering the cases where the attempted forged ciphertext is an extension of the previous ciphertexts that were

---

[5] A smaller difference is that Rogaway-Shrimpton consider associated data as an integral part of their notion, whereas we view it as an optional extension. There are also some syntactic differences between these notions, but these are of no consequence.

obtained from the encryption oracle. For the "somewhat secure" constructions, the security bound for ECB and CBC will then be roughly $O((q\ell_{max})^{\ell_{max}})$ where ℓ_{max} is the largest allowable length of any data-key (expressed in 128-bit blocks). For the masked version ECB-X and CBC-X, we get $O(q^{\ell_{max}})$, where in this case ℓ_{max} is the longest data-key that was returned by the wrapping oracle.

Input lengths that are not a multiple of the block length are handled by CTR without a problem. When using other modes, this can be handled by just padding the key. If one must preserve the length then ciphertext-stealing will also work.

Associated data. To handle associated data A, one must "hash" it together with the data-key, computing the nonce as $N \leftarrow H(\mathsf{D}, \mathsf{A})$. But as opposed to the data key, the associated data A is not random, so it should be modeled as controlled by the attacker.

For the constructions using second-preimage hashing, we must now make a stronger assumption on the hash function. Specifically, the function $H(\mathsf{D}, \mathsf{A})$ must meet the condition of "enhanced TCR" as described in [11], when D is viewed as the hashing seed: Namely an attacker that chooses A and gets a random D, should not be able to find different A', D' such that $H(\mathsf{D}, \mathsf{A}) = H(\mathsf{D}', \mathsf{A}')$. Still we note that since H can depend on a secret key then constructing such functions is easier, and in particular universal hashing satisfy even this stronger requirement.

For the constructions based on linear hashing, one way to incorporate associated data is by applying a PRF to it and then computing the nonce as $N = H(PRF(\mathsf{A})|\mathsf{D})$, where H is the same linear function from above. Practically speaking, however, if we are already using a PRF then we might as well compute $N = PRF(\mathsf{A}|\mathsf{D})$ (in which case we get back the SIV construction).

Information Theoretically Secure Multi Party Set Intersection Re-visited

Arpita Patra[*], Ashish Choudhary[**], and C. Pandu Rangan[***]

Dept of Computer Science and Engineering
IIT Madras, Chennai India 600036
arpitapatra10@gmail.com, partho_31@yahoo.co.in, prangan55@gmail.com

Abstract. We re-visit the problem of secure multiparty set intersection (MPSI) in information theoretic settings. In [15], Li et.al have proposed a protocol for MPSI with $n = 3t + 1$ parties, that provides information theoretic security, when t out of those n parties are corrupted by an active adversary having *unbounded computing power*. In [15], the authors have claimed that their protocol takes six rounds of communication and communicates $\mathcal{O}(n^4 m^2)$ field elements, where each party has a set containing m field elements. However, we show that the round and communication complexity of the protocol in [15] is much more than what is claimed in [15]. We then propose a *novel* information theoretically secure protocol for MPSI with $n \geq 3t + 1$, which significantly improves the "actual" round and communication complexity of the protocol of [15]. Our protocols employ several tools which are of independent interest.

Keywords: Multiparty Computation, Information Theoretic Security.

1   Introduction

Secure Multiparty Set Intersection (MPSI): Consider a complete synchronous network $\mathcal{N}$, consisting of n parties $\mathcal{P} = \{P_1, \ldots, P_n\}$, who are pairwise connected by a reliable and private channel. The parties do not trust each other and the distrust in the network is modeled by a centralized adversary $\mathcal{A}_t$, who has *unbounded computing* power and can actively corrupt at most t parties in Byzantine fashion, where $t < \frac{n}{3}$. A Byzantine (or actively) corrupted party is under complete control of $\mathcal{A}_t$, who may force the party to behave arbitrarily. Any protocol over $\mathcal{N}$ is assumed to operate in a sequence of rounds. In each round, a party performs some local computation, sends new messages to the other parties through the private channels and publicly *broadcasts* some information, receives the messages that were sent by the other parties in current round on the private channels and the messages that were publicly broadcast by the other parties in

[*] Financial Support from Microsoft Research India Acknowledged.
[**] Financial Support from Infosys Technology India Acknowledged.
[***] Work Supported by Project No. CSE/05-06/076/DITX/CPAN on Protocols for Secure Communication and Computation Sponsored by Department of Information Technology, Government of India.

M.J. Jacobson Jr., V. Rijmen, and R. Safavi-Naini (Eds.): SAC 2009, LNCS 5867, pp. 71–91, 2009.

current round. Here broadcast is a primitive, which allows a party to send some information *identically* to all other parties. If a physical broadcast channel is available in the system, then broadcast will take one round. Otherwise, we can simulate broadcast using a protocol among the parties in $\mathcal{P}$, which will have the same effect as a physical broadcast channel. Each party P_i has a private data-set S_i, containing m elements from a finite field $\mathbb{F}$, where $|\mathbb{F}| \geq n$. The goal of MPSI is to design a protocol that can compute the intersection of these n sets, satisfying the following properties:

1. CORRECTNESS: At the end of the protocol, each honest party correctly gets the intersection of the n sets, irrespective of the behavior of $\mathcal{A}_t$ and
2. SECRECY: The protocol should not leak any *extra* information to the corrupted parties, other than what is implied by the input of the corrupted parties (i.e., the data-sets possessed by corrupted parties) and the final output (i.e., the intersection of all the n data-sets).

MPSI problem is an interesting secure distributed computing problem and has huge practical applications such as online recommendation services, medical databases, data mining etc. [10].

Existing Literature on MPSI: The MPSI problem was first studied in *cryptographic* model in [10, 14], under the assumption that the adversary has *bounded computing power*. By representing the data-sets as polynomials, the set intersection problem is converted into the task of computing the common roots of n polynomials in [10, 14]. This is done as follows: Let $S = \{s_1, s_2, \ldots, s_m\}$ be a set of size m, where $\forall i, s_i \in \mathbb{F}$. Now set S can be represented by a polynomial $f(x)$ of degree-m, where $f(x) = \prod_{i=1}^{m}(x - s_i) = a_0 + a_1 x + \ldots + a_m x^m$. It is obvious that if an element s is a root of $f(x)$, then s is a root of $r(x)f(x)$ too, where $r(x)$ is a *random* polynomial of degree-m over $\mathbb{F}$. Now for MPSI, party P_i represents his set S_i, by a degree-m polynomial $f^{(P_i)}(x)$ and supplies $f^{(P_i)}(x)$ (i.e. its $m+1$ coefficients), as his input, in a secure manner. Then all the parties jointly and securely compute

$$F(x) = (r^{(1)}(x)f^{(P_1)}(x) + r^{(2)}(x)f^{(P_2)}(x) + \ldots + r^{(n)}(x)f^{(P_n)}(x)) \qquad (1)$$

where $r^{(1)}(x), \ldots r^{(n)}(x)$ are n secret random polynomials of degree-m over $\mathbb{F}$, jointly generated by the n parties. Note that $F(x)$ preserves all the common roots of $f^{(P_1)}(x), \ldots, f^{(P_n)}(x)$. Every element $s \in (S_1 \cap S_2 \cap \ldots \cap S_n)$ is a root of $F(x)$, i.e. $F(s) = 0$. Hence after computing $F(x)$ in a secure manner, it can be reconstructed by every party, who locally checks if $F(s) = 0$ for every s in his private set. All s's at which the evaluation of $F(x)$ is zero forms the intersection set $(S_1 \cap S_2 \cap \ldots \cap S_n)$. In [14], it has been proved formally that $F(x)$ does not reveal any *extra* information to the adversary, other than what is deduced from $(S_1 \cap S_2 \cap \ldots \cap S_n)$ and input set S_i of the corrupted parties.

Remark 1. Even though every $s \in (S_1 \cap S_2 \cap \ldots \cap S_n)$ is a root of $F(x)$, there may exist some $s' \in \mathbb{F}$, such that $F(s') = 0$, even though $s' \notin (S_1 \cap S_2 \cap \ldots \cap S_n)$. This is possible if s' happens to be the common root of all $r^{(i)}(x)$'s. However, as stated in [14], the probability of this event is negligible.

In [14], the MPSI problem is solved by securely computing $F(x)$, assuming $\mathcal{A}_t$ to be *computationally bounded*. In [15], the authors have presented the first information theoretically secure protocol for MPSI, assuming $\mathcal{A}_t$ to be *computationally unbounded* and $n \geq 3t + 1$. Specifically, the authors have shown how to securely compute $F(x)$ in the presence of a computationally unbounded $\mathcal{A}_t$. To the best of our knowledge, this is the only known information theoretically secure MPSI protocol. *Notice that, although not explicitly stated in [15], the MPSI protocol of [15] involves a negligible error probability in* CORRECTNESS. *This is due to the argument given in Remark 1.*

Our Motivation and Contribution: The authors in [15] claimed that their MPSI protocol takes *six* rounds and communicates $\mathcal{O}(n^4 m^2)$ elements from $\mathbb{F}$. However, we show that the round and communication complexity of the MPSI protocol of [15] is much more than what is claimed in [15]. We then propose a new information theoretically secure protocol for MPSI with $n \geq 3t + 1$, which significantly improves the "actual" round and communication complexity of the MPSI protocol given in [15].

2 Analysis of the MPSI Protocol of [15]

In order to securely compute $F(x)$ given in (1) against a computationally unbounded $\mathcal{A}_t$, the MPSI protocol of [15] is divided into three phases. We briefly recall the steps performed in first two phases, which are the most expensive phases in terms of round and communication complexity.

1. Input Phase: Here each party represents his private data-set as a polynomial and t-shares the coefficients of the polynomial among the n parties. To do so, the parties use a two dimensional verifiable secret sharing (VSS). A two dimensional VSS [9, 11, 13], ensures that each party (including a corrupted party) "consistently" and correctly t-shares the coefficients of his polynomial with everybody. Now, the authors in [15] claimed that this takes two rounds, where in the first round, each party does the sharing and in the second round verification is done by all parties to ensure whether everybody has received correct and consistent shares (see sec. 4.2 in [15]). However, no estimation is done for the communication complexity of this phase. Now it is well known that the minimum number of rounds taken by any VSS protocol with $n \geq 3t + 1$ is at least *three* [9, 11, 13]. Moreover, the current best three round VSS protocol with $n = 3t + 1$ requires a private communication of $\mathcal{O}(n^3)$ and broadcast of $\mathcal{O}(n^3)$ field elements [9, 13]. Now in the **Input Phase** of [15], each party executes $(m + 1)$ VSS's to share the coefficients of his secret polynomial. In addition, each party also executes $n(m + 1)$ VSS's to share the coefficients of n random polynomials, each of degree m. These polynomials are used to generate secret random polynomials $r^{(1)}(x), \ldots, r^{(n)}(x)$. So the total number of VSS done in **Input Phase** is $\mathcal{O}(n^2 m)$. Hence, the **Input Phase** will take at least three rounds, with a private communication of $\mathcal{O}(n^5 m)$ and broadcast of $\mathcal{O}(n^5 m)$ field elements. If the broadcast channel is not available, then simulation of broadcast of a single field element requires a private communication of $\mathcal{O}(n^2)$ field elements and $\Omega(t)$ rounds [16]. Thus, in the absence of broadcast channel, the **Input Phase** will require $\Omega(t)$ rounds and a communication complexity of $\mathcal{O}(n^7m)$ field elements.

2. Computation Phase: Given that the coefficients of $f^{(P_1)}(x), \ldots, f^{(P_n)}(x)$, $r^{(1)}(x), \ldots, r^{(n)}(x)$ are t-shared, in the computation phase, the parties jointly try to compute $F(x) = r^{(1)}(x)f^{(P_1)}(x) + r^{(2)}(x)f^{(P_2)}(x) + \ldots + r^{(n)}(x)f^{(P_n)}(x)$, such that the coefficients of $F(x)$ are t-shared. For this, the parties execute a sequence of steps. But we recall only first two steps, which are crucial in the communication and round complexity analysis of **Computation Phase**.

During **step 1**, the parties locally multiply the shares of the coefficients of $r^{(i)}(x)$ and $f^{(P_i)}(x)$, for $1 \leq i \leq n$. This results in $2t$-sharing of the coefficients of $f^{(P_i)}(x)r^{(i)}(x)$ for $1 \leq i \leq n$. During **step 2**, each party invokes a *re-sharing protocol* and converts the $2t$-sharing of the coefficients of $f^{(P_i)}(x)r^{(i)}(x)$ into t-sharing, for $1 \leq i \leq n$. The re-sharing protocol enables a party to generate t-sharing of an element, given the t'-sharing of the same element, where $t' > t$. In [15], the authors have given the reference of [12] for the details of re-sharing protocol and claimed that the re-sharing and other additional verifications will take *only three rounds*, with a private communication of $\mathcal{O}(n^4m^2)$ field elements (see sec. 4.2 of [15]). However, [12] presents a protocol for general secure *Multiparty Computation* (MPC), which uses "circuit based approach" to securely evaluate a function. Specifically, the MPC protocol of [12] assumes that the (general) function to be computed is represented as an arithmetic circuit over $\mathbb{F}$, consisting of addition, multiplication, random, input and output gates. The re-sharing protocol of [12] was used to evaluate a multiplication gate. But the protocol was *non-robust* in the sense that it fails to achieve its goal when at least one of the parties misbehaves, in which case the protocol outputs a pair of parties such that at least one of them is corrupted. In fact, the MPC protocol of [12] takes $\Omega(t)$ rounds in the presence of broadcast channel in the system, whereas in the absence of broadcast channel it will take $\Omega(t^2)$ rounds. The authors in [15] have not mentioned what will be the outcome of their protocol if the re-sharing protocol (whose details they have not given) fails during the **computation phase**. In fact, computing t-sharing of the coefficients of $F(x)$ by using the ideas of best known general MPC protocol with $n = 3t+1$ [2, 8, 12] will require a communication complexity of $\Omega(m^2n^2)$ field elements and round complexity of $\Omega(t)$ rounds in the presence of a broadcast channel.

To summarize, a more accurate estimation of the round complexity and communication complexity of the MPSI protocol of [15] in the presence and in the absence of a physical broadcast channel is as follows:

1. If a physical broadcast channel is available in the system, then the **Input Phase** will require a private communication of $\Omega(n^5m)$ field elements and broadcast of $\Omega(n^5m)$ field elements. Moreover, the **Computation Phase** will take $\Omega(t)$ rounds and communication complexity of $\Omega(m^2n^2)$.

2. If a physical broadcast channel is not available in the system, then the **Input Phase** will require a private communication of $\Omega(n^7m)$ field elements.

Moreover, the **Computation Phase** will take $\Omega(t^2)$ rounds and communication complexity of $\Omega(m^2 n^2)$ field elements.

3 Our Results

We propose a new, information theoretically secure MPSI protocol with $n = 3t + 1$, tolerating a *computationally unbounded* $\mathcal{A}_t$. Our protocol is based on the approach of solving the MPSI by securely computing the function given in (1). Moreover, our protocol involves a negligible error probability in correctness. However, as mentioned in Remark 1, any protocol for MPSI, based on computing the function in (1) will involve a negligible error probability. In the following tables, we compare the round complexity (RC) and communication complexity (CC) of our MPSI protocol with the *estimated* RC and CC of the MPSI protocol of [15] (as stated in previous section). In the tables, the CC is in terms of field elements. Moreover, CC/RC with (out) BC stands for communication complexity/round complexity in presence (absence) of physical broadcast channel[1].

Reference	CC with BC		RC with BC
	Private	Broadcast	
[15]	$\Omega(n^5 m + m^2 n^2)$	$\Omega(n^5 m)$	$\Omega(t)$
This Paper	$\mathcal{O}((m^2 n^3 + n^4 \log(|\mathbb{F}|))$	$\mathcal{O}(m^2 n^3 + n^4 \log(|\mathbb{F}|))$	58

Reference	CC without BC	RC without BC
	Private	
[15]	$\Omega(n^7 m)$	$\Omega(t^2)$
This Paper	$\mathcal{O}(m^2 n^5 + n^6 \log(|\mathbb{F}|))$	$\mathcal{O}(t)$

From the table, we find that our MPSI protocol significantly improves the *estimated* round and communication complexity of the MPSI protocol of [15].

3.1 Our MPSI Protocol vs. Existing General MPC Protocols

The MPSI problem is a particular variant of general secure MPC problem [20]. Informally, in MPC problem, each party P_i has a private input $x_i \in \mathbb{F}$. There is a publicly known function $f : \mathbb{F}^n \to \mathbb{F}^n$. At the end of computation of f, party P_i gets $y_i \in \mathbb{F}$, such that $(y_1, \ldots, y_n) = f(x_1, \ldots, x_n)$. The goal of any general MPC protocol is to securely compute f, where at the end of the protocol, all parties (honest) receive correct outputs, irrespective of the behavior of $\mathcal{A}_t$. Moreover, the messages seen by $\mathcal{A}_t$ during the protocol, should contain no *additional* information about the inputs and outputs of honest parties, other than what can be computed from the inputs and outputs of corrupted parties. The function f is represented as an arithmetic circuit over the finite field $\mathbb{F}$, consisting of five type of gates, namely addition, multiplication, random, input and output. The

---

[1] If a physical broadcast channel is not available, then we use the protocol of [4, 5], which takes $\mathcal{O}(t)$ rounds and private communication of $\mathcal{O}(n^2 \ell)$ bits to simulate broadcast of ℓ bit message.

number of gates of these types are denoted by c_A, c_M, c_R, c_I and c_O respectively. Any general MPC protocol tries to securely evaluate the circuit gate-by-gate, keeping all the inputs and intermediate results of the circuit as t-shared [3, 18].

The MPSI problem can be solved using any general MPC protocol. However, since a general MPC protocol does not exploit the nuances and the special properties of the problem, it is not efficient in general. Moreover, we do not know how to customize the generic MPC protocols to solve MPSI problem in an optimal fashion. However, we outline below a general approach and use the same to estimate the complexity of MPSI protocols, that could have been derived from general MPC protocols.

Suppose, we try to solve MPSI by computing the function given in (1), using general MPC protocol. The arithmetic circuit, representing the function in (1), will roughly require $c_I = n(m+1)$ input gates (every party P_i inputs $(m+1)$ coefficients of $f^{(P_i)}(x)$), $c_R = n(m+1)$ random gates (n polynomials $r^{(1)}(x), \ldots, r^{(n)}(x)$ have $n(m+1)$ random coefficients), $c_M = n(m+1)^2$ multiplication gates (computing $r^{(1)}(x)f^{(P_i)}(x)$ requires $(m+1)^2$ co-efficient multiplications) and $c_O = 2m+1$ output gates (the $2m+1$ coefficients of $F(x)$ should be output). In the following tables we give the round complexity (RC) and communication complexity (CC) of best known general MPC protocols with $n = 3t+1$, to securely compute (1) with above number of gates.

Reference	CC with BC		RC with BC
	Private	Broadcast	
[3]	$\mathcal{O}(n^5 m^2)$	$\mathcal{O}(n^5 m^2)$	$\mathcal{O}(1)$
[12]	$\mathcal{O}(n^4 m^2)$	$\mathcal{O}(n^2)$	$\mathcal{O}(n)$
[8]	$\mathcal{O}(n^2 m^2)$	$\mathcal{O}(n^2)$	$\mathcal{O}(n)$
[2]	$\mathcal{O}(n^2 m^2)$	$\mathcal{O}(n^3)$	$\mathcal{O}(n)$
This Paper	$\mathcal{O}(m^2 n^3 + n^4 \log(|\mathbb{F}|))$	$\mathcal{O}(m^2 n^3 + n^4 \log(|\mathbb{F}|))$	58

Reference	CC without BC	RC without BC
	Private	
[3]	$\mathcal{O}(n^7 m^2)$	$\mathcal{O}(n)$
[12]	$\mathcal{O}(n^6 m^2)$	$\mathcal{O}(n^2)$
[8]	$\mathcal{O}(n^4 m^2)$	$\mathcal{O}(n^2)$
[2]	$\mathcal{O}(n^4 m^2 + n^5)$	$\mathcal{O}(n^2)$
This Paper	$\mathcal{O}(m^2 n^5 + n^6 \log(|\mathbb{F}|))$	$\mathcal{O}(n)$

From the table, we find that our protocol incurs much lesser communication complexity than the protocol of [3], while keeping round complexity same. But the protocols of [2, 8, 12] provides slightly better communication complexity than ours at the cost of increased round complexity. Round complexity and communication complexity are two important parameters of any distributed protocol. If we ever hope to practically implement MPSI protocols, then we should look for a solution, which tries to *simultaneously optimize* both these parameters. In this context, our MPSI protocol fits the bill more appropriately, than the protocols mentioned in the table.

Though our main motive in this paper is to present a clean solution for MPSI, as a bi-product we have shown that our protocol simultaneously improves both communication and round complexity, whereas existing general MPC protocols (when applied to solve MPSI) improve only one of these two parameters.

3.2 Overview of Our Protocol

As mentioned earlier, our MPSI protocol tries to securely compute the function given in (1). Our protocol is divided into three phases, namely (a) Input and Preparation phase; (b) Computation Phase and (c) Output Phase. In the Input and Preparation phase, the parties t-share the coefficients of their input polynomials. Moreover, the parties jointly generate the t-sharing of the secret random $r^{(i)}(x)$ polynomials. To achieve the first task, we design a new protocol called 1DShare, which further uses a new information checking protocol (ICP) called Multi-Secret-Multi-Verifier-ICP. The second task is achieved by a sub-protocol called Random. In the Computation Phase, the parties generate the t-sharing of the coefficients of $r^{(i)}(x)f^{(i)}(x)$. For this, we use sub-protocol Mult, which is a combination of few existing ideas from the literature and few new ideas presented in this paper. Finally, in the Output Phase, the coefficients of $F(x)$ are reconstructed by each party, by using sub-protocol ReconsPublic. In the next section, we give the technical details of each of the above mentioned sub-protocols.

4 Tools Used

Here we present a number of sub-protocols each solving a specific task. Finally, we combine them to design our MPSI protocol. All the sub-protocols that are presented here are designed to *concurrently* deal with $\ell \geq 1$ values. In the literature, the sub-protocols that achieve the same functionality as ours, were designed to deal with *single* value at a time. Our sub-protocols, concurrently dealing with ℓ values, are better in terms of communication complexity, than ℓ concurrent executions of the existing sub-protocols working with single value. Thus, our sub-protocols harness the advantage offered by dealing with multiple values concurrently (this fact will be more clear in the following sections).

For convenience, we analyze the round and communication complexity of the sub-protocols assuming the existence of physical broadcast channel in the system. While presenting the sub-protocols, we assume that all computations and communications are done over a finite field $\mathbb{F}$, where $\mathbb{F} = GF(2^\kappa)$ and κ is the error parameter. Thus, each field element can be represented by $\log(\mathbb{F}) = \mathcal{O}(\kappa)$ bits. Moreover, without loss of generality, we assume that $n = \text{poly}(\kappa)$.

4.1 Information Checking Protocol and IC Signatures

Information Checking Protocol (ICP) is a tool for authenticating messages in the presence of $\mathcal{A}_t$. The notion of ICP was first introduced by Rabin et.al [18]. As described in [7, 18], an ICP is executed among three parties: a dealer D, an intermediary INT and a verifier R. The dealer D hands over a secret value $s \in \mathbb{F}$ to INT. At a later stage, INT is required to hand over s to R and convince R that s is indeed the value which INT received from D.

The basic definition of ICP involves only a *single* verifier R and deals with *only one* secret s [7, 18]. We extend this notion to *multiple* verifiers, where all the n parties in $\mathcal{P}$ act as verifiers. Thus our ICP is executed among three entities: the dealer $D \in \mathcal{P}$, an intermediary $INT \in \mathcal{P}$ and entire set $\mathcal{P}$ acting as verifiers. This will be later helpful in using ICP as a tool in our MPSI protocol. Moreover, we extend our ICP to deal with *multiple* secrets, denoted by S, which contains $\ell \geq 1$ secret values. Thus, our ICP is executed with respect to *multiple* verifiers and deals with *multiple* secrets concurrently. We call our ICP as Multi-Secret-Multi-Verifier-ICP. Now similar to the ICP defined in [7, 18], our Multi-Secret-Multi-Verifier-ICP is a sequence of following three protocols:

1. Distr$(D, INT, \mathcal{P}, S)$: is initiated by D, who hands over secret $S = (s^{(1)} \ldots s^{(\ell)})$, containing $\ell \geq 1$ elements from $\mathbb{F}$ to INT. In addition, D hands over some **authentication information** to INT and **verification information** to the individual parties (verifiers) in $\mathcal{P}$.

2. AuthVal$(D, INT, \mathcal{P}, S)$: is initiated by INT to ensure that in RevealVal, secret S held by INT will be accepted by all (honest) parties (verifiers) in $\mathcal{P}$.

3. RevealVal $(D, INT, \mathcal{P}, S)$: is carried out by INT and the verifiers. Here INT produces S, along with **authentication information** and individual verifiers in $\mathcal{P}$ produce **verification information**. Depending upon the values produced by INT and verifiers, either S is accepted or rejected by all the parties.

The **authentication information**, along with S, which is held by INT at the end of AuthVal is called D's *IC signature* on S, denoted as $ICSig(D, INT, S)$. Multi-Secret-Multi-Verifier-ICP satisfies the following properties (which are almost same as the properties, satisfied by the ICP of [7, 18]):

1. If D and INT are uncorrupted, then S will be accepted in RevealVal by each honest verifier.
2. If INT is uncorrupted, then at the end of AuthVal, INT knows an S, which will be accepted in RevealVal by each honest verifier, except with probability $2^{-\Omega(\kappa)}$.
3. If D is uncorrupted, then during RevealVal, with probability at least $1 - 2^{-\Omega(\kappa)}$, every $S' \neq S$ produced by a corrupted INT will be rejected by each honest verifier.
4. If D and INT are uncorrupted, then at the end of AuthVal, $\mathcal{A}_t$ has no information about S.

We now present our novel protocol Multi-Secret-Multi-Verifier-ICP, with $n = 3t + 1$. The high level idea of the protocol is as follows: D selects a random polynomial $F(x)$ of degree $\ell + n\kappa$ over $\mathbb{F}$, whose lower order ℓ coefficients are elements of S. In addition, D also selects a random polynomial $R(x)$ of degree $\ell + n\kappa$ over $\mathbb{F}$, which is independent of $F(x)$. D hands over $F(x)$ and $R(x)$ to INT. D then associates κ *random evaluation points* with each verifier P_i and gives the value of $F(x), R(x)$ at those evaluation points to P_i. This prevents with very high probability, a corrupted INT, to produce an incorrect $F(x)$ during RevealVal, without being un-noticed by an honest verifier P_i. This ensures third property of ICP. In order to ensure second property, an honest INT has to ensure that his $F(x)$ is consistent with the evaluation points of the honest verifiers. For this, INT and the verifiers interact in a zero knowledge fashion and check the consistency of $F(x)$ and secret evaluation points. To maintain the secrecy of S during the zero knowledge interaction, INT uses the $R(x)$ polynomial.

Multi-Secret-Multi-Verifier-ICP$(D, INT, \mathcal{P}, \ell, S = (s^{(1)}, \ldots, s^{(\ell)}))$

Distr$(D, INT, \mathcal{P}, \ell, S)$ **Round 1**: D selects a random polynomial $F(x)$ of degree $\ell + n\kappa$ over $\mathbb{F}$, whose lower order ℓ coefficients are elements of S. In addition, D selects another random polynomial $R(x)$ of degree $\ell + n\kappa$ over $\mathbb{F}$. D also selects $n\kappa$ random, non-zero, distinct evaluation points from $\mathbb{F}$, denoted by $\alpha_{i,1}, \alpha_{i,2}, \ldots, \alpha_{i,\kappa}$, for $1 \le i \le n$. D privately gives $F(x)$ and $R(x)$ to INT. To verifier $P_i \in \mathcal{P}$, D privately gives $(\alpha_{i,l}, a_{i,l}, b_{i,l})$, for $l = 1, \ldots, \kappa$, where $a_{i,l} = F(\alpha_{i,l})$ and $b_{i,l} = R(\alpha_{i,l})$.

AuthVal$(D, INT, \mathcal{P}, \ell, S)$ **Round 2**: INT chooses a random $d \in \mathbb{F} \setminus \{0\}$ and broadcasts $(d, B(x) = F(x) + dR(x))$. Parallely, each verifier $P_i \in \mathcal{P}$ broadcasts a random subset of indices $l_1, \ldots, l_{\frac{\kappa}{2}}$, the evaluation points $\alpha_{i,l_1}, \ldots, \alpha_{i,l_{\frac{\kappa}{2}}}$ and the values $a_{i,l_1}, \ldots, a_{i,l_{\frac{\kappa}{2}}}$ and $b_{i,l_1}, \ldots, b_{i,l_{\frac{\kappa}{2}}}$. Notice that each verifier randomly selects the subset of indices $l_1, \ldots, l_{\frac{\kappa}{2}}$, independent of other verifiers.

Round 3: D checks if for at least $2t+1$ verifiers P_i, it holds that $a_{i,l} + db_{i,l} = B(\alpha_{i,l})$, for all l in the set of random indices broadcasted by P_i in **Round 2**. If the above condition is not satisfied for at least $2t+1$ verifiers, then D broadcasts the polynomial $F(x)$.

Local Computation (by each party): IF $F(x)$ is broadcasted in **Round 3**, then INT replaces the $F(x)$ received from D during **Round 1**, with the $F(x)$ which is broadcasted in **Round 3**. Accordingly, each verifier P_i adjust his $a_{i,l}$ (as received in **Round 1**), for $l = 1, \ldots, \kappa$, such that $F(\alpha_{i,l}) = a_{i,l}$ holds. ELSE say that verifier P_i accepts INT if $a_{i,l} + db_{i,l} = B(\alpha_{i,l})$, for all l in the set of random indices, broadcasted by P_i in **Round 2**.

The polynomial $F(x)$ is called D's **IC signature** on $S = (s^{(1)}, \ldots, s^{(\ell)})$ given to INT, which is denoted by $ICSig(D, INT, S)$.

RevealVal$(D, INT, \mathcal{P}, \ell, S)$: (a) **Round 4**: INT broadcasts $F(x)$; (b) **Round 5**: Each verifier $P_i \in \mathcal{P}$ broadcasts all the evaluation points $\alpha_{i,l}$ which were not broadcasted during **Round 2** and $a_{i,l}$ corresponding those indices.

Local Computation (by each party): Say that verifier P_i re-accepts INT if for at least one of the newly revealed (by P_i) points, it holds that $a_{i,l} = F(\alpha_{i,l})$. If there are at least $t + 1$ verifiers who re-accepts INT, then accept the lower order ℓ coefficients of $F(x)$ as $S = (s^{(1)}, \ldots, s^{(\ell)})$. In this case, we say that D's signature on S is correct. Else reject $F(x)$ broadcasted by INT and we say that INT has failed to produce D's signature.

Lemma 1 (Property 1). *If D and INT are honest, then each honest verifier will accept S at the end of RevealVal, without any error.*

Lemma 2 (Property 2). *If D is uncorrupted, then at the end of AuthVal, INT knows an S, which will be accepted in RevealVal by each honest verifier, except with probability $2^{-\Omega(\kappa)}$.*

PROOF: If D is honest, then the proof follows from Lemma 1. So we consider the case when D is corrupted. Now there are two possible sub-cases. If D broadcasts $F(x)$ during **Round 3**, then the lemma holds trivially, without any error. So we consider the case, when D (corrupted) has not broadcasted $F(x)$ during **Round 3**. This implies that at least $2t + 1$ verifiers have *accepted INT* during AuthVal. Now, out of these $2t + 1$ verifiers, at least $t + 1$ are honest. If we can show that these honest verifiers will *re-accept INT* during RevealVal with high probability, then the proof is over. So we now proceed to prove the same.

In order that an honest P_i *accept INT* during AuthVal but does not *re-accept* it during RevealVal, it must be the case that the data (evaluation points and values) that P_i exposed during AuthVal satisfies the polynomial $B(x)$ that INT broadcasted during AuthVal, but on the other hand, out of the remaining evaluation points that are used by P_i in RevealVal, none satisfy the polynomial $F(x)$ produced by INT. That is, for the selected $\frac{\kappa}{2}$ indices $l_1, ..., l_{\frac{\kappa}{2}}$, it holds that $a_{i,l} + db_{i,l} = B(\alpha_{i,l})$, for all l in the set of indices $\{l_1, ..., l_{\frac{\kappa}{2}}\}$ and $F(\alpha_{i,l}) \neq a_{i,l}$ for all l in the *remaining* set of indices. Notice that INT chooses d independently of the values given by D. Also, P_i chooses the $\frac{\kappa}{2}$ indices randomly out of κ indices. So the probability that the above event happens is $\frac{1}{\binom{\kappa}{\kappa/2}} \approx 2^{-\Omega(\kappa)}$, which is negligible. This shows that with high probability all honest verifiers (at least $t + 1$), who have *accepted INT* during AuthVal, will *re-accept INT* during RevealVal, thus proving our lemma. $\square$

Lemma 3 (Property 3). *If D is uncorrupted, then during RevealVal, with probability at least $1 - 2^{-\Omega(\kappa)}$, every $S' \neq S$ produced by a corrupted INT will be rejected by each honest verifier.*

PROOF: If a corrupted INT produces $S' \neq S$ during RevealVal, then it implies that INT has broadcasted $F'(x) \neq F(x)$ during **Round 4**. Moreover, while broadcasting $F'(x)$, INT will have no information about the $\frac{\kappa}{2}$ random secret evaluation points (which were not broadcasted during AuthVal), corresponding to each honest verifier. Without knowing the $\frac{\kappa}{2}$ secret evaluation points of an honest verifier, say P_i, the probability that INT will be re-accepted by P_i is at most $\frac{\ell+nk}{|\mathbb{F}|}$. Thus, the total probability that any honest verifier will accept INT (who broadcasts $F'(x) \neq F(x)$) is $\frac{(\ell+nk)(2t+1)}{|\mathbb{F}|} \approx 2^{-\Omega(k)}$. $\square$

Lemma 4 (Property 4). *If D and INT are honest, then $\mathcal{A}_t$ will have no information about S at the end of AuthVal.*

PROOF: For simplicity, let the first t verifiers are corrupted. So in the **Round 1**, the adversary will know κt points on $F(x)$ and $R(x)$. In **Round 2**, the adversary will come to know about additional $\frac{k}{2}(2t+1)$ points on $F(x)$ and $R(x)$. Moreover, since D and INT are both honest, $2t + 1$ honest verifiers will accept INT and hence D will not broadcast $F(x)$ during **Round 3**. So at the end of AuthVal, adversary will know $\kappa t + \frac{\kappa}{2}(2t + 1)$ points on each of $F(x)$ and $R(x)$. However, since $n = 3t + 1$ and degree of $F(x)$ and $R(x)$ is $\ell + n\kappa$, the adversary will have no information about the lower order ℓ coefficients of $F(x)$. $\square$

Lemma 5. *Protocol Multi-Secret-Multi-Verifier-ICP takes five rounds and correctly generates IC signature on ℓ field elements, by privately communicating $\mathcal{O}((\ell + n\kappa)\kappa)$ bits and broadcasting $\mathcal{O}((\ell + n\kappa)\kappa)$ bits. The protocol works correctly, except with error probability of $2^{-\Omega(\kappa)}$.*

Important Notation: *In the rest of the paper, whenever we say that D hands over $ICSig(D, INT, S)$ to INT, we mean that Distr and AuthVal are executed in the background. Similarly, INT reveals $ICSig(D, INT, S)$ can be interpreted as INT, along with other parties, invoking RevealVal.*

Remark 2 (Comparison with Existing ICP). The current best known ICP is due to [7], which privately communicates and broadcasts $\mathcal{O}(n\kappa)$ bits, to generate IC signature on a single secret (though the ICP of [7] is designed with $n = 2t+1$, the protocol when executed with $n = 3t + 1$, will result in the same communication complexity). Had we executed ℓ times the ICP of [7], dealing with single secret, the communication complexity would turn out to be $\mathcal{O}(\ell n\kappa)$ bits (both private and broadcast). However, the communication complexity of Multi-Secret-Multi-Verifier-ICP considering all the ℓ secrets concurrently is $\mathcal{O}((\ell + n\kappa)\kappa)$ bits (both private and broadcast). This clearly shows that if ℓ is significantly large, which is the case in our MPSI protocol, then executing a *single instance* of Multi-Secret-Multi-Verifier-ICP, dealing with *multiple secrets* concurrently, is advantageous over executing *multiple instances* of ICP of [7], dealing with *single secret*. The same principle holds for other sub-protocols, which are described in the sequel.

4.2 Generating ℓ Length Random Vector

We now present a protocol called RandomVector($\mathcal{P}, \ell$), which allows the parties in $\mathcal{P}$ to jointly generate a vector, containing ℓ random elements from $\mathbb{F}$. Following the idea of [8], protocol RandomVector uses Vandermonde Matrix and its capability to extract randomness.

$$(r^{(1)}, \ldots, r^{(\ell)}) = \textbf{RandomVector}(\mathcal{P}, \ell)$$

1. Every party $P_i \in \mathcal{P}$ selects $L = \lceil \frac{\ell}{2t+1} \rceil$ random elements $r^{(1,P_i)}, \ldots, r^{(L,P_i)}$ from $\mathbb{F}$.
2. Every party $P_i \in \mathcal{P}$ as a dealer invokes Sharing Phase of four round VSS protocol of [11] with $n \geq 3t + 1$ for sharing each of the values $r^{(1,P_i)}, \ldots, r^{(L,P_i)}$.
3. For reconstructing the values $r^{(1,P_i)}, \ldots, r^{(L,P_i)}$ (shared by P_i in Sharing Phase), the Reconstruction Phase of four round VSS of [11] with $n \geq 3t+1$ is invoked for L times separately. Now corresponding to every $P_i \in \mathcal{P}$, the values $r^{(1,P_i)}, \ldots, r^{(L,P_i)}$ are public.
4. Now parties compute $(r^{(1,1)}, \ldots, r^{(1,2t+1)}) = (r^{(1,P_1)}, \ldots, r^{(1,P_n)})V$, $(r^{(2,1)}, \ldots, r^{(2,2t+1)}) = (r^{(2,P_1)}, \ldots, r^{(2,P_n)})V, \ldots, (r^{(L,1)}, \ldots, r^{(L,2t+1)}) = (r^{(L,P_1)}, \ldots, r^{(L,P_n)})V$. Here V is a $n \times (2t + 1)$ publicly known Vandermonde matrix over $\mathbb{F}$.

The values $r^{(1,1)}, \ldots, r^{(1,2t+1)}, \ldots, r^{(L,1)}, \ldots, r^{(L,2t+1)}$ constitute the elements of ℓ length random vector.

Protocol RandomVector also uses the four round perfect VSS (verifiable secret sharing) protocol of [11] (see Fig 2 of [11]) as black box. The perfect VSS (see the definition of VSS in Section 2.1 of [11]) with $n \geq 3t+1$ parties consists of two phases, namely Sharing Phase and Reconstruction Phase. The Sharing Phase takes four rounds and allows a dealer D (which can be any party from the set of n parties) to verifiably share a secret $s \in \mathbb{F}$ by privately communicating $\mathcal{O}(n^2 \log |\mathbb{F}|)$ bits and broadcasting $\mathcal{O}(n^2 \log |\mathbb{F}|)$ bits where $|\mathbb{F}| \geq n$. The Reconstruction Phase takes single round and allows all the (honest) parties to reconstruct the secret s (shared by D in Sharing Phase) by broadcasting $\mathcal{O}(n \log |\mathbb{F}|)$ bits in total. Notice that, in our context, $|\mathbb{F}| = 2^\kappa \geq n$. The VSS protocol has an important property that once D (possibly corrupted) shares a secret s during Sharing Phase, then D is *committed* to s. Later, in the Reconstruction Phase, irrespective of the behavior of the corrupted parties, the same s will be reconstructed. Thus a corrupted D will not be able to change his commitment from s to any other value, with the help of corrupted parties, during Reconstruction Phase.

Lemma 6. *Protocol RandomVector generates ℓ length random vector in five rounds and privately communicates $\mathcal{O}(\ell n^2 \kappa)$ bits and broadcasts $\mathcal{O}(\ell n^2 \kappa)$ bits.*

4.3 Unconditional Verifiable Secret Sharing and Reconstruction

Definition 1 (d-1D-sharing [1]). *: We say that a secret s is d-1D-shared, if there exists a degree-d polynomial $f(x)$, with $f(0) = s$, such that each (honest) P_i in $\mathcal{P}$ holds the ith share $f(i) = s_i$ of s. The vector $(s_1, s_2, \ldots, s_n)$ of shares is called a d-sharing of s and is denoted as $[s]_d$. We may skip the subscript d when it is clear from the context.*

If s is d-1D-shared by $D \in \mathcal{P}$, then we denote it as $[s]_d^D$. In the sequel, we describe a new protocol 1DShare, which allows a dealer $D \in \mathcal{P}$ to t-1D-share ℓ secret values $s^{(1)}, \ldots, s^{(\ell)}$, where $\ell \geq 1$, with very high probability. If D behaves correctly during the protocol, then each honest $P_i \in \mathcal{P}$ will hold i^{th} shares $s_i^{(1)}, \ldots, s_i^{(\ell)}$, of $s^{(1)}, \ldots, s^{(\ell)}$ (respectively), at the end of the protocol.

Notice that the desired sharing for each $s^{(i)}$ (separately) can be produced using a perfect (i.e., without any error) VSS protocol with $n \geq 3t + 1$ [9, 11, 13]. However, this will involve more communication complexity than 1DShare which performs the same task with less communication complexity (but with a negligible error probability). 1DShare achieves this by incorporating one of the ideas used in [8] and using Multi-Secret-Multi-Verifier-ICP as building block.

The goal of 1DShare is as follows: (a) If D is honest, then the protocol generates $[s^{(1)}]_t, \ldots, [s^{(\ell)}]_t$ with very high probability, such that the secrets $s^{(1)}, \ldots, s^{(\ell)}$ remain information theoretically secure from $\mathcal{A}_t$. (b) If D is corrupted and has not generated t-1D-sharing of secrets, then with high probability, D will be detected as corrupted during a public verification process. Moreover, every honest party accepts a pre-defined t-1D-sharing of ℓ 1's, namely $[1]_t, [1]_t, \ldots, [1]_t$ (ℓ times), on behalf of D.

Informally, the protocol works as follows: D chooses $\ell+1$ random polynomials $f^{(0)}(x), \ldots, f^{(\ell)}(x)$ over $\mathbb{F}$, each of degree t, such that $f^{(0)}(0) = r$ and $f^{(l)}(0) =$

$s^{(l)}$ for $l = 1, \ldots, \ell$. Here r is a random non-zero element from $\mathbb{F}$. D then hands over his IC signature on i^{th} points of $f^{(l)}(x)$ polynomials *concurrently* to party P_i. After this, the parties jointly produce a non-zero random value z. Now D is asked to broadcast a linear combination of the $\ell + 1$ polynomials, where the scalars of the linear combination are function of z. At the same time, each party P_i is asked to broadcast his corresponding linear combination of points. Ideally, the linear combination of points, broadcasted by the individual parties, should lie on the linear combination of the polynomial broadcasted by D. If this happens, then with very high probability, D has correctly t-1D-shared each $s^{(l)}$. Otherwise, there is a party, say P_i, for which the above condition is not satisfied. In this case, P_i is asked to reveal D's signature on the ith points of $f^{(l)}(x)$ polynomials that he has received from D. In case P_i is able to correctly produce the signature, D is detected to be corrupted and the protocol terminates, with each party assuming predefined t-1D-sharing of ℓ 1's, namely $[1]_t, [1]_t, \ldots, [1]_t$, on behalf of D.

$$([s^{(1)}]_t^D, \ldots, [s^{(\ell)}]_t^D) = \mathbf{1DShare}(D, \mathcal{P}, \ell, s^{(1)}, s^{(2)}, \ldots, s^{(\ell)})$$

1. For $l = 1, \ldots, \ell$, D picks a random polynomial $f^{(l)}(x)$ over $\mathbb{F}$ of degree-t, with $f^{(l)}(0) = s^{(l)}$. D also chooses a random polynomial $f^{(0)}(x)$ of degree-t with $f^{(0)}(0) = r$ where r is a random, non-zero element from $\mathbb{F}$. For $i = 1, \ldots, n$, let $S_i = (r_i, s_i^{(1)}, s_i^{(2)}, \ldots, s_i^{(\ell)})$, where $r_i = f^{(0)}(i)$ and $s_i^{(l)} = f^{(l)}(i)$. D hands over $ICSig(D, P_i, S_i)$ to party P_i.
2. All the parties in $\mathcal{P}$ invoke $\mathsf{RandomVector}(\mathcal{P}, 1)$ to generate a non-zero random value $z \in \mathbb{F}$.
3. D broadcasts the polynomial $f(x) = f^{(0)}(x) + \sum_{l=1}^{\ell} f^{(l)}(x)z^l = \sum_{l=0}^{\ell} f^{(l)}(x)z^l$. Parallely, every party P_i computes and broadcasts $y_i = r_i + \sum_{l=1}^{\ell} s_i^{(l)} z^l$.
4. If the polynomial $f(x)$ broadcasted by D is of degree more than t, then each party agrees that D is corrupted and outputs t-1D-sharing of ℓ 1's i.e $[1]_t, [1]_t, \ldots, [1]_t$. The protocol terminates here.
5. Every party checks whether $f(i) \stackrel{?}{=} y_i$ for all $i = 1, \ldots, n$. If yes then everybody accepts the t-1D-sharings $[s^{(1)}]_t, [s^{(2)}]_t, \ldots, [s^{(\ell)}]_t$ and the protocol terminates. Otherwise, let $P_i \in \mathcal{P}$, such that $f(i) \neq y_i$. In this case, P_i reveals $ICSig(D, P_i, S_i)$. If P_i succeeds to prove D's signature on $S_i = (r_i, s_i^{(1)}, \ldots, s_i^{(\ell)})$ and $f(i) \neq r_i + \sum_{l=1}^{\ell} s_i^{(l)} z^l$, then each party agrees that D is corrupted and outputs t-1D-sharing of ℓ 1's i.e $[1]_t, [1]_t, \ldots, [1]_t$ (ℓ times) and the protocol terminates here. We say that P_i has raised a valid **complaint** against D. But if the signature is invalid then ignore P_i's complaint against D and everybody accepts $[s^{(1)}]_t, \ldots, [s^{(\ell)}]_t$.

Lemma 7. *In protocol 1DShare, if D is honest, then t-1D-sharing of $s^{(1)}, \ldots, s^{(\ell)}$ are generated, except with error probability of $2^{-\Omega(\kappa)}$. Moreover, $\mathcal{A}_t$ will have no information about the secrets. On the other hand, if D is corrupted and any of the values $r, s^{(1)}, \ldots, s^{(\ell)}$ is not t-1D-shared, then D will be caught, except with error probability of $2^{-\Omega(\kappa)}$. The protocol takes eleven rounds, privately communicates $\mathcal{O}((\ell n + n^2 \kappa)\kappa)$ bits and broadcasts $\mathcal{O}((\ell n + n^2 \kappa)\kappa)$ bits.*

PROOF: The communication and round complexity can be checked easily by inspection. We now prove the correctness. If D is honest, then $f(i) = y_i$ will hold, corresponding to every honest P_i. However, a corrupted party P_i may broadcast incorrect $y_i' \neq y_i$, such that $y_i' \neq f(i)$. Moreover, P_i can forge honest D's IC signature on corresponding incorrect $r_i' \neq r_i$ or/and $s_i'^{(j)} \neq s_i^{(j)}$, for $j = 1 \ldots \ell$. In this case, everybody will reject the sharing done by D. However, from properties of Multi-Secret-Multi-Verifier-ICP protocol, this can happen with probability $2^{-\Omega(\kappa)}$. The secrecy of $s^{(1)}, s^{(2)}, \ldots, s^{(\ell)}$ for an honest D, follows from the fact that $\mathcal{A}_t$ will have only t shares for each $s^{(i)}, 1 \leq i \leq n$ and random r. In addition, the value $f(0)$ is blinded with a random value r, chosen by D. Thus, $\mathcal{A}_t$ will have no information about the secrets.

Next, we consider the case, when D is corrupted and the sharing of at least one of the values $r, s^{(1)}, s^{(2)}, \ldots, s^{(\ell)}$ is not a correct t-1D-sharing, i.e., the shares of the honest parties lie on a polynomial of degree higher than t. Let H be the set of honest parties in $\mathcal{P}$. Moreover, let $h^0(x), \ldots, h^\ell(x)$ denote the minimum degree polynomial, defined by the points on $f^{(0)}(x), \ldots, f^{(\ell)}(x)$ respectively, held by the parties in H. Then according to the condition, degree of at least one of the polynomials $h^0(x), \ldots, h^\ell(x)$ is more than t. Moreover, degree of $h^0(x), \ldots, h^\ell(x)$ can be at most $|H|-1$. This is because $|H|$ distinct points can define a polynomial of degree at most $|H| - 1$. Now the value y_i broadcasted by an honest P_i can be defined as $y_i = \sum_{j=0}^{\ell} z^j h^j(i)$.

We next claim that if degree of at least one of $h^0(x), \ldots, h^\ell(x)$ is more than t, then the minimum degree polynomial, say $h^{min}(x)$, defined by y_i's, corresponding to $P_i \in H$ will be of degree more than t, with very high probability. This will clearly imply that $f(x) \neq h^{min}(x)$ and hence $y_i \neq f(i)$, for at least one $P_i \in H$.

So we proceed to prove that $h^{min}(x)$ will be of degree more than t, when one of $h^0(x), \ldots, h^\ell(x)$ has degree more than t. For this, we show the following:

1. We first show that $h^{def}(x) = \Sigma_{j=0}^{\ell} z^j h^j(x)$ will of degree more than t with very high probability, if one of $h^0(x), \ldots, h^\ell(x)$ has degree more than t.
2. We then show that $h^{min}(x) = h^{def}(x)$, implying that $h^{min}(x)$ will be of degree more than t with very high probability

The first claim is easy to prove. If one of $h^0(x), \ldots, h^\ell(x)$ has degree more than t, then the linear combination of these polynomials, namely $h^{def}(x)$, can be written as $h^{def}(x) = h_1^{def}(x) + h_2^{def}(x)$. Here $h_1^{def}(x)$ contains all the coefficients of $h^{def}(x)$, having exponent more than t, while $h_2^{def}(x)$ contains all the remaining coefficients of $h^{def}(x)$. Now $h^{def}(x)$ will be of degree-t, if $h_1^{def}(x) = 0$, which can happen for at most ℓ possible values of z. Since z is generated randomly from $\mathbb{F} \setminus \{0\}$, independent of $h^0(x), \ldots, h^\ell(x)$, the probability that $h_1^{def}(x) = 0$ is at most $\frac{\ell}{|\mathbb{F}|-1} \approx 2^{-\Omega(\kappa)}$. This implies that $h^{def}(x)$ will be of degree $t_m > t$. Notice that each y_i broadcasted by an honest P_i, will lie on $h^{def}(x)$.

Now we will show that $h^{min}(x) = h^{def}(x)$ and thus $h^{min}(x)$ has degree at least t_m, which is greater than t. So consider the difference polynomial $dp(x) = h^{def}(x) - h^{min}(x)$. Clearly, $dp(x) = 0$, for all $x = i$, where $P_i \in H$. Thus $dp(x)$ will have at least $|H|$ roots. On the other hand, maximum degree of $dp(x)$ could be t_m, which is at most $|H|-1$. These two facts imply that $dp(x)$ is zero polynomial, implying that $h^{def}(x) = h^{min}(x)$ and thus $h^{min}(x)$ has degree $t_m > t$.

Since $h^{min}(x)$ has degree more than t, it implies that $h^{min}(x) \neq f(x)$ (which is of degree-t and broadcasted by D). This further imply that $f(i) \neq y_i$, for at least one $P_i \in H$. So P_i will raise a valid **complaint** against D by revealing $ICSig(D, P_i, S_i)$, where $S_i = (r_i, s_i^{(1)}, \ldots, s_i^{(\ell)})$. Since P_i is honest, the signature will be revealed successfully, except with an error probability of $2^{-\Omega(\kappa)}$ (this follows from the properties of Multi-Secret-Multi-Verifier-ICP). Moreover, everybody will publicly verify that $f(i) \neq r_i + \sum_{l=1}^{\ell} s_i^{(l)} z^l$ and hence will catch the corrupted D with very high probability.   □

Reconstruction of t-1D-Sharing: We now present a protocol called ReconsPublic, that reconstructs a secret s, given $[s]_t$. In the protocol, every party broadcasts his share of s. Now out of these n shares, at most t could be corrupted. But since $n = 3t + 1$, by applying Reed-Solomon error correction algorithm (e.g. Berlekamp Welch Algorithm [17]), s can be recovered.

$$s = \text{ReconsPublic}(\mathcal{P}, [s]_t)$$

Each party P_i broadcasts his share s_i of s. The parties apply error correction to reconstruct s from the n shares.

Lemma 8. *ReconsPublic takes one round and broadcasts $\mathcal{O}(n\kappa)$ bits.*

Important Notation: We now define few notations which are used in subsequent sections (these notations are also commonly used in the literature). By saying that parties in $\mathcal{P}$ compute (locally) $([y^{(1)}]_d, \ldots, [y^{(\ell')}]_d) = \varphi([x^{(1)}]_d, \ldots, [x^{(\ell)}]_d)$ (for any function $\varphi : \mathbb{F}^\ell \to \mathbb{F}^{\ell'}$), we mean that each P_i computes $(y_i^{(1)}, \ldots, y_i^{(\ell')}) = \varphi(x_i^{(1)}, \ldots, x_i^{(\ell)})$. Note that applying an affine (linear) function φ to a number of d-1D-sharings, we get d-1D-sharings of the outputs. So by adding two d-1D-sharings of secrets, we get d-1D-sharing of the sum of the secrets, i.e. $[a]_d + [b]_d = [a + b]_d$. However, by multiplying two d-1D-sharings of secrets, we get $2d$-1D-sharing of the product of the secrets, i.e. $[a]_d[b]_d = [ab]_{2d}$.   ◇

4.4   Upgrading t-1D-Sharing to t-2D-Sharing

Definition 2. *A value s is d-2D-shared among the parties in $\mathcal{P}$, if there exists degree-d polynomials $f, f^1, f^2 \ldots, f^n$ with $f(0) = s$ and for $i = 1, \ldots, n$, $f^i(0) = f(i)$. Moreover, every (honest) party $P_i \in \mathcal{P}$ holds a share $s_i = f(i)$ of s, the polynomial $f^i(x)$ for sharing s_i and share-share $s_{ji} = f^j(i)$ for the share s_j of every other (honest) party P_j. We denote d-2D-sharing of s as $[[s]]_d$.*

If a secret s is d-2D-shared by a party $D \in \mathcal{P}$, then we denote it as $[[s]]_d^D$. Notice that if s is d-2D-shared, then its ith share s_i is d-1D-shared. We now present a new protocol, called Upgrade1Dto2D for upgrading t-1D-sharing to t-2D-sharing. Specifically, given t-1D-sharing of ℓ secrets, namely $[s^{(1)}]_t, \ldots, [s^{(\ell)}]_t$, Upgrade1Dto2D, outputs t-2D-sharing $[[s^{(1)}]]_t, [[s^{(2)}]]_t, \ldots, [[s^{(\ell)}]]_t$, except with probability of $2^{-\Omega(\kappa)}$. Moreover, $\mathcal{A}_t$ learns nothing about the secrets during Upgrade1Dto2D. Furthermore, if a party tries to cheat during the protocol, then with very high probability, he will be caught.

Lemma 9. *Protocol Upgrade1Dto2D upgrades t-1D-sharing of ℓ secrets to t-2D-sharing, except with negligible error probability. The protocol consumes twenty eight rounds, privately communicates $\mathcal{O}((\ell n^2 + n^3 \kappa)\kappa)$ bits and broadcasts $\mathcal{O}((\ell n^2 + n^3 \kappa)\kappa)$ bits. Moreover, $\mathcal{A}_t$ learns nothing about the secrets.*

PROOF: The communication and round complexity of the protocol is easy to follow. We now prove the correctness. Provided ℓ t-1D-sharing $[s^{(1)}]_t, [s^{(2)}]_t, \ldots,$ $[s^{(\ell)}]_t$, every honest party P_i correctly t-1D-shares his shares $s_i^{(0)}, s_i^{(1)}, \ldots, s_i^{(\ell)}$. Now for every honest party P_h, the value $s_h = s_h^{(0)} + \sum_{l=1}^{\ell} r^{(l)} s_h^{(l)}$ will be reconstructed correctly, where s_h is the hth share of $s = s^{(0)} + \sum_{l=1}^{\ell} r^{(l)} s^{(l)}$. But a corrupted party P_c may share $\overline{s_c^{(0)}}, \overline{s_c^{(1)}}, \ldots, \overline{s_c^{(\ell)}}$ with $\overline{s_c^{(l)}} \neq s_c^{(l)}$ for some $l \in \{0, 1, \ldots, \ell\}$. In this case with very high probability $\overline{s_c} = \overline{s_c^{(0)}} + \sum_{l=1}^{\ell} r^{(l)} \overline{s_c^{(l)}}$ will not be equal to s_c (which is the actual c^{th} share of s) as the ℓ length vector $(r^{(1)}, \ldots, r^{(\ell)})$ is chosen uniformly at random. Hence Reed-Solomon Error correction algorithm will point $\overline{s}_c$ as a corrupted share, in which case P_c will be caught. It is easy to see that at any stage of the protocol, $\mathcal{A}_t$ learns not more than t shares for each $s^{(l)}, 1 \leq l \leq \ell$. Hence all the secrets will be secure. $\square$

$([[s^{(1)}]]_t, [[s^{(2)}]]_t, \ldots, [[s^{(\ell)}]]_t) = \mathsf{Upgrade1Dto2D}(\mathcal{P}, \ell, [s^{(1)}]_t, [s^{(2)}]_t, \ldots, [s^{(\ell)}]_t)$

1. Each $P_i \in \mathcal{P}$ invokes 1DShare$(P_i, \mathcal{P}, 1, s^{(0,P_i)})$ to generate $[s^{(0,P_i)}]_t$, where $s^{(0,P_i)} \in \mathbb{F} \setminus \{0\}$ is a random value.

2. The parties in $\mathcal{P}$ computes $[s^{(0)}]_t = \sum_{j=1}^{n} [s^{(0,P_j)}]_t$.

3. Now every P_i invokes 1DShare$(P_i, \mathcal{P}, \ell + 1, s_i^{(0)}, s_i^{(1)}, \ldots, s_i^{(\ell)})$ to generate $[s_i^{(0)}]_t, [s_i^{(1)}]_t, \ldots, [s_i^{(\ell)}]_t$, where $s_i^{(0)}, s_i^{(1)}, \ldots, s_i^{(\ell)}$ are the ith shares of secrets $s^{(0)}, s^{(1)}, \ldots, s^{(\ell)}$ respectively.

4. Now to detect the parties P_k (at most t), who have generated $[\overline{s_k^{(0)}}]_t, [\overline{s_k^{(1)}}]_t, \ldots, [\overline{s_k^{(\ell)}}]_t$ such that $\overline{s_k^{(l)}} \neq s_k^{(l)}$ for some $l \in \{0, 1, \ldots, \ell\}$, the parties in $\mathcal{P}$ jointly generate an ℓ length random vector $(r^{(1)}, \ldots, r^{(\ell)})$ by invoking Protocol RandomVector$(\mathcal{P}, \ell)$. Now all the parties publicly reconstruct $s_i = s_i^{(0)} + \sum_{l=1}^{\ell} r^{(l)} s_i^{(l)}$ and $s = s^{(0)} + \sum_{l=1}^{\ell} r^{(l)} s^{(l)}$ by executing following steps:

   (a) The parties in $\mathcal{P}$ compute $[s_i]_t = [s_i^{(0)}]_t + \sum_{l=1}^{\ell} r^{(l)} [s_i^{(l)}]_t$ and invoke ReconsPublic$(\mathcal{P}, [s_i]_t)$ to publicly reconstruct s_i, for $i = 1, \ldots, n$.

   (b) Every party apply Reed-Solomon error correction algorithm (e.g. Berlekamp Welch Algorithm [17]) to $s_1, s_2, \ldots, s_n$, to recover s. Reed-Solomon error correction algorithm also points out the corrupted shares. Hence if s_i is pointed as a corrupted share, then $[s_i^{(0)}]_t, [s_i^{(1)}]_t, \ldots, [s_i^{(\ell)}]_t$ are ignored by every party.

5. Output $[[s^{(1)}]]_t, [[s^{(2)}]]_t, \ldots, [[s^{(\ell)}]]_t$.

Remark 3 (Comparison with Existing Protocols). In [1], the authors have given a protocol to upgrade d-1D-Sharing to d-2D-Sharing, where $n = 2t+1$. However, the protocol is non-robust. That is, if all the n parties behave honestly, then the protocol will perform the upgradation. Otherwise, the protocol will fail to do the upgradation, but will output a pair of parties, of which at least one is corrupted.

4.5 Proving $c = ab$

Given t-1D-sharing of ℓ pairs, $([a^{(1)}]_t^D, [b^{(1)}]_t^D), \ldots, ([a^{(\ell)}]_t^D, [b^{(\ell)}]_t^D)$, let $c^{(l)} = a^{(l)}b^{(l)}$ for $l = 1, \ldots, \ell$. $D \in \mathcal{P}$ now wants to generate $[c^{(1)}]_t^D, \ldots, [c^{(\ell)}]_t^D$ such that the (honest) parties in $\mathcal{P}$ know that the shared $c^{(l)}$ values satisfy $c^{(l)} = a^{(l)}b^{(l)}$ for $l = 1, \ldots, \ell$. If D is honest, then during this process all $a^{(l)}$, $b^{(l)}$ and $c^{(l)}$ values should remain secure.

We propose a protocol ProveCeqAB to achieve the above task. The idea of the protocol is inspired from [7], where a protocol for the same purpose is proposed, with *a single* pair of values, namely (a, b). Our protocol concurrently deals with ℓ pairs, which leads to a gain in communication complexity. Our protocol uses 1DShare as a building block.

$([c^{(1)}]_t^D, \ldots, [c^{(\ell)}]_t^D) = \mathsf{ProveCeqAB}(D, \mathcal{P}, \ell, [a^{(1)}]_t^D, [b^{(1)}]_t^D, \ldots, [a^{(\ell)}]_t^D, [b^{(\ell)}]_t^D)$

1. D chooses a random non-zero ℓ length tuple $(\beta^{(1)}, \ldots, \beta^{(\ell)}) \in \mathbb{F}^\ell$. D then invokes $\mathsf{1DShare}(D, \mathcal{P}, \ell, c^{(1)}, \ldots, c^{(\ell)})$, $\mathsf{1DShare}(D, \mathcal{P}, \ell, \beta^{(1)}, \ldots, \beta^{(\ell)})$ and $\mathsf{1DShare}(D, \mathcal{P}, \ell, b^{(1)}\beta^{(1)}, \ldots, b^{(\ell)}\beta^{(\ell)})$ to verifiably t-1D-share $(c^{(1)}, \ldots, c^{(\ell)})$, $(\beta^{(1)}, \ldots, \beta^{(\ell)})$ and $(b^{(1)}\beta^{(1)}, \ldots, b^{(\ell)}\beta^{(\ell)})$ respectively. If in any of these 1DShare protocol, D is found to be corrupted, then every party conclude that D fails in this protocol and hence this protocol terminates.
2. Now all the parties in $\mathcal{P}$ invoke $\mathsf{RandomVector}(\mathcal{P}, 1)$ to generate a random value $r \in \mathbb{F}$.
3. For every $l \in \{1, \ldots, \ell\}$, all parties locally compute $[Y^{(l)}]_t = (r[a^{(l)}]_t + [\beta^{(l)}]_t)$ and invoke $\mathsf{ReconsPublic}(\mathcal{P}, [Y^{(l)}]_t)$ to reconstruct $Y^{(l)}$. Parallely, D broadcasts the values $Z^{(1)} = (ra^{(1)} + \beta^{(1)}), \ldots, Z^{(\ell)} = (ra^{(\ell)} + \beta^{(\ell)})$. All the parties check whether $Z^{(l)} \overset{?}{=} Y^{(l)}$. If not then every party concludes that D fails in this protocol and hence the protocol terminates.
4. For every $l \in \{1, \ldots, \ell\}$, the parties locally compute $[X^{(l)}]_t = \left(Y^{(l)}[b^{(l)}]_t - [b^{(l)}\beta^{(l)}]_t - r[c^{(l)}]_t \right)$ and invoke $\mathsf{ReconsPublic}(\mathcal{P}, [X^{(l)}]_t)$ to reconstruct $X^{(l)}$. The parties then check $X^{(l)} \overset{?}{=} 0$. If not then every party concludes that D fails in this protocol and hence the protocol terminates. Otherwise D has proved that $c^{(l)} = a^{(l)}b^{(l)}$.

Lemma 10. *In protocol* ProveCeqAB, *if D does not fail, then $(a^{(l)}, b^{(l)})$, $c^{(l)}$ satisfies $c^{(l)} = a^{(l)}b^{(l)}$ for $l = 1, \ldots, \ell$, except with negligible error probability.* ProveCeqAB *takes eighteen rounds, privately communicates $\mathcal{O}((\ell n + n^2\kappa)\kappa)$ bits and broadcasts $\mathcal{O}((\ell n + n^2\kappa)\kappa)$ bits. Moreover, if D is honest then $\mathcal{A}_t$ learns no information about $a^{(l)}, b^{(l)}$ and $c^{(l)}$, for $1 \leq l \leq \ell$.*

4.6 Multiplication

Given t-1D-sharing of ℓ pairs of secrets, say $([a^{(1)}]_t, [b^{(1)}]_t), \ldots, ([a^{(\ell)}]_t, [b^{(\ell)}]_t)$, we now present a protocol called Mult which allows the parties to compute t-1D-sharing $[c^{(1)}]_t, \ldots, [c^{(\ell)}]_t$ such that $c^{(l)} = a^{(l)}b^{(l)}$ for $l = 1, \ldots, \ell$. Our protocol is motivated from the protocol of [7], which deals with *a single* pair of t-1D-sharing. However, our protocol concurrently deals with ℓ pairs of t-1D-sharing. This leads to a gain in communication complexity.

Lemma 11. *Except with negligible error probability, Mult produces $[c^{(1)}]_t, \ldots,$ $c^{(\ell)}]_t$ from ℓ pairs $([a^{(1)}]_t, [b^{(1)}]_t), \ldots, ([a^{(\ell)}]_t, [b^{(\ell)}]_t)$. The protocol takes 46 rounds, privately communicates $\mathcal{O}((\ell n^2 + n^3 \kappa)\kappa)$ bits and broadcasts $\mathcal{O}((\ell n^2 + n^3 \kappa)\kappa)$ bits. Moreover, $\mathcal{A}_t$ learns nothing about $c^{(l)}, a^{(l)}$ and $b^{(l)}$, for $1 \le l \le \ell$.*

$$([c^{(1)}]_t, \ldots, [c^{(\ell)}]_t) = \mathsf{Mult}(\mathcal{P}, \ell, ([a^{(1)}]_t, [b^{(1)}]_t), \ldots, ([a^{(\ell)}]_t, [b^{(\ell)}]_t))$$

1. All the parties invoke $\mathsf{Upgrade1Dto2D}(\mathcal{P}, \ell, [a^{(1)}]_t, \ldots, [a^{(\ell)}]_t)$ and $\mathsf{Upgrade1Dto2D}(\mathcal{P}, \ell, [b^{(1)}]_t, \ldots, [b^{(\ell)}]_t)$ to upgrade t-1D-sharings of 2ℓ values to t-2D-sharings, i.e., to generate $[[a^{(1)}]]_t, \ldots, [[a^{(\ell)}]]_t$ and $[[b^{(1)}]]_t, \ldots, [[b^{(\ell)}]]_t$ respectively.

2. Let $(a_1^{(l)}, \ldots, a_n^{(l)})$ and $(b_1^{(l)}, \ldots, b_n^{(l)})$ denote the 1D sharings of $a^{(l)}$ and $b^{(l)}$ respectively. Since $a^{(l)}$ and $b^{(l)}$ is t-2D-shared, their i^{th} shares $a_i^{(l)}$ and $b_i^{(l)}$ are t-1D-shared (see the definition of t-2D-sharing). The parties in $\mathcal{P}$ locally compute $[c^{(l)}]_{2t} = [a^{(l)}]_t[b^{(l)}]_t$ for $l = 1, \ldots, \ell$ where $[c^{(l)}]_{2t} = (a_1^{(l)} b_1^{(l)}, \ldots, a_n^{(l)} b_n^{(l)})$.

3. Each party P_i has in his possession i^{th} share of $[c^{(l)}]_{2t}$ i.e. $a_i^{(l)} b_i^{(l)}$ for $l = 1, \ldots, \ell$ where both $a_i^{(l)}$ and $b_i^{(l)}$ are already t-1D-shared by P_i during Protocol $\mathsf{Upgrade1Dto2D}$ executed in step 1 of this protocol. Now each party P_i invokes $\mathsf{ProveCeqAB}(P_i, \mathcal{P}, \ell, [a_i^{(1)}]_t^{P_i}, [b_i^{(1)}]_t^{P_i}, \ldots, [a_i^{(\ell)}]_t^{P_i}, [b_i^{(\ell)}]_t^{P_i})$ to produce $[c_i^{(1)}]_t^{P_i}, \ldots, [c_i^{(\ell)}]_t^{P_i}$ such that $c_i^{(l)} = a_i^{(l)} b_i^{(l)}$ for $l = 1, \ldots, \ell$. At most t (corrupted) parties may fail to execute $\mathsf{ProveCeqAB}$. For simplicity assume first $2t + 1$ parties are successful in executing $\mathsf{ProveCeqAB}$.

4. Now for each $l \in \{1, \ldots, \ell\}$, first $(2t + 1)$ parties have produced $[c_1^{(l)}]_t^{P_1}, \ldots, [c_{(2t+1)}^{(l)}]_t^{P_{(2t+1)}}$. So for $l = 1, \ldots, \ell$, parties in $\mathcal{P}$ compute $[c^{(l)}]_t$ as follows: $[c^{(l)}]_t = r_1[c_1^{(l)}]_t^{P_1} + \ldots + r_{2t+1}[c_{(2t+1)}^{(l)}]_t^{P_{(2t+1)}}$. Here $r_i = \prod_{j=1, j \neq i}^n \frac{-j}{i-j}$. The vector $(r_1, \ldots, r_{2t+1})$ is called recombination vector [6] which is public and known to every party.

4.7 Generating Random t-1D-Sharing

We now present a protocol called $\mathsf{Random}(\mathcal{P}, \ell)$, which allows the parties in $\mathcal{P}$ to jointly generate ℓ random t-1D-sharings, $[r^{(1)}]_t, \ldots, [r^{(\ell)}]_t$, where each $r^{(i)} \in \mathbb{F}$.

$$\mathsf{Random}(\mathcal{P}, \ell)$$

Every party $P_i \in \mathcal{P}$ invokes $\mathsf{1DShare}(P_i, \mathcal{P}, \ell, r^{(1,P_i)}, \ldots, r^{(\ell,P_i)})$ to verifiably t-1D-share ℓ random values $r^{(1,P_i)}, \ldots, r^{(\ell,P_i)}$ from $\mathbb{F}$. Now all the parties in $\mathcal{P}$ jointly computes $[r^{(l)}]_t = \sum_{i=1}^n [r^{(l,P_i)}]_t$ for $l = 1, \ldots, \ell$

Lemma 12. *With overwhelming probability, Random generates ℓ random t-1D-sharing $[r^{(1)}]_t, \ldots, [r^{(\ell)}]_t$ in 11 rounds, by privately communicating $\mathcal{O}((\ell n^2 + n^3 \kappa)\kappa)$ bits and broadcasting $\mathcal{O}((\ell n^2 + n^3 \kappa)\kappa)$ bits.*

5 Unconditionally Secure MPSI Protocol with $n = 3t + 1$

We now present our unconditionally secure MPSI protocol with $n = 3t + 1$.

Remark 4. In any MPSI protocol that computes the intersection of the sets using the function given in (1), $\mathcal{A}_t$ may disrupt the security of the protocol by forcing a corrupted party to input a zero polynomial representing his set [14, 15]. To avoid this, the authors of [14, 15] specified the following trick. They noticed that the coefficient of mth degree term in every P_j's polynomial $f^{(P_j)}(x)$ is 1 always. Hence, every party assumes a predefined $[1]_t$ on behalf of the mth coefficient of every $f^{(P_j)}(x)$ (instead of allowing individual parties to t-1D-share the mth coefficient). This stops the corrupted parties to commit a zero polynomial.

Input and Preparation Phase

1. Every $P_i \in \mathcal{P}$ represents his set $S_i = \{e_i^{(1)}, e_i^{(2)}, \ldots, e_i^{(m)}\}$ by a polynomial $f^{(P_i)}(x)$ of degree m such that $f^{(P_i)}(x) = (x - e_i^{(1)}) \ldots (x - e_i^{(m)}) = a^{(0,P_i)} + a^{(1,P_i)}x + \ldots + a^{(m,P_i)}x^m$. P_i then invokes $\mathsf{1DShare}(P_i, \mathcal{P}, m, a^{(0,P_i)}, \ldots, a^{(m-1,P_i)})$ to generate $[a^{(0,P_i)}]_t, \ldots, [a^{(m-1,P_i)}]_t$. Since $a^{(m,P_i)} = 1$ always, every party in $\mathcal{P}$ assumes a predefined t-1D-sharing for 1, namely $[1]_t$ on behalf of $a^{(m,P_i)}$ (see Remark 4).

2. The parties in $\mathcal{P}$ invoke n times $\mathsf{Random}(\mathcal{P}, m+1)$ parallely, where i^{th} invocation of $\mathsf{Random}(\mathcal{P}, m+1)$ generates $m+1$ t-1D-sharings $[b^{(0,i)}]_t, \ldots, [b^{(m,i)}]_t$. Now the parties assume that $r^{(i)}(x) = b^{(0,i)} + b^{(1,i)}x + \ldots + b^{(m,i)}x^m$ for $i = 1, \ldots, n$. This step can be executed parallely with step 1.

Computation Phase

1. Let $F^{(i)}(x) = r^{(i)}(x)f^{(P_i)}(x) = c^{(0,i)} + c^{(1,i)}x + \ldots + c^{(2m,i)}x^{2m}$ for $i = 1, \ldots, n$. For $i = 1, \ldots, n$, to generate $[c^{(0,i)}]_t, \ldots, [c^{(2m,i)}]_t$, the parties in $\mathcal{P}$ do the following:
   (a) The parties invoke $\mathsf{Mult}(\mathcal{P}, (m+1)^2, ([a^{(0,i)}]_t, [b^{(0,i)}]_t), ([a^{(0,i)}]_t, [b^{(1,i)}]_t), \ldots, ([a^{(m,i)}]_t, [b^{(m-1,i)}]_t), ([a^{(m,i)}]_t, [b^{(m,i)}]_t))$ with $(m+1)^2$ pairs (every coefficient of $r^{(i)}(x)$ should be multiplied with every coefficient of $f^{(P_i)}(x)$) to produce $(m+1)^2$ t-1D-sharings $[a^{(0,i)}b^{(0,i)}]_t, [a^{(0,i)}b^{(1,i)}]_t, \ldots, [a^{(m,i)}b^{(m-1,i)}]_t, [a^{(m,i)}b^{(m,i)}]_t$.
   (b) The parties compute $[c^{(0,i)}]_t = [a^{(0,i)}b^{(0,i)}]_t$, $[c^{(1,i)}]_t = [a^{(0,i)}b^{(1,i)}]_t + [a^{(1,i)}b^{(0,i)}]_t, \ldots, [c^{(2m,i)}]_t = [a^{(m,i)}b^{(m,i)}]_t$.

2. Let $F(x) = \sum_{i=1}^{n} F^{(i)}(x) = d^{(0)} + d^{(1)}x + \ldots + d^{(2m)}x^{2m}$. To generate $[d^{(0)}]_t, \ldots, [d^{(2m)}]_t$, the parties compute $[d^{(j)}]_t = \sum_{i=1}^{n}[c^{(j,i)}]_t$ for $j = 0, \ldots, 2m$.

Output Phase

1. The parties invoke $\mathsf{ReconsPublic}(\mathcal{P}, [d^{(j)}]_t)$ to publicly reconstruct $d^{(j)}$ for $j = 0, \ldots, 2m$. Thus now parties have reconstructed $F(x)$.

2. Each P_i with his private set $S_i = \{e_i^{(1)}, \ldots, e_i^{(m)}\}$ locally checks whether $F(e_i^{(l)}) \overset{?}{=} 0$ for $l = 1, \ldots, m$. If $F(e_i^{(l)}) = 0$, the P_i adds $e_i^{(l)}$ in a set IS_i (initially $IS_i = \emptyset$). P_i outputs IS_i as the intersection set $S_1 \cap S_2 \ldots, \cap S_n$.

Theorem 1. *MPSI protocol with $3t+1$ takes 58 rounds, privately communicates $\mathcal{O}((m^2n^3 + n^4\kappa)\kappa)$ and broadcasts $\mathcal{O}((m^2n^3 + n^4\kappa)\kappa)$ bits, when physical broadcast channel is available in the system. In the absence of a physical broadcast channel, the protocol takes $\mathcal{O}(t)$ rounds and privately communicates $\mathcal{O}((m^2n^5 + n^6\kappa)\kappa)$ bits. The protocol solves MPSI problem with very high probability.*

6   Open Problem

Designing efficient information theoretically secure MPSI protocol with optimal resilience (i.e., with $n = 2t + 1$) is left as an open problem.

References

1. Beerliová-Trubíniová, Z., Hirt, M.: Efficient Multi-party Computation with Dispute Control. In: Halevi, S., Rabin, T. (eds.) TCC 2006. LNCS, vol. 3876, pp. 305–328. Springer, Heidelberg (2006)
2. Beerliová-Trubíniová, Z., Hirt, M.: Perfectly-Secure MPC with Linear Communication Complexity. In: Canetti, R. (ed.) TCC 2008. LNCS, vol. 4948, pp. 213–230. Springer, Heidelberg (2008)
3. Ben-Or, M., Goldwasser, S., Wigderson, A.: Completeness Theorems for Non-Cryptographic Fault Tolerant Distributed Computation. In: 20th ACM Symposium on Theory of Computing, pp. 1–10. ACM Press, New York (1988)
4. Berman, P., Garay, J.A., Perry, K.J.: Bit Optimal Distributed Consensus. Comp. Sci. Research, 313–322 (1992)
5. Carter, L., Wegman, M.N.: Universal Classes of Hash Functions. J. of Comp. and Sys. Sci. 18(4), 143–154 (1979)
6. Cramer, R., Damgård, I.: Multiparty Computation: An Introduction: Contemporary Cryptography. Birkhäuser, Basel (2005)
7. Cramer, R., Damgård, I., Dziembowski, S., Hirt, M., Rabin, T.: Efficient Multiparty Computations Secure Against an Adaptive Adversary. In: Stern, J. (ed.) EUROCRYPT 1999. LNCS, vol. 1592, pp. 311–326. Springer, Heidelberg (1999)
8. Damgård, I., Nielsen, J.B.: Scalable and Unconditionally Secure Multiparty Computation. In: Menezes, A. (ed.) CRYPTO 2007. LNCS, vol. 4622, pp. 572–590. Springer, Heidelberg (2007)
9. Fitzi, M., Garay, J., Gollakota, S., Pandu Rangan, C., Srinathan, K.: Round-Optimal and Efficient Verifiable Secret Sharing. In: Halevi, S., Rabin, T. (eds.) TCC 2006. LNCS, vol. 3876, pp. 329–342. Springer, Heidelberg (2006)
10. Freedman, M.J., Nissim, K., Pinkas, B.: Efficient Private Matching and Set Intersection. In: Cachin, C., Camenisch, J.L. (eds.) EUROCRYPT 2004. LNCS, vol. 3027, pp. 1–19. Springer, Heidelberg (2004)
11. Gennaro, R., Ishai, Y., Kushilevitz, E., Rabin, T.: The Round Complexity of Verifiable Secret Sharing and Secure Multicast. In: 33rd ACM Symposium on Theory of Computing, pp. 580–589. ACM Press, New York (2001)
12. Hirt, M., Maurer, U., Przydatek, B.: Efficient Secure Multi-party Computation. In: Okamoto, T. (ed.) ASIACRYPT 2000. LNCS, vol. 1976, pp. 143–161. Springer, Heidelberg (2000)
13. Katz, J., Koo, C.Y., Kumaresan, R.: Improving the Round Complexity of VSS in Point-to-Point Networks. In: Aceto, L., Damgård, I., Goldberg, L.A., Halldórsson, M.M., Ingólfsdóttir, A., Walukiewicz, I. (eds.) ICALP 2008, Part II. LNCS, vol. 5126, pp. 499–510. Springer, Heidelberg (2008)
14. Kissner, L., Song, D.: Privacy-Preserving Set Operations. In: Shoup, V. (ed.) CRYPTO 2005. LNCS, vol. 3621, pp. 241–257. Springer, Heidelberg (2005)
15. Li, R., Wu, C.: An Unconditionally Secure Protocol for Multi-Party Set Intersection. In: Katz, J., Yung, M. (eds.) ACNS 2007. LNCS, vol. 4521, pp. 226–236. Springer, Heidelberg (2007)

16. Lynch, N.A.: Distributed Algorithms. Morgan Kaufmann, San Francisco (1996)
17. MacWilliams, F.J., Sloane, N.J.A.: The Theory of Error Correcting Codes. North-Holland Publishing Company, Amsterdam (1978)
18. Rabin, T., Ben-Or, M.: Verifiable Secret Sharing and Multiparty Protocols with Honest Majority. In: 21st ACM Symposium on Theory of Computing, pp. 73–85. ACM Press, New York (1989)
19. Srinathan, K., Narayanan, A., Pandu Rangan, C.: Optimal Perfectly Secure Message Transmission. In: Franklin, M. (ed.) CRYPTO 2004. LNCS, vol. 3152, pp. 545–561. Springer, Heidelberg (2004)
20. Yao, A.C.: Protocols for Secure Computations. In: 23rd IEEE Symposium on Foundations of Computer Science, pp. 160–164. IEEE Press, Los Alamitos (1982)

Real Traceable Signatures

Sherman S.M. Chow

Department of Computer Science
Courant Institute of Mathematical Sciences
New York University, NY 10012, USA
schow@cs.nyu.edu

Abstract. Traceable signature scheme extends a group signature scheme with an enhanced anonymity management mechanism. The group manager can compute a tracing trapdoor which enables anyone to test if a signature is signed by a given misbehaving user, while the only way to do so for group signatures requires revealing the signer of all signatures. Nevertheless, it is not tracing in a strict sense. For all existing schemes, T tracing agents need to recollect all N' signatures ever produced and perform RN' "checks" for R revoked users. This involves a high volume of transfer and computations. Increasing T increases the degree of parallelism for tracing but also the probability of "missing" some signatures in case some of the agents are dishonest.

We propose a new and efficient way of tracing – the tracing trapdoor allows the reconstruction of tags such that each of them can uniquely identify a signature of a misbehaving user. Identifying N signatures out of the total of N' signatures ($N << N'$) just requires the agent to construct N small tags and send them to the signatures holder. N here gives a trade-off between the number of unlinkable signatures a member can produce and the efforts for the agents to trace the signatures. We present schemes with simple design borrowed from anonymous credential systems. Our schemes are proven secure respectively in the random oracle model and in the common reference string model (or in the standard model if there exists a trusted party for system parameters initialization).

Keywords: traceable signatures, efficient tracing, group signatures, anonymity management, bilinear groups, standard model.

1   Introduction

Group signature is one of the important privacy enhancing cryptographic primitives. Each group member can sign a message on behalf of a group such that anyone can verify that the group signature is produced by someone enrolled to the group, but not exactly whom. In other words, one can give a proof of group-membership without revealing the true identity.

Unconditional anonymity may be abused by misbehaving users, and timely identification of "bad" signatures is of the utmost importance. For example, consider the use of group signature in the Vehicle Safety Communications (VSC)

M.J. Jacobson Jr., V. Rijmen, and R. Safavi-Naini (Eds.): SAC 2009, LNCS 5867, pp. 92–107, 2009.

system from the Department of Transportation in the U.S. [19], any wrong traffic information purported by a misbehaving driver or a compromised car should be identified to avoid possibly traffic accident which may cost human life.

Group signatures come with a mechanism which allows the group manager (GM) to "open" a signature and reveal the true signer by the GM's decision. To identity any signatures previously generated by a misbehaving user, the GM is required to open all signatures. This incurs three problems – it penalizes the privacy of all other good users, and imposes a high computational overhead on the GM. Besides, these signatures may be distributed in various locations (e.g. in the VSC scenario) and the GM needs to re-collect all these signatures, which may delay the identification of bad signatures. In view of these shortcomings, Kiayias, Tsiounis and Yung [16] proposed the concept of traceable signatures.

Traceable signature is a group signature with an enhanced anonymity management mechanism. Opening of the signatures is no longer the only option. The GM can compute a user-specific tracing trapdoor which enables anyone to test if a signature is signed by a given user. In this way, the objective of identifying all the signatures produced by a misbehaving user can be achieved, without compromising the privacy of all other good users.

Nevertheless, we found that the latter two problems remain unsolved. The tracing mechanism of the existing schemes [10,13,16,18] actually does not trace the signatures. Instead, it *checks* whether a signature was issued by a given user. The GM may delegate the trapdoor to many tracing agents (TA's) to check in parallel, but the TA's still need to recollect all N' signatures ever produced and perform RN' invocations of the "tracing" algorithm for R revoked users in total. This involves a high volume of transfer and computations. There is also a trade-off between the degrees of parallelism and the trust on the TA's. The more TA's employed, the higher chance that a TA may "miss" some signatures, either intentionally or accidentally.

In this paper, we propose a new and efficient way of tracing – the user-specific tracing trapdoor allows the reconstruction of tags such that each of them can uniquely identify a signature of a misbehaving user. Identifying N signatures just requires the agent to invoke N tag-reconstruction and send these N small tags to the signatures holder, instead of requiring the transfer of $N' >> N$ signatures in the traditional approach. Our new traceable signatures still enjoy the original applications mentioned in [16], namely, transforming an anonymous system to one with "fair privacy", a mix-net application where originators of messages can be opened, and open-bid auctions.

We present schemes with simple design borrowed from existing anonymous credential systems. In particular, [8] has briefly mentioned that their compact e-cash system can be viewed as a bounded group signature scheme supporting efficient tracing. Our schemes are proven secure respectively in the random oracle (RO) model and in the common reference string (CRS) model (or in the standard model if there exists a trusted party for system parameters initialization). The former is more efficient while the latter gives a more modular design and higher security guarantees.

2 Design of Traceable Signatures and Building Blocks

Before we delve into the formal definition of traceable signature and our constructions, we first talk about its high-level design, which motivates the discussion of several building blocks. Since traceable signature is an enhancement of group signature, we start by the latter.

2.1 High Level Designs

Group Signatures. When a user joins a group, the GM gives this new member a signature. The user presents this signature to a verifier to show the membership. However, a verifier who got the same signature can claim the membership too. This means the GM should sign on something that is related to some valuable secrets of the users, such that they would not share it with other easily. To sign on behalf of the group, the user should generate another signature as well. This latter signature can be given by the member's own private key. The signature of the GM and the member's own private key form the credential of a member.

We assume the private key is valuable, and a user does not want to leak this private key to any one, including the GM (for exculpability). It can be "hidden" in the form of a *commitment*. The GM can then give a *signature on the commitment*. The GM may also store part of the communication with this user in an archive. This concludes the joining stage.

A group member wants to preserve anonymity in signing. There should be a *protocol for proof of knowledge (PoK) of a signature*. Another feature of group signature is that it can be opened to reveal the true signer. Thus, it should contain an *encryption* of some information that uniquely identifies a user, such that only a designated party (e.g. the GM) can decrypt it. There should be a way to let any verifier to know that this encryption has been done correctly, so another *zero-knowledge protocol for showing the correctness of the encryption* is needed. All these proofs by the signer should be verifiable by everyone, hence they must be non-interactive. The proof should also be *witness-indistinguishable* such that it is generated equally likely by each possible credential (the witness). This concludes our discussions on the idea of group signature.

Traceable Signatures. The traditional way of tracing only tells if a signature is given by a particular user. We know from the existing schemes that a function of the user's private key can serve as this "decisional" trapdoor which supports an efficient detection mechanism.

For our new way of tracing, we found that it is easier for the GM to generate the user-specific tracing trapdoor instead, which is also stored in the archive. To make sure the signature of a member should be related to this chosen value, the GM should give a *signature on a block of messages*, i.e. on the trapdoor and a commitment of the user's private key. This may not be required for the typical group signature schemes or the traditional traceable signature schemes.

To enable a tracing agent TA to uniquely identify every signatures produced by a particular member, everyone is required to compute a deterministic tag based on a seed given by the GM. A TA can then reproduce the tag to identify the signature. The tag is generated by a *pseudorandom function* taking the secret seed and the counter value as its input, which makes the signatures produced by the same user remain unlinkable to any one without the trapdoor.

Now we have a function that is only computable with the help of a secret seed, but what should be its input? We cannot afford an exponentially-large domain here since it will be time-consuming for a TA to re-generate all possible tags. We thus need to confine the domain. The verifier should ensure that this input value is an integer less than a limit N pre-selected by the GM. This can be done with the help of a zero-knowledge *range proof*.

The above idea has actually been employed by compact e-cash systems [4,8] to support "join once, spend many". In our case, we use the deterministic recovery nature to support efficient tracing. A weakness of this approach is that, N gives a trade-off between the number of unlinkable signature a member can produce and the efforts to trace the signatures.

2.2 Number-Theoretic Preliminaries

A mapping $\hat{e} : \mathbb{G}_1 \times \mathbb{G}_2 \to \mathbb{G}_T$ is a bilinear pairing if

- $\mathbb{G}_1$ and $\mathbb{G}_2$ are cyclic multiplicative groups of prime order p.
- g, h are generators of $\mathbb{G}_1$ and $\mathbb{G}_2$ respectively.
- $\psi : \mathbb{G}_2 \to \mathbb{G}_1$ is a computable isomorphism from $\mathbb{G}_2$ to $\mathbb{G}_1$, with $\psi(h) = g$.
- Each group element has a unique binary representation.
- (Bilinear) $\forall x \in \mathbb{G}_1$, $y \in \mathbb{G}_2$ and $a, b \in \mathbb{Z}_p$, $\hat{e}(x^a, y^b) = \hat{e}(x, y)^{ab}$.
- (Non-degenerate) $\hat{e}(g, h) \neq 1$.

$\mathbb{G}_1$ and $\mathbb{G}_2$ can be the same group or different groups. We say that $(\mathbb{G}_1, \mathbb{G}_2)$ is a bilinear group pair if the group action in $\mathbb{G}_1$, $\mathbb{G}_2$, ψ and $\hat{e}$ are all efficiently computable. We name $(p, \mathbb{G}_1, \mathbb{G}_2, \mathbb{G}_T, \hat{e}, g, h)$ as bilinear map context *params$_{BM}$*.

Definition 1 (Decisional Diffie-Hellman (DDH)). *The DDH problem in $\mathbb{G}$ is, on input a quadruple $(g, g^a, g^b, g^c) \in \mathbb{G}^4$, output 1 if $c = ab$ and 0 otherwise.*

Definition 2 (eXternal Diffie-Hellman (XDH)). *The XDH problem in a bilinear group pair $(\mathbb{G}_1, \mathbb{G}_2)$ with trace map ψ is to solve the DDH problem in $\mathbb{G}_1$. If XDH is hard, there exists no efficiently computable isomorphism $\psi' : \mathbb{G}_1 \to \mathbb{G}_2$.*

Definition 3 (Decisional Linear (DLin)). *The DLin problem in $\mathbb{G} = \langle g \rangle$ is defined as follow: On input a sextuple $(u, v, g, u^a, v^b, g^c) \in \mathbb{G}^6$, decide if $c = a+b$.*

Definition 4 (q-Decisional Diffie-Hellman Inversion (q-DDHI)). *The q-DDHI problem in prime order group $\mathbb{G} = \langle g \rangle$ is defined as follow: On input a $(q+2)$-tuple $(g, g^x, g^{x^2}, \ldots, g^{x^q}, g^c \in \mathbb{G}^{q+2})$, decide if $c = 1/x$.*

Definition 5 (q-Strong Diffie-Hellman (q-SDH)). *The q-SDH problem in a bilinear group pair $(\mathbb{G}_1, \mathbb{G}_2)$ with an efficient (computable in polynomial time) trace map $\psi : \mathbb{G}_2 \to \mathbb{G}_1$ is, on input a $(q+2)$-tuple $(g, h, h^\gamma, h^{\gamma^2}, \ldots, h^{\gamma^q}) \in \mathbb{G}_1 \times \mathbb{G}_2^{q+1}$ where $g = \psi(h)$, output a pair $(B, e) \in \mathbb{G}_1 \times \mathbb{Z}_p^*$ such that $B^{(\gamma+e)} = g$.*

We say that an X assumption holds if no probabilistic polynomial time algorithm has non-negligible advantage (over random guessing if X is decisional) in solving problem X. The q-SDH assumption in $(\mathbb{G}_1, \mathbb{G}_2)$ with a trace map $\psi : \mathbb{G}_2 \to \mathbb{G}_1$ is shown to be true [5] in the generic group model.

2.3   Cryptographic Building Blocks

Signature with Efficient Protocols. A signature scheme with efficient protocols refers to a signature scheme with two protocols for the following purposes.

1. The signer only needs a commitment of a block of messages $(m_1, \ldots, m_L)$ but not the messages themselves to give a signature on $(m_1, \ldots, m_L)$;
2. A signature holder can prove the knowledge of a signature on some block of messages without revealing the signature nor the block of messages.

Examples include BBS+ signature [2] (a variant of BBS signature in [6] as outlined in [9]), and P-signature [3,4]. The latter supports non-interactive zero-knowledge proofs in the common reference string model, and the construction in [4] supports $L \geq 1$.

In a P-signature, PSigSetup setups the global parameters used by all other algorithms to be described below. A signer uses PSigKG to generate a pair of signing / verification key. Any user can use an associated commitment scheme Com to make a commitment of the message(s) to be signed and run PObtain, which interacts with the algorithm PIssue executed by the signer. As a result, the user obtains a P-signature on the message(s). If the privacy of the message(s) is not a concern, the signer can simply use the PSign algorithm. The possession of a signature can be shown using the PProve algorithm, which can then be verified by anyone using the PVer algorithm. Details can be found in [4].

For a BBS+ signature on (m_1, m_2), the global parameters contain (g, g_1, g_2, h). Using a signing key μ, the signer picks a random e and gives the signature as $\varsigma = (gg_1^{m_1}g_2^{m_2})^{1/(\mu+e)}$, which can be verified under the verification key $Z = h^\mu$ by checking if $\hat{e}(\varsigma, Zh^e) = \hat{e}(gg_1^{m_1}g_2^{m_2}, h)$. A computational zero-knowledge proof of signature (for a single message block) has been given in [6]. A perfect zero-knowledge proof (since the signature is not encrypted) for multiple message blocks has been given in [2]. These can be made non-interactive by using Fiat-Shamir heuristics in the random oracle model.

Pseudorandom Function, Weakly-Secure Signature and Strong One-Time Signature. We employed a variant of a pseudorandom function (PRF) due to Dodis and Yampolskiy [11], defined as $PRF_{g,s}(x) : x \mapsto g^{\frac{1}{s+x}}$ where $\mathbb{G}_p = \langle g \rangle$ is a cyclic group of prime order p, $s \in_R \mathbb{Z}_p$ is the secret seed and $x \in \mathbb{Z}_p$ is the input. We use it in our constructions for the tag-generation. Its pseudorandomness relies on the q-DDHI assumption, see [8,4] for details.

This PRF function appeared in a short signature scheme proposed by Boneh and Boyen [5]. The secret key is the seed s and the input x encodes the message. The signature is the PRF value. Verification of signature is possible if we use a bilinear group pair instead of $\mathbb{G}_p$. (On the other hand, the PRF is pseudorandom only if the DDH is hard in $\mathbb{G}_p$.) We will use this signature in the range proof and the user signing part of our CRS-based construction. It is unforgeable under a non-adaptive chosen-message attack under the q-SDH assumption. Since it is deterministic, it is also strongly-unforgeable.

For the "anonymity against CCA attack", we use a strong one-time signature in our CRS-based construction, which informally means that the adversary can ask for the signature on a chosen message, but can neither create a different signature on that message nor forge a signature on a different message.

Non-Interactive Proofs for Bilinear Groups. Groth and Sahai [15] proposed an efficient non-interactive zero-knowledge (NIZK) or non-interactive witness-indistinguishable (NIWI) proof system for statements of the form

$$\prod_{q=1}^{Q} \hat{e}(a_q \prod_{m=1}^{M} x_m^{\alpha_{q,m}}, b_q \prod_{n=1}^{N} y_n^{\beta_{q,n}}) = t$$

where $t \in \mathbb{G}_T, \{a_q\} \subset \mathbb{G}_1, \{b_q\} \subset \mathbb{G}_2, \{\alpha_{q,m}\}, \{\beta_{q,n}\}, \subset \mathbb{Z}_p, \{x_m\} \subset \mathbb{G}_1, \{y_n\} \subset \mathbb{G}_2$ when given $\{C_m\}$ – commitments of $\{x_m\}$, $\{D_m\}$ – commitments of $\{y_n\}$, and a CRS $param_{BM}$. This is also referred to as pairing product equation.

The proof system can be instantiated by the subgroup decision assumption in composite order groups, the DLin assumption or the XDH assumption. However, the associated commitment scheme based on the first assumption is only binding over one of the prime order subgroups, which gives different PRF values for two identically distributed commitments of the same value. Hence, the e-cash scheme in [4] and our CRS-based construction employ either one of the latter two assumptions. It gives a NIZK protocol for Dodis-Yampolskiy PRF, a NIZK PoK for a tag-based encryption [17] (to be described), and a NIWI PoK for Boneh-Boyen signature in [14] It can be seen that the "structures" in all these primitives conformed to the pairing product equations.

Range Proof. Proving a secret value is within a public range can be done in this way – the verifier gives signatures on each value in the range, the prover then makes a commitment of the secret value and proves the knowledge of a signature that signs on the committed value. This idea appeared in the k-times anonymous authentication system in [20], and is used in [2,4] and our CRS-based construction. Camenisch, Chaabouni and shelat [7] gave a generalization of this approach. By writing the secret value in base-D and commit to these D-ary digits, this yields a proof of size $O(k/(\log k - \log\log k))$ instead of $O(k)$, for proving the secret lies in $[0, 2^{k+1} - 1]$.

Linear Encryption and Tag-Based Encryption. Linear encryption proposed in [6] is a natural extension of ElGamal encryption based on the decision linear assumption, which is secure even in groups where DDH problem is easy. The encryption key is $(u, v, g_0) \in \mathbb{G}_1^3$ where $u^a = v^b = g_0$, and the decryption key is (a, b). An encryption of a message $M \in \mathbb{G}_1$ is $(T_1, T_2, T_3) = (u^\alpha, v^\beta, Mg_0^{\alpha+\beta})$, where $\alpha, \beta \in_R \mathbb{Z}_p^*$; which can be decrypted by $T_3/(T_1^a \cdot T_2^b)$. The scheme is secure against chosen-plaintext attacks (CPA) under the DLin assumption.

Kiltz [17] extended this linear encryption to a tag-based encryption which is secure against selective-tag weak chosen-ciphertext attacks (CCA), under the same assumption. The encryption key is $(u, v, g_0, U, V) \in \mathbb{G}_1^5$ where $u^a = v^b = g_0$, and the decryption key is (a, b). To encrypt a message $M \in \mathbb{G}_1$ under a tag (or a label) $t \in \mathbb{Z}_p^*$, picks $\alpha, \beta \in_R \mathbb{Z}_p^*$ and returns $(T_1, T_2, T_3, T_4, T_5) = (u^\alpha, v^\beta, Mg_0^{\alpha+\beta}, (g_0^t U)^\alpha, (g_0^t V)^\beta)$, which can be decrypted by $T_3/(T_1^a \cdot T_2^b)$ if $\hat{e}(u, T_4) = \hat{e}(T_1, g_0^t U)$ and $\hat{e}(v, T_5) = \hat{e}(T_2, g_0^t V)$ hold. The latter check can also be done without pairing if the discrete logarithm of U, V with respect to u, v respectively are kept. We will call the tag used in tag-based encryption as "label" to avoid any confusion with the tracing tag in the traceable signature.

3 Framework

3.1 Syntax

Our new model of traceable signature is based on the original framework in [16]. A traceable signature involves three kinds of entities, namely, the group manager (GM), the users (U_i) and the tracing agents (TA). It consists of nine polynomial time algorithms or protocols. The following enumerates the syntax.

- Setup. On input a security parameter 1^λ for $\lambda \in \mathbb{N}$, a trusted party executes this algorithm to output the system parameters *params*. For simplicity of the framework, we assume that Setup also outputs the group public/private key (gpk, gsk), and the opening agent public/private key pair (opk, osk). For brevity, all algorithms below take $(params, gpk, opk)$ implicitly as inputs.
- Join. A (prospective) user U_i joins the group and obtains a member public/private key pair (pk_i, sk_i) as a result of the interaction with the GM via this protocol. The GM also adds U_i's identification and part of the transcript of the protocol to the membership archive $\mathcal{DB}$, which is kept private.
- Sign. Given a message m and a member private key sk_i, user U_i uses this algorithm to give a signature σ on m on behalf of the group gpk.
- Verify. Given a signature σ and a message m, anyone can use this algorithm to verify if σ is a valid signature on m signed by a member of the group gpk.
- Reveal. On input of the member archive $\mathcal{DB}$ and a user's identification U_i, the GM outputs a trapdoor s_i for tracing the signatures produced by U_i.
- Trace. Anyone can use Trace with the trapdoor s_i generated by Reveal to output a set of tags which can uniquely identify each of the signatures of U_i.
- Open. Given a valid signature σ, the GM uses the opening secret key osk to output some information ς_i which enables the retrieval of the user's identification information U_i in the membership archive $\mathcal{DB}$.
- Claim. Given the member private key sk_i and a valid signature σ, user U_i can give an evidence z that proves the original authorship of σ.
- ClaimVer. Given a message m, a valid signature σ and an evidence z produced by Claim, anyone can verify whether σ is originated from user U_i holding sk_i.

3.2   Requirements

Definition 6. *A traceable signature scheme (of security parameter λ) is correct if the four conditions below are satisfied (with overwhelming probability in λ).*

- *Sign-Correctness. For all messages m, and all sk_i obtained from the Join protocol, $\mathsf{Verify}(\mathsf{Sign}(m, sk_i), m) = \top$.*
- *Open-Correctness. For all messages m, all sk_i obtained from the Join protocol of user U_i, and all membership archives $\mathcal{DB}$ which contain the information for user U_i, $\mathsf{Open}(\mathsf{Sign}(m, sk_i), osk, \mathcal{DB}) = U_i$.*
- *Trace-Correctness. For all messages m, all sk_i obtained from the Join protocol of user U_i, and all membership archives $\mathcal{DB}$ which contain the information for user U_i, $\mathsf{Sign}(m, sk_i) \subseteq \mathsf{Trace}(\mathsf{Reveal}(\mathcal{DB}, U_i))$, where σ is in the set S when a specific component s of σ is in the set S.*
- *Claim-Correctness. For all messages m and all sk_i obtained from the Join protocol, $\mathsf{ClaimVer}(m, \sigma, \mathsf{Claim}(sk_i, \sigma)) = \top$, where $\sigma = \mathsf{Sign}(m, sk_i)$.*

We briefly recall the security concerns. Formal definition can be found in [16].

- *Identification Security.* Any subset of colluded users and tracing agents cannot output a valid signature which cannot be opened to anyone in this collusion group or cannot be traced (by the trapdoors produced by an honest execution of Reveal algorithm) to one of them.
- *Anonymity.* No collusion of users and tracing agents can distinguish between the signatures of two honest group members. (Note that the tracing agents are not given the user-specific trapdoor of these two honest members.)
- *Non-Frameability.* There are two different ways an honest user may be framed. A conspiration of the GM and any subset of colluded users may construct a signature that opens or trace to an innocent user outside this group, or may claim a signature that was generated by an honest user as their own.

Due to the new traceability feature we introduce, our schemes can only be secure against a weaker variant of anonymity attack described below.

- *N-Anonymity.* Same as Anonymity Attack, but the adversary can only see at most N (determined in Setup) signatures from each of the honest members.

Remarks. Since Open is considered as an internal operation (which is different from Trace), the adversary in the first two attacks are not allowed to query an "open" oracle. Nevertheless, our CRS-based construction achieves "CCA" anonymity, i.e. the adversary has an "open" oracle in breaking anonymity, under the natural constraint that it cannot be queried with the challenge signature.

Given the deterministic nature of the tracing tag, the GM (or a tracing agent who "colludes" with the GM) may launch a misidentification or framing attack by giving a signature with a "legitimate" tracing tag [1]. However, a user can dispute if the self-claiming component is deterministically determined by the tracing tag and part of the membership private key which is only known to the user. Specifically, the existence of two valid signatures with exactly the same tracing tag but different self-claiming components means that the GM "reused" the same seed in issuing "different" membership credentials.

4   Constructions

We first give our construction in the common reference string model. This can be seen as a concrete realization of the design we gave in Section 2. Then we will present a more efficient construction in the random oracle model.

4.1   Construction in the Common Reference String Model

This somewhat generic and moderately efficient construction is mostly based on the building blocks we presented in Section 2.3, except we have instantiated the signature in the range proof by Boneh-Boyen signature and the PRF by Dodis-Yampolskiy PRF. It is largely based on the compact e-cash scheme in [4]. We added a tag-based encryption [17] of the user's identity and the user self-claiming component, but removed the double-spending detection.

Setup. This algorithm setups all the building blocks. Namely, it runs PSigSetup and returns the P-signature parameters *params*, PSigSetup needs to run the setup of the Groth-Sahai proof system to get its parameters $params_{GS}$, which in turn contain the bilinear map context $params_{BM} = (p, \mathbb{G}_1, \mathbb{G}_2, \mathbb{G}_T, \hat{e}, g, h)$. Let H be a collision-free hash function which maps to $\mathbb{Z}_p^*$. All these should be determined by the CRS, or executed by a trusted initializer.

For credential issuing, runs PSigSetup to generate a key pair (gpk, gsk).

For opening, setups a key pair (opk, osk) by tag-based encryption's TEncSetup.

For tracing, manages a list of triple (U_i, pk_i, s_i), which is initially empty. For the range proof system, the GM picks a number N which is polynomial in λ, runs PSigSetup($params$) again to generate another pair of signing key (pk_r, sk_r), generates and publishes the signatures $\Sigma_i = \mathsf{PSign}(i, sk_r), \forall i \in [1, 2, \cdots, N]$.

Join. User U_i obtains a credential from the GM through the interactions below.

1. User U_i randomly selects $x_i \in_R \mathbb{Z}_p^*$, computes a public key $pk_i = g^{x_i}$, and a commitment $comm_{sk} = \mathsf{Com}(x_i, open_{x_i})$. U_i sends $comm_{sk}$ to the GM, proves in zero-knowledge the knowledge of $open_{x_i}$, and that $comm_{sk}$ corresponds to the secret key used for computing pk_i.
2. The GM verifies the proof, randomly selects $s_i \in_R \mathbb{Z}_p^*$, computes $comm_{seed} = \mathsf{Com}(s_i, open_{s_i})$ and sends $(s_i, open_{s_i})$ to the user. The tracing trapdoor for this user will be s_i and the GM should ensure it is unique.
3. The user and the GM run the algorithms $\mathsf{PObtain}(gpk, (x_i, s_i), (open_{x_i}, open_{s_i}))$
   $\leftrightarrow$ $\mathsf{PIssue}(gsk, (comm_{sk}, comm_{s_i}))$ respectively. The user obtains a P-signature ς_i on (x_i, s_i), and stores (ς_i, x_i, s_i) as the member private key.
4. The GM adds the entry (U_i, pk_i, s_i) to the membership archive.

Sign(m). User U_i manages a counter n_i on the number of signatures produced.

- U_i generates a one-time signature key pair (pk_o, sk_o).
- U_i signs on pk_o by $\sigma = g^{\frac{1}{x_i + H(pk_o)}}$.
- U_i computes the tracing tag $S = g^{\frac{1}{s_i + n_i}}$ and the self-claiming tag $R = S^{x_i}$.
- U_i encrypts pk_i by computing $\mathfrak{C} = \mathsf{TEnc}_{opk}(pk_i, H(pk_o))$, where $H(pk_o)$ serves as the label of the encryption.
- U_i proves in non-interactive zero-knowledge manner the relations (1) - (6):
    1. U_i is in possession of a P-signature ς_i from the GM on (x_i, s_i).
    2. U_i generated a commitment C_{sig} of $\sigma = g^{\frac{1}{x_i + H(pk_o)}}$, a signature on pk_o.
    3. $\mathfrak{C}$ is a tag-based encryption of pk_i with the label $H(pk_o)$.
    4. S is $PRF_{g,s_i}(n_i)$, that is, $S = g^{\frac{1}{s_i + n_i}}$.
    5. $R = S^{x_i}$.
    6. $0 \le n_i < N$, i.e. U_i is in possession of a P-signature ς_i under pk_r on n_i.
- U_i uses sk_o to give a signature σ_{ots} on m concatenated with the above proofs.

All these proofs need to be done non-interactively by Groth-Sahai proof system or non-interactive P-signature (which utilizes the former). Specifically, U_i

1. runs PProve on ς_i and gpk to obtain commitments and proof $(C_{pk}, C_{seed}, \pi_1)$
   $\leftarrow$ PProve($params, gpk, \varsigma_i, (x_i, s_i)$) for secret key x_i and seed s_i respectively.
2. runs PProve on Σ_{n_i} and pk_r to obtain commitment and proof for counter
   n_i, i.e. $(C_{ctr}, \pi_2) \leftarrow$ PProve($params, pk_r, \Sigma_{n_i}, (n_i)$).
3. uses the Groth-Sahai proof system to construct proofs showing that the values $(R, S, \mathfrak{C}, C_{pk}, C_{seed}, C_{ctr}, C_{sig})$ are indeed well formed. This involves the proofs $\pi_S, \pi_R, \pi_O, \pi_C$ of the following languages:
    - $\mathcal{L}_S = \{C_s, C_n, y | \exists n, s, open_n, open_s$ such that
      $C_s = \mathsf{Com}(s, open_s) \land C_n = \mathsf{Com}(n, open_n) \land y = PRF_{g,s}(n)\}$
    - $\mathcal{L}_R = \{C_x, Y, y | \exists x, open_x$ such that $C_x = \mathsf{Com}(s, open_x) \land Y = y^x\}$
    - $\mathcal{L}_O = \{C_x, C_\sigma, pk_o | \exists x, open_x, open_\sigma$ such that
      $C_x = \mathsf{Com}(x, open_x) \land C_\sigma = \mathsf{Com}(\sigma, open_\sigma) \land \sigma = g^{\frac{1}{x + H(pk_o)}}\}$
    - $\mathcal{L}_C = \{C_x, \mathfrak{C} | \exists x, open_x$ such that
      $C_x = \mathsf{Com}(x, open_x) \land pk = g^x \land \mathfrak{C} = \mathsf{TEnc}_{opk}(pk)\}$

The signature is $(R, S, T, \mathfrak{C}, pk_o, \sigma_{ots}, C_{pk}, C_{seed}, C_{ctr}, C_{sig}, \pi_1, \pi_2, \pi_S, \pi_R, \pi_O, \pi_C)$. $\mathcal{L}_R$ is relatively simple. $\mathcal{L}_S$ can be found in [4, Section 4.2]. $\mathcal{L}_O, \mathcal{L}_C$ are somewhat simplified variants of the proofs in [14, Section 7].

Verify. To verify a signature, returns true if all of the following checks succeed:

1. PVer($params, gpk, \pi_1, (C_{pk}, C_{seed})$) = $\top$.
2. PVer($params, pk_r, \pi_2, C_{ctr}$) = $\top$.
3. σ_{ots} is a valid signature on $(m||\pi_1||\pi_2||\pi_S||\pi_R||\pi_O||\pi_C)$ under pk_o.
4. π_S is a valid proof on (C_{seed}, C_{ctr}, S).
5. π_R is a valid proof on (C_{pk}, R, S).
6. π_O is a valid proof on (C_{pk}, C_{sig}, pk_o).
7. π_C is a valid proof on $(C_{pk}, \mathfrak{C})$.

Open. Given a valid signature $(\cdots, \mathfrak{C}, pk_o, \cdots)$, anyone (the GM, or an opening agent) who holds the decryption key osk recovers $pk' = \mathsf{TDec}(\mathfrak{C}, H(pk_o))$. From the membership archive $\{(\mathsf{U}_i, pk', s_i)\}$, the GM outputs the corresponding U_i.

Reveal. From the membership archive, the GM retrieves s_i of the ith user.

Trace. Given s_i, the TA computes $\{S_j = g^{\frac{1}{s_i+j}}\}_{0 \le j < N}$. If a given signature has the S component inside this set, the TA concludes that user i is its originator.

Claim. U_i who gave the signature $\sigma = (R, S, \cdots)$ generates a non-interactive proof of knowledge π_R of the value x_i such that $R = S^{x_i}$ as a proof of authorship.

ClaimVer. Verify the proof π_R given by Claim.

4.2   Construction in the Random Oracle Model

Our second scheme assumes random oracle and employs CPA linear encryption instead of weak CCA tag-based encryption for better efficiency. The design is similar to the traditional-style traceable signature in [10] which is extended from [6]. However, we have moved the component for tracing component (which also helps in self-claiming) from $\mathbb{G}_T$ to $\mathbb{G}_p$. Since $\mathbb{G}_T$ is usually a subgroup of $\mathbb{Z}_{q^\alpha}$, it is vulnerable to sub-exponential discrete logarithm attacks and needs very large representation. For example, for 128-bit security, $|\mathbb{G}_T| \ge 3072$ bits. This can also been as a variant of [2], with a verifiable encryption and the self-claiming component added and double-spending detection removed.

Our scheme relies on the DLin assumption in $\mathbb{G}_1$, the q-SDH assumption in $(\mathbb{G}_1, \mathbb{G}_2)$, and the q-DDHI assumption in $\mathbb{G}_p$. We describe the scheme in $(\mathbb{G}_1, \mathbb{G}_2)$ but these two groups can be the same (since we instantiate the PRF in another DDH-hard group $\mathbb{G}_p$). Our scheme does not rely on the XDH assumption, this gives more flexibility in the choices of the underlying elliptic curve. If we are willing to make the XDH assumption, the signature can be made shorter since the linear encryption can be replaced by ElGamal encryption [6].

Setup. Let $(\mathbb{G}_1, \mathbb{G}_2)$ be a bilinear group pair with computable isomorphism ψ as discussed such that $\mathbb{G}_1 = \langle g \rangle$, $\mathbb{G}_2 = \langle h \rangle$ and $|g| = |h| = p$ for some prime p of λ bits. Let $\mathbb{G}_p = \langle f \rangle$ be a cyclic group of order p where the DDH assumption holds. Let $\mathfrak{g}_1, g_1, \mathfrak{g}_2, g_2$ be random elements in $\mathbb{G}_1$, which are for the zero-knowledge PoK protocols. Let H be a collision-free hash function which maps to $\mathbb{Z}_p^*$.

For credential issuing, the GM randomly selects $\mu \in_R \mathbb{Z}_p^*$ and computes $Z = h^\mu$ as the public key gpk of the group. The GM keeps $gsk = \mu$ in secret.

For opening, the GM setups an encryption key pair by picking $g_0 \in_R \mathbb{G}_1$, $osk = (a, b) \in_R (\mathbb{Z}_p^*)^2$. The public key is $opk = (u, v, g_0)$ where $u^a = v^b = g_0$.

For tracing, the GM manages a list of triple $(\mathsf{U}_i, s_i, \varsigma_i)$, which is initially empty; and picks a number N which is polynomial in λ. The GM also setup a range proof system, e.g. [7] to be discussed in Appendix A.

Join. User U_i obtains a credential from the GM through the interaction below.

1. U_i selects $x_i \in_R \mathbb{Z}_p^*$, sends $y_i = g_2^{x_i}$ to the GM with the proof $\mathsf{SPK}_0\big\{(x_i) : y_i = g_2^{x_i}\big\}$. This can be done non-interactively by Schnorr signature.
2. The GM verifies the proof, picks $s_i, e_i \in_R \mathbb{Z}_p^*$, computes $\varsigma_i = (gg_1^{s_i}y_i)^{\frac{1}{\mu+e_i}}$ and sends (ς_i, e_i, s_i) to the user. The GM also stores the triple (U_i, s_i, ς_i). The tracing trapdoor for U_i will be s_i and the GM should ensure it is unique.
3. U_i checks if $\hat{e}(\varsigma_i, Zh^{e_i}) = \hat{e}(gg_1^{s_i}g_2^{x_i}, h)$. The member public key and secret key are (y_i, e_i) and (x_i, ς_i, s_i) respectively.

Sign(m). User U_i manages a counter n_i on the number of signatures produced.

- U_i computes the tracing tag $S = f^{\frac{1}{s_i+n_i}}$ and the self-claiming tag $R = S^{x_i}$.
- U_i encrypts ς_i in $(T_1 = u^\alpha, T_2 = v^\beta, T_3 = \varsigma_i g_0^{\alpha+\beta})$ where $\alpha, \beta \in_R \mathbb{Z}_p^*$.
- U_i proves in non-interactive zero-knowledge manner such that (T_1, T_2, T_3) is a linear encryption of ς_i, where (ς_i, e_i) is a BBS + signature from the GM on (s_i, x_i), $S = f^{\frac{1}{s_i+n_i}}$, $R = S^{x_i}$ and $0 \le n_i < N$. This can be abstracted as

$$
\mathsf{SPK}_1\Big\{ (\varsigma_i, e_i, s_i, x_i, n_i, \alpha, \beta) :
$$
$$
\begin{aligned}
\hat{e}(\varsigma_i, Zh^{e_i}) &= \hat{e}(gg_1^{s_i}g_2^{x_i}, h) \quad &\wedge \\
(T_1, T_2, T_3) &= (u^\alpha, v^\beta, \varsigma_i g_0^{\alpha+\beta}) \quad &\wedge \\
(S, R) &= (f^{\frac{1}{s_i+n_i}}, f^{\frac{x_i}{s_i+n_i}}) \quad &\wedge \\
0 \le \quad n_i \quad &< N \quad &\Big\}(m)
\end{aligned}
$$

To conduct SPK_1, U_i computes $\mathfrak{A}_1 = \mathfrak{g}_1^{e_i}\mathfrak{g}_2^{\rho_1}$, $\mathfrak{A}_2 = \mathfrak{g}_1^{x_i}\mathfrak{g}_2^{\rho_2}$, $\mathfrak{A}_3 = \mathfrak{g}_1^{n_i}\mathfrak{g}_2^{\rho_3}$ for $\rho_1, \rho_2, \rho_3 \in_R \mathbb{Z}_p^*$. Next, U_i computes the following two SPK's.

$$
\mathsf{SPK}_{1A}\Big\{ (e_i, s_i, x_i, n_i, \alpha, \beta, \rho_1, \rho_2, \rho_3, \gamma_1, \gamma_2) :
$$
$$
\begin{aligned}
\mathfrak{A}_1 = \mathfrak{g}_1^{e_i}\mathfrak{g}_2^{\rho_1} \wedge \mathfrak{A}_2 = \mathfrak{g}_1^{x_i}\mathfrak{g}_2^{\rho_2} \wedge \mathfrak{A}_3 = \mathfrak{g}_1^{n_i}\mathfrak{g}_2^{\rho_3} \quad &\wedge \\
T_1 = u^\alpha \quad \wedge \quad T_2 = v^\beta \quad &\wedge \\
T_1^{e_i} = u^{\gamma_1} \quad \wedge \quad T_2^{e_i} = v^{\gamma_2} \quad &\wedge \\
f = S^{s_i}S^{n_i} \quad \wedge \quad R = S^{x_i} \quad &\wedge \\
\hat{e}(T_3, Zh^{e_i}) = \hat{e}(gg_1^{s_i}g_2^{x_i}, h)\hat{e}(g_0, Z^{\alpha+\beta}h^{\gamma_1+\gamma_2}) \quad &\Big\}(m)
\end{aligned}
$$
$$
\mathsf{SPK}_{1B}\Big\{ (n_i, \rho_3) : \mathfrak{A}_3 = \mathfrak{g}_1^{n_i}\mathfrak{g}_2^{\rho_3} \wedge 0 \le n_i < N \Big\}(m)
$$

We show how to instantiate SPK_{1A} below and SPK_{1B} in Appendix A.

(Commitment.) U_i picks $r_{e_i}, r_{s_i}, r_{x_i}, r_{n_i}, r_\alpha, r_\beta, r_{\rho_1}, r_{\rho_2}, r_{\rho_3}, r_{\gamma_1}, r_{\gamma_2} \in_R \mathbb{Z}_p^*$ and computes

$$
\mathfrak{T}_1 = \mathfrak{g}_1^{r_{e_i}}\mathfrak{g}_2^{r_{\rho_1}}, \ \mathfrak{T}_2 = \mathfrak{g}_1^{r_{x_i}}\mathfrak{g}_2^{r_{\rho_2}}, \ \mathfrak{T}_3 = \mathfrak{g}_1^{r_{n_i}}\mathfrak{g}_2^{r_{\rho_3}}, \ \mathfrak{T}_4 = u^{r_\alpha}, \ \mathfrak{T}_5 = v^{r_\beta} \text{ in } \mathbb{G}_1,
$$
$$
\mathfrak{T}_6 = \hat{e}(T_3, h)^{r_{e_i}}\hat{e}(g_1, h)^{-r_{s_i}}\hat{e}(g_2, h)^{-r_{x_i}}\hat{e}(g_0, Z)^{-r_\alpha-r_\beta}\hat{e}(g_0, h)^{-r_{\gamma_1}-r_{\gamma_2}} \text{ in } \mathbb{G}_T,
$$
$$
\mathfrak{T}_7 = T_1^{r_{e_i}}u^{-r_{\delta_3}}, \ \mathfrak{T}_8 = T_2^{r_{e_i}}v^{-r_{\delta_4}} \text{ in } \mathbb{G}_1, \ \mathfrak{T}_9 = S^{r_{s_i}}S^{r_{n_i}}, \ \mathfrak{T}_{10} = S^{r_{x_i}} \text{ in } \mathbb{G}_p.
$$

(Challenge and Response) Let $\mathfrak{T} = (\mathfrak{T}_1||\dots||\mathfrak{T}_{10})$. For a challenge $c = H(m||R||S||T_1||T_2||T_3||\mathfrak{T})$, U_i computes, in $\mathbb{Z}_p$, $z_{e_i} = r_{e_i} - ce_i$, $z_{s_i} = r_{s_i} - cs_i$, $z_{x_i} = r_{x_i} - cx_i$, $z_{n_i} = r_{n_i} - cn_i$, $z_{\rho_1} = r_{\rho_1} - c\rho_1$, $z_{\rho_2} = r_{\rho_2} - c\rho_2$, $z_{\rho_3} = r_{\rho_3} - c\rho_3$, $z_{\gamma_1} = r_{\gamma_1} - c\alpha e_i$, $z_{\gamma_2} = r_{\gamma_2} - c\beta e_i$, U_i sets $\mathfrak{z} = (z_{e_i}, z_{s_i}, z_{x_i}, z_{n_i}, z_{\rho_1}, z_{\rho_2}, z_{\rho_3}, z_{\gamma_1}, z_{\gamma_2})$.

Verify. To verify a signature $(S, T_1, T_2, T_3, \pi_{SPK})$, where π_{SPK} denotes the commitments (e.g. $\mathfrak{A}_1, \mathfrak{A}_2, \mathfrak{A}_3, \mathfrak{T}$) and the responses (e.g. $\mathfrak{z}$) generated by the above PoK protocol, this algorithm executes the verification algorithm of SPK_1.

In particular, SPK_{1A} can be verified by first computing

$$\mathfrak{T}_1' = \mathfrak{A}_1^c \mathfrak{g}_1^{z_{e_i}} \mathfrak{g}_2^{z_{\rho_1}} \qquad \mathfrak{T}_2' = \mathfrak{A}_2^c \mathfrak{g}_1^{z_{x_i}} \mathfrak{g}_2^{z_{\rho_2}} \qquad \mathfrak{T}_3' = \mathfrak{A}_3^c \mathfrak{g}_1^{z_{n_i}} \mathfrak{g}_2^{z_{\rho_3}}$$

$$\mathfrak{T}_4' = T_1^c \mathfrak{T}_4 \qquad\qquad \mathfrak{T}_5' = T_2^c \mathfrak{T}_5$$

$$\mathfrak{T}_7' = T_1^{z_{e_i}} u^{-z_{\gamma_1}} \qquad\quad \mathfrak{T}_8' = T_2^{z_{e_i}} v^{-z_{\gamma_2}}$$

$$\mathfrak{T}_9' = f^c S^{z_{s_i}} S^{z_{n_i}} \qquad\quad \mathfrak{T}_{10}' = R^c S^{z_{x_i}}$$

$$\mathfrak{T}_6' = \hat{e}(T_3, h^{z_{e_i}}/Z^c) \hat{e}(g^c g_1^{-z_{s_i}} g_2^{-z_{x_i}}, h) \hat{e}(g_0, Z^{-(z_\alpha + z_\beta)} h^{-(z_{\gamma_1} + z_{\gamma_2})})$$

and checking if $c \overset{?}{=} H(m \| R \| S \| T_1 \| T_2 \| T_3 \| \mathfrak{T}_1' \| \ldots \| \mathfrak{T}_{10}')$. If the equation holds, (R, S) and (T_1, T_2, T_3) are well-formed, thus outputs $\top$, $\bot$ otherwise.

Open. Given a valid signature $(m, S, T_1, T_2, T_3, \pi_{SPK})$, anyone who holds the decryption key (a, b) computes $\varsigma' = T_3/(T_1^a \cdot T_2^b)$. From the membership archive, the GM outputs U_i of the entry with the ς_i component matches with ς'.

Reveal. From the membership archive, the GM retrieves s_i of the ith user.

Trace. Given s_i, the TA computes $\{S_j = f^{\frac{1}{s_i + j}}\}_{0 \le j < N}$. If a given signature has the S component inside this set, the TA concludes that user i is its originator[1].

Claim. U_i who gave the signature $\sigma = (R, S, \cdots)$ can provide a proof of authorship π by generating a proof of knowledge π of the value x_i such that $R = S^{x_i}$.

ClaimVer. Verify the proof π given by Claim.

4.3   Security Analysis

Since both of our constructions are based on the same design, below we give an overall picture of the security analysis. The security of both schemes is hinged upon DLin, q-DDHI and q-SDH assumptions. Our RO-based construction relies on the DDH assumption, while our CRS-based one relies on the XDH assumption and any others required for the security of the multi-block P-signature [4]. Details for the RO-based construction can be found in [2,6,10] and those for the CRS-based construction follow from [4,14].

- *Identification Security.* Misidentification is an attack by a subset of colluded users and TA's. To model the former requires the signing oracle of the underlying signature. For our CRS-based construction, the underlying signing protocol (the credential issuing protocol in our case) of the P-signature has

---

[1] The TA may recover f^{x_i} from (R, S), but it does not help in forging a signature. Nevertheless, the GM can simply send $\{S_j = f^{\frac{1}{s_i + j}}\}_{0 \le j < N}$ to the TA to avoid this.

guaranteed that computing these signatures in such an interactive manner reveals nothing else about the secret key (by the fact that the protocol can be simulated by blackbox access of the signing oracle). For the TA's, they are just *given* the seeds (but they cannot choose it), which can be easily simulated. In fact, the seeds are just part of the messages to be signed by the underlying signature scheme, and their secrecy is not relevant here.

There are two attacks goals in misidentification attack. The first one is to output a valid signature which cannot be opened to anyone outside this collusion group. With the soundness of the ZK proof for the encryption, this translates to giving an encryption of a credential which the GM never issues. If this happens, the simulator S can extract this credential and returns it as the forgery of the signature scheme used for credential issuing.

The second attack goal is to produce a tag that cannot be computed by the Trace algorithm. There are two possibilities. The adversary A used an entirely new seed that is never "certified" by the GM, or A used a seed from the GM but produced something that cannot be produced by the Reveal algorithm. For the first case, S can decrypt the ciphertext and obtain a forgery of the underlying signature scheme. The second case will break either the soundness of the ZK proof for the PRF or that of the range proof.

- *N-Anonymity.* The compromised parties controlled by an adversary A is the same as those in misidentification attack. N-anonymity goes by a series of transformation such that a signature produced by an honest member is eventually transformed to one produced by another honest member, and argue that each transformation is computationally indistinguishable to A.

  Firstly, we change the parameter for the ZK proofs to a simulated one such that the commitments are perfect hiding and the ZK proofs involved in a signature will be "faked" by a simulation instead of giving a real proof. The adversary cannot notice this change by the security of the ZK proofs. Now the proofs are faked, we can change the elements that can differentiate one honest user from another in the signatures. We replace the tracing tag S with random element, then replace it with the tracing tag of another user. As long as these honest users produced less than N signatures, the indistinguishability of these changes are guaranteed by the pseudorandomness of the PRF. We then change the self-claiming tag R, by the DDH assumption (in $\mathbb{G}_p$ or in $\mathbb{G}_1$ – XDH). Finally, we encrypt a random element instead of the signature, by the indistinguishability of the encryption.

- *Non-Frameability.* The group member gives a signature based on the secret key x_i. The simulator S does not know x_i, but it gives out many signatures by the simulators of the underlying signature scheme and the ZK proofs (and manipulating the random oracle in our RO-based scheme). Eventually, the adversary A produces a valid forgery. In our CRS-based construction, S extracts the knowledge of σ by the extractability of the underlying proof system, which is a solution of the q-SDH problem as long as $H(pk_o)$ does not match with those public keys of a one-time signature scheme pre-selected by the simulator in the simulation of the signing oracle. In our RO-based construction, S can use the standard rewinding technique to extract x_i.

5 Conclusion

We found that the original idea of tracing signatures is nice but may not fully solve the problems of the group signature as expected. We propose a new and efficient way of tracing by borrowing the idea from compact e-cash. Our notion gives an alternative when timely tracing is important and when the signatures are scattered around in an anonymous system.

References

1. Au, M.H.: Personal communication (2009)
2. Au, M.H., Susilo, W., Mu, Y.: Constant-Size Dynamic k-TAA. In: De Prisco, R., Yung, M. (eds.) SCN 2006. LNCS, vol. 4116, pp. 111–125. Springer, Heidelberg (2006)
3. Belenkiy, M., Chase, M., Kohlweiss, M., Lysyanskaya, A.: P-signatures and Noninteractive Anonymous Credentials. In: Canetti, R. (ed.) TCC 2008. LNCS, vol. 4948, pp. 356–374. Springer, Heidelberg (2008)
4. Belenkiy, M., Chase, M., Kohlweiss, M., Lysyanskaya, A.: Compact E-Cash and Simulatable VRFs Revisited. In: Boyen, X., Waters, B. (eds.) Pairing 2009. LNCS, vol. 5671, pp. 114–131. Springer, Heidelberg (2009)
5. Boneh, D., Boyen, X.: Short Signatures Without Random Oracles and the SDH Assumption in Bilinear Groups. J. Cryptology 21(2), 149–177 (2008)
6. Boneh, D., Boyen, X., Shacham, H.: Short Group Signatures. In: Franklin [12], pp. 41–55
7. Camenisch, J., Chaabouni, R., Shelat, A.: Efficient Protocols for Set Membership and Range Proofs. In: Pieprzyk, J. (ed.) ASIACRYPT 2008. LNCS, vol. 5350, pp. 234–252. Springer, Heidelberg (2008)
8. Camenisch, J., Hohenberger, S., Lysyanskaya, A.: Compact E-Cash. In: Cramer, R. (ed.) EUROCRYPT 2005. LNCS, vol. 3494, pp. 302–321. Springer, Heidelberg (2005)
9. Camenisch, J., Lysyanskaya, A.: Signature Schemes and Anonymous Credentials from Bilinear Maps. In: Franklin [12], pp. 56–72
10. Choi, S.G., Park, K., Yung, M.: Short Traceable Signatures Based on Bilinear Pairings. In: Yoshiura, H., Sakurai, K., Rannenberg, K., Murayama, Y., Kawamura, S.-i. (eds.) IWSEC 2006. LNCS, vol. 4266, pp. 88–103. Springer, Heidelberg (2006)
11. Dodis, Y., Yampolskiy, A.: A Verifiable Random Function with Short Proofs and Keys. In: Vaudenay, S. (ed.) PKC 2005. LNCS, vol. 3386, pp. 416–431. Springer, Heidelberg (2005)
12. Franklin, M. (ed.): CRYPTO 2004. LNCS, vol. 3152. Springer, Heidelberg (2004)
13. Ge, H., Tate, S.R.: Traceable Signature: Better Efficiency and Beyond. In: Gavrilova, M., Gervasi, O., Kumar, V., Tan, C.J.K., Taniar, D., Laganà, A., Mun, Y., Choo, H. (eds.) ICCSA 2006. LNCS, vol. 3982, pp. 327–337. Springer, Heidelberg (2006)
14. Groth, J.: Fully Anonymous Group Signatures Without Random Oracles. In: Kurosawa, K. (ed.) ASIACRYPT 2007. LNCS, vol. 4833, pp. 164–180. Springer, Heidelberg (2007)
15. Groth, J., Sahai, A.: Efficient Non-interactive Proof Systems for Bilinear Groups. In: Smart, N.P. (ed.) EUROCRYPT 2008. LNCS, vol. 4965, pp. 415–432. Springer, Heidelberg (2008)

16. Kiayias, A., Tsiounis, Y., Yung, M.: Traceable Signatures. In: Cachin, C., Camenisch, J.L. (eds.) EUROCRYPT 2004. LNCS, vol. 3027, pp. 571–589. Springer, Heidelberg (2004)
17. Kiltz, E.: Chosen-Ciphertext Security from Tag-Based Encryption. In: Halevi, S., Rabin, T. (eds.) TCC 2006. LNCS, vol. 3876, pp. 581–600. Springer, Heidelberg (2006)
18. Nguyen, L., Safavi-Naini, R.: Efficient and Provably Secure Trapdoor-Free Group Signature Schemes from Bilinear Pairings. In: Lee, P.J. (ed.) ASIACRYPT 2004. LNCS, vol. 3329, pp. 372–386. Springer, Heidelberg (2004)
19. IEEE P1556 Working Group. VSC Project. Dedicated short range communications, DSRC (2003)
20. Teranishi, I., Sako, K.: k-Times Anonymous Authentication with a Constant Proving Cost. In: Yung, M., Dodis, Y., Kiayias, A., Malkin, T.G. (eds.) PKC 2006. LNCS, vol. 3958, pp. 525–542. Springer, Heidelberg (2006)

A   Signature-Based Range Proof

To implement a range proof system with D signatures [7], the GM setups a signing key $Z' = h^\nu$ and gives D signatures $\{\sigma_i = g^{1/\nu+i}\}$ for each $i \in \mathbb{Z}_D$.

To conduct SPK_{1B} in Section 4.2, i.e. to prove a secret value $t = n_i$ lies in $[0, N = D^\ell)$, the prover U_i writes t in base u where $t = \sum_j (t_j D^j)$ to obtain ℓ elements $\{t_j\}$, then picks τ_j, computes $\mathfrak{V}_j = \sigma_{t_j}^{\tau_j}$ for $j \in \mathbb{Z}_\ell$. Finally, U_i computes:

$$\mathsf{SPK}_{1C}\Big\{\, (\{t_j\}, \{\tau_j\}, \rho) : \mathfrak{A} = (\textstyle\prod_j \mathfrak{g}_1^{D_j})^{t_j} \mathfrak{g}_2^\rho \wedge (\wedge_j \mathfrak{V}_j = \sigma_j^{\tau_j}) \Big\}(m)$$

which can be instantiated by

(Commitment.) U_i picks $r_{t_1}, \ldots, r_{t_\ell}, r_{\tau_1}, \ldots, r_{\tau_\ell}, r_\rho \in_R \mathbb{Z}_p^*$ and computes $\{\mathfrak{U}_j = \hat{e}(\sigma_j, h)^{-r_{t_j}} \hat{e}(g, h)^{r_{\tau_j}}\}$ and $\mathfrak{U}^* = \prod_j (\mathfrak{g}_1^{D_j r_{t_j}}) \mathfrak{g}_2^{r_\rho}$.

(Challenge and Response) Let $\mathfrak{U} = (\mathfrak{U}_1||\ldots||\mathfrak{U}_\ell||\mathfrak{U}^*)$. For a challenge $c = H(m||\mathfrak{U})$, U_i computes, in $\mathbb{Z}_p$, $z_{t_1} = r_{t_1} - ct_j, \ldots, z_{t_\ell} = r_{t_\ell} - ct_\ell, z_{\tau_1} = r_{\tau_1} - c\tau_j, \ldots, z_{\tau_\ell} = r_{\tau_\ell} - c\tau_\ell, z_\rho = r_\rho - c\rho$ and U_i sets $\mathfrak{z} = (z_{t_1}, \ldots, z_{t_\ell}, z_{\tau_1}, \ldots, z_{\tau_\ell}, z_\rho)$.

To verify, compute $\mathfrak{U}'_j = \hat{e}(\mathfrak{V}_j, Z'^c h^{-z_{t_j}}) \hat{e}(g, h)^{z_{\tau_j}}, \forall j \in [1, 2, \cdots, \ell], \mathfrak{U}'' = \mathfrak{A}^c \prod_j (\mathfrak{g}_1^{D_j z_{t_j}}) \mathfrak{g}_2^{z_\rho}$ and check if $c \stackrel{?}{=} H(m||\mathfrak{U}')$ where $\mathfrak{U}' = (\mathfrak{U}'_1||\ldots||\mathfrak{U}'_\ell||\mathfrak{U}'')$.

Cryptanalysis of Hash Functions with Structures

Dmitry Khovratovich

University of Luxembourg
dmitry.khovratovich@uni.lu

Abstract. Hash function cryptanalysis has acquired many methods, tools and tricks from other areas, mostly block ciphers. In this paper another trick from block cipher cryptanalysis, the structures, is used for speeding up the collision search. We investigate the memory and the time complexities of this approach under different assumptions on the round functions. The power of the new attack is illustrated with the cryptanalysis of the hash functions Grindahl and the analysis of the SHA-3 candidate Fugue (both functions as 256 and 512 bit versions). The collision attack on Grindahl-512 is the first collision attack on this function.

Keywords: cryptanalysis, hash functions, SHA-3, truncated differentials, Grindahl, Fugue, structures.

1 Introduction

Since 1990 the MD family of hash functions and its successor SHA family have been most widely used data integrity primitives. In contrast with few cryptanalytic results in 90s recent attacks on MD5 [19], SHA-0 [14], and SHA-1 [5] encouraged the cryptographic community to look for more reliable components and then motivated the recent SHA-3 competition [16]. The Merkle-Damgård approach [8,15] to build hash functions from compression functions, has lost a part of credit due to such generic attacks as multicollisions [11] and second-preimage search with expandable messages [12].

In contrast to generic attacks like multicollisions, which are applicable to hash functions with Merkle-Damgard strengthening, attacks on lower level *compression functions* are highly dependent on a particular proposal and can rarely be extended to other functions. Common ideas are mostly related to the notion of differentials since the fact that two different messages produce the same hash value (a *collision*) can be expressed as a zero difference in the output.

The idea of differentials comes from block cipher cryptanalysis and pioneering papers by Biham and Shamir [3]. As high probability differential characteristics were exploited in attacks on block ciphers as high probability zero-ending differential *trails* are used to find collisions for compression functions.

Since block cipher cryptanalysis is a highly developed topic, many cryptanalysts try to use the most efficient methods and tools in attacks on hash functions. However, due to stronger requirements on the results of an attack only few of them were applied. The use of truncated differentials [17] is an example.

M.J. Jacobson Jr., V. Rijmen, and R. Safavi-Naini (Eds.): SAC 2009, LNCS 5867, pp. 108–125, 2009.
© Springer-Verlag Berlin Heidelberg 2009

In this paper we investigate another tool from block cipher cryptanalysis: *structures*. A structure is originally a set of plaintexts that pairwise have some property (e.g., zero difference in particular bytes). Since the number of pairs with desired properties in a structure is much larger than the structure size, such constructions are widely used in order to save memory and time in attacks on block ciphers [4,6].

Intuitively, structures might have been used in attacks on the hash functions built on block ciphers [18]. However, the authors are aware of only one such attack: a recent attack on Snefru [2], though Snefru does not directly fit the constructions from [18].

We have found that structures are especially useful in attacks on stream-based hash functions, where parts of a message can independently be controlled. We analyze the hash functions GRINDAHL and FUGUE. For GRINDAHL-256, we improve the best known attack by Peyrin [17], while for GRINDAHL-512 this paper presents the first known collision attack. The hash function FUGUE [10] is a strengthened successor to GRINDAHL, so we did not manage to break its security claims. However, our attack is substantially faster than a trivial internal-collision attack.

This paper is organized as follows. First we briefly explain how the use of structures reduce the cost of collision search. Then we investigate how structures follow the differential trail and collapse to pairs in some step (Section 3) so that the standard differential approach can be applied afterwards. We also derive the memory complexity of the attack.

Then we attack GRINDAHL and FUGUE with structures (Section 4). The number of computations required ro find a collision for GRINDAHL-256 is reduced compared to the attack by Peyrin [17]. We also present the first collision attack on GRINDAHL-512 and the first external analysis of FUGUE . In the Appendix the time complexity of the attack is estimated under different assumptions.

2   Idea in Brief

In many attacks on compression functions a cryptanalyst deals with a set of pairs that are to follow a particular differential trail. Here the *trail* is a sequence of differences in the internal state of the hash function. (see [1] for a more formal approach). At some steps an adversary may vary a message part to be injected thus increasing the number of pairs that follow the trail (the attack by De Cannière and Rechberger on SHA-1 [5] is an example). If there is not enough freedom to satisfy round conditions, the number of candidate pairs tends to decrease. We show that this effect can be postponed if a differential trail allows to incorporate pairs into structures.

In order to distinguish the approach when a cryptanalytic deals with pairs from our approach we call the former one the *standard differential attack*. It is also known as the *trail backtracking* [1]. Our attack is later called *the structural approach*, or the *structural attack*.

Now assume that the trail deals with truncated differentials, and the possible differences form a linear space R of differences. Then if a pair of states (S_1, S_2)

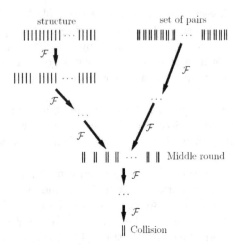

Fig. 1. Comparative view on the structural and the differential approaches. $\mathcal{F}$ is a round function. In the first case the number of states remains stable till structures collapse to pairs (*middle round*).

fits the trail, and a pair (S_2, S_3) fits the trail, then the pair (S_1, S_3) fits the trail too. Such a group of states is called a *structure*.

Suppose at some step a structure of size Q enters the round with probability P. Then every state S_i will have a desired difference with PQ other states thus composing a smaller structure. Therefore the initial structure splits into $1/P$ smaller structures. If the structure collapses into separate pairs then the differential attack is launched.

Suppose there is now freedom from the message injection, i.e., for a pair of messages (M, M') there are V possibilities for M and D more possibilities for M' (VD pairs at all). So if a round differential has probability P then of T pairs about $V \cdot D \cdot P \cdot T$ pairs survive. See Figure 1 for the outline of the situation.

When we work with structures, the value of the injected message can be chosen freely only for the first state of a new structure, or the *leader state*; the messages injected to the other states should have a desired difference with a first one. Consequently, the message freedom results in structures of size $\approx D \cdot P \cdot Q$. The number of states remains the same; however, it can be increased if we take other states as leading ones.

3   Analysis of Structure Fission

In order to benefit from the number of pairs in a structure an adversary should keep the size of the structure as big as possible. Let us estimate the size as a function of the round probability and the freedom given by message injections. Denote by $\sim$ the desired binary relation between two states, which can be also interpreted as the fact that the difference in the state satisfies the trail conditions.

The particular relation is usually clear from the context. Then a *structure* is a set of internal states such that any two states of the set satisfy the binary relation $\sim$. In our attacks on GRINDAHL and FUGUE (Section 4) the relation is of form "bytes (respectively, words) $i_1, i_2, \ldots, i_k$ are equal".

No freedom in the message injection. Suppose a structure of size $Q = 2^q$ states enters a round with probability 2^{-p}. First consider the case where there is no freedom in message injection due to the differential trail or the message schedule. After the application of the round function any state has a desired difference with 2^{q-p} other states, which form a structure of size 2^{q-p}. Therefore, the initial structure splits into about 2^p smaller structures of average size 2^{q-p}.

It is easy to prove that the partition of a structure into structures of equal size gives a lower bound on the overall number of pairs. If p is high enough then the structure collapses to separate pairs. Since there were 2^{2q-1} pairs in the structure, about 2^{2q-1-p} pairs come out of one round. Then the pairs are processed by the standard differential attack.

Value freedom in the message injection. Suppose now there is some freedom in message injections but the differential trail does not allow to introduce a difference, or the value of the difference is fixed. Then every state can be transformed to at most one element of a new structure, so the structures do not grow in this case. However, one may increase the number of structures and thus the number of considered states.

The latter approach increase the memory complexity so we do not use it except for the round when all structures collapse to pairs. Given $V = 2^v$ possibilities for an injected message we get $2^{2q-1+v-p}$ pairs after the round.

Difference freedom in the message injection. Assume that 2^d possible differences can be injected, and they form a linear space. Then we get larger structures because we have more freedom in steering a state into the structure.

Suppose that state S_i has already transformed to state S_i' by message v_i: $S_i[v_i] = S_i'$. Let us compute the probability that a randomly chosen state S_j can be transformed to some state with the desired difference with S_i' by some message m', which follows the trail too. The probability can be expressed as $\mathbb{P}(\exists m' : S_j[m'] \sim S_i', m' \sim m \mid S_i \sim S_j)$.

Assuming that the events for particular messages are independent we obtain the following expression:

$$\mathbb{P}(\exists m' : S_j[m'] \sim S_i', m' \sim m \mid S_i \sim S_j) =$$
$$= 1 - \prod_{m' \sim m} \mathbb{P}(S_j[m'] \nsim S_i' \mid S_i \sim S_j) = 1 - (1 - 2^{-p})^{2^d} \approx 2^{d-p}. \quad (1)$$

Consequently, one structure splits into structures of average size $Q' = 2^{q+d-p}$. Analogously, if $q + d - p < 0$ the structure collapse to pairs. Since $2^{2q-1+v+d}$ pairs can be composed about $2^{2q-1+v+d-p}$ come out of the round.

Size of the initial structure. By *degrees of freedom* we understand the base 2 logarithms of the number of admissible values. Suppose that at round i there are v_i degrees of freedom in the values of the injected message, d_i degrees of freedom in the differences in injected messages, and p_i (bit) conditions to be satisfied. In the standard differential attack we start with 2^c pairs and leave with one pair in the end. Therefore, we obtain the following equation:

$$c + \sum_{i=1}^{T}(v_i + d_i - p_i) = 0.$$

Here T stands for the number of rounds covered by the trail. We also denote by $c(t)$ the logarithm of the number of pairs after t-th round:

$$c(t) = c + \sum_{i=1}^{t}(v_i + d_i - p_i) = c + \underbrace{\sum_{i=1}^{t} v_i}_{v(t)} + \underbrace{\sum_{i=1}^{t} d_i}_{d(t)} - \underbrace{\sum_{i=1}^{t} p_i}_{p(t)}.$$

Suppose we start with a structure of size 2^q, which collapse to pairs after $l + 1$ rounds, $l < T$. The structure splits to $2^{p(l)-d(l)}$ smaller structures after l rounds. Each structure is of size about $2^{q+d(l)-p(l)}$. Therefore, about $2^{2q+d(l)-p(l)-1}$ pairs come out of round l.

In order to continue the collision search and obtain one pair in the end the following equation should hold:

$$2q + d(l) - p(l) - 1 = c(l) \;\Leftrightarrow\; 2q = c + v(l) + 1 \;\Leftrightarrow\; q = \frac{c + v(l) + 1}{2}.$$

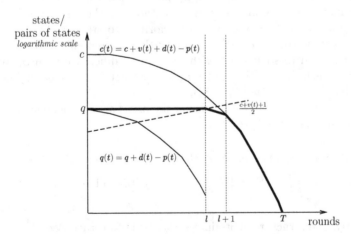

Fig. 2. Memory complexity of the collision search with structures

The memory complexity is thus determined by the maximum of 2^q and $2^{c(l+1)+1}$. It can be finally expressed as

$$\min_{0 \le l < T} \max(2^{\frac{c+v(l)+1}{2}}, 2^{c(l+1)+1}). \tag{2}$$

The plot of the memory complexity of the attack with structures compared to a standard differential attack is drawn in Figure 2. There c stands for the logarithm of the number of pairs required by the differential attack, q stands for the logarithm of the size of the structure that is used for the structural attack.

4   Concrete Attacks

4.1   How to Construct a Trail

The trails used in our attacks on GRINDAHL and FUGUEhave been obtained by a simple backtracking process. The idea is to start with zero-difference state and step back with introducing differences by all message injections. The differences spread to the internal state till every byte (or another building block) contains the difference. The number of steps is subject to the diffusion properties of the internal transformations.

4.2   Grindahl-256

Description. GRINDAHL is a family of hash functions proposed by Knudsen, Rechberger and Thomsen at FSE 2007 [13] as a stream-based hash function. The round function of GRINDAHL uses the design components of AES [7]: SUBBYTES and the MIXCOLUMNS operation. Since the internal state of GRINDAHL is wider than that of AES (GRINDAHL-256 can be viewed as a byte matrix of 4 rows and 13 columns) it uses a modified SHIFTROWS transformation in order to obtain better diffusion. The other message-independent transformation is ADDCONSTANT, which adds a constant to a particular byte.

In GRINDAHL-256 the message injection is just the overwriting of the first column with 4-byte message block. The round function is defined as the following composition of transformations:

$$P(\alpha, M) = \text{MIXCOLUMNS} \circ \text{SHIFTROWS} \circ \text{SUBBYTES} \circ$$
$$\circ \text{ADDCONSTANT} \circ \text{INJECTMESSAGE}(\alpha, M).$$

Here α denotes the state to be iterated, and M the message block to be injected. Every message block is used only once.

In order to obtain a hash value the state filled with zeros is iterated till the message is ended. Then eight blank rounds (no message injection) are applied and the resulting state is truncated to 256 bits, which is the hash value.

Security. The designers of GRINDAHL-256 claimed the security level of 2^{128} operations against both collision and second-preimage attack. Peyrin in [17] found a differential trail, which leads to a full collision in an internal state before the blank rounds are applied. The trail deals with two values of byte differences: non-zero and zero. It starts with a pair of states that differ in all bytes and after 9 message injections leads to a collision. Following our notation, he had 55 byte conditions, 21 byte degrees of value freedom, and 20 byte degrees of difference freedom thus obtaining complexity $2^{(55-21-20)\cdot 8} = 2^{112}$ message pairs. In early steps there was more freedom that is required by the trail so there was no clear difference between the value freedom and the difference freedom. However, the structural approach benefits from the difference freedom so we first exploit the latter one. There is also an attack on the prefix-MAC built on GRINDAHL [9].

Although GRINDAHL-256 is already broken, the goal of our attack is not only the illustration of structural technique. Peyrin provided some ad-hoc observations on the fact that his attack is one of the best dealing with the truncated-differential approach, and 2^{104} is the lower bound on the complexity of such attack. Our attack breaks this bound.

Attack. In order to apply the structural approach we first have to modify a bit the class of truncated differentials. Here and later we consider two-valued byte-difference: $*$ (random difference, including 0) and 0 (bytes coincide). They are marked as grey and white cells in Figure 3, respectively. One can easily check that this not only barely affect the probability of the trail and the complexity of the collision search but also simplify computations.

The second barrier is that the trail used by Peyrin for collision search is badly suited for the structural approach due to the distribution of probabilities among the iterations, which helps the standard differential attack but does not provide the best results for the structural attack. Table 1 (a) shows that we would have to start with a structure if $2^{12.5\cdot 8} = 2^{100}$ states, which does not offer enough advantage against Peyrin's attack.

The better complexity is provided by the second trail from [17], which was proposed for the second-preimage search. However, there is a mistake in Peyrin's paper: the byte C inserted before the k-th iteration does not affect column 11 in the $k+1$-th iteration. As a result, the complexity of a simple truncated differential attack is $2^{21\cdot 8} = 2^{168}$ pairs. However, the structural attack needs only a set of $2^{10.5\cdot 8 + 0.5} = 2^{84.5}$ states (Table 1 (b)).

The attack works as follows. Iterate GRINDAHL-256 for 10 rounds with randomly chosen messages and obtain a structure with $2^{84.5}$ states. Then we keep the size of the strcture after the first iteration thanks to the 4-byte difference freedom. After the second round the structure collapses to 2^{72} pairs, and only one pair comes out of the next iteration.

Time complexity of the attack. Since some message bytes pass several SUB-BYTES transformations it is not clear how costly the steps when we deal with structures are. A trivial upper bound is $2^{q+\max(d_i)} \approx 2^{116}$. We propose some optimizations, which lead to a complexity about 2^{100} operations though the

Table 1. Parameters of differential trails for GRINDAHL-256. Measurement in bytes.

i	v_i	d_i	p_i	$c(i)$	$\frac{c+v(i)}{2}$	$q(i)$
Start	–	–	–	14	7	12.5
1	0	2	2	14	7	12.5
2	3	4	7	14	8.5	9.5
3	4	3	7	14	10.5	5.5
4	4	2	7	13	**12.5**	1.5
5	4	3	9	11	14.5	–
6	4	4	14	5	16.5	–
7	2	2	9	0	17.5	–
8 − 9	0	0	0	0	17.5	–

(a)

i	v_i	d_i	p_i	$c(i)$	$\frac{c+v(i)}{2}$	$q(i)$
Start	–	–	–	21	10.5	10.5
1	0	4	4	21	**10.5**	10.5
2	4	4	20	9	12	–
3	4	3	15	0	14	–
4 − 5	0	0	0	0	14	–

(b)

technique can probably be improved. The reader may refer to Table 8 for better understanding.

In the first step there are 4 bytes of difference freedom and 4 bytes where the difference should be canceled. The leader state of a new structure is defined by iterating the round function with a random message block. For each next state S in the structure we must find the message bytes (A, B, C, D) to be injected (we keep this notation in the further text) that lead to a state colliding in particular 4 bytes with the leader state. First consider column 7 before the MIXCOLUMNS transformation in the second iteration. Three bytes of column 7 are not affected by the message injection and can be derived explicitly. On the other hand, one byte after the MIXCOLUMNS transformation is known because a collision there is needed. Thus, compute both the input and the output of the MIXCOLUMNS transformation of column 7 and thus derive the value of D and the value of second byte in column 9 in the next iteration.

Then try all the values of C. For each value derive one more byte in column 9 in the third iteration. As a result, two bytes in column 9 are known before the MIXCOLUMNS transformation and two bytes are known due to the fact of collision. As a result, derive the values of A and B and check the MIX-COLUMNS transformation in column 3 of the second iteration with the latter two values. On average, 2^7 trials of C are required.

The second step is actually the bottleneck of our attack, though we believe that the complexity may be reduced. First vary B and C for each state thus obtaining $2^{100.5}$ states. Then the 16 bytes in the third iteration where zero difference is desired are fully determined by 6 bytes that are affected by A and D. This gives us $16 - 6 = 10$ byte conditions, which can be used to divide the set of states into structures. One more condition we get from the second iteration, where the byte was affected by just fixed B. Therefore, we obtain $2^{(10+1)\cdot 8} = 2^{88}$ blocks each of size $2^{12.5}$. In every block we have 6 variables and 6 conditions; the other conditions are provided by constants. Since we process the blocks independently, the memory complexity is not increased.

Then consider the unknown bytes in columns 3, 5 and 11 that are affected only by A. Consider two random states in a block and denote by x_A and x'_A the message byte A after the SUBBYTES transformation. Then the fact of zero difference in those columns can be expressed as the following system of equations:

$$
\begin{cases}
a_{12}S(a_{21}x_A + c_1) + c_2 = a_{12}S(a_{21}x'_A + c'_1) + c'_2; \\
a_{11}S(a_{31}x_A + c_3) + c_4 = a_{11}S(a_{31}x'_A + c'_3) + c'_4; \\
a_{14}S(a_{41}x_A + c_5) + c_6 = a_{14}S(a_{41}x'_A + c'_5) + c'_6.
\end{cases}
$$

Here a_{ij} are coefficients of the MIXCOLUMNS matrix and c_i are state-dependent constants. Due to properties of the AES S-box x_A and x'_A are uniquely determined (if there is a solution) by constants $\bar{c} = (c_1, c_2, c_3, c_4, c_5, c_6)$ and $\bar{c}' = (c'_1, c'_2, c'_3, c'_4, c'_5, c'_6)$. Furthermore, this property is transitive, so that we precompute the function $f : \bar{c} \rightarrow x_A$.

As a result, a block of 2^{12} states splits to 2^8 blocks with 2^4 states each where A, B and C are fixed. In order to obtain the value of D repeat the same trick in columns 7, 11, and 12 thus getting one pair per 2^4 states, or 2^{96} pairs at all. Only 2^{72} pairs of them pass through 3 conditions in column 9 in the fourth iteration.

In the last third step we have to pass 15 byte conditions given 6 byte degrees of freedom. Since we deal with separate pairs, the filtering process be maintained with precomputations (see [17]).

4.3   Grindahl-512

The hash function GRINDAHL-512 is defined similarly to GRINDAHL-256, but the internal state is twice as big as that of GRINDAHL-512: it has 8 rows and 13 columns. Each injection of a message block substitutes the first column with 8 bytes of a message. The row offset values are defined by the following expression:

$$
c_i = i + 1; \ 0 \le i \le 7.
$$

The MixColumns matrix is also redefined but the exact coefficients are irrelevant to our attack. The only property we use is that this matrix is MDS with branch number 9.

So far there is no collision attack on GRINDAHL-512 though a weakness of using GRINDAHL-512 as the base of prefix-MAC was shown [9].

Attack. We use a 3-round differential trail, which is shown at Figure 4. The trail is obtained by iterating GRINDAHL-512 backwards from the zero-difference state. It is assumed that the last truncation (before the injection) deletes a column with 6 byte differences, while the first two truncations delete the full-difference column. The parameters of the trail are listed in Table 2 at the left. However, the second step becomes so time-consuming that the resulting complexity overcomes the brute-force one. The reason is that structures are too large to be quickly recomposed into pairs. On the other hand, if we test all the possible injections, the time complexity increases as well.

Table 2. Parameters of the differential trail for GRINDAHL-512. The second table is obtained by splitting the second step into two substeps. Measurement in bytes.

i	v_i	d_i	p_i	$c(i)$	$q(i)$
	–	–	–	48	28
1	8	8	21	43	15
2	8	8	49	10	–
3	2	2	14	0	–

i	v_i	d_i	p_i	$c(i)$	$q(i)$
	–	–	–	48	28
1	8	8	21	43	15
2 - I	0	8	21	36	2
2 - II	8	0	28	10	–
3	2	2	14	0	–

We choose to decompose the second round into two sub-rounds with only slight increase of the complexity. The idea is as follows. We first process the zeros that are the result of the second MixColumns transformation and that are affected by the second message injection. These are 21 zeros in columns 1–8. For any two states that follow the trail before the second injection the condition of having zero difference in these positions is equivalent to 21 linear equations with the differences in the internal state after the S-box application as variables. Since the message injection can be equivalently swapped with the S-box transformation we obtain that the 21 equations are 21 linear conditions on 8 differences in the message block.

Therefore, $2^{13 \cdot 8}$ structures of size $2^{15 \cdot 8}$ split into $2^{(13+21-8) \cdot 8} = 2^{26 \cdot 8}$ structures of size $2^{2 \cdot 8}$. These structures collapse to pairs and are partly filtered out due to the remaining 28 byte conditions though 8 byte degrees of freedom are still available. Then we compose all possible $2^{30 \cdot 8}$ pairs and filter them out. The desired values to be injected can be derived from pre-computed tables, which are applicable since we already deal with pairs. The resulting complexity is 2^{240} computations and still 2^{224} memory. The complexity of the last step is negligible. We also modify the memory complexity table taking into account the considerations discussed above (Table 2).

4.4 Fugue

Hash family FUGUE [10] has been recently submitted to the SHA-3 contest [16], and has been recently chosen to the second round. It was designed by a group of researchers in IBM. The design of FUGUE resembles that of GRINDAHL with several improvements, that should have increased the security. However, FUGUE is slower than GRINDAHL, which can be a serious disadvantage during the competition.

We analyze FUGUE with the structural approach and show that its security is much higher than that of GRINDAHL. Though we do not break the FUGUE security claims, the our attack is significantly faster than a trivial internal-collision attack.

Description

FUGUE-*256*. FUGUE-256 has internal state, denoted by S, of 120 bytes, which is viewed as a 4×30 array. We denote by S_i ($i = 0 \ldots 29$) the i-th column of S. A message, appropriately padded, is split to 4-byte blocks. Each block I is an input to the round transformation of S, which is defined in pseudo-code as follows:

- TIX(I);
- Repeat 2 times:
  - ROR3;
  - CMIX;
  - SMIX;

TIX, ROR3 and CMIX are linear transformations. TIX consists of the following steps:

$$S_{16}+ = S_0; \quad S_0 = I; \quad S_8+ = S_0; \quad S_1+ = S_{24},$$

where + stands for XOR. CMIX is linear as well:

$$S_0+ = S_4; \quad S_1+ = S_5; \quad S_2+ = S_6; \quad S_{15}+ = S_4; \quad S_{16}+ = S_5; \quad S_{17}+ = S_6.$$

ROR3 rotates the state three columns to the right.

SMIX is a more complicated transformation. It process bytes in columns S_0–S_3. First, the AES S-box is applied to those 16 bytes. Then they are composed into a 16-byte vector, that is multiplied by matrix N, which is an almost-MDS matrix with branch number 16.

After all the blocks have been processed, the final round transformation is applied, and then eight columns of S are taken as hash output. Since we produce a collision before the final round, we skip its description (see full details in [10]).

FUGUE-*512*. FUGUE-512 follows the same philosophy, but has a stronger design: 36 columns (instead of 30) and twice as many operations as FUGUE-256 per round:

- TIX'(I);
- Repeat 4 times:
  - ROR3;
  - CMIX';
  - SMIX;

The CMIX' and TIX' operations have more column additions compared to FUGUE-256, and column indices are different. TIX':

$$S_{22}+ = S_0; \quad S_0 = I; \quad S_8+ = S_0; \quad S_1+ = S_{24}; \quad S_4+ = S_{27}; S_7+ = S_{30}.$$

CMIX':

$$S_0+ = S_4; \quad S_1+ = S_5; \quad S_2+ = S_6; \quad S_{18}+ = S_4; \quad S_{19}+ = S_5; \quad S_{20}+ = S_6.$$

Table 3. Column dependencies in FUGUE-256 and FUGUE-512. Value $-i$ for column j means that before r-th round the last message block that affected column j is M_{r-i}.

Column	0–6	7–12	13	14–17	18–23	24–29
Depend on	-1	-2	-3	-1	-2	-3

Column	0–12	13–17	18–26	27–35
Depend on	-1	-2	-1	-2

Properties of Internal Transformations. We consider truncated differentials, where difference in one byte may be either zero or random. We assume that two columns have equal differences with probability 2^{-32}, so every column addition in CMIX and TIX operations costs us 2^{32} if producing a zero column from two random ones. The SMIX transformation is more complicated. The matrix N is not MDS but is so called *almost MDS* with the branch number equal to 13. As a result, when constructing a trail in the backward direction, we get no benefit from having few active S-boxes in the input of S-Mix so we always assumed that any active S-Mix output was produced by the input where all the 16 bytes are active. We certainly assume that this approach may not be optimal though we do not see any properties of the S-Mix transformation which may lead to other possibilities.

The designers provide several arguments for the resistance of FUGUE to pure and truncated differential attacks and even provide lower bounds for several attack modes, which unfortunately do not cover the mode that we use. We only point out that the complexity of the trivial internal collision attack on FUGUE is about $2^{29 \cdot 8 \cdot 2} = 2^{464}$ for FUGUE-256 and 2^{560} for FUGUE-512.

Analysis of Fugue-256. The optimal trail that we found for FUGUE-256 is a 6-round trail depicted in Table 5. Although differences in round $r + 2$ can theoretically be managed with a message injection in round r, this is not the case for this trail. We use the r-th message injection to get proper differences in only rounds r and $r + 1$ (mostly in round r).

We start with a structure of internal states of size $2^{44 \cdot 8} = 2^{352}$. It splits into 2^{320} structures of 2^{32} states each after three rounds (Table 4). About $2^{24 \cdot 8} = 2^{192}$ pairs come out of the next round, and we get one colliding pair after two more

Table 4. Parameters of differential trails for FUGUE-256 and FUGUE-512

i	v_i	d_i	p_i	$c(i)$	$q(i)$
Start	–	–	–	80	**44**
-6	0	4	4	80	**44**
-5	4	4	16	72	32
-4	4	4	32	48	4
-3	4	4	32	24	–
-2	4	4	32	0	–
-1	0	4	4	0	–

i	v_i	d_i	p_i	$c(i)$	$q(i)$
Start	–	–	–	116	**60**
-4	4	4	28	96	36
-3	4	4	56	48	–
-2	4	4	56	0	–
-1	0	4	4	0	–

Table 5. Trail for FUGUE-256

R\C	0	1	2	3	4	5	6	7	8	9	10	11	12	13	14	15	16	17	18	19	20	21	22	23	24	25	26	27	28	29
−5	*	*	*	*	*	*	*	*	*	*	*	*	*	*	*	*	*	*	*	*	*	*	*	*	*	*	*	*	*	*
−4	*	*	*	*	*	—	—	*	*	*	*	—	*	*	*	*	*	*	*	*	*	*	*	*	*	*	*	*	*	*
−3	*	*	*	—	*	—	—	—	*	*	*	—	—	*	—	—	—	—	*	*	*	*	*	*	*	*	*	*	*	*
−2	*	*	—	—	*	—	—	—	*	—	*	—	—	—	—	—	—	—	—	—	—	—	—	*	*	*	*	*	*	*
−1	*	—	—	—	—	—	—	—	—	*	—	—	—	—	—	—	—	—	—	—	—	—	—	—	—	—	—	—	—	—
0	—	—	—	—	—	—	—	—	—	—	—	—	—	—	—	—	—	—	—	—	—	—	—	—	—	—	—	—	—	—

Table 6. Summary of our attacks on concrete hash functions

Hash function	Attack	Memory complexity	Time complexity
GRINDAHL-256	Truncated differential [17]	2^{32}	2^{112}
	Structural	2^{84}	2^{100}
GRINDAHL-512	**Structural**	2^{224}	2^{240}
FUGUE-256	Internal collision	2^{464}	-
	Structural	2^{352}	2^{352}
FUGUE-512	Internal collision	2^{560}	-
	Structural	2^{480}	2^{480}

Table 7. Optimal trail for FUGUE-512

R\C	0	1	2	3	4	5	6	7	8	9	10	11	12	13	14	15	16	17	18	19	20	21	22	23	24	25	26–30	31–35
-4	*	*	*	*	*	*	*	*	*	*	*	*	*	*	*	*	*	*	*	*	*	*	*	*	*	*	* * * * *	* * * * *
-3	*	*	-	-	*	-	*	*	*	*	*	-	*	*	*	*	*	*	*	*	*	*	*	*	*	*	* * * * *	* * * * *
-2	*	*	-	-	*	-	-	*	*	-	*	-	*	-	-	-	-	-	-	-	-	-	-	*	*	*	* * * * *	* * * * *
-1	*	-	-	-	-	-	-	-	-	*	-	-	-	-	-	-	-	-	-	-	-	-	-	-	-	-	-----	-----
0	-	-	-	-	-	-	-	-	-	-	-	-	-	-	-	-	-	-	-	-	-	-	-	-	-	-	-----	-----

rounds. Due to big memory complexity of the attack, we assume that we are allowed to run much precomutation and store the results in tables. We thus assume that we spend negligible time complexity per each state and each pair, so the resulting time complexity should be about 2^{352} as well. This complexity is clearly much larger than the birthday bound (2^{128}) though it is at the same time much smaller than a birthday bound for the internal collision (2^{448}). We would also like to point out that we have not found any non-trivial differential attack with a comparable complexity.

Analysis of Fugue-512. The optimal trail that we found for FUGUE-512 is a 5-round trail depicted in Table 7. Here we use a message injection to get proper differences in the same round. We start with a structure of internal states of size $2^{60.8} = 2^{480}$. It splits into 2^{192} structures of 2^{288} states in the next round (Table 4), and collapse to 2^{352} pairs after two rounds. Following the same observation, we again assume that we spend negligible time complexity per each state and each pair, so the resulting time complexity should be about 2^{480}, which is still much larger than the birthday bound (2^{256}) and smaller than a birthday bound for the internal collision (2^{560}).

5   Conclusions and Future Work

We showed how the organization of internal states into structures can drastically reduce the complexity of collision search providing an appropriate differential trail. The exact formulas for memory complexity and estimates on time complexity of the attack with structures have been provided. We successfully combined our approach with simply obtained differential trails and presented the best known attacks on GRINDAHL and the only external analysis of FUGUE. The results are summarized in Table 6.

We conclude that FUGUE is much more resistant to attacks with truncated differentials, that were successfully used for the cryptanalysis of GRINDAHL. This is mostly due to a better diffusion and a larger internal state, which prevents from this style of attacks. We believe that our attacks can be further improved with other differential trails or better optimization of the maintenance of structures. The complexity of the attack is now determined by the bottleneck step when structures collapse to pairs. It is likely that the plot of the complexity function can be significantly smoothed for some hash functions.

Acknowledgement. I greatly thank anonymous reviewers for their valuable comments, which helped to improve the paper. I am supported by the PRP "Security & Trust" grant of the University of Luxembourg.

References

1. Bertoni, G., Daemen, J., Peeters, M., Van Assche, G.: Radiogatun, a belt-and-mill hash function (2006), http://radiogatun.noekeon.org/
2. Biham, E.: New techniques for cryptanalysis of hash functions and improved attacks on Snefru. In: Nyberg, K. (ed.) FSE 2008. LNCS, vol. 5086, pp. 444–461. Springer, Heidelberg (2008)
3. Biham, E., Shamir, A.: Differential cryptanalysis of DES-like cryptosystems. In: Menezes, A., Vanstone, S.A. (eds.) CRYPTO 1990. LNCS, vol. 537, pp. 2–21. Springer, Heidelberg (1991)
4. Biham, E., Shamir, A.: Differential Cryptanalysis of the Data Encryption Standard. Springer, Heidelberg (1993)
5. De Cannière, C., Rechberger, C.: Finding SHA-1 characteristics: General results and applications. In: Lai, X., Chen, K. (eds.) ASIACRYPT 2006. LNCS, vol. 4284, pp. 1–20. Springer, Heidelberg (2006)
6. Daemen, J., Knudsen, L.R., Rijmen, V.: The block cipher SQUARE. In: Biham, E. (ed.) FSE 1997. LNCS, vol. 1267, pp. 149–165. Springer, Heidelberg (1997)

7. Daemen, J., Rijmen, V.: The Design of Rijndael. AES — the Advanced Encryption Standard. Springer, Heidelberg (2002)
8. Damgård, I.: A design principle for hash functions. In: Brassard, G. (ed.) CRYPTO 1989. LNCS, vol. 435, pp. 416–427. Springer, Heidelberg (1990)
9. Gorski, M., Lucks, S., Peyrin, T.: Slide attacks on a class of hash functions. In: Pieprzyk, J. (ed.) ASIACRYPT 2008. LNCS, vol. 5350, pp. 143–160. Springer, Heidelberg (2008)
10. Halevi, S., Hall, W.E., Jutla, C.S.: The hash function fugue. Submission to NIST (2008)
11. Joux, A.: Multicollisions in iterated hash functions. Application to cascaded constructions. In: Franklin, M. (ed.) CRYPTO 2004. LNCS, vol. 3152, pp. 306–316. Springer, Heidelberg (2004)
12. Kelsey, J., Schneier, B.: Second preimages on n-bit hash functions for much less than 2^n work. In: Cramer, R. (ed.) EUROCRYPT 2005. LNCS, vol. 3494, pp. 474–490. Springer, Heidelberg (2005)
13. Knudsen, L.R., Rechberger, C., Thomsen, S.S.: The Grindahl hash functions. In: Biryukov, A. (ed.) FSE 2007. LNCS, vol. 4593, pp. 39–57. Springer, Heidelberg (2007)
14. Manuel, S., Peyrin, T.: Collisions on sha-0 in one hour. In: Nyberg, K. (ed.) FSE 2008. LNCS, vol. 5086, pp. 16–35. Springer, Heidelberg (2008)
15. Merkle, R.C.: One way hash functions and DES. In: Brassard, G. (ed.) CRYPTO 1989. LNCS, vol. 435, pp. 428–446. Springer, Heidelberg (1990)
16. NIST. Cryptographic hash algorithm competition, http://www.nist.gov/hash-competition
17. Peyrin, T.: Cryptanalysis of Grindahl. In: Kurosawa, K. (ed.) ASIACRYPT 2007. LNCS, vol. 4833, pp. 551–567. Springer, Heidelberg (2007)
18. Preneel, B., Govaerts, R., Vandewalle, J.: Hash functions based on block ciphers: A synthetic approach. In: Stinson, D.R. (ed.) CRYPTO 1993. LNCS, vol. 773, pp. 368–378. Springer, Heidelberg (1994)
19. Wang, X., Yu, H.: How to break MD5 and other hash functions. In: Cramer, R. (ed.) EUROCRYPT 2005. LNCS, vol. 3494, pp. 19–35. Springer, Heidelberg (2005)

A   Analysis of Complexity

The time complexity analysis is much harder because we have to arrange states into structures as fast as possible.

We consider only one structure, because this process is independently applied for all structures. To obtain the whole complexity one should multiply the derived values by the current number of structures.

No freedom in message injection. States S'_i and S'_j belong to the same structure if $S'_i \sim S'_j$. On the other hand, if $\sim$ defines set R of linear differences then the condition can be expressed in terms of projections to space $R^\perp$ that is orthogonal to R:

$$S'_i \sim S'_j \iff \mathrm{pr}_{R^\perp} S'_i = \mathrm{pr}_{R^\perp} S'_j.$$

As a result, we compute the ordered set of projections of structures and use binary search to derive the structure a state belongs to. Assuming the sorting and search costs are negligible comparing to the round iteration we derive the complexity roughly equal to the number of states.

Value freedom in message injection. The fact that there is the value freedom in message injection does not affect the complexity of the attack if structures do not collapse into pairs yet. The exact value of the complexity in this case is just the number of states after the round iteration.

Consider the case where structures are collapsed to pairs and single states. As mentioned in Section 3, the structure contains 2^{2q-1+v} pairs. The further steps depend on whether we can exploit properties of the round function.

- If the round function is viewed as a black box, we just derive pairs for each possible message m. The complexity is about 2^{q+v}.
- If we can quickly find solutions m for the equation

$$\mathrm{pr}_{R^\perp} S_i[m] = \mathrm{pr}_{R^\perp} S_j[m] \tag{3}$$

  then it is solved for all possible pairs about 2^{2q-1} times.
- If there exist not only a fast algorithm for solution (3) but also function f such that (3) has a solution iff $f(S_i) = f(S_j)$. Then we compose the ordered set of $f(S)$ and for each new state look for a pair with negligible cost. The complexity would be equal to the maximum of the size of the initial structure (2^q) and the number of resulting pairs $(2^{2q-1+v-p})$.

Difference freedom in message injection. Again, first, we investigate the case where structures do not collapse to pairs. Suppose states $S'_1, \ldots, S'_i$ have been already distributed into just created structures. We also require that every leader state is obtained by the same injected message m_0. A state S_{i+1} can be distributed to the structure with the leader state S' if there exist a message m_{i+1} such that $m_0 \sim m_{i+1}$ and $\mathrm{pr}_{R^\perp} S' = \mathrm{pr}_{R^\perp} S_{i+1}[m_{i+1}]$. Denote by S the set of all such states $S_{i+1}[m_{i+1}]$. Then the question is whether $\mathrm{pr}_{R^\perp} S'$ belongs to $\mathrm{pr}_{R^\perp} \mathsf{S}$.

If S is an affine space, and the linear space does not depend on S_{i+1} then we can easily compute the projection and find the corresponding structure using the ordered set approach. The complexity would be equal to the number of states. If S is not an affine space but can be represented as a union of affine spaces then we compute the projection for each space. In the worst case the complexity is equal to 2^d multiplied by the number of states.

Now consider the case where a structure collapse to pairs. This is actually the most complicated case and can be considered as a bottleneck. Indeed, about $2^{2q-1+v+d}$ pairs are composed from a structure with 2^q states. About $2^{2q-1+v+d-p}$ pairs come out of the round iteration. The possible approaches are similar to the case where there is no freedom in difference. If the round function is a black box, the complexity varies from 2^{q+v} to 2^{q+v+d}. If there exists a function f such that (3) has a solution iff $f(S_i) = f(S_j)$, then the complexity is between 2^q and $2^{2q-1+v+d-p}$).

B   Trails

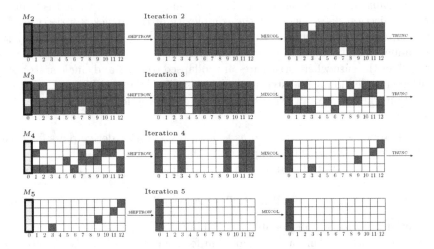

Fig. 3. Differential trail for GRINDAHL-256 (Table 1 (b))

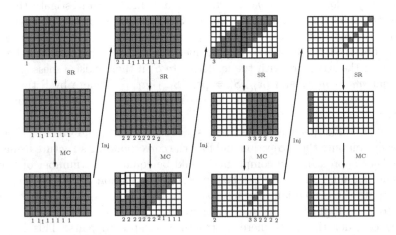

Fig. 4. Differential trail for GRINDAHL-512 (Table 2 (b))

Table 8. Dependencies of the message block in the differential trail for GRINDAHL-256

It	Col	Cost	1				2				3			
2	2	1					B							
	3	1	B	A										
	7	1			D									
3	1	2									A			
	2	2										B		
	3	3					B	A						
	5	3					C		A					
	6	3						C	B					
	7	2								D				
	8	2							C					
	9	2	C	D	AC	BD								
	10	2												D
	11	2					D		A					
	12	1						D	B					
4	3	3									B	A		
	9	3					C	D	AC	BD				
	11	3									D			A
	12	3										B		

Message bytes

Cryptanalysis of the LANE Hash Function

Shuang Wu, Dengguo Feng, and Wenling Wu

State Key Lab of Information Security, Institute of Software
Chinese Academy of Sciences
Beijing 100190, China
{wushuang,feng,wwl}@is.iscas.ac.cn

Abstract. The LANE[4] hash function is designed by Sebastiaan Indesteege and Bart Preneel. It is now a first round candidate of NIST's
SHA-3 competition. The LANE hash function contains four concrete
designs with different digest length of 224, 256, 384 and 512.

The LANE hash function uses two permutations P and Q, which
consist of different number of AES[1]-like rounds. LANE-224/256 uses
6-round P and 3-round Q. LANE-384/512 uses 8-round P and 4-round
Q. We will use LANE-n-(a,b) to denote a variant of LANE with a-round
P, b-round Q and a digest length n.

We have found a semi-free start collision attack on reduced-round
LANE-256-(3,3) with complexity of 2^{62} compression function evaluations
and 2^{69} memory. This technique can be applied to LANE-512-(3,4) to
get a semi-free start collision attack with the same complexity of 2^{62} and
2^{69} memory. We also propose a collision attack on LANE-512-(3,4) with
complexity of 2^{94} and 2^{133} memory.

Keywords: hash function, collision attack, rebound attack, LANE,
SHA-3 candidates.

1 Introduction

The SHA-3 competition hosted by NIST aims to find a new cryptographic hash
standard as a replacement of SHA-2. 51 of the 64 submitted designs are accepted
to entered the first round. The LANE hash function is one of the first round
candidates.

The attacks on widely used hash standards such as MD5[2] and SHA-1[3]
are based on differential analysis. Many of the first round candidates of SHA-
3 competition use AES[1]-like SPN structures and claim to resist differential
attacks.

Florian Mendel et al. have proposed a new tool of "Rebound"[5] attack for
cryptanalysis of AES-based designs. The main idea of rebound attack is to take
advantage of weakness implied by S-box's optimal non-linearity. Random input
and output differences of an S-box match with surprisingly high probability of
1/2 and at least two values can be selected for each S-box. The complexity of
one round in the traditional truncated differential path can be totally eliminated

M.J. Jacobson Jr., V. Rijmen, and R. Safavi-Naini (Eds.): SAC 2009, LNCS 5867, pp. 126–140, 2009.
© Springer-Verlag Berlin Heidelberg 2009

at the cost of exhausting degrees of freedom of the active state values. In this paper, we analyze reduced LANE with rebound techniques. There are other parallel works of improved rebound techniques and applications in [7,8,9].

This paper is organized as follows. In section 2, we briefly describe the LANE hash function. In section 3, we discuss inner collisions of only two lanes in the first layer P. Then semi-free start collision attacks on reduced LANE-256 are described in section 4. Section 5 describes the attacks on reduced LANE-512. Section 6 is the conclusion.

2   Description of LANE Hash Function

The LANE hash function uses iterative MD structure with counters and output transformation. Digest values of LANE-224 and LANE-384 are truncated from LANE-256 and LANE-512 separately. Details of padding rules and output transformation are omitted here since they do not influence this attack.

In this section we briefly describe the compression function of LANE-256 and LANE-512. LANE-256/512 is an iterative hash function, whose compression function $f(H_{i-1}, M_i, C_i)$ processes a 512/1024-bit message block, a 512/1024-bit chaining value, a 64-bit counter, and outputs 256/512-bit digest length. The chaining state H_{i-1} and the message block M_i are expanded to six 256/512-bit blocks. Each block enters a different lane of P. Different lanes use different constants and counters. Output of the first three lanes and the last three lanes are XORed separately as input of the Q permutations. At last, output of both Q permutations are XORed as the next chaining value H_i. The structure of compression function in LANE is shown in Figure 1.

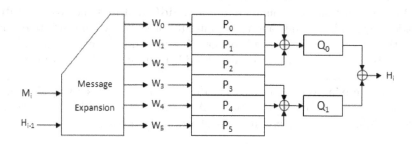

Fig. 1. The compression function of LANE

2.1   Message Expansion

The compression function of LANE-256 expands chaining value $H_{i-1} = h_0 \| h_1$ and message block $M_i = m_0 \| m_1 \| m_2 \| m_3$ to six 256-bit blocks $W_0, ... W_5$ as shown in equation 1.

$$
\begin{aligned}
W_0 &= h_0 \oplus m_0 \oplus m_1 \oplus m_2 \oplus m_3 \parallel h_1 \oplus m_0 \oplus m_2 \\
W_1 &= h_0 \oplus h_1 \oplus m_0 \oplus m_2 \oplus m_3 \parallel h_0 \oplus m_1 \oplus m_2 \\
W_2 &= h_0 \oplus h_1 \oplus m_0 \oplus m_1 \oplus m_2 \parallel h_0 \oplus m_0 \oplus m_3 \\
W_3 &= h_0 \parallel h_1 \\
W_4 &= m_0 \parallel m_1 \\
W_5 &= m_2 \parallel m_3
\end{aligned}
\tag{1}
$$

The message expansion in LANE-512 is analogous. The only difference is that all blocks are double-sized.

2.2 Permutations P and Q

P and Q in Figure 1 are permutations with AES-like state update rounds. LANE-256 uses 6-round P, 3-round Q and LANE-512 uses 8-round P and 4-round Q. Each round contains five steps. One round of state update operation in LANE-256 is shown in Figure 2. The difference in LANE-512 is that all operations are on four 4×4 matrices.

In this paper, we use S_i to denote the state value after the i-th round. Between S_i and S_{i+1}, the state values are denoted as S'_i, S''_i, S'''_i and S''''_i consecutively. $\triangle S_i$ is used to denote the XOR difference of state S_i.

The five steps of one round are:

- **SB**: the non-linear operation **SubBytes** applies an S-Box to each byte of the state. The S-box is the same as the one used in AES[1].
- **SR**: the cyclical permutation **ShiftRows** rotates the bytes of the i-th row leftwards by i positions.
- **MC**: the diffusion layer **MixColumns** multiplies each column by a MDS matrix which is the same as the one in AES.
- **AC**: the constants and counter additions **AddConstants** and **AddCounter** add the round constants and the counter to the states. We use **AC** to denote both of them. The last rounds of both P and Q don't have AC operations. Details of the AC operations are omitted here since they have nothing to do with our attacks.
- **SC**: the mixing operation between different 4×4 states **SwapColumns** reorders the columns in the state. LANE-256 and LANE-512 use different SwapColumns operations which are shown in Figure 3.

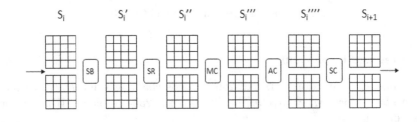

Fig. 2. One round of state update operation in LANE-256

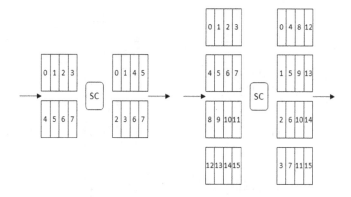

Fig. 3. SwapColumns operations used in LANE-256 and LANE-512

3   Construct Inner Collisions Using Rebound Techniques

The LANE hash function has six lanes in the first layer P. In this section, we are trying to construct collisions between only two lanes, namely the inner collisions. It's easy to see that two simultaneous inner collisions could directly lead to a full collision.

3.1   Optimal Differential Pattern for LANE

The message expansion used in LANE is based on a linear $(6,3,4)$-code over GF_4, which means for any possible differential path, there are at least four active lanes in the first layer P. Once the difference enters layer Q, there would be more active S-Boxes. So we want to eliminate all differences before they enter layer Q.

This is the best differential pattern for LANE with four active lanes P_1, P_2, P_4 and P_5. Two inner collisions in layer P ensure no difference enter layer Q as shown in Figure 4.

Let $\triangle m_0 = \triangle m_2 \neq 0$ and $\triangle h_0 = \triangle h_1 = \triangle m_1 = \triangle m_3 = 0$, we have four active lanes P_1, P_2, P_4 and P_5 and the differences in W are in the form of $(\triangle, 0)$ and $(0, \triangle)$. Differential paths with initial difference of $(\triangle, 0)$ and $(0, \triangle)$ behave

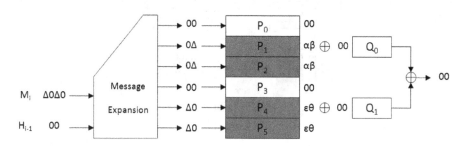

Fig. 4. Optimal differential pattern for LANE

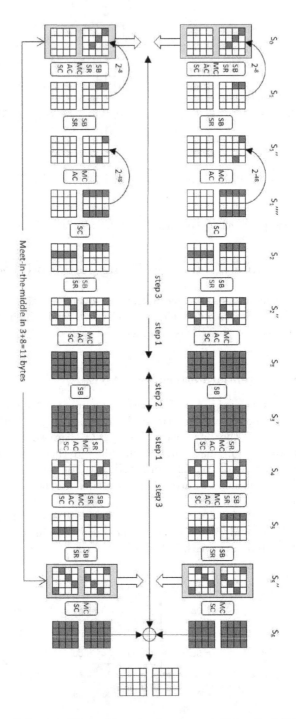

Fig. 5. Rebound differential path of an inner collision for LANE-256

in a similar way. So we only need to consider one type of differential path in the final attack. We will talk about this in section 4.2.

3.2 Rebound Differential Path of Inner Collision

In this section, we only consider an inner collision of two lanes. Using rebound techniques proposed by Florian Mendel et al. in [5], we can easily attack round 6 of layer P in LANE-256 with a complexity of 2^{100}. The differential path is shown in Figure 5.

In traditional truncated differential path, difference propagates from initial state to hash value in forward direction. In a rebound attack, we search for an inbound differential path in internal states first. Then the outbound part can be considered as two truncated differential paths in different directions - forward and backward. Since complexity of the inbound phase can be eliminated, we only need to consider probability of the outbound phase.

Here, we briefly describe the attack of inner collision. This is similar to the attack on Grøstl[5]. For more details of rebound attack, please refer to the original paper.

Step 1: We start from choosing random differences in both S_2'' and S_4. Then compute $\triangle S_3$ from $\triangle S_2''$ and $\triangle S_3'$ from $\triangle S_4$. These difference propagations $\triangle S_2'' \rightarrow \triangle S_3$ and $\triangle S_3' \leftarrow \triangle S_4$ hold with probability of 1 because all operations between them are linear transformations SR, MC, AC and SC.

Step 2: We expect to find a match of possible differential character at the S-box in the third round with probability of 2^{-32}, because random difference in input and output of an S-Box matches with probability of $1/2$ and there are 32 active S-boxes. Once we have found a match, we get 2^{32} staring points (attempts) for the outbound phase, since we have at least two values for each S-Box match. So we can generate 2^{32} attempts with complexity of 2^{32}. For any $x \geq 32$ we can generate 2^x attempts with complexity of 2^x.

Step 3: Each starting point (attempt) can lead to our demanded differential pattern in S_0 with a probability of $2^{-48} \times 2^{-8} = 2^{-56}$. In other words, we can generate a successful attempt in one lane with complexity of 2^{56}.

Step 4: In order to find a match in the three bytes of difference in S_0 and eight bytes in S_2'' between two lanes, we need $2^{8 \times (3+8)/2} = 2^{44}$ successful attempts in both lanes. So the complexity is $2^{56} \times 2^{44} = 2^{100}$ for a inner collision with the same initial difference in both lanes. The memory requirements of step 4 is $2 \times 2^{44} = 2^{45}$.

4 Semi-free Start Collision Attack on LANE-256-(3,3)

Even if we have successfully found two inner collisions of four lanes in layer P, we can not get a collision of full LANE. The problem is the message expansion since rebound attack require a full control of the state values. Four initial state values of the two inner collisions will probably lead to a contradiction since we have a degree of freedom for only three states, namely (h_0, h_1), (m_0, m_1) and (m_2, m_3).

More precisely, from two inner collisions we get the exact values of W_1, W_2, W_4 and W_5. Recall equation (1), and we can see that W_4 and W_5 can determine values of m_0, m_1, m_2 and m_3. By selecting the values of h_0 and h_1, we can change the value of $(h_0 \oplus h_1 \oplus m_0 \oplus m_2 \oplus m_3, h_0 \oplus m_1 \oplus m_2)$ to W_1 which we have got from the first inner collision. Since all degrees of freedom are used, we have to leave W_2 satisfied by chance.

There are 256 bits left in W_2 along with the 24-bit initial difference. We need $2^{(256+24)/2} = 2^{140}$ inner collisions in both P_1, P_2 and P_4, P_5 to find a match in $256 + 24 = 280$ bits. So in both lanes of one inner collision, we need $2^{140/2} = 2^{70}$ times more attempts. The complexity of semi-free start collision attack on LANE-256 is $2^{100} \times 2^{70} = 2^{170} > 2^{128}$ which exceeds the birthday bound of 256-bit hash functions and this attack fails.

4.1   Rebound Differential Path with Partially Fixed State Values

We are inspired by Dmitry Khovratovich et al. of their meet-in-the-middle attacks on several SHA-3 candidates[6]. The idea is to fix values of certain bits to get an actually smaller size in the meet-in-the-middle part of the target state and lower the complexity.

If we fix some bytes in an AES state, they would be affected by other bytes in at most two rounds. We have got an observation that diffusion in LANE is not as efficient as in AES. Fixed values in certain positions of the initial state can proceed to the third round in both LANE-256 and LANE-512.

Combining this small observation and rebound techniques, we have found a solution for LANE-256-(3,3) as shown in Figure 6.

In this figure, one byte with a mark of "X" means its value can be pre-computed and fixed during the attack. In our attack, we let all the X bytes in S_0 to be zeros and calculate values of the following ones. When we choose differences in S_1' and S_2'', we also set the values of fixed bytes in S_1' and S_2'' to what we have pre-computed.

In the four active lanes of this attack, the round constants and counters are different. So the exact values of fixed bytes in S_1' and S_2'' are different in four lanes. But they would all lead to zero values in the certain positions of initial states.

We also let the values of marked bytes in h_0 and h_1 to be zeros. So when we have got W_1, W_2, W_4 and W_5 from two inner collisions, we calculate the values of the non-zero bytes of h_0, h_1, m_0, m_1, m_2 and m_3 from W_1, W_4 and W_5. Then there are only 128 bits of state values left unsatisfied in W_2 instead of 256 bits, since all zero bytes are already satisfied in advance.

4.2   Details of the Attack

In this attack, We use two inner collision differential paths with initial differences of the patterns $(\triangle, 0)$ and $(0, \triangle)$ separately. If we change the position of two 4×4 matrices in the initial state of one path, the differential path don't change

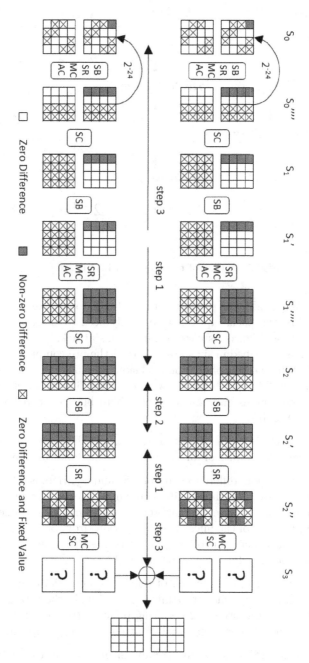

Fig. 6. Rebound differential path for LANE-256-(3,3)

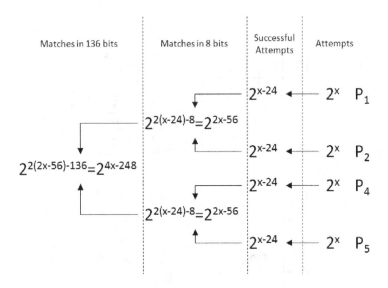

Fig. 7. Outline of the attack on LANE-256-(3,3)

substantially. Especially, the positions of fixed bytes don't change. So we consider these two differential paths equivalent and only need to analyze one of them in the following steps. Figure 7 shows outline of this attack.

This attack is described in six steps. Step 1 is the pre-computation. Steps 2 to 4 are the details in one lane of an inner collision. Step 5 is the meet-in-the-middle step of the initial difference between two lanes. Step 6 is the meet-in-the-middle step of the state values and the difference byte between two inner collisions except for the X bytes.

Step 1: Set all fixed bytes in the initial states in four lanes to zeros and compute the consecutive exact values of all fixed bytes in the following states.

Step 2: Choose random differences in both S_1' and S_2''. Here is a little difference from the attack above. We choose differences in S_2'' to be the same in both lanes of one inner collision. These two differences will remain the same when they proceed to S_3 because of the linear transformations from S_2'' to S_3. Even though we don't know the exact value of $\triangle S_3$ in both lanes, they must be the same and will offset each other before they enter layer Q.

Step 3: We expect to find a match of possible differential character at the S-box in the second round with probability of 2^{-32} as in the attack above. Once we have found a match, we get 2^{32} staring points (attempts) for the outbound phase. Now assume that we generated 2^x attempts with complexity of 2^x.

Step 4: We can find a successful attempt in one lane in every 2^{24} attempts. With 2^x attempts, we expect to find 2^{x-24} successful ones.

Step 5: Now we have 2^{x-24} successful attempts in both lanes, so we can find $2^{2(x-24)-8} = 2^{2x-56}$ matches in the only one byte difference in S_0. So we have got 2^{2x-56} inner collisions in P_1, P_2 and the same number of inner collisions in P_4, P_5. This step requires $4 \times 2^{x-24}$ memory.

Step 6: After we select the values of h_0 and h_1, there are 128 bits in W_2 and 8 bits in the initial difference unsatisfied. So we expect $2^{2(2x-56)-136} = 2^{4x-248}$ matches in these 136 bits. This step requires $2 \times 2^{2x-56}$ memory.

If $x = 62$, we expect to find a final match. Memory requirements of step 5 and step 6 are 2^{40} and 2^{69}. So the semi-free start collision attack on LANE-256-(3,3) has an overall complexity of 2^{62} and requires about 2^{69} memory.

5   Applications to LANE-512

We can also use rebound techniques to find inner collisions for LANE-512. By fixing certain bytes in the state values, we can find semi-free collision and collision attacks on LANE-512-(3,4).

5.1   Inner Collision of LANE-512

For LANE-512, we can proceed to round 8 of P in an inner collision attack of two lanes. The differential path will be shown in Figure 8 as an appendix. Details of this attack is similar to inner collision attack on LANE-256 in section 3.2.

For any given attempt, it is successful with probability of $2^{-24} \times 2^{-96} = 2^{-120}$, which means we can generate one successful attempt with complexity of 2^{120}. Then we have to match in $8 \times (16 + 16) = 256$ bits, and we need $2^{256/2} = 2^{128}$ successful attempts in both lanes. The complexity of inner collision on LANE-512 is $2^{120} \times 2^{128} = 2^{248}$ and the memory requirement of meet-in-the-middle step is $2 \times 2^{128} = 2^{129}$.

5.2   Semi-free Start Collision Attack and Collision Attack on LANE-512-(3,4)

Using the fixed bytes techniques, we can find a semi-free start collision of reduced LANE-512-(3,4) with a differential path shown in Figure 9 as an appendix. This attack is almost the same as the attack in section 4.2 with the same complexity of 2^{62} and 2^{69} memory.

As you can see in Figure 9, we can fix more bytes in the 3-round path for LANE-512. If we don't use the degrees of freedom in the initial chaining values h_0 and h_1, we have 16 more bytes in the final meet-in-the-middle part. The difference is that fixed bytes in W_1 and W_2 are not set to zeros in the marked positions. Recall equation 1, since now values of h_0 and h_1 are fixed, if we set fixed bytes of W_4 and W_5 to zeros, values of W_1 and W_2 in the marked positions are determined by the value of standard $IV = (h_0, h_1)$.

Assume that we have generated 2^x attempts in each lanes, we expect 2^{2x-56} inner collisions in both P_1, P_2 and P_4, P_5. The difference is now we have to match

$256 + 8 = 264$ bits. So we expect $2^{2\times(2x-56)-264} = 2^{4x-376}$ final matches with memory requirement of $2 \times 2^{2x-56}$. If $x = 94$, we expect to find one final match.

So, we have found a collision attack on LANE-512-(3,4). The complexity of collision attack on LANE-512-(3,4) is 2^{94} and memory requirement is 2^{133}.

5.3   Semi-free Start Collision Attack on LANE-512-(4,4)

If we want to attack more than three round in P, we can no longer use fixed values, since fixed values can only proceed to the third round. Without fixed values, we can attack LANE-512-(4,4) with a differential path shown in Figure 10 which is part of the one shown in Figure 8.

Assume that we have generated 2^x attempts in each lanes, and only 2^{x-120} of them will be successful ones. Then we expect $2^{2(x-120)-8} = 2^{2x-248}$ inner collisions in both P_1, P_2 and P_4, P_5. Here, we have to match $512 + 8 = 520$ bits. So we expect $2^{2\times(2x-248)-520} = 2^{4x-1016}$ final matches with memory requirement of $2 \times 2^{2x-248}$. If $x = 254$, we expect to find one final match.

So, the complexity of semi-free start collision attack on LANE-512-(4,4) is 2^{254} and memory requirement is 2^{261}. Though computational complexity is less than birthday bound, memory requirement of this attack is more than 2^{256}. This attack can be considered unsuccessful.

6   Conclusion

In this paper, we analyzed the LANE hash function using rebound and meet-in-the-middle techniques. We give several attacks on reduced variants of LANE-256 and LANE-512. Table 1 shows all the results of these attacks. Notation " † " in this table means the attack can be considered unsuccessful.

The memory requirements of all these attacks come from the meet-in-the-middle steps. But the memoryless variants seem not easy to be implemented in our attacks.

We can hardly attack more than three rounds of P with method of fixing certain bytes, since the fixed values can only proceed to the third round. Our attacks on reduced variants do not hurt collision resistance of full LANE.

Acknowledgments. The authors would like to thank the anonymous referees for their valuable comments. Furthermore, this work is supported by the National

Table 1. Results of collision attacks in this paper

hash function	P rounds	Q rounds	collision type	complexity	memory
LANE-256	6	-	inner collision of P	2^{100}	2^{45}
	3	3	semi-free start collision	2^{62}	2^{69}
LANE-512	8	-	inner collision of P	2^{248}	2^{129}
	3	4	semi-free start collision	2^{62}	2^{69}
			collision	2^{94}	2^{133}
	4	4	semi-free start collision	2^{254}	2^{261}†

High-Tech Research and Development 863 Plan of China (No. 2007AA01Z470), the National Natural Science Foundation of China (No. 60873259), and the National Grand Fundamental Research 973 Program of China (No. 2004CB318004).

References

1. National Institute of Standards and Technology: FIPS PUB 197, Advanced Encryption Standard (AES). Federal Information Processing Standards Publication 197, U.S. Department of Commerce (November 2001)
2. Wang, X., Yu, H.: How to break MD5 and other hash functions. In: Cramer, R. (ed.) EUROCRYPT 2005. LNCS, vol. 3494, pp. 19–35. Springer, Heidelberg (2005)
3. De Cannière, C., Rechberger, C.: Finding SHA-1 characteristics: General results and applications. In: Lai, X., Chen, K. (eds.) ASIACRYPT 2006. LNCS, vol. 4284, pp. 1–20. Springer, Heidelberg (2006)
4. Indesteege, S., Preneel, B.: The LANE hash function, http://www.cosic.esat.kuleuven.be/lane/
5. Mendel, F., Rechberger, C., Schläffer, M., Thomsen, S.S.: The Rebound Attack: Cryptanalysis of Reduced Whirlpool and Grøstl. In: Dunkelman, O. (ed.) FSE 2009. LNCS, vol. 5665, pp. 260–276. Springer, Heidelberg (2009)
6. Khovratovich, D., Nikolić, Weinmann, R.: Meet-in-the-Middle Attacks on SHA-3 Candidates. In: Dunkelman, O. (ed.) FSE 2009. LNCS, vol. 5665, pp. 228–245. Springer, Heidelberg (2009)
7. Mendel, F., Peyrin, T., Rechberger, C., Schläffer, M.: Improved Cryptanalysis of the Reduced Grøstl Compression Function, ECHO Permutation and AES Block Cipher. In: Jacobson, M.J., Rijmen, V., Safavi-Naini, R. (eds.) SAC 2009. LNCS, vol. 5867, pp. 16–35. Springer, Heidelberg (2009)
8. Lamberger, M., Mendel, F., Rechberge, C., Rijmen, V., Schläffer, M.: Rebound Distinguishers: Results on the Full Whirlpool Compression Function. In: Matsui, M. (ed.) ASIACRYPT 2009. LNCS, vol. 5912. Springer, Heidelberg (to appear, 2009)
9. Matusiewicz, K., Naya-Plasencia, M., Nikolić, I., Sasaki, Y., Schläffer, M.: Rebound Attack on the Full LANE Compression Function. In: Matsui, M. (ed.) ASIACRYPT 2009. LNCS, vol. 5912. Springer, Heidelberg (to appear, 2009)

Appendix

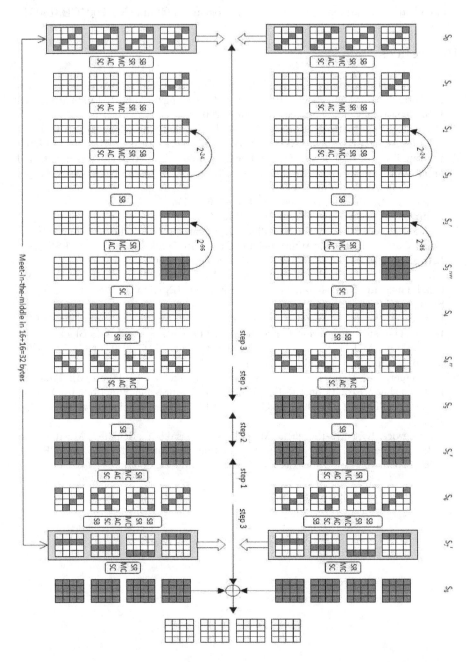

Fig. 8. Rebound differential path of inner collision for LANE-512

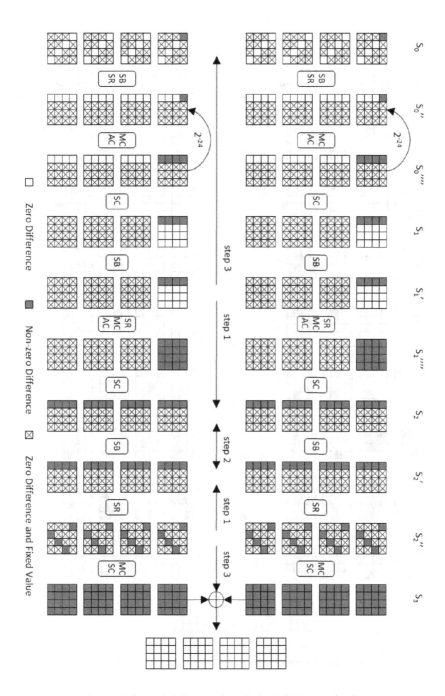

Fig. 9. Rebound differential path for LANE-512-(3,4)

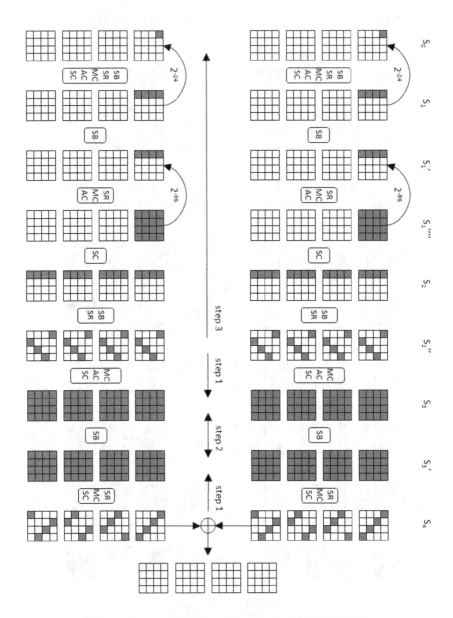

Fig. 10. Rebound differential path for LANE-512-(4,4)

Practical Pseudo-collisions for Hash Functions ARIRANG-224/384

Jian Guo[1,*], Krystian Matusiewicz[2], Lars R. Knudsen[2], San Ling[1], and Huaxiong Wang[1]

[1] Division of Mathematical Sciences,
School of Physical and Mathematical Sciences,
Nanyang Technological University, Singapore
{guojian,lingsan,hxwang}@ntu.edu.sg
[2] Department of Mathematics,
Technical University of Denmark, Denmark
{K.Matusiewicz,Lars.R.Knudsen}@mat.dtu.dk

Abstract. In this paper we analyse the security of the SHA-3 candidate ARIRANG. We show that bitwise complementation of whole registers turns out to be very useful for constructing high-probability differential characteristics in the function. We use this approach to find near-collisions with Hamming weight 32 for the full compression function as well as collisions for the compression function of ARIRANG reduced to 26 rounds, both with complexity close to 2^0 and memory requirements of only a few words. We use near collisions for the compression function to construct pseudo-collisions for the complete hash functions ARIRANG-224 and ARIRANG-384 with complexity 2^{23} and close to 2^0, respectively. We implemented the attacks and provide examples of appropriate pairs of H, M values. We also provide possible configurations which may give collisions for step-reduced and full ARIRANG.

Keywords: practical, pseudo-collision, ARIRANG, hash function.

1 Introduction

ARIRANG [1] is one of the first-round candidates in the SHA-3 competition organized by NIST. It is an iterated hash function that uses a variant of the Merkle-Damgård mode augmented by a block counter. The compression function is a dedicated design that iterates a step transformation that can be seen as a target-heavy unbalanced Feistel network [9]. Its construction seems to be influenced by an earlier design called FORK-256 [4] with the important difference of using a bijective function based on a layer of S-boxes and an MDS mapping as the source of non-linearity. This prevents attacks similar to the ones developed for FORK-256 [7,6,2] from working on ARIRANG. A single sequence of 40 steps rather than four parallel branches makes it immune to meet-in-the-middle attacks [8].

* The paper was partly done during the author's visit to Technical University of Denmark and was partly supported by a DCAMM grant there.

M.J. Jacobson Jr., V. Rijmen, and R. Safavi-Naini (Eds.): SAC 2009, LNCS 5867, pp. 141–156, 2009.
© Springer-Verlag Berlin Heidelberg 2009

Related Work. To the best of our knowledge, the only published previous work on ARIRANG is a step-reduced preimage attack by Hong *et al* [3]. Based on the meet-in-the-middle preimage attack framework developed by Sasaki *et al*, Hong *et al* were able to find [3-33] step-reduced pseudo-preimages with complexity 2^{241} and 2^{481} for ARIRANG-256 and ARIRANG-512, respectively.

Our Contributions. In this paper we report results of our security assessment of ARIRANG. The initial observation that motivated our analysis was the fact that differences created by complementing (flipping) all bits in a register propagate quite nicely through the function due to a particular interaction of the layer of S-boxes and an MDS mapping. We were able to exploit this fact to derive a range of attacks on the compression function and extend some of them to attacks on the complete hash function.

After a short description of ARIRANG given in section 2 we explain in details our ideas of managing all-ones differences in section 3 and show how to find conforming messages in section 4. After that, we describe two attacks on ARIRANG. In section 5 we show how to find collisions for 26 out of 40 steps of the compression function with complexity close to the cost of computing a single hash value of ARIRANG. Next, we show in Section 6 that by injecting all-ones difference in one of the chaining values we can easily (with complexity close to one evaluation) obtain 32-bit (resp. 64-bit) near collisions for the full compression function of ARIRANG-256 (resp. ARIRANG-512). We use the freedom of selecting in which chaining register we want to have differences to convert those near-collisions for the compression function to pseudo-collisions for the full hash functions ARIRANG-224 and ARIRANG-384 which we can obtain with complexity 2^{23} and close to 2^0 respectively. Finally, we discuss some open problems and conclude in Section 8. Our results are summarized in Table 1.

Table 1. Summary of the results of this paper

Compression function		
Result	Complexity	Example
32-bit near-collision for full ARIRANG-256 compress	1	Y
64-bit near-collision for full ARIRANG-512 compress	1	Y
26-step collision for ARIRANG-256/512	1	Y
Hash function		
Result	Complexity	Example
pseudo-collision for full ARIRANG-224/384 hash	2^{23} / 1	Y

2   Brief Description of ARIRANG

We start with providing a minimal description of ARIRANG necessary to understand our attacks. More details can be found in the original submission document.

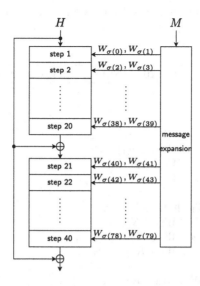

Fig. 1. Compression function of ARIRANG

Compression Function. The fundamental building block of the hash function ARIRANG-256 (ARIRANG-512) is the compression function that takes 256-bit (512-bit) chaining value and 512-bit (1024-bit) message block and outputs a new 256-bit (512-bit) chaining value. The function, depicted in Fig. 1, consists of two main parts: the message expansion process and the iteration of the step transformation.

The message expansion function takes as input 16 words of the message $M_0, \ldots, M_{15}$ and produces 80 expanded message words in two stages. First, 32 words W_i are generated according to the procedure described in Alg. 1, where K_i are word constants and r_i are fixed rotation amounts. Our attacks do not depend on their actual values. Next, these 32 words are used 80 times, two in

Table 2. Ordering σ of expanded message words W_i used in step transformations

i	$\sigma(i)$	i	$\sigma(i)$	i	$\sigma(i)$	i	$\sigma(i)$
0, 1	16, 17	20, 21	20,21	40, 41	24, 25	60, 61	28, 29
2, 3	0, 1	22, 23	3, 6	42, 43	12, 5	62, 63	7, 2
4, 5	2, 3	24, 25	9,12	44, 45	14, 7	64, 65	13, 8
6, 7	4, 5	26, 27	15, 2	46, 47	0, 9	66, 67	3, 14
8, 9	6, 7	28, 29	5, 8	48, 49	2, 11	68, 69	9, 4
10, 11	18, 19	30, 31	22,23	50, 51	26, 27	70, 71	30, 31
12, 13	8, 9	32, 33	11,14	52, 53	4, 13	72, 73	15, 10
14, 15	10, 11	34, 35	1, 4	54, 55	6, 15	74, 75	5, 0
16, 17	12, 13	36, 37	7,10	56, 57	8, 1	76, 77	11, 6
18, 19	14, 15	38, 39	13, 0	58, 59	10, 3	78, 79	1, 12

each step transformation, in the order defined by the function σ described in Table 2.

Algorithm 1. Generation of expanded message words in ARIRANG.

for $i = 0, \ldots, 15$ **do**
    $W_i \leftarrow M_i$
end for
$W_{16} \leftarrow (W_9 \oplus W_{11} \oplus W_{13} \oplus W_{15} \oplus K_0) \lll r_0$
$W_{17} \leftarrow (W_8 \oplus W_{10} \oplus W_{12} \oplus W_{14} \oplus K_1) \lll r_1$
$W_{18} \leftarrow (W_1 \oplus W_3 \oplus W_5 \oplus W_7 \oplus K_2) \lll r_2$
$W_{19} \leftarrow (W_0 \oplus W_2 \oplus W_4 \oplus W_6 \oplus K_3) \lll r_3$

$W_{20} \leftarrow (W_{14} \oplus W_4 \oplus W_{10} \oplus W_0 \oplus K_4) \lll r_0$
$W_{21} \leftarrow (W_{11} \oplus W_1 \oplus W_7 \oplus W_{13} \oplus K_5) \lll r_1$
$W_{22} \leftarrow (W_6 \oplus W_{12} \oplus W_2 \oplus W_8 \oplus K_6) \lll r_2$
$W_{23} \leftarrow (W_3 \oplus W_9 \oplus W_{15} \oplus W_5 \oplus K_7) \lll r_3$

$W_{24} \leftarrow (W_{13} \oplus W_{15} \oplus W_1 \oplus W_3 \oplus K_8) \lll r_0$
$W_{25} \leftarrow (W_4 \oplus W_6 \oplus W_8 \oplus W_{10} \oplus K_9) \lll r_1$
$W_{26} \leftarrow (W_5 \oplus W_7 \oplus W_9 \oplus W_{11} \oplus K_{10}) \lll r_2$
$W_{27} \leftarrow (W_{12} \oplus W_{14} \oplus W_0 \oplus W_2 \oplus K_{11}) \lll r_3$

$W_{28} \leftarrow (W_{10} \oplus W_0 \oplus W_6 \oplus W_{12} \oplus K_{12}) \lll r_0$
$W_{29} \leftarrow (W_{15} \oplus W_5 \oplus W_{11} \oplus W_1 \oplus K_{13}) \lll r_1$
$W_{30} \leftarrow (W_2 \oplus W_8 \oplus W_{14} \oplus W_4 \oplus K_{14}) \lll r_2$
$W_{31} \leftarrow (W_7 \oplus W_{13} \oplus W_3 \oplus W_9 \oplus K_{15}) \lll r_3$

The iterative part uses the step transformation to update the state of 8 chaining registers, a, b, $\ldots$, h. First, the input chaining values $H[0], \ldots, H[7]$ are loaded into chaining registers $a, \ldots, h$. Then, the step transformation is applied 20 times. After 20 steps, the initial chaining values are XOR-ed to the current chaining values and the computation is carried on for another 20 steps. At the end, the usual feed-forward is applied by XOR-ing initial chaining values to the output of the iteration.

The step transformation updates chaining registers using two expanded message words $W_{\sigma(2t)}$, $W_{\sigma(2t+1)}$ as follows

$$
\begin{aligned}
T_1 &\leftarrow \mathbf{G}^{(256)}(a_t \oplus W_{\sigma(2t)}), & T_2 &\leftarrow \mathbf{G}^{(256)}(e_t \oplus W_{\sigma(2t+1)}), \\
b_{t+1} &\leftarrow a_t \oplus W_{\sigma(2t)}, & f_{t+1} &\leftarrow e_t \oplus W_{\sigma(2t+1)}, \\
c_{t+1} &\leftarrow b_t \oplus T_1, & g_{t+1} &\leftarrow f_t \oplus T_2, \\
d_{t+1} &\leftarrow c_t \oplus (T_1 \lll 13), & h_{t+1} &\leftarrow g_t \oplus (T_2 \lll 29), \\
e_{t+1} &\leftarrow d_t \oplus (T_1 \lll 23), & a_{t+1} &\leftarrow h_t \oplus (T_2 \lll 7).
\end{aligned}
$$

This transformation is illustrated in Fig. 2. In ARIRANG-256, it uses a function $\mathbf{G}^{(256)}$ which splits 32-bit input value into 4 bytes, transforms them using AES

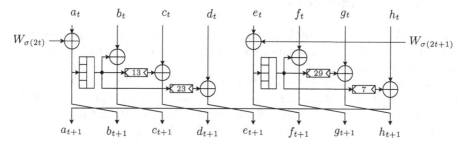

Fig. 2. Step transformation of ARIRANG updates the state of eight chaining registers

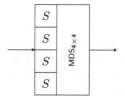

Fig. 3. Function $\mathbf{G}^{(256)}$ of ARIRANG-256 uses four AES S-Boxes followed by AES MDS mapping

S-Box and feeds the result to the AES MDS transformation, as presented in Fig. 3. ARIRANG uses the same finite field as AES, defined by the polynomial $x^8 + x^4 + x^3 + x + 1$. MDS mapping for 256 bit variant is defined as

$$
MDS_{4\times4} = \begin{bmatrix} z & z+1 & 1 & 1 \\ 1 & z & z+1 & 1 \\ 1 & 1 & z & z+1 \\ z+1 & 1 & 1 & z \end{bmatrix}.
$$

In ARIRANG-512, an analogous function $\mathbf{G}^{(512)}$ is defined using a layer of 8 S-boxes and an appropriate 8×8 MDS matrix.

Hash Function. The hash function ARIRANG is an iterative construction closely following the original Merkle-Damgård mode. The message is first padded by a single '1' bit followed by an appropriate number of zero bits and a 64-bit field containing the length of the original message. After padding and appending block length field, the message is divided into 512-bit blocks and the compression function is applied to process each of the blocks one by one. The construction has one additional variable compared to the plain Merkle-Damgård mode. A new variable that stores the current message block index is introduced and its value is XOR-ed into chainings before each application of the compression function. However, this does not affect our attacks.

3   All-One Differences

From the description of ARIRANG-256, it is clear that it uses only three essential building blocks: XORs, bit rotations and the function $\mathbf{G}^{(256)}$, which is the only part non-linear over $\mathbb{F}_2$.

Let us focus on the function $\mathbf{G}^{(256)}$ first. First, note that for the AES S-Box input difference of 0xff maps to output difference 0xff with probability 2^{-7}, the two values x for which $S(x) \oplus S(x \oplus \text{0xff}) = \text{0xff}$ are 0x7e, 0x81.

The second observation is that for the 256-bit MDS mapping all the vectors of the form (a, a, a, a) are fixed points since $a \cdot z + a(z + 1) + a + a = a$.

This means all-one difference will map to all-one difference through $MDS_{4 \times 4}$. In turns, there are 16 32-bit values x such that

$$\mathbf{G}^{(256)}(x) \oplus \mathbf{G}^{(256)}(x \oplus \text{0xffffffff}) = \text{0xffffffff}$$

and the probability of such a differential is 2^{-28}.

This means we can consider a differential that uses only all-one differences in active registers. The big advantage of such differences is that they are rotation invariant, so we can easily model differentials like that by replacing all the rotations and function $\mathbf{G}^{(256)}$ with identity.

MDS mapping for ARIRANG-512 is different and all-ones is not its fixed-point, but after combining S-box layer with MDS, we get the differential of the same type with probability 2^{-56}, so the same principle applies to the larger variant as well.

To minimize the complexity of the attack, we need to use as few active $\mathbf{G}^{(256)}$-functions as possible in the part of the function where we cannot control input values to them. Since there are only 2^{16} possible combinations of all-one differences in message words and 2^{24} combinations including chaining registers $H[0], \ldots, H[7]$, it is easy to enumerate them all using a computer search.

We note that all-one differences trick is also used in [5].

4   Message Adjustments

The method used to find messages that make the differences in the actual function to follow the differential can be called a message adjustment strategy.

We have full control over the message words $W_0, \ldots, W_{15}$. Through combinations of the message words, we can still control some of the messages W_i for $16 \leq i \leq 31$. We can modify the messages used in the first 4 steps freely, yet leaving the output chaining values of 4-th step unchanged by modifying the corresponding input chaining values $H[0], \ldots, H[7]$.

For example, changing W_2 and $H[6]$ by the same amount ($\oplus$ both with a same value) will keep the output of step 3 stable. Beyond step 4, if we change the value of W_6 in step 5, we still make the output of step 5 stable by changing the $H[4]$ by a same amount. However this change will be propagated by the right G function in step 1, we can fix this by changing the $H[5], H[6]$ and $H[7]$ by proper values, respectively. This method applies to W_7 in step 5 similarly. In step 6, if W_{19} is changed, we can still keep the output after step 6 stable. We achieve this by $\oplus$ with $H[7]$ by the same amount of the change. Note that this difference will be propagated through the left G function in step 2 (Note we can only do this when the left G in step 2 is not active). We can fix this by $\oplus$ with $H[0], H[1], H[2]$ by proper values, respectively. Then the change in $H[0]$ will be propagated through the G function in step 1. We then fix this by $\oplus$ with $H[0], H[1], H[2]$ by proper values. Similar method applies to W_{18} in step 6.

5   Collisions for Reduced Round Compression Function

A search for collision configuration that minimizes the overall number of active $\mathbf{G}^{(256)}$ functions shows that the best strategy is to flip all message words. Then throughout the whole compression function only 16 out of 80 $\mathbf{G}^{(256)}$ are active. When we restrict the attention to steps 20-40 (the part which almost certainly is beyond any message-modification techniques) we can find a configuration with only 5 active $\mathbf{G}^{(256)}$ and in fact only 3 in steps 22-40. Details of minimal paths are summarized in Table 3. The second characteristic with probability 2^{-140} in steps 21-40 shows that the claim made in [1, section 6.2, page 37] that *"there is no collision producing characteristics which has a probability higher than 2^{-256} in the last two rounds"* is based on assumptions that do not hold in practice.

Table 3. Results of search for collision characteristics in ARIRANG-256

type	minimize	min. value	diffs in message words
collisions	total active G	16	0,...,15 (all)
collisions	active G rounds 20-40	5	2,3,7,8,9,13

Even though using all-one differences does not seem to allow for finding good collision differentials for the full compression function, one can use them to mount an attack on its reduced-round variants. In the rest of this section we illustrate it with a method that instantly finds collisions for 26 steps of ARIRANG-256.

5.1   Finding Step Reduced Collision Differential

To find the optimal path for reduced-round attack, we searched the all-one differentials using the following criteria.

1. We count the number of active G from step 11, as we have a complete control over the first 10 steps,
2. there are only differences in message words, not in chaining values,
3. the differential should give round reduced collision,
4. the differential should have minimum number of active G,
5. preferably, the active G-s should appear as early as possible.

Table 4. 26-step reduced collision characteristics in ARIRANG

Step	W (left)	Active G (left)	W (right)	Active G (right)
1	W_{16}		W_{17}	
2	W_0		W_1	
3	W_2		W_3	
4	W_4	✓	W_5	
5	W_6	✓	W_7	✓
6	W_{18}	✓	W_{19}	
7	W_8		W_9	✓
8	W_{10}	✓	W_{11}	✓
9	W_{12}	✓	W_{13}	✓
10	W_{14}	✓	W_{15}	✓
11	W_{20}		W_{21}	✓
12	W_3		W_6	
13	W_9	✓	W_{12}	
14	W_{15}		W_2	
15	W_5	✓	W_8	

The search result[1] shows a differential with differences in message words M_4, M_6, M_8, M_{10} and the corresponding active G is shown in Table 4, steps after 16 are not shown because there is no active G between step 16 and step 26 and we do not consider steps after step 26.

5.2   Finding Step Reduced Collisions

To find the example of the 26-step reduced collision, we need to deal with all those active G so that the input to the active G are one of those all-one difference pairs. As our algorithm runs in a deterministic way, we actually force the input to a chosen pair $(\gamma, \bar{\gamma}) = $ (81818181, 7E7E7E7E). In the first 10 steps, whenever there is an active G, we can fix the input by modifying the immediate message word. After step 10, we follow the algorithm below:

1. For active G in step 11, we change W_{21} to the proper value by modifying W_1 and W_3 by the same amount so that W_{18} does not change, we compensate the change of W_1 and W_3 using the method in section 4.
2. For active G in step 13, we modify the message word W_6, which is used one step before. We modify W_2 also by a same amount so that W_{19} is constant, and then compensate the changes.
3. For active G in step 15, we modify W_5 directly. We compensate the change of W_5 and W_{18}.

As we can see the algorithm is deterministic, so the complexity is 1 with no memory requirements. An example of the chaining values and a pair of messages obtained using this procedure is shown in Table 5.

---

[1] Active G may not be paired with active messages, as the differences in message may be canceled by differences from preceding steps.

Table 5. 26-step reduced collision for ARIRANG-256 with differences in M only

input H	C0E5A81E	952A32CB	730C4EB7	78730E23	757D7CAC	00000000	D69B0F52	D69B0F52
M	D69B0F52	78730E23	D69B0F52	730C4EB7	E3E3E3E3	952A32CB	1A1A1A1A	49494949
	00000000	02020202	D3DCBDB8	D9BDE3CB	562D250E	9B9F0611	662E4BD8	E75B0B2F
M'	D69B0F52	78730E23	D69B0F52	730C4EB7	1C1C1C1C	952A32CB	E5E5E5E5	49494949
	FFFFFFFF	02020202	2C234247	D9BDE3CB	562D250E	9B9F0611	662E4BD8	E75B0B2F
step 26	B4931778	F1615E8C	0E3756B9	93ED3536	4EBCBBFE	86C9ADD8	34334617	340155F6

6   Pseudo-collisions for ARIRANG-224 and ARIRANG-384

If we relax the condition of no difference at the output of the compression function we can find much better differentials. A near-collision attack for the complete compression function makes use of the three particular features of the compression function of ARIRANG. The first one is the existence of all-ones differentials. The second element that enables our attack is the fact that in the first steps we can manipulate chaining values and message words to adjust input values of G-functions, similarly to the message modification strategy. Finally, we exploit the double-feed-forward feature of the compression function (cf. Fig. 1) to restrict the differences to only first half of the steps.

Once we have such near-collisions for the compression function, we can use them to construct pseudo-collisions for the complete hash function ARIRANG-224 and ARIRANG-384. This is possible thanks to the details of message padding and the way the final digest is produced. Because the final hash value is just a truncated chaining value, we can introduce the chaining differences in the register which is going to be truncated when producing the digest. Also, the padding and appending the length information does not use a separate message block but rather a few last words of a block. This means we need to deal with only one message block with the last three words determined by the padding scheme and the message length.

We will talk about ARIRANG-224, however our attack is not specific to it, so it also works for ARIRANG-384.

6.1   Finding Near Collision Differential

Based on the same idea and model as used for searching the collision, we did the search for finding near collisions and we observed an interesting phenomenon. With input differences in a single chaining variable, we could get differentials that go through the first twenty steps and collapse back to the same register at step 20. Then after the middle feed-forward, there is no difference in chaining registers and nothing happens until the final feed-forward. Only then the initial difference is injected again and results in an output difference restricted to only one register, 32 bits in case of ARIRANG-256. Actually all configurations with differences in chaining variables behaves similarly, we can treat them as combinations of single difference.

With difference in H[7], we find it is easy to find the appropriate chaining values and messages. And advantage of this differential is, H[7] of the final output

Table 6. Active G functions in $H[7]$ near collision characteristics for ARIRANG

Step	W (left)	Active G (left)	W (right)	Active G (right)
1	W_{16}		W_{17}	
2	W_0	✓	W_1	
3	W_2		W_3	✓
4	W_4	✓	W_5	✓
5	W_6		W_7	
6	W_{18}		W_{19}	
7	W_8		W_9	✓
8	W_{10}		W_{11}	✓
9	W_{12}		W_{13}	
10	W_{14}		W_{15}	
11	W_{20}		W_{21}	
12	W_3	✓	W_6	
13	W_9		W_{12}	✓
14	W_{15}	✓	W_2	✓
15	W_5		W_8	
16	W_{22}		W_{23}	
17	W_{11}		W_{14}	✓
18	W_1		W_4	✓

is discarded for ARIRANG-224 and ARIRANG-384, hence instead of near collision, it gives collisions. The differential with corresponding active G is listed in Table 6 and the detailed picture of it can be found in Fig 4. There is no active G after step 18, and there is no difference in the output before the final feed-forward. Steps after 18 are not listed in Table 6.

6.2 Finding Chaining Values and Messages

The algorithm used to solve the near collision starts with setting all messages and chaining values to be a random value, here we make use of 0. To get pseudo-collisions for the complete hash function, we need to consider the message padding and the encoding of the block length. In ARIRANG, the message padding is performed by appending '1' followed by as many zeros as necessary and the message length is encoded in the last two words. To accommodate for this, we use 13 word long message which we can manipulate freely and fix $M_{13} = 10 \cdots 0_2$ and M_{14}, M_{15} to contain encoded length (which is $13 \cdot 32$ for ARIRANG-224 and $13 \cdot 64$ for ARIRANG-384). Thanks to that, the input to the compression function is consistent with the definition of the hash function and we still have a complete control over 13 message words $M_0, \ldots, M_{12}$. Now we can focus on finding a message pair that follows the differential in the compression function and we proceed as follows.

1. Steps 1-9, whenever there is an active G, we force the input to the G to γ $((\gamma, \bar{\gamma})$ is one of good input pairs to $\mathbf{G}^{(256)})$ by modifying the immediate W values.

Table 7. Collision Example for ARIRANG-224

input H	969F43DE	781BBD62	E6E7CEC7	075AF1AC	EE30CDD2	670D94E4	7AD337C6	60026A7A
input H'	969F43DE	781BBD62	E6E7CEC7	075AF1AC	EE30CDD2	670D94E4	7AD337C6	9FFD9585
M	43F40822	00000000	22EE1F96	30B48FFB	AD6E028F	958F43D5	5819FFF7	00000000
	00000000	34B65233	00000000	C16DE896	00000000	80000000	00000000	000001A0
output H	CBF6A53B	0D7EB2CB	ACFD326A	2BA6E962	4C2087AA	2ABD938A	221AED0E	
output H'	CBF6A53B	0D7EB2CB	ACFD326A	2BA6E962	4C2087AA	2ABD938A	221AED0E	
H ⊕ H'	00000000	00000000	00000000	00000000	00000000	00000000	00000000	

2. Step 12, we modify W_3. Note that W_3 is also used in step 3 and 6 (W_{18}), we can compensate this change using the method described before.

3. Step 13, we modify W_{20} through W_0, we also modify W_2 so that W_{19} keeps stable. We compensate the change of W_0 and W_2 again using the described method.

4. Step 14, left active G can be dealt with using W_6 and W_2.

5. Step 15, right active G can be choosing a random W_9, we compensate the change of W_9 used in step 7 by modifying $H[6]$. However the input to the left G in step 3 changes, we compensate this using W_{19} in step 6, $H[0]$ and $H[1]$ in step 1. Again input to left G in step 1 changes as $H[0]$ changes, we compensate as done for change of W_7. Note W_{19} can only be changed indirectly, here we use W_2 and then compensate using $H[6]$. We repeat this step until we find the right active G in step 14 is good. Note we can do the compensation work only after a good value is found.

6. step 17, we modify W_5 which is used in step 15. Then we compensate the change of W_5 and W_{18}

7. Step 18, the active G is dealt with by using W_4 and W_0.

The only active G left is the one in step 15. We leave this to a chance by looping over different W_9. This requires 2^{28} tries, which is equivalent to around 2^{23} (2^{51} for ARIRANG-384) calls to the compression function as we only need to compute two G functions in the loop and there are 80 such computations in the compression function. Examples shown in Table 7 can be found in few seconds on a standard computer, and the the algorithm has no memory requirements apart from a few words used for intermediate variables.

6.3 Collisions for **ARIRANG-384**

We can find collisions for ARIRANG-384 the same way as done for ARIRANG-224. However, the corresponding complexity of 2^{51} is too high for a standard computer to handle. To get over this difficulty, we can use the fact that the final transform for ARIRANG-384 is done by discarding the last two chaining values, i.e. $H[6]$ and $H[7]$. So besides $H[7]$-differential, we can also consider $H[6]$-differential and $H[6-7]$-differential (Indeed this also gives near collisions with outputs differ in $H[6]$ and $H[7]$). Thanks to a different positions of active G-functions, it turns out that the $H[6]$-differential can be solved with complexity 1. Table 8 lists the active G for this differential. Note that this differential works for all instances of

Table 8. Active G functions in $H[6]$ near collision characteristics for ARIRANG

Step	W (left)	Active G (left)	W (right)	Active G (right)
1	W_{16}		W_{17}	
2	W_0		W_1	
3	W_2	✓	W_3	
4	W_4		W_5	✓
5	W_6	✓	W_7	✓
6	W_{18}		W_{19}	
7	W_8		W_9	
8	W_{10}		W_{11}	✓
9	W_{12}		W_{13}	✓
10	W_{14}		W_{15}	
11	W_{20}		W_{21}	
12	W_3		W_6	
13	W_9	✓	W_{12}	
14	W_{15}		W_2	✓
15	W_5	✓	W_8	✓
16	W_{22}		W_{23}	
17	W_{11}		W_{14}	
18	W_1		W_4	✓
19	W_7		W_{10}	✓

ARIRANG. So this also gives another solution for finding 224/256 near collision for ARIRANG-256 with complexity 1.

Referring to table 8, we can solve this differential (finding chaining values and messages) using the following procedure:

1. Step 1-9 can be handled as usual.
2. Step 13, we modify W_6 in step 12. We compensate the change of W_6 and W_{19}
3. Step 14, we modify W_2 directly and then compensate the change of W_2 and W_{19}
4. Step 15, for the left active G, we modify W_5 and compensate; for the right active G, we modify W_8. Note that the change of W_8 can be compensated similarly as done for W_{19}.
5. Step 18, we modify W_4 and W_0 simultaneously.
6. Step 19, we modify W_1 as used in step 18 and W_7 simultaneously.

As shown above, every step in the algorithm is deterministic, hence it gives complexity close to 1. Experiments also support the result, collisions can be found in terms of μs. An example of collision for ARIRANG-384 is shown in Table 9, note it is also 448/512 near collision for ARIRANG-512.

6.4   Pseudo-preimages

It is possible to further extend the pseudo-near-collision attack to pseudo-preimages of ARIRANG. Take the configuration $H = (0, 1, 0, 0, 0, 0, 0, 0)$ for

Table 9. Pseudo-collision example for ARIRANG-384

input H	BA36BCB93BFD8D20	6B951DB399EB2EDC	1950E807876279AE	AF16B3C9901076DC
	62372888DECEB1E5	939957A5F4B4EE05	AA31DB9CB0EF684C	49B72A01D8C86B6F
input H'	BA36BCB93BFD8D20	6B951DB399EB2EDC	1950E807876279AE	AF16B3C9901076DC
	62372888DECEB1E5	939957A5F4B4EE05	55CE24634F1097B3	49B72A01D8C86B6F
M	B5127D606F0860D8	3E2BD987F6626D29	4EF941810127832F	0000000000000000
	B5127D606F0860D8	A8FF942B50A3F3F8	A99E61F4B41D9347	F6E3114F3EAAA5E1
	AFE28E981D9AE700	0000000000000000	C80D9570708720C3	AD8760D00E4D14C8
	0000000000000000	8000000000000000	0000000000000000	0000000000000340
output H	5939B28C23F6435F	BFA7FC0F59F0BFF7	FBF8D1923EED2060	AE79BE18FC078E32
	F4CE359791C979E7	543F7F214A45D0A9	193A61B727F9BC5A	3E8CFA173B9D48B2
output H'	5939B28C23F6435F	BFA7FC0F59F0BFF7	FBF8D1923EED2060	AE79BE18FC078E32
	F4CE359791C979E7	543F7F214A45D0A9	E6C59E48D80643A5	3E8CFA173B9D48B2

example, we are able to solve it in time 1 and it gives a near collision with all-one difference in $H[1]$ of final output. Note that once one such near collision pair is found, we are able to find 2^{32} pairs by trying different values for W_1 (W_7, W_0, and W_4 are changed accordingly) and compensate at the beginning. To find exact values, we need to compute steps 18 – 40 only, so the complexity to find one pair is reduced to about 2^{-1}. Given a target t, any match with t or $t \oplus 0^{32}1^{32}0^{192}$ will give us a pseudo-preimage. So we are able to find a match by finding 2^{255} different values, and finding each value costs 2^{-1}. The overall complexity for finding a pseudo-preimage is 2^{254} for ARIRANG-256. Similarly, we can find pseudo-preimage for ARIRANG-512 within 2^{510}. However this does not give a preimage attack, as converting pseudo-preimage to preimage requires the complexity to be less than 2^{n-2} in general.

7   Possible Extensions

With the similar method above, we can see that it is reasonable to count the active G from step 21, as most of the time, we can handle the first 20 steps using the message adjustment with low complexity. We did the search and found two interesting configurations ($M = (0, 1, 1, 0, 0, 1, 0, 1, 0, 0, 1, 0, 0, 1, 0, 1)$ and $M = (0, 0, 0, 0, 0, 1, 0, 0, 1, 0, 0, 0, 0, 0, 0, 0)$, where i-th bit of the configuration indicates whether there is a difference in $M[i]$) which gives 29-step reduced and 34-step reduced collisions with 1 and 2 active Gs, respectively. These two configurations may give step-reduced collisions with complexity less than birthday bound. With configuration $M = (1, 1, 0, 0, 1, 1, 0, 0, 1, 1, 0, 0, 0, 0, 1, 0), H = (1, 0, 0, 0, 0, 1, 1, 1)$ we may find [2-37] step reduced pseudo-collision as there are only 4 active G after step 20 and the active G in step 21 seems easy to deal with. With configuration $M = (0, 0, 1, 1, 0, 0, 0, 1, 1, 1, 0, 0, 0, 1, 0, 0)$, we may find semi-free-start collision for full ARIRANG as there are 5 active Gs after step 20 and seems those 3 active Gs in step 21 and 23 can be dealt with by modifying the chaining values.

Some investigation shows that similar idea of message adjustment can be used to find collisions based on semi-free-start collision. Note that when messages are modified, chaining values are modified in accordingly. We can do the reverse: modify the chaining values to those we required, and change the messages accordingly. However we need to be careful to ensure that active Gs are not affected.

8   Conclusions

We presented a range of attacks on ARIRANG. They all use the same type of differential based on flipping all bits in a register and the fact that all-one differences propagate with non-zero probability through the non-linear function $G^{(256)}$ and are not affected by all the other building blocks of the function.

This approach allowed us to find collisions for step-reduced compression function and pseudo-collisions for the hash function. Even though this method seems to be effective when looking for collisions for up to around 30 steps, we do not see a way to extend it to a collision attack on the full hash function at the moment.

A possible alternative approach would be to consider other types of differences. Note that we can get high-probability local collision patterns by having only one S-box active inside of $G^{(256)}$ and canceling the (dense) output differences in later steps by appropriate differences in message words. With this approach we can have up to 18 S-boxes active in the part of the function beyond our message-modification control to beat the birthday bound. The main difficulty seems to find a superposition of such local patterns that agrees with the message expansion process.

One could also think about ways to "patch" the design to defend against our attacks. It seems that the double feed-forward is not a good idea as it enabled us to skip half of the steps of the function in our pseudo-collision attack. Moreover, it should not be possible to use all-one differences that easily. To this end, one could either break the symmetry of rotations somewhere (perhaps in the message expansion process as seen in SHA-256 that uses also shifts in addition to rotations) or modify the MDS mapping to make sure that none of the possible output differences of the layer of S-boxes obtained for all-one input difference maps to all-ones difference through the MDS. However, all those fixes are quite ad-hoc and address only one particular attack strategy exploited in this paper.

Acknowledgements

The work in this paper was supported in part by the National Research Foundation of Singapore under Research Grant NRF-CRP2-2007-03 and the Singapore Ministry of Education under Research Grant T206B2204.

Krystian Matusiewicz was supported by grant 274-07-0246 from the Danish Research Council for Technology and Production Sciences.

The authors would like to thank Christian Rechberger, Praveen Gauravaram and the anonymous reviewers for the helpful comments and Wei Lei for his shell script.

References

1. Chang, D., Hong, S., Kang, C., Kang, J., Kim, J., Lee, C., Lee, J., Lee, J., Lee, S., Lee, Y., Lim, J., Sung, J.: ARIRANG: SHA-3 Proposal. NIST SHA-3 candidate, http://csrc.nist.gov/groups/ST/hash/sha-3/Round1/documents/ARIRANG.zip
2. Contini, S., Matusiewicz, K., Pieprzyk, J.: Extending FORK-256 attack to the full hash function. In: Qing, S., Imai, H., Wang, G. (eds.) ICICS 2007. LNCS, vol. 4861, pp. 296–305. Springer, Heidelberg (2007)
3. Hong, D., Kim, W.-H., Koo, B.: Preimage attack on arirang. Cryptology ePrint Archive, Report 2009/147 (2009), http://eprint.iacr.org/2009/147
4. Hong, D., Sung, J., Lee, S., Moon, D., Chee, S.: A new dedicated 256-bit hash function. In: Robshaw, M.J.B. (ed.) FSE 2006. LNCS, vol. 4047, pp. 195–209. Springer, Heidelberg (2006)
5. Indesteege, S., Mendel, F., Rechberger, C., Schläffer, M.: Practical Collisions for SHAMATA. In: Jacobson, M.J., Rijmen, V., Safavi-Naini, R. (eds.) SAC 2009. LNCS, vol. 5867, pp. 1–15. Springer, Heidelberg (2009)
6. Matusiewicz, K., Peyrin, T., Billet, O., Contini, S., Pieprzyk, J.: Cryptanalysis of FORK-256. In: Biryukov, A. (ed.) FSE 2007. LNCS, vol. 4593, pp. 19–38. Springer, Heidelberg (2007)
7. Mendel, F., Lano, J., Preneel, B.: Cryptanalysis of reduced variants of the FORK-256 hash function. In: Abe, M. (ed.) CT-RSA 2007. LNCS, vol. 4377, pp. 85–100. Springer, Heidelberg (2007)
8. Saarinen, M.-J.: A Meet-in-the-Middle collision attack against the new FORK-256. In: Srinathan, K., Rangan, C.P., Yung, M. (eds.) INDOCRYPT 2007. LNCS, vol. 4859, pp. 10–17. Springer, Heidelberg (2007)
9. Schneier, B., Kesley, J.: Unbalanced Feistel networks and block cipher design. In: Gollmann, D. (ed.) FSE 1996. LNCS, vol. 1039, pp. 121–144. Springer, Heidelberg (1996)

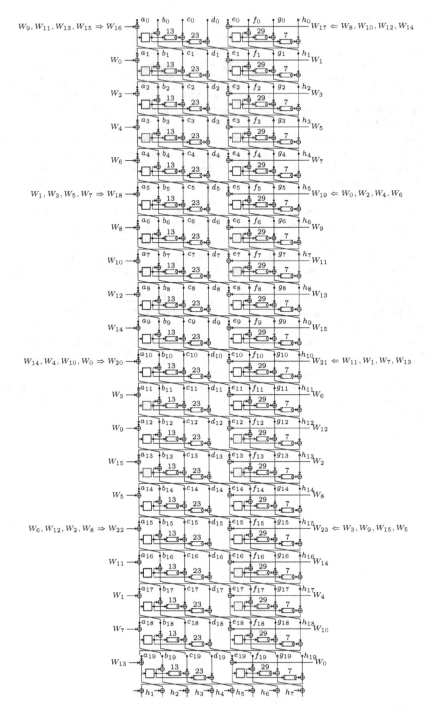

Fig. 4. Differential path in steps 1-20 used to find near-collisions in the compression function. There are no differences in steps 21-40.

A More Compact AES

David Canright[1] and Dag Arne Osvik[2]

[1] Naval Postgraduate School, Monterey CA 93943, USA
dcanright@nps.edu
[2] École Polytechnique Fédérale de Lausanne
dagarne.osvik@epfl.ch

Abstract. We explore ways to reduce the number of bit operations required to implement AES. One way involves optimizing the composite field approach for entire rounds of AES. Another way is integrating the Galois multiplications of MixColumns with the linear transformations of the S-box. Combined with careful optimizations, these reduce the number of bit operations to encrypt one block by 9.0%, compared to earlier work that used the composite field only in the S-box. For decryption, the improvement is 13.5%. This work may be useful both as a starting point for a bit-sliced software implementation, where reducing operations increases speed, and also for hardware with limited resources.

Keywords: AES, tower field, composite Galois field, bitslice.

1 Introduction

There have been many implementations of the Advanced Encryption Standard, optimized for various criteria, for different applications. Some approaches seek to minimize circuitry, e.g., [1,2,3,4,5]. For this goal, Rijmen[6] suggested using subfield arithmetic in the crucial step of computing an inverse in the Galois Field of 256 elements. [Note: strictly speaking, the operation is not $x \rightarrow x^{-1}$ but rather $x \rightarrow x^{254}$ so $0 \rightarrow 0$, but we will refer to this as the inverse for convenience; similarly for subfields.] Rudra et al.[1] gave a detailed implementation using that subfield approach. This idea was further extended by Satoh et al.[2], using sub-subfields (the "tower-field" representation of Paar[7], also called the "composite-field" approach), along with other innovative optimizations, which resulted in the smallest AES circuit at that point. The S-box architecture of Satoh was improved by Canright[8], mainly through carefully chosen normal bases, resulting in the most compact S-box to date. This S-box has been used in bit-sliced software implementations of AES, by Rebeiro et al.[9], and (slightly improved by [10]) by Käsper and Schwabe[11].

The present work seeks to further reduce the size of AES, in terms of the number of bit operations. While [8] showed that normal bases gave a more compact Galois inverter for the S-box, the specific basis chosen did not yield compact Galois multiplications by the constants used in the MixColumns step; hence that composite basis was used *only* for the S-box. Here we reconsider the approach

M.J. Jacobson Jr., V. Rijmen, and R. Safavi-Naini (Eds.): SAC 2009, LNCS 5867, pp. 157–169, 2009.
© Springer-Verlag Berlin Heidelberg 2009

of maintaining the composite-field representation throughout the rounds of encryption, as in Rudra et al.[1]. We find that a different choice of basis than in [8] does indeed give a smaller AES implementation with this approach, in part through combining the linear transformations of the S-box with the constant multiplications (or "scalings") of MixColumns. Moreover, applying optimization software to certain portions of the logic further reduces the number of operations. Together, these improvements give a 9.0% reduction in the number of bit operations needed to encrypt one block with a 128-bit key.

First we briefly review the AES algorithm in Section 2, then detail our method in Section 3, including choices of basis for the tower field and integration of the scalings of MixColumns with the linear transformations of the S-box. Finally, we summarize our results in Section 4 and briefly discuss conclusions in Section 5.

2   AES Algorithm

The Rijndael algorithm, as adopted for the Advanced Encryption Standard, is a symmetric block cipher with 128-bit blocks and three key sizes: 128, 192, or 256 bits[12]. Here, we give just enough detail to explain our method below.

For encryption, each block of 16 bytes is processed by several rounds: 10, 12, or 14, depending on key size. From the initial key, the key schedule generates a different round key for each round. Each round comprises the following steps.

1. *SubBytes* subjects each byte independently to a nonlinear function, often called the S-box, and substitutes the result for the original byte. The S-box function consists of two sequential operations:

   (a) first, *inversion* treats the byte as an element of $GF(2^8)$, where the bits are coefficients of a polynomial, and polynomial arithmetic is modulo the irreducible polynomial $q(x) = x^8 + x^4 + x^3 + x + 1$; each nonzero byte is replaced by its multiplicative inverse in this field, while a zero byte remains unchanged.

   (b) then an *affine transformation* is applied: treating the byte as a vector of bits, the byte is multiplied by a constant bit matrix M and then a constant byte b is added (with bit arithmetic in $GF(2)$, where multiplication is AND and addition is XOR), so $x \rightarrow M x + b$.

   In software, the S-box is often implemented as a table lookup.

2. *ShiftRows* considers the 16 bytes as a 4×4 array and rotates each row to the left by its position, so row #0 does not move, row #1 moves 1, etc.

3. *MixColumns* operates independently on each column of the 4×4 array: the column, as a vector of four bytes, is multiplied by a constant byte matrix, where the byte arithmetic is in $GF(2^8)$ as in the S-box inversion:

$$\begin{pmatrix} x_0 \\ x_1 \\ x_2 \\ x_3 \end{pmatrix} \rightarrow \begin{pmatrix} 2\ 3\ 1\ 1 \\ 1\ 2\ 3\ 1 \\ 1\ 1\ 2\ 3 \\ 3\ 1\ 1\ 2 \end{pmatrix} \begin{pmatrix} x_0 \\ x_1 \\ x_2 \\ x_3 \end{pmatrix}$$

4. *AddRoundKey* bitwise adds (XOR) a 128-bit round key to the 128-bit state.

The first round is preceded by an *AddRoundKey* step, and the last round skips the *MixColumns* step.

For decryption, the whole process is reversed, using the inverse operation for each step in the reverse order. AddRoundKey is its own inverse, and the inverse of ShiftRows rotates rows to the right instead of left. The inverse of MixColumns just multiplies each column by the inverse of the constant byte matrix, so

$$
\begin{pmatrix} x_0 \\ x_1 \\ x_2 \\ x_3 \end{pmatrix} \rightarrow \begin{pmatrix} E & B & D & 9 \\ 9 & E & B & D \\ D & 9 & E & B \\ B & D & 9 & E \end{pmatrix} \begin{pmatrix} x_0 \\ x_1 \\ x_2 \\ x_3 \end{pmatrix}
$$

where the constant values are in hexadecimal (the leading 4 bits of each are 0). For the inverse of SubBytes, first the inverse affine transformation is applied, so $x \rightarrow M^{-1}(x + b)$, then the Galois inversion is its own inverse operation.

Some reordering of the steps in each round is possible. SubBytes commutes with ShiftRows, and MixColumns and AddRoundKey can be swapped by modifying the key schedule appropriately; similarly with the inverse operations for decryption. Commonly, fast software implementations, e.g., those of Bernstein and Schwabe[13], combine the S-box function with the Galois multiplications of MixColumns, using each input byte to index a table of 4-byte columns, as suggested in the Rijndael proposal[14], which called them "T-tables."

3   Method

Our goal was to develop an implementation of AES with a minimal number of bit operations. The result could be useful for a bit-sliced software implementation, or for hardware with limited resources. (Our original inspiration was considering a bit-sliced AES for the CellBE processor[15].) Our starting point, and baseline for comparison, was the compact AES of [2], with the improved S-box of [8].

These prior works used a tower-field representation of $GF(2^8)$ so that the Galois inversion in the S-box could be calculated compactly. But where [8] optimized the choice of basis for the S-box only, we sought to find the best basis for whole rounds of AES, as [1] did for a different composite-field representation.

One reason *not* to do this, i.e., to change back to the standard basis after the S-box, is that the Galois multiplication by the constant 2 byte, as required in MixColumns, is very compact in the standard basis: three bitwise XORs. Nonetheless, we found that overall the advantages of our approach overcame this disadvantage.

One way we reduced operations is by combining the constant Galois multiplications of MixColumns with the linear part of the affine transformation of the S-box. We tried different places to put these combined transformations, either earlier as parts of the S-box or later as parts of MixColumns, as we will describe in subsection 3.2.

Another way was by finding the tower-field basis that would be most compact not just for the inverter, but for an entire round of encryption or decryption.

There are actually many different tower-field representations possible; here we only need to examine a small subset of those considered in [16], as discussed in subsection 3.1 below.

Lastly, we applied state-of-the-art optimizing software to the transformation matrices and to parts of the inverter operation. The optimizing software employs heuristics to arrive at very efficient implementations.

3.1 Basis Choices

In [8], 432 different choices of basis were considered for the tower-field representation of $GF(2^8)$, where $GF(2^8)$ is considered as a quadratic extension of $GF(2^4)$, which in turn is considered as a quadratic extension of $GF(2^2)$. We will use the notation $GF(2^8)/GF(2^4)/GF(2^2)$ to indicate such a tower-field representation; of course, all representations of $GF(2^8)$ are isomorphic. Such a representation really involves three bases: one each for $GF(2^2)/GF(2)$, for $GF(2^4)/GF(2^2)$, and for $GF(2^8)/GF(2^4)$, where each basis consists of two elements linearly independent over the subfield.

Only polynomial bases (of the form $[r, 1]$) and normal bases (of the form $[r^q, r]$, where $q = 2^1, 2^2,$ or 2^4 is the size of the subfield, and r, r^q are conjugates) were considered in [8]; other types are generally less efficient. And only choices with a trace of unity $\tau = r + r^q = 1$ were considered, that is, where the minimal polynomial has the form $x^2 + x + \nu$ and $\nu = r \times r^q$ is the norm of r, since this choice eliminates some operations. Some other special forms would also eliminate some operations, such as where the norm is unity $\nu = 1$ or where the trace and norm are equal $\tau = \nu$, but these turned out to be less efficient for the Galois inverter of the S-box. Normal bases were shown to have a definite advantage for the inverter, since more factors are shared in the lower level operations. And one particular choice, #4 of the 432 in [16, App. E], gave the smallest optimized transformation matrices, and hence the smallest merged S-box (where encryption and decryption share a Galois inverter), as well as the smallest S-box for encryption only or for decryption only.

But basis #4 did not give a compact form for the Galois multiplications needed in MixColumns, so the tower-field representation was only used for the S-box, with the rest of each round using the standard basis. In particular, for encryption, MixColumns requires multiplying bytes by the constants 2 and 3 (in the standard representation), where multiplying by 2 only requires three bitwise XORs. (In the standard representation, "2" represents a root of the irreducible polynomial $q(x) = x^8 + x^4 + x^3 + x + 1$, and "3" = "2" + 1, where 1 is the multiplicative identity as usual. So multiplication of a byte by 2 involves shifting left one bit, and if the msb was 1, then XOR with 0x1B.)

We explored whether a different approach might give a more compact implementation: using a tower-field representation throughout the rounds of encryption, similar to the approach of [1]. We sought an optimum basis for both the S-box and MixColumns steps; ShiftRows is just a re-ordering of bytes, and AddRoundKeys works the same in any basis. To keep the optimization tractable, we limited consideration to encryption only, or separately for decryption only;

we did *not* consider a merged encrypt/decrypt architecture as in [2]. And to keep the Galois inverter of the S-box small, we only looked at normal bases with unit trace.

This left 16 possibilities out of the 432 in [16, App. E]: basis numbers 1, 4, 19, 22, 37, 40, 55, 58, 73, 76, 91, 94, 109, 112, 127, and 130. It turns out that all these cases give a Galois inverter of the same size; though the specifics of the operations change, the total number of bit operations in the optimized inverter is the same.

Besides the inverter, the S-box includes the affine transformation, and Mix-Columns requires Galois multiplication by 2 and 3, or by four different constants for decryption. We will use the term "scaling" to indicate such Galois multiplying of a byte by a specified constant byte. Then both scaling and the affine part of the S-box (ignoring the additive constant for now; see subsection 3.3) can each be represented as a linear transformation: multiplication of an 8-bit vector by a bit matrix (with all bit arithmetic modulo 2). These transformations can be combined by simply multiplying the bit matrices. To see more precisely which matrices would be required in each round, we needed to consider how to implement MixColumns. Then we could choose the basis that gave the most compact versions of those matrices.

3.2 MixColumns

Satoh[2] gave an elegant implementation that combined MixColumns with its inverse, for the architecture with both encryption and decryption merged (with a selector signal). For just MixColumns, each column, as a vector of four bytes, is multiplied by a 4×4 matrix, where scalar multiplication is in $GF(2^8)$. Satoh effectively decomposed the matrix as below:

$$
\begin{pmatrix} 2\,3\,1\,1 \\ 1\,2\,3\,1 \\ 1\,1\,2\,3 \\ 3\,1\,1\,2 \end{pmatrix} = 2 \times \begin{pmatrix} 1\,1\,0\,0 \\ 0\,1\,1\,0 \\ 0\,0\,1\,1 \\ 1\,0\,0\,1 \end{pmatrix} + \begin{pmatrix} 0\,0\,1\,1 \\ 0\,0\,1\,1 \\ 1\,1\,0\,0 \\ 1\,1\,0\,0 \end{pmatrix} + \begin{pmatrix} 0\,1\,0\,0 \\ 1\,0\,0\,0 \\ 0\,0\,0\,1 \\ 0\,0\,1\,0 \end{pmatrix}
$$

This decomposition allowed reuse of certain combinations of bytes[2, (6)], and each byte multiplication by 2 took three XORs, so altogether each 4-byte column took 108 XORs.

For decryption, the inverse MixColumns matrix of [2] came from adding more terms to the MixColumns matrix:

$$
\begin{pmatrix} E\,B\,D\,9 \\ 9\,E\,B\,D \\ D\,9\,E\,B \\ B\,D\,9\,E \end{pmatrix} = \begin{pmatrix} 2\,3\,1\,1 \\ 1\,2\,3\,1 \\ 1\,1\,2\,3 \\ 3\,1\,1\,2 \end{pmatrix} + 4 \times \begin{pmatrix} 1\,0\,1\,0 \\ 0\,1\,0\,1 \\ 1\,0\,1\,0 \\ 0\,1\,0\,1 \end{pmatrix} + 8 \times \begin{pmatrix} 1\,1\,1\,1 \\ 1\,1\,1\,1 \\ 1\,1\,1\,1 \\ 1\,1\,1\,1 \end{pmatrix}
$$

where the 4 and 8 came from repeated multiplications of common terms by 2, at 3 XORs each. With the reuse of common terms, altogether each 4-byte column took 195 XORs.

We considered similar decompositions that could re-use some byte sums, but with the constant scaling combined with the affine transformation of the S-box. Let T_2 be the matrix (given in subsection 3.3) below that performs the "times 2" operation, and similarly for other constants. (Note: $T_1 = I$, the identity matrix.) So with the matrix M of the affine transformation, the combined transformations needed for encryption are M, $T_2 M$, and $T_3 M$.

Our approach uses the decomposition below:

$$\begin{pmatrix} 2 & 3 & 1 & 1 \\ 1 & 2 & 3 & 1 \\ 1 & 1 & 2 & 3 \\ 3 & 1 & 1 & 2 \end{pmatrix} = \begin{pmatrix} 2 & 3 & 0 & 0 \\ 0 & 3 & 2 & 0 \\ 0 & 0 & 2 & 3 \\ 2 & 0 & 0 & 3 \end{pmatrix} + \begin{pmatrix} 0 & 0 & 1 & 1 \\ 1 & 1 & 1 & 1 \\ 1 & 1 & 0 & 0 \\ 1 & 1 & 1 & 1 \end{pmatrix}$$

Using the common terms in the last matrix, this approach has 11 byte additions (88 XORs) per column.

One way to do the transformations is for half the bytes, after the inverter of the prior S-box, to get transformed with both M and $T_2 M$ separately, and the other half with M and $T_3 M$; no byte needs both $T_2 M$ and $T_3 M$. Another way is to do half the bytes with just $T_2 M$, the other half with $T_3 M$, and later apply M only to two common terms: sums of untransformed bytes 0 & 1 and 2 & 3. While the latter ("later" transformations) way has fewer transformations overall, in the former ("early" transformations) way, pairs of transformations apply to each byte, allowing additional optimizations of the pairs. We explored both, and it turned out the early transformation approach was slightly better.

For decryption, the inverse MixColumns matrix has four different constants (hexadecimal E, B, D, & 9), linearly independent over $GF(2)$, so is more expensive. These scalings can be combined with the inverse affine transformation for the following inverse S-box, since inverse MixColumns is a linear operation.

We considered a direct, early approach, where after AddRoundKey, each byte is transformed by the four transformations $M^{-1} T_E$, $M^{-1} T_B$, $M^{-1} T_D$, and $M^{-1} T_9$, followed by the 12 byte additions (96 XORs) for inverse MixColumns. We also considered a decomposition:

$$\begin{pmatrix} E & B & D & 9 \\ 9 & E & B & D \\ D & 9 & E & B \\ B & D & 9 & E \end{pmatrix} = \begin{pmatrix} 3 & 2 & 0 & 0 \\ 0 & 3 & 2 & 0 \\ 0 & 0 & 3 & 2 \\ 2 & 0 & 0 & 3 \end{pmatrix} + D \times \begin{pmatrix} 1 & 0 & 1 & 0 \\ 0 & 1 & 0 & 1 \\ 1 & 0 & 1 & 0 \\ 0 & 1 & 0 & 1 \end{pmatrix} + 9 \times \begin{pmatrix} 0 & 1 & 0 & 1 \\ 1 & 0 & 1 & 0 \\ 0 & 1 & 0 & 1 \\ 1 & 0 & 1 & 0 \end{pmatrix}$$

Each byte gets transformed with both $M^{-1} T_2$ and $M^{-1} T_3$, and later the two common expressions, sums of untransformed bytes 0 & 2 and 1 & 3, each get both $M^{-1} T_D$ and $M^{-1} T_9$. This still has 12 byte additions but fewer transformations. Again, we tried both, and this time the later approach was better.

3.3   Transformation Matrices

In the ShiftRows, S-box, and MixColumns steps of a normal encryption round, each byte is routed to the correct position in a column, is inverted in a Galois inverter, then goes through the affine transformation along with 2 or 3 times that result (as shown above). The affine transformation on a byte x looks like $y = M\,x + b$, or in detail

$$
\begin{pmatrix} y_7 \\ y_6 \\ y_5 \\ y_4 \\ y_3 \\ y_2 \\ y_1 \\ y_0 \end{pmatrix} = \begin{pmatrix} 1\,1\,1\,1\,1\,0\,0\,0 \\ 0\,1\,1\,1\,1\,1\,0\,0 \\ 0\,0\,1\,1\,1\,1\,1\,0 \\ 0\,0\,0\,1\,1\,1\,1\,1 \\ 1\,0\,0\,0\,1\,1\,1\,1 \\ 1\,1\,0\,0\,0\,1\,1\,1 \\ 1\,1\,1\,0\,0\,0\,1\,1 \\ 1\,1\,1\,1\,0\,0\,0\,1 \end{pmatrix} \begin{pmatrix} x_7 \\ x_6 \\ x_5 \\ x_4 \\ x_3 \\ x_2 \\ x_1 \\ x_0 \end{pmatrix} + \begin{pmatrix} 0 \\ 1 \\ 1 \\ 0 \\ 0 \\ 0 \\ 1 \\ 1 \end{pmatrix}
$$

where bit #7 is the most significant and all bit operations are modulo 2.

To do the same operation in a different basis, we need to apply a similarity transformation to this matrix M (to account for the change of basis on both input and output vectors). Let X refer to the 8×8 bit matrix that converts a byte *from* the tower-field basis *to* the standard basis, and let u and v be the tower-field representations of x and y, respectively (so $x = X\,u$ and $y = X\,v$). Then the affine transformation becomes

$$ v = \left(X^{-1} M\,X \right) u + c \qquad \text{where} \qquad c = X^{-1} b $$

or equivalently

$$ v = \left(X^{-1} M\,X \right) (u + d) \qquad \text{where} \qquad d = X^{-1} M^{-1} b $$

Galois multiplication by the constants 2 and 3 can also be done by matrices; let T_2 and T_3 respectively be these matrices with respect to the standard representation. Then

$$
2 \times x = T_2\,x = \begin{pmatrix} 0\,1\,0\,0\,0\,0\,0\,0 \\ 0\,0\,1\,0\,0\,0\,0\,0 \\ 0\,0\,0\,1\,0\,0\,0\,0 \\ 1\,0\,0\,0\,1\,0\,0\,0 \\ 1\,0\,0\,0\,0\,1\,0\,0 \\ 0\,0\,0\,0\,0\,0\,1\,0 \\ 1\,0\,0\,0\,0\,0\,0\,1 \\ 1\,0\,0\,0\,0\,0\,0\,0 \end{pmatrix} \begin{pmatrix} x_7 \\ x_6 \\ x_5 \\ x_4 \\ x_3 \\ x_2 \\ x_1 \\ x_0 \end{pmatrix}
$$

and $T_3 = T_2 + I$. To get the scaling matrices for decryption, let $T_4 = (T_2)^2$, $T_8 = T_4 T_2$, $T_C = T_8 + T_4$; then $T_E = T_C + T_2$, $T_B = T_8 + T_3$, $T_D = T_C + I$, $T_9 = T_8 + I$. Again, to do these same operations in the tower-field basis, we would apply a similarity transformation to these matrices. Or, if we combine with the affine transformation, for encryption we get

$$ 2 \times v = \left(X^{-1} T_2 M\,X \right) (u + d) \qquad \text{and} \qquad 3 \times v = \left(X^{-1} T_3 M\,X \right) (u + d) $$

So for a given byte, with the early transformation strategy, first we apply the Galois inverter, then apply two transformations, either affine and 2×affine, or affine and 3×affine, depending on in which row of a column it ends up. Thus for a given basis X we need to optimize the matrix *pairs* $[(X^{-1} M X), (X^{-1} T_2 M X)]$ and $[(X^{-1} M X), (X^{-1} T_3 M X)]$, where each pair is considered as a single 16×8 matrix. For the later transformation strategy, the three separate matrices $(X^{-1} M X)$, $(X^{-1} T_2 M X)$ and $(X^{-1} T_3 M X)$ would be optimized.

The additive constant c can usually be included by simply replacing some XORs by XNORs, or it may be incorporated into the key schedule. Note that, because the row sum of the constants in the MixColumns matrix (or its inverse) is 1, then c really only needs to be added to any *one* of the four terms in the row.

For each of the 16 different normal bases, we applied our optimization software to minimize (smallest number of XORs) the two 16×8 bit matrices (each would apply to a pair of bytes per column) of the early transformation strategy, and also the three 8×8 matrices (again for a pair of input bytes) for the later strategy.

Basis #127 was the winner, with an early strategy optimized total of $17 + 18 = 35$ XORs, barely beating the later strategy at $11 + 13 + 12 = 36$ XORs. Here are the basis change matrices for basis #127:

$$X = \begin{pmatrix} 0\,0\,1\,0\,0\,1\,0\,0 \\ 0\,1\,1\,0\,0\,0\,1\,1 \\ 1\,1\,0\,1\,1\,0\,1\,1 \\ 0\,1\,0\,1\,0\,1\,1\,0 \\ 0\,0\,1\,0\,1\,1\,1\,0 \\ 1\,0\,1\,1\,0\,1\,1\,1 \\ 1\,1\,0\,1\,1\,1\,0\,1 \\ 1\,1\,0\,0\,0\,0\,1\,0 \end{pmatrix}, \quad X^{-1} = \begin{pmatrix} 0\,0\,1\,0\,1\,1\,0\,1 \\ 0\,1\,0\,1\,0\,1\,0\,1 \\ 1\,1\,0\,1\,1\,0\,1\,1 \\ 0\,1\,1\,0\,0\,1\,1\,1 \\ 1\,1\,1\,1\,0\,0\,0\,1 \\ 0\,1\,0\,1\,1\,0\,1\,1 \\ 0\,1\,1\,1\,1\,0\,0\,1 \\ 1\,0\,1\,1\,0\,1\,1\,1 \end{pmatrix}$$

Besides the matrix transformations in normal rounds, this approach also uses two others. Before the first round, each byte must be transformed from the standard representation to the tower-field basis, by the matrix (X^{-1}) above; the optimized version requires 15 XORs per byte, which we treat as part of round 0, the initial AddRoundKey. And the last round of encryption skips MixColumns and needs to end up in the standard representation, so after the last inverter, the affine transformation is combined with the basis change in the matrix $(M X)$; this requires 13 XORs per byte (with the constant b incorporated into the last round key; otherwise a NOT is needed).

For decryption, again for each of the 16 normal bases, we optimized the single 32×8 bit matrix (four transformations) for the early transformation strategy, and also the two 16×8 matrices for the later strategy. In ranking the results, recall that the early approach applies the 32×8 matrix to each input byte; the later approach applies one 16×8 matrix to each input byte but applies the other only to the shared sums, half as many bytes. This time, basis #94 won, with the best later strategy result at $19 + \frac{1}{2} \times 18 = 28$ XORs per byte, better than the best early strategy result of #58, at 31 XORs. Here are the basis change matrices for basis #94:

$$
X = \begin{pmatrix} 0\,1\,1\,1\,0\,1\,0\,0 \\ 0\,1\,1\,1\,0\,0\,1\,0 \\ 1\,0\,0\,0\,1\,0\,1\,1 \\ 1\,0\,1\,1\,1\,1\,0\,1 \\ 1\,1\,1\,1\,0\,1\,1\,0 \\ 0\,0\,1\,1\,1\,0\,1\,0 \\ 0\,0\,1\,0\,0\,0\,1\,0 \\ 0\,0\,1\,0\,0\,1\,1\,0 \end{pmatrix}, \quad X^{-1} = \begin{pmatrix} 0\,1\,0\,0\,1\,0\,1\,1 \\ 0\,1\,1\,1\,0\,0\,1\,1 \\ 1\,1\,0\,0\,0\,0\,0\,1 \\ 0\,0\,1\,1\,0\,0\,0\,1 \\ 0\,0\,1\,1\,0\,1\,1\,1 \\ 0\,0\,0\,0\,0\,0\,1\,1 \\ 1\,1\,0\,0\,0\,0\,1\,1 \\ 1\,0\,0\,1\,1\,1\,1\,1 \end{pmatrix}
$$

For the first round of decryption, with no MixColumns, we need the transformation $X^{-1} M^{-1}$, which takes 13 XORs. After the last decryption round, before the additional AddRoundKey corresponding to encryption round #0, we need to switch back to the standard basis with X, at 13 XORs.

3.4 Galois Inverter

For the inverter of basis #127, applying (by hand) the OR gate substitution reduced the inverter size to 56 XORs, 30 ANDs, 10 ORs. This is the same number of bit operations as the inverter for basis #4 of [8], which, like that for basis #94, is 56 XORs, 34 ANDs, 6 ORs.

But in the tower-field representation, each 8-bit Galois inverter includes a 4-bit Galois inverter in the subfield. The $GF(2^4)$ inverter performs a bijection function, as does a 4×4 S-box, and hence is a natural target for optimization with methods like those in [17]. The result reduced the 4-bit inverter from 9 XORs, 8 ANDs, 2 ORs down to 8 XORs, 5 ANDs, 2 ORs, a savings of 4 bit operations.

3.5 Round Keys

The AddRoundKey step of each round is a simple bitwise XOR in any basis. In our approach, each round key must be represented in the tower-field basis. One way to do this would be to pre-compute the usual round keys by some means, then apply the X^{-1} matrix transformation to each byte. Another approach is to do the whole key schedule in the tower-field representation. First the initial key needs to be transformed into the tower field. For the last round the round key needs to be transformed, either back to the standard representation and added after the data block is transformed back, or transformed by the inverse affine transformation and added before the data is transformed back.

In our comparisons below we do not consider the cost of the key schedule that generates the round keys. We assume the round keys have been pre-computed, including their tower-field representations. This is appropriate to using our approach for bit-sliced software, where the round keys can be stored and applied to many blocks. But this assumption is less appropriate to compact hardware implementations, comparable to that of [2], where storing keys in registers is expensive, so typically keys are computed on the fly.

3.6   Validation

We implemented our approach in the form of Verilog (hardware description language) code. This code was written mainly for testing purposes, and defines one module for each kind of round, including the key schedule. We successfully tested this implementation by compiling and running it on a FPGA in a SRC 6e computer system, with correct results. (The FPGA implementation was only to check correctness; true optimization for an FPGA would need to exploit their Look-Up-Table structure.)

4   Results

We have described our methods to reduce the number of bit operations needed for AES. Our original motivation was to explore bit-slice techniques to implement AES in software. For that, reducing the number of operations essentially translates into increasing the speed. Then an appropriate measure is the total number of bit operations needed to encrypt a block, as shown in Table 1. The reduction in operations we achieved might also be useful for compact hardware implementations, where area is limited. We do not propose a specific hardware design here.

Our baseline for comparison is a compact encryption-only (or decryption-only) AES implementation using the S-box of [8] and the MixColumns of [2], shown above in subsection 3.2, and our units of comparison are bit operations: XOR, AND, OR, NOT. These two implementations are compared in detail in Table 1.

One normal round of encryption took 155 ops/byte in the baseline; our new approach needs only 139.5 ops/byte, smaller by 10.0%. However, our approach requires an initial transformation into the composite field (15 ops/byte), which adds on to the cost of round #0, the initial AddRoundKey (8 ops/byte). The

Table 1. Results. The number of bit operations per byte is given for various operations up to the full 10-round AES, comparing our approach with a baseline compact implementation

	encryption		decryption	
	baseline	new approach	baseline	new approach
Galois inverter	96	92	96	92
initial transformation	0	15	0	13
round transformations	24	17.5	25	28
last transformation	24	13	25	13
MixColumns	27	22	48.75	24
AddRoundKey	8	8	8	8
round #0	8	23	8	21
normal round	155	139.5	177.75	152
last round	128	113	129	113
10-round AES	1531	1391.5	1736.75	1502

last round skips MixColumns: the baseline version takes 128 ops/byte; ours takes 113 ops/byte.

For a bit-sliced software approach, a reasonable basis for comparison is the total number of bit operations. Altogether, for 128-bit keys (10 rounds), the baseline requires 24496 bit operations to encrypt one block, while ours requires 22264, which is 9.0% smaller. For 256-bit keys, our approach is 9.4% smaller. For decryption, the improvement is even greater: 13.5% for 128-bit keys and 13.8% for 256-bit keys.

For a compact hardware approach, comparison is less clear, depending on the specific architecture. Suppose we assume an encryption-only version of the compact design in [2]. There, the 32-bit data path goes through four S-boxes including transformations, a MixColumns operation, and AddRoundKey. Selectors are used to skip MixColumns on the last round and also skip the S-box for round 0. Data register connections do the ShiftRows. To simply plug in our approach, the basic round has 10% fewer operations, but the paths for round 0 and the last round would need different transformations added; the result is actually 8.1% *larger* than the baseline (based only on the bit operations in the rounds, excluding selectors and registers). A different architecture would be needed in order to take advantage of our approach, such as one where just the initial transformation into the tower-field basis, and the last tranformation to the standard representation, have 8-bit data paths instead; this approach would make those byte-serial rounds slower, but would reduce the total operations for rounds by 5.5%.

5   Conclusions

We have reduced the number of bit operations for 10-round AES by 9.0%. We achieved this reduction partly through finding a tower-field representation that compactly calculates both the Galois inversion and the constant scaling of Mix-Columns (when combined with the affine transformation). The other contribution to increased efficiency comes from very effective optimization, both of the 4-bit inverter (within the 8-bit Galois inverter) and of the various transformation matrices.

Our more compact AES approach may be useful for software bit-slice implementations, or for hardware with limited resources. Of course, in developing a bit-sliced program for a specific target processor, then parallelism and register constraints need to be taken into account, as well as the cost of slicing and unslicing. In fact, soon next-generation Intel and AMD processors will include single instructions to perform whole AES rounds[18], which may render bit-sliced implementations uncompetitive on such targets. However, current and older Intel and AMD processors may be promising targets for some time, as well as other possibilities such as the CellBE processor[15]. And further optimization may be possible for a particular processor, using a suitable slicing arrangement and taking advantage of specific instructions; for example, on a CellBE processor, the $GF(2^4)$ inverter takes only 8 instructions, compared to the 15 bit operations mentioned in subsection 3.4. Also, Intel's coming AVX technology, with 256-bit registers and RISC-style SSE instructions, may make bit-slicing competitive with the native AES instructions.

Future work includes developing a bitsliced AES implementation for the CellBE processor, and possibly for others.

Acknowledgements

We would like to thank the reviewers for several helpful suggestions, including pointing out some recent relevant work.

References

1. Rudra, A., Dubey, P.K., Jutla, C.S., Kumar, V., Rao, J.R., Rohatgi, P.: Efficient Rijndael encryption implementation with composite field arithmetic. In: Koç, Ç.K., Naccache, D., Paar, C. (eds.) CHES 2001. LNCS, vol. 2162, pp. 171–184. Springer, Heidelberg (2001)
2. Satoh, A., Morioka, S., Takano, K., Munetoh, S.: A compact Rijndael hardware architecture with S-box optimization. In: Boyd, C. (ed.) ASIACRYPT 2001. LNCS, vol. 2248, pp. 239–254. Springer, Heidelberg (2001)
3. Wolkerstorfer, J., Oswald, E., Lamberger, M.: An ASIC implementation of the AES S-boxes. In: Preneel, B. (ed.) CT-RSA 2002. LNCS, vol. 2271, pp. 67–78. Springer, Heidelberg (2002)
4. Chodowiec, P., Gaj, K.: Very compact FPGA implementation of the AES algorithm. In: Walter, C.D., Koç, Ç.K., Paar, C. (eds.) CHES 2003. LNCS, vol. 2779, pp. 319–333. Springer, Heidelberg (2003)
5. Feldhofer, M., Wolkerstorfer, J., Rijmen, V.: AES implementation on a grain of sand. In: IEE Proceedings on Information Security, IEE, vol. 152, pp. 13–20 (2005)
6. Rijmen, V.: Efficient implementation of the Rijndael S-box (2001), http://www.esat.kuleuven.ac.be/~rijmen/rijndael/sbox.pdf
7. Paar, C.: Efficient VLSI Architectures for Bit-Parallel Computation in Galois Fields. PhD thesis, Institute for Experimental Mathematics, University of Essen, Germany (1994)
8. Canright, D.: A very compact S-box for AES. In: Rao, J.R., Sunar, B. (eds.) CHES 2005. LNCS, vol. 3659, pp. 441–455. Springer, Heidelberg (2005)
9. Rebeiro, C., Selvakumar, D., Devi, A.: Bitslice implementation of AES. In: Pointcheval, D., Mu, Y., Chen, K. (eds.) CANS 2006. LNCS, vol. 4301, pp. 203–212. Springer, Heidelberg (2006)
10. Boyar, J., Peralta, R.: New logic minimization techniques with applications to cryptology. Cryptology ePrint Archive, Report 2009/191 (2009), http://eprint.iacr.org/
11. Käsper, E., Schwabe, P.: Faster and timing-attack resistant aes-gcm. Cryptology ePrint Archive, Report 2009/129 (2009), http://eprint.iacr.org/
12. NIST: Specification for the Advanced Encryption Standard (AES), FIPS PUB 197 (2001)
13. Bernstein, D.J., Schwabe, P.: New aes software speed records. In: Chowdhury, D.R., Rijmen, V., Das, A. (eds.) INDOCRYPT 2008. LNCS, vol. 5365, pp. 322–336. Springer, Heidelberg (2008)

14. Daemen, J., Rijmen, V.: AES proposal: Rijndael (1999),
    http://csrc.nist.gov/archive/aes/rijndael/Rijndael-ammended.pdf
15. IBM: Introduction to the Cell Broadband Engine (2005),
    http://www-01.ibm.com/chips/techlib/techlib.nsf/techdocs/
    D21E662845B95D4F872570AB0055404D
16. Canright, D.: A very compact Rijndael S-box. Technical Report NPS-MA-05-001,
    Naval Postgraduate School (2005)
17. Osvik, D.A.: Speeding up Serpent. In: AES Candidate Conference, pp. 317–329
    (2000)
18. Intel: Advanced encryption standard (AES) instructions set, rev. 2 (2009),
    http://software.intel.com/en-us/articles/
    advanced-encryption-standard-aes-instructions-set/

Optimization Strategies for Hardware-Based Cofactorization

Daniel Loebenberger[1] and Jens Putzka[2]

[1] b-it
D-53113 Bonn
daniel@bit.uni-bonn.de
http://www.b-it-center.de
[2] MPI für Mathematik
D-53111 Bonn
putzka@mpim-bonn.mpg.de
http://www.mpim-bonn.mpg.de

Abstract. We use the specific structure of the inputs to the cofactorization step in the general number field sieve (GNFS) in order to optimize the runtime for the cofactorization step on a hardware cluster. An optimal distribution of bitlength-specific ECM modules is proposed and compared to existing ones. With our optimizations we obtain a speedup between 17% and 33% of the cofactorization step of the GNFS when compared to the runtime of an unoptimized cluster.

Keywords: General Number Field Sieve (GNFS), Elliptic Curve Method (ECM), hardware cluster, cofactorization step.

1 Introduction

Factoring natural numbers using the elliptic curve method (ECM) is based on the seminal work of Hendrik Lenstra (Lenstra 1987), which is a natural adaption of Pollard's $(p-1)$-method (Pollard 1974) to elliptic curves. In recent implementations of the general number field sieve (GNFS), the ECM is used to factor intermediate sieving results (this is the so called cofactorization step). For example in the record factorization of Franke & Kleinjung (2005) the sieving step produced intermediate numbers of length up to 128 bits. Adapting this to the factorization problem of the number RSA-768 (RSA Laboratories 2007) results in the task of factoring roughly $2 \cdot 10^{12}$ numbers of length up to 140 bit using the ECM.

Since cofactorization is a costly part of the GNFS, it is natural to think about highly specialized hardware realizations of this step, to improve the performance of the GNFS considerably. In particular, since the task consists of many very similar steps, a realization as a hardware *cluster* is suitable. On such a cluster one has many computational units running in parallel that are able to process inputs up to a certain bitlength. The question remains how many of those bitlength-specific modules should be implemented, regardless of the concrete implementation of the corresponding ECM modules. A straightforward approach

M.J. Jacobson Jr., V. Rijmen, and R. Safavi-Naini (Eds.): SAC 2009, LNCS 5867, pp. 170–181, 2009.

would be to construct only modules capable of factoring inputs of any size from the GNFS. It is clear, however, that this approach is a great waste of logical resources and that a detailed study of the bitlength-structure of the inputs to the cofactorization step results in much better performance than the naïve approach. Furthermore we quantify the gain we achieve using our optimized construction and generalize our result to arbitrary clusters.

2   The General Number Field Sieve

In this section we give a brief overview of the GNFS in the version which was used by Franke et al. in their record factorization of RSA-640. The GNFS is asymptotically the best known factorization algorithm for large integers. For a more detailed explanation, see for example Lenstra & Lenstra (1993). In this section we will always consider pairs of object, which are indexed by the variable $i \in \{1,2\}$.

Polynomial Selection: Find good polynomials $F_i(X, Y)$ (see Kleinjung (2006)).

Sieving: Choose two bounds L_i and two bounds B_i. The task is to find many coprime pairs of integers (a, b) with $b > 0$ such that both $F_i(a, b)$ are L_i-smooth. This means that $F_i(a, b)$ decomposes into prime factors smaller than L_i. These pairs (a, b) are called relations. In general it is more than enough to find $\pi(L_1) + \pi(L_2)$ relations. In practice, however, one takes usually some more. We can write for each pair (a, b)

$$F_i(a, b) = R_i(a, b)S_i(a, b)$$

where $R_i(a, b)$ is B_i-rough, i.e. has no factor $< B_i$ and $S_i(a, b)$ is B_i-smooth.
Sieve: Approximation of $\log R_i(a, b)$. This can be done using a lattice sieve (Franke & Kleinjung 2006).
Find Candidates: Take $(R_1(a, b), R_2(a, b))$ for pairs (a, b) if the approximately computed $\log R_i(a, b)$ are below a given bound. These pairs are called candidates. Remove the remaining ones.
Trial Division: For all candidates find the $S_i(a, b)$ (using trial divisions) and calculate the $R_i(a, b)$.
Remove Candidate: If $R_i(a, b) > L_i$ do a fast compositeness test and remove the candidate if $R_i(a, b)$ is pseudoprime.
Apply Strategy: One can precompute a list with pairs of bitlengths which have the property that integers of that size can be factorized in the next step with high probability. For example pairs where both $R_i(a, b)$ are large in some sense can be removed (Kleinjung 2004).
Cofactorization: Find the factors of $R_i(a, b)$ using ECM or MPQS (see for example Cohen (1997)). In our case this should be done using a hardware cluster which uses ECM to find the factors.

Simplification: The relations define a sparse matrix. One now uses some elementary column/row transformations to reduce the size.

Linear Algebra: Solve the resulting system of linear equations.

Computing Square Roots: To be able to find the factor one needs to calculate a square root in a number field.

3   Modelling the Cluster System

Our goal is a model of a hardware cluster (e.g. a COPACOBANA, see Kumar *et al.* (2006), using Virtex4 XC4VSX35 FPGAs). In our specific example the cluster has 16 slots, each containing 8 FPGAs (in the following called chips). Each chip can run several ECM-processes in parallel depending on the size of the corresponding ECM-module. We assume that each chip can only be filled with ECM modules of a particular size. This requirement is from a theoretical point of view unnecessary, but for the concrete realization we have in mind we actually have to require this, since the device controlling all the chips is in our case not able to perform otherwise. Of course modules constructed for a given bitlength can also factor shorter integers. If one wants to factor a number using the cluster, the number is forwarded to a module suitable for its bitlength. The corresponding module then attempts to find a nontrivial factor of the input number. If this succeeds after a certain number of trials (each being a separate run of the ECM with a different elliptic curve), the factor is sent back to the controlling host computer, otherwise the number is discarded. If the factor that is sent back or the remaining cofactor is still composite, another factoring attempt is made. We assume for our estimates that the effort for these additional factorizations is negligible when compared to the first factorization attempt.

The first question we have to answer is the following: From an engineering point of view it is unrealistic to build arbitrary sized ECM modules. What is the smallest bitlength $g \in \mathbb{N}$ for which such a construction is practical? We call this g the *granularity* of the implementation. Of course one cannot give a general answer to this question. The answer heavily depends on the type of the chips one is using and the concrete implementation one has in mind. In our example, we will have $g = 17$ due to the design of the Virtex4 XC4VSX35 FPGAs.

Another question is: How can we get rid of modules for which the numbers of integers having that bitlength is very small? In other words if for a particular bitlength there are only very few numbers to factor, it would be better to factor such numbers using modules capable of factoring larger integers. This would ensure that we would not waste any resources on the cluster, resulting in a better runtime of the cofactorization step.

We describe now the model of the cluster: Let N denote the number of chips on the cluster, e.g. $N = 128$ in our concrete example, and let $\mathcal{D}$ denote the set of inputs to the cofactorization step with $M := \#\mathcal{D}$. For $d \in \mathcal{D}$ let $\text{len}(d)$ denote the bitlength of the number d, i.e. $\text{len}(d) := \lfloor \log_2(d) \rfloor + 1$. Each of the input numbers can be handled by specific modules suitable for their bitlength. The size for which the modules are designed is always a multiple of g. We denote by n_i the number of parallel ECM modules for an integer having $i \cdot g$ bits and by c_i the average runtime of such an integer on the corresponding chips. We are now going to model the classes the numbers may fall into. In general, if we are given an interval $\mathcal{I} := [x, y]$ with $x, y \in \mathbb{N}$ and $x \leq y$, a *partition* of $\mathcal{I}$ is a sequence $\mathcal{C} := (\mathcal{C}_0, \mathcal{C}_1, \ldots, \mathcal{C}_k) \in \mathbb{N}^k$ for some $k \in \mathbb{N}$, with $x = \mathcal{C}_0 < \mathcal{C}_1 < \cdots < \mathcal{C}_k = y$. We call k the *size* of the partition $\mathcal{C}$. The interval $(\mathcal{C}_{i-1}, \mathcal{C}_i]$ is called the *i-th subinterval* of $\mathcal{C}$. If now $\mathcal{C}^1$ and $\mathcal{C}^2$ are partitions of $\mathcal{I}$, we say that $\mathcal{C}^2$ is a *refinement* of $\mathcal{C}^1$ if for any $0 \leq i \leq k$ there is some j, such that $\mathcal{C}_i^1 = \mathcal{C}_j^2$. In other words that means that we have subdivided the subintervals of $\mathcal{C}^1$ into smaller pieces without changing already existing cuts and we write $\mathcal{C}^1 \preceq \mathcal{C}^2$. Conversely, $\mathcal{C}^1$ is called a *coarsening* of $\mathcal{C}^2$. For our purposes we only consider partitions $\mathcal{C}$ of the interval $\mathcal{I} = [x, y]$ where $x := \lfloor \min(\text{len}(d) \mid d \in \mathcal{D}) \rfloor_g$ and $y := \lceil \max(\text{len}(d) \mid d \in \mathcal{D}) \rceil_g$, where the notation $\lfloor . \rfloor_g$ ($\lceil . \rceil_g$) means that the rounding is done down to (up to) the next multiple of g. Additionally we require that for any $0 \leq i < \#\mathcal{C}$ the number $\mathcal{C}_i$ is a multiple of g. We will call such partitions *g-partitions* of the intervall induced by $\mathcal{D}$. In particular the finest partition we will consider is the g-partition $\mathcal{C}^f := (x, x + g, x + 2g, \ldots, y)$ and the possible partitions we may have at the end are always coarsenings of $\mathcal{C}^f$.

For the following, fix a data set $\mathcal{D}$ and define $K := \#\mathcal{C}^f - 1 = (y - x)/g$. Now given any $\mathcal{C} \preceq \mathcal{C}^f$ of size k, let $a_i(\mathcal{C}) \in \mathbb{N}$ be the number of occurrences in the i-th subinterval of $\mathcal{C}$, i.e. $a_i(\mathcal{C}) := \#\{d \in \mathcal{D} \mid \text{len}(d) \in (\mathcal{C}_{i-1}, \mathcal{C}_i]\}$. For later use we define the input distribution

$$\alpha(\mathcal{C}) := \left(\frac{a_1(\mathcal{C})}{M}, \ldots, \frac{a_k(\mathcal{C})}{M} \right) \in \mathbb{R}^k.$$

If we consider the ith subinterval of $\mathcal{C}$ the average cost of factoring such a number is $c_{\mathcal{C}_i/g}$. The space used for such a module is roughly $1/n_{\mathcal{C}_i/g}$. Thus the area-time product for class i is given by

$$\vartheta_i(\mathcal{C}) := \frac{c_{\mathcal{C}_i/g}}{n_{\mathcal{C}_i/g}}.$$

A *layout* of the cluster is given by an ordered partition $\ell \vdash_k N$ of the N chips into k summands, one for each class. Thus we have

$$\ell \vdash_k N :\Longleftrightarrow \ell = (\ell_1, \ldots, \ell_k) \in \{1, \ldots, N\}^k \wedge \sum_{1 \leq i \leq k} \ell_i = N,$$

with $\ell_i > 0$, implying $N \geq k$. That means we assume that the number of chips is always greater than the number of classes, which is also reasonable. Note that we have indeed two different notions of partitions here: First a partition of an interval and second an additive ordered partition of a natural number. This could of course be unified, but for our work it is preferable to have these two different notions, since for the former notion we emphasize on the variable number of subintervals while for the latter we assume a fixed number of summands.

Write $\mathcal{C}|_j$ for the restriction of $\mathcal{C}$ on its first j subintervals. The minimal runtime for $\mathcal{C}|_j$ is given by

$$\mu_{\mathcal{C}}(N, j) := \min_{\ell \vdash_j N} \max_{1 \leq i \leq j} \frac{\vartheta_i(\mathcal{C}|_j) \cdot a_i(\mathcal{C}|_j)}{\ell_i} \tag{1}$$

The value $\mu_{\mathcal{C}}(N, j)$ is indeed a time measurement, since c_i is given in seconds, n_i has unit $1/$ chip and ℓ_i has unit chip. We will use the following convention: If we write $\mu_{\mathcal{C}}(N)$ we actually mean $\mu_{\mathcal{C}}(N, \#\mathcal{C} - 1)$. Further we define

$$\tau(N) := \min_{\mathcal{C} \preceq \mathcal{C}^f} \mu_{\mathcal{C}}(N) \tag{2}$$

Equation (1) and (2) actually depend on the data set $\mathcal{D}$ and we write $\mu_{\mathcal{D}, \mathcal{C}}(N, j)$ and $\tau_{\mathcal{D}}(N)$, respectively, if there is more than one data set under consideration. In the following we will show how one can compute $\mu_{\mathcal{C}^f}(N)$ efficiently, namely with $\mathcal{O}(N \cdot K)$ arithmetic operations. Note that the imprecision of considering arithmetic operations only is in our case not a problem, since the size of the numbers is bounded from above by a constant.

We can compute Equation (1) easily using Bellman's dynamic programming. To do so, we need to handle two things:

1. The solutions for the boundaries have to be computed (i.e. for the case $j = 1$):

$$\mu_{\mathcal{C}}(N, 1) = \frac{\vartheta_1(\mathcal{C}|_1) \cdot a_1(\mathcal{C}|_1)}{N} \tag{3}$$

2. We need a recursion formula for $\mu_{\mathcal{C}}(N, j)$. Assume we know $\mu_{\mathcal{C}}(N', j-1)$ for all $N' < N$. Then we have

$$\mu_{\mathcal{C}}(N, j) = \min_{N' < N} \max \left(\mu_{\mathcal{C}}(N', j-1), \frac{\vartheta_j(\mathcal{C}|_j) \cdot a_j(\mathcal{C}|_j)}{N - N'} \right) \tag{4}$$

The function $\mu_{\mathcal{C}}(N, j)$ can thus be computed with $\mathcal{O}(N \cdot j)$ arithmetic operations.

Let us now compute the function $\tau(N)$. The total number of classes $\mathcal{C} \preceq \mathcal{C}^f$ is $2^K/4$. Since K will be small in all our examples of the GNFS, a straightforward algorithm would just compute $\mu_{\mathcal{C}}(N)$ for all $\mathcal{C} \preceq \mathcal{C}^f$ and select the classes with minimal runtime. Employing such an algorithm for the computation of $\tau(N)$ will use $\mathcal{O}(NK2^K)$ arithmetic operations.

We will now describe a greedy approach which will find in many cases the optimal classes using only $\mathcal{O}(K)$ evaluations of the function $\mu_{\mathcal{C}}(N)$ for various $\mathcal{C} \preceq \mathcal{C}^f$, i.e. compute $\tau(N)$ with $\mathcal{O}(N \cdot K^2)$ arithmetic operations: Let $\mathcal{C} := [\mathcal{C}_0, \mathcal{C}_1, \ldots, \mathcal{C}_k]$ be any partition of the interval $\mathcal{I} = [x, y]$.

For $p \in [1, K-1]$ denote by $\mathcal{C}^{(p)}$ the refinement of $\mathcal{C}$ at position $g \cdot p$. Our algorithm will work as follows: Starting from the partition (x, y), we successively refine (x, y) until the optimal partition is found. In particular if we are given in step r a partition $\mathcal{C}$, we compute $\mu_{\mathcal{C}^{(p)}}(N)$ for all p and take in the next round the partition $\mathcal{C}^{(p)}$ with the smallest runtime $\mu_{\mathcal{C}^{(p)}(N)}$. If there are two positions p_1, p_2 with the same minimal runtime, we select one of the partitions randomly for the next step. This approach is indeed greedy, since we take in every round the best subdivision. The algorithm terminates if for all p the value $\mu_{\mathcal{C}(p)}(N)$ is not strictly smaller than $\mu_{\mathcal{C}}(N)$. In this case the partition $\mathcal{C}$ is returned. Observe that this algorithm will in general *not* find the optimal classes, since we cannot guarantee that the algorithms terminates in a local minimum. In our experiments, however, this heuristic indeed computed $\tau(N)$ in all our examples.

In order to measure the advantage of our optimization, we compare the estimated runtime of the cluster using our construction with the runtime of a naïvely constructed cluster, i.e. a cluster only containing bitlength-specific modules for numbers having y bits. On such a cluster the runtime for a data set $\mathcal{D}$ of M numbers is bounded from below by the following expression:

$$\sigma_{\mathcal{D}}^-(N) := \frac{1}{N \cdot n_K} \sum_{1 \leq i \leq K} c_i a_i \tag{5}$$

and bounded from above by

$$\sigma_{\mathcal{D}}^+(N) := \frac{M c_K}{N n_K} \tag{6}$$

with $K := \#\mathcal{C}^f - 1$ as above. The first estimate is a bit optimistic since the runtime of a module does not only depend on the input but also on the arithmetic built into the module. Further the second estimate is too pessimistic, since a module running on smaller input numbers will also run faster on average.

We use the functions

$$\gamma_{\mathcal{D}}^-(N) := \frac{\sigma_{\mathcal{D}}^-(N) - \tau_{\mathcal{D}}(N)}{\sigma_{\mathcal{D}}^-(N)}$$

and

$$\gamma_{\mathcal{D}}^+(N) := \frac{\sigma_{\mathcal{D}}^+(N) - \tau_{\mathcal{D}}(N)}{\sigma_{\mathcal{D}}^+(N)}$$

as lower and upper bounds, respectively, to measure the runtime gain we achieve with our optimized cluster. This expression is exactly the runtime gain achieved by the optimization (having runtime $\tau_{\mathcal{D}}(N)$) in contrast to the naïvely constructed cluster (having runtime between $\sigma_{\mathcal{D}}^-(N)$ and $\sigma_{\mathcal{D}}^+(N)$).

4   Concrete Statistical Analyses

We will now perform a rigorous statistical analysis of six concrete runs of the GNFS up to the cofactorizations step for the number RSA-768 using Franke and Kleinjung's implementation, and study the function $\tau(N)$ for these particular inputs: Each data set $\mathcal{D}$ consists of many $(2 \cdot 10^8)$-rough composite numbers of bitlength between 58 and 160, each $\mathcal{D}$ being a specific output of the sieving step of the GNFS for different choices of a polynomial pair and the sieving region of the lattice siever. Following von zur Gathen *et al.* (2007), we estimate the

Table 1. Number of parallel ECM-modules per chip depending on the bitlength

Bitlength $17i$	17	34	51	68	85	102	119	136	153	170
Processes n_i	32	26	22	18	15	12	10	9	8	7

Table 2. Average runtime of the ECM on a Virtex4 XC4VSX35 FPGA

Bitlength $17i$	17	34	51	68	85
Cost c_i (in μs)	491.49125	673.9225	856.35375	1038.785	1221.21625

Bitlength $17i$	102	119	136	153	170
Cost c_i (in μs)	1403.6475	1586.07875	1768.51	1950.94125	2133.3725

Table 3. Relative frequencies of the input data

Bitlength	$0-68$	$69-85$	$86-102$	$103-119$	$120-136$	$137-153$
$\mathcal{D}_1$	0.0015	0.0553	0.4540	0.0886	0.2826	0.1181
$\mathcal{D}_2$	0.0007	0.0547	0.4493	0.0889	0.2823	0.1241
$\mathcal{D}_3$	0.0008	0.0540	0.4533	0.0881	0.2836	0.1203
$\mathcal{D}_4$	0.0009	0.0567	0.4440	0.0874	0.2902	0.1209
$\mathcal{D}_5$	0.0011	0.0518	0.4306	0.0875	0.2992	0.1299
$\mathcal{D}_6$	0.0009	0.0461	0.4340	0.0834	0.3031	0.1326
Mean	0.0010	0.0531	0.4442	0.0873	0.2902	0.1243
Stdev.	0.0003	0.0038	0.0099	0.0020	0.0091	0.0058

number of parallel ECM modules and the runtime on the Virtex4 XC4VSX35 FPGAs according to Table 1 and 2, respectively. In the implementation that was used only modules for $17i$ bit integers were build. Note that such a module will also be capable of factoring samller integers.

Let us have a look at the distribution $\alpha(\mathcal{C}^f)$ of the input data for the various data sets (see Table 3). Note the low standard deviation of the corresponding entries. In Figure 1 a histogram as well as the distribution on the classes $\mathcal{C}^f$ is given for data set $\mathcal{D}_1$.

We now employ our model to find an optimal layout for the cluster and compute the runtime gain we achieved with our optimization. Let the notation be as in Section 3. In the case of the COPACOBANA we will have $N = 8 \cdot 16 = 128$. There are 351306039 ordered partitions of the number 128 in not more than 6 parts. The total number of layouts of the cluster, including the choice of the classes is in our example 402858941.

After having computed the function $\tau_{\mathcal{D}}(128)$ for all data sets $\mathcal{D}$ we obtain for every set an optimal layout (consisting of the interval partition $\mathcal{C}$ and the distribution of chips ℓ). If we take the result of the optimization for data set $\mathcal{D}_1$, for example, we will have 47 modules for integers of up to 102 bit, 58 for integers up to 136 bit and 23 for the remaining integers (up to 153 bit). The size of the first class is in this case 102 bit, the size of the second one 34 bit and of the third class 17 bit. The results are summarized in Table 4 and 5.

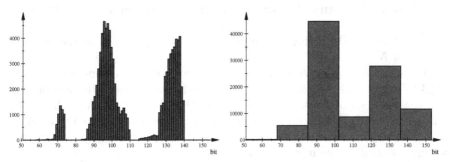

Fig. 1. Left: Histogram of data set $\mathcal{D}_1$. Right: Distribution onto specific modules.

Table 4. Optimal partitions for the data sets $\mathcal{D}_1$, $\mathcal{D}_2$ and $\mathcal{D}_3$

	$\mathcal{D}_1$	$\mathcal{D}_2$	$\mathcal{D}_3$
$(\mathcal{C}_{i+1} - \mathcal{C}_i)/g$	(3,2,1)	(1,2,2,1)	(3,1,1,1)
ℓ	(47, 58, 23)	(1, 46, 57, 24)	(48, 11, 45, 24)
$\tau_\mathcal{D}$ (μs)	124966.936	96137.13955	126309.5441
$\#\mathcal{D}$	98322	75013	99488
$\tau_\mathcal{D}/\#\mathcal{D}$	1.271	1.2816	1.2696

Table 5. Optimal partitions for the data sets $\mathcal{D}_4$, $\mathcal{D}_5$ and $\mathcal{D}_6$

	$\mathcal{D}_4$	$\mathcal{D}_5$	$\mathcal{D}_6$
$(\mathcal{C}_{i+1} - \mathcal{C}_i)/g$	(3,1,1,1)	(3,1,1,1)	(3,2,1)
ℓ	(47, 11, 46, 24)	(45, 11, 47, 25)	(44, 59, 25)
$\tau_\mathcal{D}$ (μs)	113592.0763	37653.16612	65015.11716
$\#\mathcal{D}$	90141	29719	50273
$\tau_\mathcal{D}/\#\mathcal{D}$	1.2602	1.267	1.2932

Table 6. Performance gain for data set $\mathcal{D}_1$ (in percent) of the optimized cluster

	$\mathcal{D}_1$	$\mathcal{D}_2$	$\mathcal{D}_3$	$\mathcal{D}_4$	$\mathcal{D}_5$	$\mathcal{D}_6$
$\gamma_\mathcal{D}^-$	17.47	16.97	17.66	18.38	18.4	16.88
$\gamma_\mathcal{D}^+$	33.29	32.73	33.36	33.86	33.5	32.12

In order to measure the advantage of our optimization, we use the estimates from Section 3. We have here at maximum 153 bit numbers and use the values in the tables above. The result of our optimization is shown in Table 6.

5   Generalizations to an Arbitrary Number of Clusters

Fix one data set $\mathcal{D}$. In this section we analyze the behaviour of the function $\gamma^-(N)$ for $N \to \infty$.

In practice a growing N would mean that we employ not only one COPA-COBANA, but a whole collection of these, running simultaneously, and optimize over the whole set of chips. We will now show that the runtime gain achieved by this collection of clusters converges to roughly 21% when compared to a collection of naïvely constructed clusters. It is clear that the actual gain however will strongly depend on the input data $\mathcal{D}$.

Now let's say we are going to build m clusters and we wish to optimize the number of bitlength specific ECM modules as above. The formulae in Section 3 are still valid, except that we will have $N = 128m$ chips in a collection of m clusters instead of $N = 128$ as above.

We wish to compute $\lim_{N \to \infty} \gamma^{\pm}(N)$. To do so, we first need to compute $\tau(N)$ for $N \to \infty$. Unfortunately, the dynamic programming approach used above is only useful if we consider fixed N, but does not tell us anything about the limit. In Figure 2 the value of $\gamma^-(N)$ is plotted for the case of $m \in \{1, \ldots, 100\}$ clusters using data set $\mathcal{D}_1$. Note that this observation follows also our intuition, since with an increasing number of clusters one cannot expect more runtime gain.

Assume we are given classes $\mathcal{C} \preceq \mathcal{C}^f$. Set $k := \#\mathcal{C} - 1$. In order to be able to compute the limit, we look at the problem of computing $\mu_{\mathcal{C}}(N)$ over the reals, i.e. we will have $\ell \in \mathbb{R}^k$. With this simplifications it is clear that the expression

$$\max_{1 \leq i \leq k} \frac{\vartheta_i(\mathcal{C}) \cdot a_i(\mathcal{C})}{\ell_i}$$

is minimal if and only if

$$\frac{\vartheta_i(\mathcal{C}) \cdot a_i(\mathcal{C})}{\ell_i} = \frac{\vartheta_j(\mathcal{C}) \cdot a_j(\mathcal{C})}{\ell_j} \text{ for all } i, j \in \{1, \ldots, k\}$$

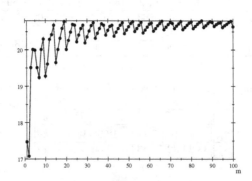

Fig. 2. Lower bound on the runtime gain for an increasing number m of clusters

Write $\vartheta_i'(\mathcal{C}) := \vartheta_i(\mathcal{C}) \cdot a_i(\mathcal{C})$. We end up in solving the following system of equations:

$$\ell_1 + \cdots + \ell_k = N$$
$$\vartheta_1'(\mathcal{C}) \cdot \ell_2 = \vartheta_2'(\mathcal{C}) \cdot \ell_1$$
$$\vdots \quad \vdots$$
$$\vartheta_1'(\mathcal{C}) \cdot \ell_k = \vartheta_k'(\mathcal{C}) \cdot \ell_1$$

This system of k equations is linear in the k unknowns $\ell_1, \ldots, \ell_k$, having the solution

$$\ell_i = \frac{\vartheta_i'(\mathcal{C})N}{\vartheta_1'(\mathcal{C}) + \cdots + \vartheta_k'(\mathcal{C})}$$

We could have used this approach also for our computation of $\mu_C(n)$ in Section 3. There we would have computed the approximate partition of N (being a vector of reals) and would then have rounded the results appropriately. To find the minimum we would have then to round 2^k times resulting in an algorithm that would have used $\mathcal{O}(k \cdot 2^k)$ arithmetic operations, which is of course preferable if k is small compared to N. Back to our question of computing the limit we have

$$\lim_{N \to \infty} \mu_C(N) = \lim_{N \to \infty} \frac{1}{N} \sum_{1 \leq i < \#\mathcal{C}} \vartheta_i'(\mathcal{C}) \quad \text{and} \quad \lim_{n \to \infty} \tau(N) = \min_{\mathcal{C} \preceq \mathcal{C}^f} \lim_{N \to \infty} \mu_C(N).$$

Furthermore

$$\lim_{N \to \infty} \sigma^-(N) = \lim_{N \to \infty} \frac{1}{N \cdot n_K} \sum_{1 \leq i \leq K} a_i \cdot c_i \quad \text{and} \quad \lim_{N \to \infty} \sigma^+(N) = \lim_{N \to \infty} \frac{M c_K}{N n_K}$$

Together

$$\lim_{N \to \infty} \gamma^-(N) = \min_{\mathcal{C} \preceq \mathcal{C}^f} 1 - \frac{n_K \sum_{1 \leq i < \#\mathcal{C}} \vartheta_i'(\mathcal{C})}{\sum_{1 \leq i \leq K} c_i \cdot a_i}$$

and

$$\lim_{N \to \infty} \gamma^+(N) = \min_{\mathcal{C} \preceq \mathcal{C}^f} 1 - \frac{n_K \sum_{1 \leq i < \#\mathcal{C}} \vartheta_i'(\mathcal{C})}{M c_K}.$$

Table 7 shows the results for our six test sets. We observe again that the corresponding values for the different data sets are very similar. Thus it seems that only the distribution of the inputs is crucial for the outcome of the optimization.

Table 7. Bounds on the limit of the runtime gain (in percent) for the various data sets

	$\mathcal{D}_1$	$\mathcal{D}_2$	$\mathcal{D}_3$	$\mathcal{D}_4$	$\mathcal{D}_5$	$\mathcal{D}_6$
$\lim_{N\to\infty} \gamma_{\mathcal{D}}^-$	20.81	20.58	20.70	20.56	20.00	19.81
$\lim_{N\to\infty} \gamma_{\mathcal{D}}^+$	35.99	35.66	35.82	35.63	34.80	34.51

6   Conclusion

We have described a mathematical model of a hardware cluster like the COPA-COBANA. Using this model we were able to compute the optimal distribution of bitlength specific modules on such a cluster efficiently, independent of which concrete ECM implementation was used. For our optimization it is necessary to have an estimate of the expected input distribution. This is in the case of the GNFS a nontrivial question (given some fixed parameter set), but it seems that the outputs of the GNFS always follow a certain distribution. To study this distribution in general is a challenging task and requires a deep understanding of the number theoretical properties of the inputs for the cofactorization step. Results in this direction are reserved for a forthcoming publication. The methods that were used are standard and were well studied in the 1960th and the 1970th. Nonetheless our optimization gives a speedup between 17% and 33% for the cofactorization step of the GNFS. As far as we know such a mathematical optimization was never done before for a hardware cluster like the COPACOBANA. Additionally our results are applicable for any scalable problem, when one wants to implement it efficiently on a dedicated hardware cluster.

Acknowledgements

Both authors were funded by the b-it foundation, the state of Northrhine-Westfalia and the German Federal Office for Information Security (BSI). The second author was additionally funded by the MPI and the Hausdorff-Center for Mathematics in Bonn. We want to express our thanks to Thorsten Kleinjung for helpful discussions on the output of the sieving step. Additional thanks go to Jérémie Detrey.

References

1. Bellman, R.: Dynamic Programming. Princeton University Text (1957)
2. Cohen, H.: A course in computational algebraic number theory. Springer, Berlin (1997)
3. Franke, J., Kleinjung, T.: RSA 640 (2005), http://www.crypto-world.com/announcements/rsa640.txt
4. Franke, J., Kleinjung, T.: Continued Fractions and Lattice Sieving (Unpublished) (2006), http://www.math.uni-bonn.de/people/thor/confrac.ps
5. von zur Gathen, J., Güneysu, T., Kargl, A., Loebenberger, D., Paar, C., Putzka, J.: Faktorisierung großer Zahlen: Hardware für Elliptische Kurven Faktorisierung. Technical report, HGI Bochum, b-it Bonn & Siemens AG München (2007)

6. Kleinjung, T.: Cofactorisation Strategies for the Number Field Sieve and an Estimate for the Sieving Step for Factoring 1024-bit Integers (Unpublished) (2004), http://www.math.uni-bonn.de/people/thor/cof.ps

7. Kleinjung, T.: On Polynomial Selection for the General Number Field Sieve. Mathematics of Computation 75(256), 2037–2047 (2006), http://dx.doi.org/10.1090/S0025-5718-06-01870-9

8. Kumar, S., Paar, C., Pelzl, J., Pfeiffer, G., Schimmler, M.: Breaking ciphers with COPACOBANA –A Cost-Optimized Parallel Code Breaker. In: Goubin, L., Matsui, M. (eds.) CHES 2006. LNCS, vol. 4249, pp. 101–118. Springer, Heidelberg (2006), http://dx.doi.org/10.1007/11894063_9

9. Lenstra, A.K., Lenstra Jr., H.W. (eds.): The development of the number field sieve. Lecture Notes in Mathematics, vol. 1554. Springer, Berlin (1993)

10. Lenstra Jr., H.W.: Factoring integers with elliptic curves. Annals of Mathematics 126, 649–673 (1987)

11. Pollard, J.M.: Theorems on factorization and primality testing. Proceedings of the Cambridge Philosophical Society 76, 521–528 (1974)

12. RSA Laboratories. The RSA Challenge Numbers (2007)

More on the Security of Linear RFID Authentication Protocols

Matthias Krause and Dirk Stegemann

Theoretical Computer Science
University of Mannheim
Mannheim, Germany

Abstract. The limited computational resources available in RFID tags implied an intensive search for lightweight authentication protocols in the last years. The most promising suggestions were those of the HB-familiy (HB$^+$, HB$^\#$, TrustedHB, ...) initially introduced by Juels and Weis, which are provably secure (via reduction to the Learning Parity with Noise (LPN) problem) against passive and some kinds of active attacks. Their main drawbacks are large amounts of communicated bits and the fact that all known HB-type protocols have been proven to be insecure with respect to certain types of active attacks. As a possible alternative, authentication protocols based on choosing random elements from L secret linear n-dimensional subspaces of $GF(2)^{n+k}$ (so called CKK-protocols) were introduced by Cichoń, Klonowski, and Kutyłowski. These protocols are special cases of (linear) (n, k, L)-protocols which we investigate in this paper. We present several active and passive attacks against (n, k, L)-protocols and propose $(n, k, L)^{++}$-protocols which we can prove to be secure against certain types of active attacks. We obtain some evidence that the security of (n, k, L)-protocols can be reduced to the hardness of the *learning unions of linear subspaces* (LULS) problem. We then present a learning algorithm for LULS based on solving overdefined systems of degree L in Ln variables. Under the hardness assumption that LULS-problems cannot be solved significantly faster, linear (n, k, L)-protocols (with properly chosen n, k, L) could be interesting for practical applications.

Keywords: Lightweight Cryptography, RFID Authentication, Algebraic Attacks, HB$^+$, CKK, CKK^2.

1 Introduction

In lightweight cryptography one tries to solve the problem of determining the minimal amount of computational resources which have to be invested for reaching certain security goals. This problem implies a lot of interesting and nontrivial theoretical questions. Since weak computational devices (e.g., mobile devices, RFIDs) are used in practice to a rapidly growing extent, results in lightweight cryptography are highly desired also from a practical point of view.

M.J. Jacobson Jr., V. Rijmen, and R. Safavi-Naini (Eds.): SAC 2009, LNCS 5867, pp. 182–196, 2009.

RFID (radio frequency identification) tags are small devices that are equipped with only little memory and computational power. Their main application is the identification of objects. In order to prevent cloning and tracing attacks and to preserve the tagged object's privacy, RFID tags should reveal their identities only to legitimate readers. Since most practically relevant RFID tags are too weak to execute standard authentication protocols, alternative measures are necessary. Besides technical approaches based on blocking or disturbing the communication, lightweight authentication protocols and corresponding security models are intensively discussed (see, e.g., [13,15]).

One of the most promising proposals was the HB$^+$ protocol due to Juels and Weis [14], which is provable secure (via reduction to the learning parity with noise (LPN) problem) with respect to passive and some kinds of active attacks. A severe drawback of the protocol is that presumably secure parameter combinations imply large amounts of transmitted data. Together with the small available bandwidth in RFID communication, this may add up to authentication times that are unacceptable for many applications. Another disadvantage is that HB$^+$ and all its variants suggested so far have been broken by man-in-the-middle (MITM) attacks. Particularly, the HB$^+$-protocol was broken by Gilbert, Robshaw and Sibert in [12], the HB$^{\#}$-protocol introduced by Gilbert, Robshaw and Seurin in [11] was recently broken by Ouafi, Overbeck and Vaudenay in [16], and Trusted-HB introduced by Bringer and Chabanne in [3] was broken by Frumkin and Shamir in [10].

As a possible alternative to HB-type protocols, another class of lightweight authentication protocols (so called CKK-protocols) were introduced by Cichoń, Klonowski, and Kutyłowski [4]. These protocols can be generalized to linear (n, k, L)-protocols, in which the secret key (the identification information in the RFID tag) consists of the specification of L n-dimensional linear subspaces $V_1, \ldots, V_L$ of GF$(2)^{n+k}$, while the identification is performed by collaboratively generating an element $v \in V_l$ for a random $l \in \{1, \ldots, L\}$. In [4], the CKK^2-protocol, a special linear $(n, k, 2)$-protocol, and the CKK$^{\sigma, L}$-protocol, a special linear (n, k, L)-protocol, were suggested for practical application.

Compared with HB-type protocols, the advantages of (n, k, L)-protocols are that fewer bits have to be communicated, computational effort and memory requirements are lower on the prover's side (essentially, the prover has to generate random elements from L different n-dimensional subspaces of GF$(2)^{n+k}$), and that (n, k, L)-protocols seem to be more resistant against active attacks. The drawback is that we can not yet prove the security of (n, k, L)-protocols by reduction to a well-established problem like the LPN-problem. However, in this paper we show that, similarly to HB-type protocols, the security of (n, k, L)-protocols can be related to the hardness of a certain learning problem, in this case the *Learning Unions of L linear subspaces* (LULS) problem.

Our strategy for designing a lightweight authentication protocol is the same as in the context of HB-type protocols and consists of the following two steps.

1. Define an appropriate lightweight symmetric encryption function $E : X \times K \longrightarrow Y$, the basis operation, which guarantees that that the basic

E-protocol is secure against a passive adversary. Hereby, X denotes an appropriate input-, K an appropriate key-, and Y an appropriate output space.

2. Define a protocol structure P over E such that the security of P with respect to active adversaries can be reduced to the security of the basic E-protocol against a passive adversary.

The basic E-protocol is defined as follows: Alice and Bob share a secret key $k \in K$. In one round, the verifier Alice sends *hello* to the prover Bob. Receiving *hello*, Bob chooses a random element $x \in X$, which is distributed according to a publicly known probability distribution $\Pr_B$, and sends $E_k(x)$ back to Alice. After a predefined number of rounds, Alice decides about accepting or rejecting by applying a verification operation to the messages sent by Bob. The definition of the verification operation depends on the definition of E. A passive adversary has only passive access to the insecure channel between Alice and Bob, i.e., she has to reach her goal on the basis of a set of observations $F_k(x_1), \ldots, F_k(x_m)$, where for all $i = 1, \ldots, m$, x_i is randomly and independently choosen according to $\Pr_B$.

Note that for HB-type protocols, $K = \mathrm{GF}(2)^n$, $X = Y = \mathrm{GF}(2)^n \times \mathrm{GF}(2)$, $y = \mathrm{GF}(2)$, and the basis operation is defined by

$$E((x, \nu), k) = (x, y) \ ,$$

where $y = x \cdot k \oplus \nu$. Bob chooses x with respect to the uniform distribution and sets the noise bit ν to one with probability $p < 0.5$. Alice accepts if the number of rounds in which $y_i = x_i \cdot k$ is satisfied exceeds a certain threshold.

Obviously, basic E-protocols are vulnerable to replay attacks. In both cases, HB-type- and linear protocols, the basic E-protocol is also vulnerable to active key recovery attacks (see [14] and the attack described in subsection 2.4, respectively). Consequently, solving challenge (2.) is an important task, which could not be done in a satisfactory way so far in the case of HB-type protocols.

Our results and the outline of this paper are as follows. In Subsect. 2.1 we define the basis operation of linear protocols and specify the adversary models. In Subsect. 2.2 we take a look at CKK-protocols [4], the first type of linear protocols occuring in the literature. We present a fast passive (polynomial time) attack against the CKK^2-protocol which allows to recover the secret key for the proposed parameters $(n, k) = (128, 30)$ in less than a second on a standard PC, while an earlier (exponential time) attack on CKK^2 published in [7] requires a couple of hours on comparable hardware.

In Subsect. 2.3 we describe special active key recovery attacks against linear protocols, so called equality attacks, and show that the basic linear (n, k, L)-protocol and the linear $(n, k, L)^+$-protocol (which is based on the same design principle as HB$^+$) are vulnerable to these attacks.

In Subsect. 2.4 we introduce $(n, k, L)^{++}$-protocols and prove their security against equality attacks.

In Subsect. 2.5 we list some generic attacks against linear protocols. Moreover, we introduce the *Learning Unions of L linear subspaces* (LULS) problem. The complexity of the LULS-problem characterizes the security of linear protocols with respect to passive adversaries. We give a generic exponential time algorithm to solve this problem, and we show that active adversaries that are able to efficiently solve the LULS-problem can break the $(n, k, L)^+$-protocol.

In Sect. 3 we present a nontrivial learning algorithm for the LULS-problem. We outline the algorithm in all details for the case $L = 2$ and describe how the ideas can be generalized to the case $L > 2$. The algorithm is based on generating and solving $\frac{k}{s}$ special overdefined systems of degree-L equations over $GF(2^s)$ for appropriate $s \leq k$. Our hardness assumption is that the running time of this learning algorithm characterizes the complexity of the LULS-problem and the complexity of actively attacking $(n, k, L)^{++}$-protocols.

In Sect. 4 we discuss some aspects of the practical use of $(n, k, L)^{++}$-protocols. General (n, k, L)-protocols have a huge keylength of $L \cdot n \cdot n + k$. One idea could be to use $\mathsf{CKK}^{\sigma, L}$-protocols (see [4]), a special $(n, 1, L)$-protocol which is still unbroken. Other ideas for reducing the keylength in similar cases were discussed in the literature, e.g., using keys defined by Toeplitz matrices instead of random matrices [11], or defined by special Toeplitz matrices generated by Linear Feedback Shift Registers (LFSRs) [3]. The security analysis of the corresponding types of special (n, k, L)-protocols remains a matter of further research.

We have experimentally confirmed the correctness and efficiency of our attacks and algorithms with the computer algebra system Magma [2].

2 Linear (n, k, L)-Protocols

2.1 The Basis Operation and the Adversary Models

In a linear (n, k, L)-protocol, Alice (the verifier, e.g., an RFID reader) and Bob (the prover, e.g., an RFID tag) share a common secret information (the tag's ID) from a certain keyspace. As usual, we assume that the secret key is hardwired in the RFID tag, while Alice has legal access to a database containing Bob's secret information.

We define now the basis operation of linear (n, k, L)-protocols and denote for a positive integer N the set $\{1, \ldots, N\}$ by $[N]$.

The secret keys of the protocols consist of the specifications of L n-dimensional injective linear functions $F_1, \ldots, F_L : GF(2)^n \longrightarrow GF(2)^{n+k}$. The inputs are pairs (x, l), where $x \in GF(2)^n$ and $l \in [L]$.

Let us denote by $V_1, \ldots, V_L$ the n-dimensional linear subspaces of $GF(2)^{n+k}$ corresponding to the images of $F_1, \ldots, F_L$, respectively.

In the basic linear protocol, Alice accepts a message $w \in GF(2)^{n+k}$ coming from Bob if $w \in V_l$ for some $l \in [L]$.

We analyze the security of (n, k, L)-protocols with respect to passive and active adversaries. A passive adversary is able to read the messages exchanged by Alice and Bob. His aim is (partial) key recovery, i.e., to try to compute nontrivial information about the secret key from a set of messages produced by the honest parties Alice and Bob.

An active adversary has the additional abilities

- To corrupt or to replace messages sent from Alice to Bob,
- To corrupt or to replace messages sent from Bob to Alice,
- To retrieve the information whether a (possibly corrupted) transcript has been accepted or rejected by Alice.

We assume that neither of the adversaries is able to read nor modify the keybits nor the inner state bits nor the private random bits of Alice or Bob.

2.2   The CKK-Protocols

The protocols CKK^1, CKK^2 and $\mathsf{CKK}^{\sigma,L}$ suggested by Cichoń, Klonowski and Kutyłowski in [4] are restricted types of (n, k, L)-protocols.

The protocol $\mathsf{CKK}^{\sigma,L}$ is an (n, k, L)-protocol with the restriction $F_l(u) = \sigma^l(u\|f(u))$ for all $l \in [L]$, where σ denotes a secret permutation $\sigma \in S_{n+k}$ and f a secret linear function $f : \mathrm{GF}(2)^n \longrightarrow \mathrm{GF}(2)^k$. Hence, the secret keys have the form (f, σ).

The protocol CKK^2 is an $(n + k, k, 2)$-protocol with the additional properties that $F_1(u, a) = (u, f(u), a)$ and $F_2(u, a) = (u, a, f(u))$ for all $u \in \mathrm{GF}(2)^n$ and $a \in \mathrm{GF}(2)^k$, where f denotes a secret linear function $f : \mathrm{GF}(2)^n \longrightarrow \mathrm{GF}(2)^k$.

CKK^2 and $\mathsf{CKK}^{\sigma,L}$ protocols were suggested for practical application in [4], with the parameters $n = 128$ and $k = 30$.

So far, the only nontrivial cryptanalytic result concerning linear (n, k, L)-protocols is due to Gołebiewski, Majcher and Zagórski [7]. They present an attack against the CKK^2-protocol, which cannot be applied to the general case. Its running time is proportional to $\sum_{s=0}^{k-1} \binom{n}{s}$, i.e., of order $n^{\Theta(k)}$.

We now describe a very fast attack against the CKK^2-protocol with parameters (n, k) whose running time is dominated by the effort required for inverting k $(n \times n)$-matrices.

Let $f : \mathrm{GF}(2)^n \longrightarrow \mathrm{GF}(2)^k$ denote the secret key, recall that

$$V_1 = \{(v, f(v), a),\ v \in \mathrm{GF}(2)^n, a \in \mathrm{GF}(2)^k\}\ ,$$
$$V_2 = \{(v, a, f(v)),\ v \in \mathrm{GF}(2)^n, a \in \mathrm{GF}(2)^k\}\ .$$

Let the functions $f^1, \dots, f^k : \mathrm{GF}(2)^n \longrightarrow \mathrm{GF}(2)$ denote the component functions of the secret function f, i.e., $f(v) = (f^1(v), \dots, f^k(v))$ for all $v \in \mathrm{GF}(2)^n$. The attack is based on the simple fact that if an observation (v, a, b) satisfies $a_r = b_r$ for some $r \in [k]$, which is true with probability $1/2$, then we know that $f^r(v) = a_r = b_r$.

The attack works as follows.

1. Let $e_1, \dots, e_n$ denote the standard basis of $\mathrm{GF}(2)^n$.
2. **FOR** $r \in [k]$
    2.1 Consider a set of messages produced by Bob and extract from it a set $O_r = ((v_{r,1}, a_{r,1}, b_{r,1}), \dots, (v_{r,n}, a_{r,n}, b_{r,n}))$ such that $v_{r,1}, \dots, v_{r,n}$ form a basis of $\mathrm{GF}(2)^n$ and $a_{r,i}(r) = b_{r,i}(r) = f^r(v_{r,i})$ for all $i \in [n]$.
    2.2 Derive $f^r(e_1), \dots, f^r(e_n)$ from O_r.

Table 1. Performance of the passive attack on CKK^2

(n,k)	approx. number of observations	approx. attack time
$(128, 30)$	311	0.3 s
$(1024, 256)$	2197	179 s

The correctness of the attack follows straightforwardly from the definitions. The expected number of messages needed for constructing O_r can be estimated based on the following experiment.

1. Set $B := \emptyset$.
2. **REPEAT**
   2.1 Choose a random $v \in \mathrm{GF}(2)^n$ (w.r.t. the uniform distribution).
   2.2 $V := V \cup \{v\}$.
3. **UNTIL** V is a generating system of $\mathrm{GF}(2)^n$.

Let $p(n)$ denote the probability that the experiment stops after n iterations (i.e., V is a basis of $\mathrm{GF}(2)^n$), and $E(n)$ denote the expected number of iterations of the experiment. It is known that $p(n) \approx 0.2887$ and $E(n) \approx n + 1.6067$ (see, e.g., [7]). Hence, an estimate for the expected number of messages for constructing O_r is $2 \cdot E(n) \approx 2n + 3.2134$. For the parameter choices proposed for practical applications, the attack is very efficient already on standard PC hardware (Magma V2.15-9 [2] on a 3.4 GHz Intel Pentium IV with 4 GB RAM), see Table 1.

2.3 Basic Protocol Types and Equality Attacks

In the basic linear protocol, Alice starts the communication by sending some signal triggering Bob to compute a proof w of his identity. In particular, Bob computes $w = F_l(u)$ for randomly (independently and uniformly) chosen $l \in [L]$ and $u \in \mathrm{GF}(2)^n$. Alice accepts a proof $\tilde{w}$ if there is an $l \in [L]$ such that $\tilde{w} \in V_l$ (see Fig. 1).

Obviously, this protocol is vulnerable to replay attacks, since an adversary can store a number of proofs and then impersonate Bob by presenting these proofs to Alice.

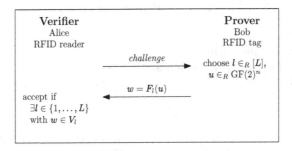

Fig. 1. Basic Communication Mode

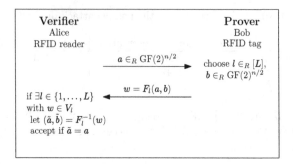

Fig. 2. $(n, k, L)^+$ Communication Mode

Moreover, an active adversary can successfully recover the key as follows.

1. Collect a set of messages $O = \{v^1, \ldots, v^s\}$ sent by Bob. The parameter s should be chosen in such a way that O contains a basis for V_l for all $l \in [L]$ with high probability (This can be achieved for $s \in \Theta(L \cdot E(n)) = \Theta(Ln)$, see Sect. 2.2.)

2. Construct an $s \times s$-matrix M over $\{0, 1\}$, where $M_{i,j} = 1$ iff Alice accepts $v^i \oplus v^j$.

Note that if v^i and v^j belong to the same subspace V_l, the probability for $M_{i,j} = 1$ is one. If $\{v^i, v^j\} \nsubseteq V_l$ for all $l \in [L]$ then the probability that $M_{i,j} = 1$ equals the probability that $v_i \oplus v_j \in \bigcup_{l=1}^{L} V_l$, which is at most $(L - 2)2^{-k}$. Hence, it is possible to efficiently compute specifications of $V_1, \ldots, V_L$ and to impersonate Bob by replying with $w \in V_l$ for arbitrary $l \in [L]$.

To prevent this kind of attack we consider the following distributed communication mode, which, analogously to the HB$^+$-protocols, defines $(n, k, L)^+$-protocols. Alice starts by sending a random $a \in GF(2)^{n/2}$ to Bob. Bob chooses random values $b \in GF(2)^{n/2}$ and $l \in [L]$ and sends $w = F_l(a, b)$ to Alice. Alice accepts $w \in GF(2)^{n+k}$ if there is some $l \in [L]$ with $w \in V_l$ and the prefix of length $n/2$ of $F_l^{-1}(w)$ equals a (see Fig. 2).

However, also $(n, k, L)^+$-protocols can be broken by an MITM attack:

1. Fix $a_1 \neq 0$ in $GF(2)^{n/2}$.
2. Send a_1 to Bob and receive $w_1 \in V_l$ for some $l \in [L]$.
3. **FOR** $r = 2, \ldots, s$
   3.1 **REPEAT**
       3.1.1 Catch a from Alice.
       3.1.2 Send $a' := a \oplus a_1$ to Bob and receive w'.
       UNTIL Alice accepts $w' \oplus w_1$ (which happens with probability at least $1/L$).
   3.2 Define $a_r := a'$ and $w_r := w'$.

The parameter s is chosen such that $\{w_1, \ldots, w_s\}$ contains a basis of V_l with high probability (see Sect. 2.2). This procedure will be repeated until specifications of $V_1, \ldots, V_L$ have been computed.

In the next subsection we propose linear $(n, k, L)^{++}$-protocols, a slightly modified version of $(n, k, L)^{+}$-protocols, and show that they are secure against a certain type of MITM-attack.

2.4 Linear $(n, k, L)^{++}$-Protocols and Provable Security against MITM-Attacks

The parameters n, k, L as well as V_l, F_l for $l \in [L]$ are defined as above. Let $n = 2N$. The $(n, k, L)^{++}$-protocol works similarly to the $(n, k, L)^{+}$-protocol, but uses an additional publicly known invertible function $f : \mathrm{GF}(2)^n \longrightarrow \mathrm{GF}(2)^n$, which we call connection function (see Fig. 3).

1. Alice chooses a random $a \in \mathrm{GF}(2)^N$, $a \neq \mathbf{0}$, moves to the inner state a and sends a to Bob.
2. Bob chooses random values $b \in \mathrm{GF}(2)^N$ and $l \in [L]$ and sends $w = F_l(f(a, b))$ back to Alice.
3. Alice accepts a message $w \in \mathrm{GF}(2)^n$ in inner state a if
   - $w \neq \mathbf{0}$, and
   - $\exists l \in [L]$ such that $w \in V_l$, and
   - $f^{-1}(F_l^{-1}(w))$ has the form (a, b) for some $b \in \mathrm{GF}(2)^N$.

Note that choosing f to be the identity yields the $(n, k, L)^{+}$-protocol.

We construct now a connection function f which yields provable security of $(n, k, L)^{++}$-protocols with respect to a certain type of MITM-attack which we call (x, y)-equality attack.

The aim of an (x, y)-equality attacker Eve is to generate two messages $w \neq w' \in \mathrm{GF}(2)^{n+k}$ and to efficiently test by MITM-access to the protocol if w and $w \oplus w'$ belong to the same linear subspace V_l for some $l \in [L]$. As described above, such an attack can be used to efficiently compute specifications of the subspaces $V_1, \ldots, V_L$.

Eve works in three phases:

1. Send a message $y \in \mathrm{GF}(2)^N$ to Bob and receive $w' = F_l(f(y, b'))$.
2. Observe a challenge $a \in \mathrm{GF}(2)^N$ sent by Alice.

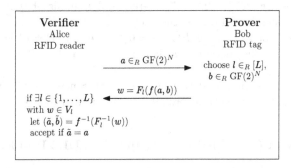

Fig. 3. $(n, k, L)^{++}$ Communication Mode

3. Compute a value $x = x(y, w', a) \in \mathrm{GF}(2)^N$, send it to Bob, receive the message $w = F_r(f(x, b))$ and send $w \oplus w'$ to Alice.

The success probability of the attack is given by the probability that Alice accepts $w \oplus w'$ given that $l = r$.

Note that if f is GF(2)-linear (as in the $(n, k, L)^+$-protocol), then setting $x = a \oplus y$ yields an attack with success probability one.

We define now a connection function which yields provable security against (x, y)-equality attacks. In the following we identify $\{0, 1\}^N$ with the finite field $K = \mathbf{F}_{2^N}$ and denote by $+, \cdot$ the addition and multiplication in K. Let the function value $f(a, b)$ for all $a, b \in K$ be defined by

$$f(a, b) = (ab, ab^3) \ .$$

Thus, Alice accepts a message w with $F_l^{-1}(w) = (u, v) \in K^2$ in inner state $a \in K^*$ if $(a^{-1}u)^3 = a^{-1}v$, which is equivalent to $u^3 = a^2 v$.

Theorem 1. *The success probability of an (x, y)-equality attacker against the $(n, k, L)^{++}$-protocol is at most $\frac{3}{2^N - 1}$.*

Proof. For given $y, a \in K^*$, Eve has to choose an element $x \in K^*$ such that $w + w' = (u, v) \in K \times K$ will be accepted by Alice in inner state a, where $w = F_l(x, b)$ and $w' = F_l(y, b')$ for some $l \in [L]$, and $b, b' \in K^*$. Note that Eve has no information about b, b', and that $u = xb + yb'$ and $v = xb^3 + yb'^3$.

Consequently, Eve's choice for the value x has to satisfy

$$(xb + yb')^3 = a^2(xb^3 + yb'^3) \ .$$

This is equivalent to

$$(x + yc)^3 = a^2(x + yc^3) \ ,$$

where $c = b'(b^{-1})$, which is equivalent to $P(x, c) = 0$, where the polynomial $P(x, d)$ is for all $d \in K^*$ defined as

$$P(x, d) = x^3 + (yd)x^2 + (y^2 d^2 + a^2)x + d^3(y^3 + y^2 a^2) \ .$$

Note that there are $|K^*| = 2^N - 1$ different polynomials of type $P(x, d)$ with respect to the variable x (Look at the coefficient yd of x^2).

For all $x \in K^*$ let $P(x) = \{d, P(x, d) = 0\}$. Note that $P(x, d)$ is a polynomial of degree 3 also in the unknown d. This implies that for all $x \in K^*$ it holds $|P(x)| \le 3$.

Eve has to choose an x that satisfies $c \in P(x)$. Since she does not have any information about c, her success probability is bounded by $\frac{3}{2^N - 1}$.   □

2.5   Security of (n, k, L)-Protocols and the LULS-Problem

There are several exhaustive search strategies for computing specifications of the secret subspaces $V_1, \ldots, V_L$, see, e.g., the search-for-a-basis heuristic described in Appendix A. The parameters (n, k) should be chosen in such a way that these attacks become infeasible. Moreover, k should be large enough such that the probability p of a random $v \in GF(2)^{n+k}$ belonging to $\bigcup_{l=1}^{L} V_l$ is negligibly small. Note that $p < L2^{-k}$.

The subspaces $V_1, \ldots, V_L$ should have the property $V_i \oplus V_j = GF(2)^{n+k}$ for all $i \neq j \in [L]$, otherwise the effective keylength would be reduced. This implies $n \geq k$.

The Learning Unions of L Linear Subspaces (LULS) Problem refers to the following communication game between a learner and an oracle. The oracle holds the specifications of L n-dimensionial linear subspaces $V_1, \ldots, V_L$ of $GF(2)^{n+k}$. The learner can send requests *hello* to the oracle. If the oracle receives *hello*, it chooses randomly and uniformly an $l \in [L]$ and $v \in V_l$ and sends the (positive) example v to the learner. The aim of the learner is to compute specifications of $V_1, \ldots, V_L$ from a sufficiently large set $v^1, \ldots, v^s$ of examples produced by the oracle. Note that this corresponds to a passive key recovery attack against (n, k, L)-type protocols. As described above, a possible strategy is the search-for-a-basis heuristic, which we outline in Appendix A together with implied suggestions on how to choose n and k.

An active adversary who is able to solve the LULS-problem efficiently can break the $(n, k, L)^+$-protocol. In particular, knowing specifications of the secret subspaces $V_1, \ldots, V_L$, he can generate specifications of the subspaces $V_l(a)$ (i.e., the image of $F_l(a, \cdot)$), for arbitrary $a \in GF(2)^{n/2}$ and $l \in [L]$ by repeatedly sending a to Bob. Then the adversary uses $N = n/2$ subspaces $V_l(a_1), \ldots, V_l(a_N)$ for $\{a_1, \ldots, a_N\}$ linearly independent to forge a response for a challenge $a = \sum_{i=1}^{N} \alpha_i a_i$ by computing

$$w = \sum_{i=1}^{N} \alpha_i v_i \text{ with } v_i \in_R V_l(a_i)$$

$$= \sum_{i=1}^{N} \alpha_i F_l(a_i, b_i)$$

$$= F_l(a, b') \text{ with } b = \sum_{i=1}^{N} b_i .$$

In the case of the $(n, k, L)^{++}$-protocol, the adversary cannot just return a random $w \in V_l(a)$, but has to make sure that the first half of $f^{-1}(F_l^{-1}(w))$ corresponds to a. How such a w can be found efficiently (possibly based on the specifications of the subspaces $V_l(a)$) is a matter of further research.

In Sect. 3 we present and discuss an algebraic learning algorithm for LULS.

3   On Solving the LULS-Problem

3.1   A Learning Algorithm for the LULS-Problem

Recall that the LULS-problem with parameters n, k, L consists in computing specifications of L secret n-dimensional linear subspaces of $GF(2)^{n+k}$ from positive examples v produced by an oracle which chooses randomly and uniformly $l \in [L]$ and $v \in V_l$. In this paper we treat the case $L = 2$ and consider the special case that $V_l = \{(v, f(v)), v \in \mathrm{GF}(2)^n\}$, $l \in \{1, 2\}$ for secret linear functions $f_1, f_2 : \mathrm{GF}(2)^n \longrightarrow \mathrm{GF}(2)^k$. Our algorithm computes for all $i \in [k]$ specifications of the i-th component functions $f_1^i, f_2^i : \mathrm{GF}(2)^n \longrightarrow \mathrm{GF}(2)$ separately, i.e., it suffices to consider the case $k = 1$. The learning algorithm is based on the following reasoning.

1. Take a set $O = \{(v^1, w_1), \ldots, (v^n, w_n)\} \subseteq \mathrm{GF}(2)^{n+1}$ of examples such that $B = \{v^1, \ldots, v^n\}$ forms a basis of $\mathrm{GF}(2)^n$. For all $i \in [n]$ let x_i and y_i denote the variables corresponding to $f_1(v^i)$ and $f_2(v^i)$, respectively.
2. For $b \in \{0, 1\}$ let $I_b = \{i \in [n], w_i = b\}$.
3. For all $i \in [n]$ let $t_i = x_i \oplus y_i$, and for all $i < j \in [n]$ let $t_{i,j} = x_i y_j \oplus x_j y_i$.
4. Observe that for all $i \in [n]$ the equality $(w_i \oplus x_i)(w_i \oplus y_i) = 0$ holds. This implies

$$x_i y_i = 0 \text{ if } i \in I_0 \text{ and } x_i y_i = 1 \oplus t_i \text{ if } i \in I_1 \ . \tag{1}$$

5. Observe that for each example $(v, w) \in \mathrm{GF}(2)^{n+1}$, $v \notin B$, the following holds: If $v = \bigoplus_{i \in I} v_i$, (i.e., $I \subseteq [n]$ defines the unique representation of v w.r.t. B), then

$$\left(w \oplus \bigoplus_{i \in I} x_i\right)\left(w \oplus \bigoplus_{i \in I} y_i\right) = 0 \ . \tag{2}$$

Observe that relation (2) can be rewritten as a relation $T_B(I, w)$ in the variables t_i and $t_{i,j}$ in the following way. If $w = 0$ then relation (2) is equivalent to $\bigoplus_{i \in I} x_i y_i \oplus \bigoplus_{i < j \in I} t_{i,j} = 0$. Together with relation (1) this implies $\bigoplus_{i \in I_1 \cap I}(t_i \oplus 1) \oplus \bigoplus_{i < j \in I} t_{i,j} = 0$ for $w = 0$. Consequently, for $w = 0$ we define $T_B(I, w)$ as

$$\bigoplus_{i \in I \cap I_1} t_i \oplus \bigoplus_{i < j \in I} t_{i,j} = \begin{cases} 0 & \text{if } |I \cap I_1| \text{ is even} \\ 1 & \text{if } |I \cap I_1| \text{ is odd} \end{cases} \ .$$

If $w = 1$ then relation (2) is equivalent to $1 \oplus \bigoplus_{i \in I} t_i \oplus \bigoplus_{i \in I \cap I_1}(t_i \oplus 1) \oplus \bigoplus_{i < j \in I} t_{i,j} = 0$. Hence, for $w = 1$ we define $T_B(I, w)$ as

$$\bigoplus_{i \in I \cap I_0} t_i \oplus \bigoplus_{i < j \in I} t_{i,j} = \begin{cases} 0 & \text{if } |I \cap I_1| \text{ is odd} \\ 1 & \text{if } |I \cap I_1| \text{ is even} \end{cases} \ .$$

Note that a relation similar to relation (2) was also exhibited in [1] for designing an algebraic attack against so-called F_f-protocols.

The learning algorithm now proceeds as follows.

1. Let initially the system LES of linear equations in the $\frac{1}{2}(n^2 + n)$ variables t_i ($i \in [n]$) and $t_{i,j}$ ($i < j \in [n]$) be empty.
2. **REPEAT**
   2.1 Choose an observation (v, w), $v \notin B \cup \{0\}$, and compute the unique subset $I \subseteq [n]$ with $v = \bigoplus_{i \in I} v^i$.

2.2 Enlarge the system LES by the linear equation $T_B(I, w)$.

3. **UNTIL** the system LES has $\frac{1}{2}(n^2 + n)$ linearly independent equations.

4. Compute by Gaussian elimination the unique solution θ of the system LES.

5. Compute from θ the unique correct assignments to x_i, y_i for all $i \in [n]$.

The correct assignments to the x_i and y_i variables (step 5 of the algorithm) can be computed from $\theta = (\theta_i)_{i \in [n]} (\theta_{i,j})_{i < j \in [n]}$ as follows.

For $b = 0, 1$ let K_b denote the set $K_b = \{i \in [n], \theta_i = b\}$. We know that for all $i \in K_0$ it holds that $x_i = y_i = w_i$, and for all $i \in K_1$ it holds that $y_i = x_i \oplus 1$. This implies that for all $i < j$ in K_1, $\theta_{i,j}$ satisfies

$$\theta_{i,j} = x_i(x_j \oplus 1) \oplus x_j(x_i \oplus 1) = x_i \oplus x_j .$$

This yields a system LES^* of $1/2|K_1|(|K_1| - 1)$ linear equations in the variables x_i, $i \in K_1$, of rank $|K_1| - 1$. As it does not matter which of the two secret linear subspaces we denote by V_1 and which by V_2, we have the freedom to set $x_k = 0$ for some fixed $k \in K_1$. The system LES^* together with $x_k = 0$ yields a system of full rank and allows to compute the correct assigment to the x_i-variables by Gaussian elimination.

3.2   Analysis and Experimental Results

The background for the fact that the repeat cycle of the algorithm is left after a finite number of rounds is that the following $(2^n - (n + 1)) \times (n(n + 1)/2)$-matrix $M(n)$ over GF(2) has full row rank (which is not hard to show). The row indices of $M(n)$ are all subsets $I \subseteq [n]$ with $|I| \geq 2$, the column indices are $[n] \cup \{(i, j), 1 \leq i < j \leq n\}$. We have $M(n)_{I,i} = 1$ iff $i \in I$ and $M(n)_{I,(i,j)} = 1$ iff $\{i, j\} \subseteq [n]$.

We do not give here a theoretical analysis of the expected number of rounds of the repeat cycle. Our experiments show that the algorithm needs only slightly more than $\frac{1}{2}(n^2 + n) + n$ observations to compute the secret functions f_1 and f_2. Particularly for $n = 128$, we need approx. 8390 examples and 4 minutes on a 3.4 GHz Intel Pentium IV with 4 GB RAM and Magma V2.15-9 [2].

How severe is the restriction that the secret subspaces have the special form $V = \{(v, f(v)), v \in GF(2)^n\}$ for some surjective linear mapping $f : GF(2)^n \longrightarrow GF(2)^k$? Let us consider the general case $V = \{A \circ v, v \in GF(2)^n\}$ for an $(n + k) \times n$ matrix A. V can be written in the special form iff the first n rows of A are linearly independent. For randomly chosen A this is true with probability $p(n) \approx 0.2887$ (see Sect. 2.2).

We have seen that we could solve the LULS-problem with parameters $(n, k, 2)$ by solving k LULS-problems with parameters $(n, 1, 2)$.

For the special LULS-problem with parameters $(n, 1, L)$, $L > 2$, we can define a similar system LES consisting of degree-L equations in the variables x_i^l, $i \in [n]$, $l \in [L]$, induced as above by equations of the form

$$\left(w \oplus \bigoplus_{i \in I} x_i^1 \right) \ldots \left(w \oplus \bigoplus_{i \in I} x_i^L \right) = 0 . \tag{3}$$

The problem is that for $L > 2$ the equations have several symmetries such that the system can not be solved uniquely. The way out is to

- Choose an appropriate parameter $s < k$ which divides k, let $k = s \cdot p$,
- Write vectors $w \in \mathrm{GF}(2)^k$ as vectors $w \in \mathrm{GF}(2^s)^p$, and
- Solve the corresponding p LULS-problem with parameters $(n, 1, L)$ over $\mathrm{GF}(2^s)$.

How to find the best choices of s is a matter of further theoretical and experimental research.

We are convinced that there is no faster way to solve an (n, k, L)-LULS-problem other than solving a system of degree-L equations in Ln variables (if n, k, L are appropriately chosen). Such a system is defined over at least $\Phi(n, L) = \binom{n}{L} + 2\sum_{k=1}^{L-1} \binom{n}{k}$ different monomials, i.e., solving it by linearization means to solve a system of linear equations of size $\Phi(n, L)$. This will cost $\mathcal{O}(\Phi(n, L)^3)$ operations, which can be considered infeasible already for $(n, L) \in \{(128, 5), (256, 4)\}$, since $\Phi(128, 5) \approx 2^{28}$ and $\Phi(256, 4) \approx 2^{27}$.

4   Summary

We have seen that the secret key of CKK^2-protocols can be computed very quickly from a sufficiently large set of messages sent by Bob. This kind of protocol should not be used in practice.

The parameters of $(n, k, L)^{++}$-protocols have to be chosen in such a way that solving the LULS problem with parameters $(\frac{n}{2}, k, L)$ is infeasible. We recommend to use $n = 256$, $k = 64$ and $L = 5$.

Another interesting question is to search for simpler nonlinear connection functions f, for which a security proof can be found. In our proposal, for computing $f(a, b)$ Bob has to perform three multiplications in the finite field of order $2^{n/2}$.

It is another interesting open question whether the very symmetrically structured systems of degree-L equations arising in our LULS-algorithm in Sect. 3 can be more efficiently solved by more advanced techniques like the F4- or F5-algorithm or cube attacks [8,9,5,6]. If one could generate convincing evidence that such algorithms cannot beat our linearization attack, then $(n, k, L)^{++}$-protocols with the above parameters could be seriously considered for practical use.

A problem of (n, k, L)-protocols is the large key length in the case that random mappings $F_1, \ldots, F_L$ are used. It is an important task to look for secure and efficient ways to generate pseudorandom keys. In this context, the (still unbroken) $\mathsf{CKK}^{\sigma, L}$-protocols could become interesting. However, we conjecture that $\mathsf{CKK}^{\sigma, L}$-protocols can be efficently broken.

Interesting suggestions for keylength reductions have been made in [11] and [3]. Adapting these ideas to (n, k, L)-protocols would mean

- To consider special forms of secret subspaces $V_l = \{(A_l \circ v), v \in \mathrm{GF}(2)^n\}$, where A_l denotes a secret $(n + k) \times n$ Toeplitz matrix [11], and

– To define the Toeplitz matrix A_l to be generated by a secret Linear Feedback Shift Register [3].

Checking the feasibility and security of these constructions should be a matter of further research.

Acknowledgement

We are very thankful to Mirek Kutyłowski for the introduction to the topic, and to Matthias Hamann, Stefan Lucks, Willi Meier, Frederik Armknecht, and Erik Zenner for helpful discussions.

References

1. Blass, E.-O., Kurmus, A., Molva, R., Noubir, G., Shikfa, A.: The F_f-family of protocols for RFID-privacy and authentication, http://eprint.iacr.org/2008/476
2. Bosma, W., Cannon, J., Playoust, C.: The magma algebra system. i. the user language. J. Symbolic Comput. 24, 235–265 (1997)
3. Bringer, J., Chabanne, H.: Trusted-HB: A low cost version of HB$^+$ secure against a man-in-the-middle attack. IEEE Trans. Inform. Theor. 54, 4339–4342 (2008)
4. Cichoń, J., Klonowski, M., Kutyłowski, M.: Privacy protection for RFID with hidden subset identifiers. In: Indulska, J., Patterson, D.J., Rodden, T., Ott, M. (eds.) PERVASIVE 2008. LNCS, vol. 5013, pp. 298–314. Springer, Heidelberg (2008)
5. Dinur, I., Shamir, A.: Cube attacks on tweakable black box polynomials. Cryptology ePrint Archive, Report 2008/385 (2008), http://eprint.iacr.org
6. Dinur, I., Shamir, A.: Cube attacks on tweakable black box polynomials. In: Joux, A. (ed.) EUROCRYPT 2009. LNCS, vol. 5479, pp. 278–299. Springer, Heidelberg (2009)
7. Gołębiewski, Z., Majcher, K., Zagórski, F.: Attacks on CKK family of RFID authentication protocols. In: Coudert, D., Simplot-Ryl, D., Stojmenovic, I. (eds.) ADHOC-NOW 2008. LNCS, vol. 5198, pp. 241–250. Springer, Heidelberg (2008)
8. Faugère, J.-C.: A new efficient algorithm for computing Gröbner bases (F4). J. Pure Appl. Algebra 139, 61–68 (1999)
9. Faugère, J.-C.: A new efficient algorithm for computing Gröbner basis without reduction to zero (F5). In: Mora, T. (ed.) ISSAC 2002, pp. 75–83. ACM Press, New York (2002)
10. Frumkin, D., Shamir, A.: Untrusted-HB: Security vulnerabilities of Trusted-HB. Cryptology ePrint Archive, Report 2009/044 (2009), http://eprint.iacr.org
11. Gilbert, H., Robshaw, M.J.B., Seurin, Y.: HB$^\#$: Increasing the security and efficiency of HB$^+$. In: Smart, N.P. (ed.) EUROCRYPT 2008. LNCS, vol. 4965, pp. 361–378. Springer, Heidelberg (2008)
12. Gilbert, H., Robshaw, M.J.B., Sibert, H.: Active attack against HB$^+$: A provable secure lightweight authentication protocol. Electronic Letters 41, 1169–1170 (2005)
13. Juels, A.: RFID privacy: A technical primer for the non-technical reader. In: Strandburg, K., Raicu, D.S. (eds.) Privacy and Technologies of Identity: A Cross-Disciplinary Conversation. Springer, Heidelberg (2005)
14. Juels, A., Weis, S.A.: Authenticating pervasive devices with human protocols. In: Shoup, V. (ed.) CRYPTO 2005. LNCS, vol. 3621, pp. 293–308. Springer, Heidelberg (2005)

15. Langheinrich, M.: A survey of RFID privacy approaches. J. Personal and Ubiqui-
    tous Comp. 13, 413–421 (2009)
16. Ouafi, K., Overbeck, R., Vaudenay, S.: On the security of HB$^{\#}$ against a man-in-
    the-middle attack. In: Pieprzyk, J. (ed.) ASIACRYPT 2008. LNCS, vol. 5350, pp.
    108–124. Springer, Heidelberg (2008)

A   The Search-for-a-Basis Heuristic

The search-for-a-basis heuristic tries to construct a set Q of examples which form
a basis of V_l for some $l \in L$. For all linearly independent sets Q of n examples
let $p(Q)$ denote the probability that an example coming from the oracle belongs
to the linear span $< Q >$ of Q. It is quite obvious that $p(Q)$ is maximal if Q
is a basis of V_l for some $l \in L$. If $p(Q)$ is not too small, we can compute an
approximation $\tilde{p}(Q)$ of $p(q)$ by testing for $w \in < Q >$ for a sufficiently large
number of examples w.

For $v \in Q$ and $w \notin Q$ we denote by $Q(v, w)$ the set obtained by replacing v
by w in Q.

The idea of the heuristic is to start with an arbitrary linear independent set
Q of n examples and to try to improve this set by finding $v \in Q$ and $w \notin Q$ such
that $\tilde{p}(Q) < \tilde{p}(Q(v, w))$. Iterating this at most n times yields a basis for V_l for
some $l \in [L]$.

This kind of heuristic is infeasible if the following condition is fulfilled. For a
random linear independent set Q of n examples the probability $p(Q)$ is negligibly
small with probability $1 - \epsilon$, ϵ negligibly small. The parameters n, k should be
chosen in such a way that this condition is guaranteed.

We estimate the probability $p(Q)$ for the case $L = 2$. For a linear independent
set Q of n examples let $Q = Q_1 \cup Q_2$, where $Q_1 \subseteq V_1$ and $Q_2 \subseteq V_2 \setminus V_1$. W.l.o.g.
let $|Q_1| = n/2 + s$ and $|Q_2| = n/2 - s$. The event $w \in < Q >$ happens iff
$w \in V_1 \cap < Q_1 >$ or $w \in V_2$ and $w \in V_2 \cap < Q_1 >$, i.e.,

$$p(Q) \leq \frac{1}{2}\left(2^{s-n/2} + 2^{-k}\right) \ .$$

(Note that $\dim(V_1 \cap V_2) = n - k$ for random n-dimensional subspaces V_1, V_2).
If n, k are chosen in such a way that 2^{-k}, $2^{-n/4}$ and the probability that $|v| \notin$
$[n/4, 3n/4]$ are negligibly small, then the above condition is fulfilled (note that
the expected value of s is $2^{-k}n/2$).

Differential Fault Analysis of Rabbit

Aleksandar Kircanski and Amr M. Youssef

Concordia Institute for Information Systems Engineering
Concordia University
Montreal, Quebec, H3G 1M8, Canada
{a_kircan,youssef}@ciise.concordia.ca

Abstract. Rabbit is a high speed scalable stream cipher with 128-bit key and a 64-bit initialization vector. It has passed all three stages of the ECRYPT stream cipher project and is a member of eSTREAM software portfolio. In this paper, we present a practical fault analysis attack on Rabbit. The fault model in which we analyze the cipher is the one in which the attacker is assumed to be able to fault a random bit of the internal state of the cipher but cannot control the exact location of injected faults. Our attack requires around $128 - 256$ faults, precomputed table of size $2^{41.6}$ bytes and recovers the complete internal state of Rabbit in about 2^{38} steps.

1 Introduction

The ECRYPT stream cipher project, also known as eSTREAM, is a project that aimed to identify new promising stream ciphers. The first call for stream cipher submissions was made in 2004 and it consisted of profile 1 and profile 2: software oriented ciphers and hardware oriented ciphers. The ciphers were put through a three-phase elimination process, finalizing in 2008, when four software oriented ciphers, including Rabbit, and three hardware oriented ciphers were selected as members of the eSTREAM portfolio.

After passing all three phases, Rabbit [4] (also see RFC 4503) has become a member of profile 1 eSTREAM portfolio. While originally designed with high software performance in mind, Rabbit turns out to be also very fast and compact in hardware. Fully optimized software implementations achieve an encryption speed of up to 3.7 clock cycles per byte (CPB) on a Pentium 3, and of 9.7 CPB on an ARM7. It uses a 128-bit secret key, 64-bit IV and generates 128 pseudo-random bits as keystream output at each iteration. The size of the secret internal state amounts to 513 bits, consisting of two sets of 8 32-bit words and one additional 1-bit value.

The security of Rabbit has been thoroughly investigated in a series of white papers published by the crypto lab at Cryptico A/S. These papers include analysis of the key setup function [9], analysis of IV-setup [13], mod n cryptanalysis [14], algebraic cryptanalysis [8] and periodic properties [11]. Also, a distinguishing attack requiring 2^{247} 128-bit samples was reported in [2]. The bias utilized in this attack was resulting from the bias in the Rabbit core function where it was

M.J. Jacobson Jr., V. Rijmen, and R. Safavi-Naini (Eds.): SAC 2009, LNCS 5867, pp. 197–214, 2009.

shown that images of the Rabbit core function, g, have significantly less zeros than ones at each offset and this was used to show that there exists a bias in the least significant bit of certain keystream subblocks. This work was extended in [19], where the probability distribution of several keystream bits together was calculated by means of Fast Fourier Transform, using the techniques described in [22]. The complexity of the latter attack is 2^{158}. The authors also presented an attack in which the $2^{51.5}$ instantiations of the cipher are analyzed based on the first three keystream output blocks of each instantiation. The additional assumption is that certain differences expressed in terms of XOR among these $2^{51.5}$ internal states are known. This attack recovers all $2^{51.5}$ keys and requires 2^{32} precomputation steps, 2^{32} memory, and $2^{97.5}$ steps. According to the authors, the attack is given under an unusual cryptanalytic assumption. This attack was considered the first known key recovery attack on Rabbit.

In this paper we present a fault analysis attack on Rabbit. The fault model adopted is the one in which an attacker is assumed to be able to cause a bit-flip at a random location in the internal state of the cipher. However, the exact position of the flipped bit is unknown to the attacker. The attacker is also assumed to be able to reinitialize the cipher sufficient amount of times, iterate and obtain keystream words. The proposed attack requires around $128 - 256$ faults, an off-line precomputed table of size $2^{41.6}$ bytes and recovers the complete internal state of Rabbit in about 2^{38} steps.

2   Fault Analysis

Cryptanalytic attacks can be broadly classified into two classes. In the first class of attacks, the attacker tries to exploit any weakness in the underlying mathematical structure of the cipher. This type includes, for example, differential cryptanalysis, linear cryptanalysis and algebraic attacks. The second class of attacks are implementation dependent attacks, which include side channel attacks, such as timing analysis [18] and power analysis [17], and fault analysis attacks. In fault analysis attacks [5], the attacker applies some kind of physical influence on the internal state of the cryptosystem, such as ionizing radiation which flips random bits in devices' memory. By examining the results of cryptographic operations under such faults, it is often possible to deduce information about the secret key or the secret internal state of the cipher.

Fault attacks were first introduced by Boneh et al. [5] in 1996 where they described attacks that targeted the RSA public key cryptosystem by exploiting a faulty Chinese remainder theorem computation to factor the modulus n. Subsequently, fault analysis attacks were extended to symmetric systems such as DES [3] and later to AES [15] and other primitives. Fault analysis attacks became a more realistic serious threat after cheap and low-tech methods of applying faults were presented (e.g., [1,23]).

Hoch and Shamir [16] showed that fault analysis attacks present a powerful tool against stream ciphers as well. Stream ciphers based on LFSRs, LILI-128 and SOBER-t32 as well as RC4 were shown to be insecure in a fault analysis model in which the attacker does not have the ability to choose the exact location of the induced fault. In the case of RC4, the key recovery attack requires 2^{16} faults and 2^{26} keystream words. In [6], RC4 was assessed using a different fault model in which the attacker may specify the location at which the fault is induced but can not specify the value of injected faults. The attack requires 2^{16} induced faults. Another more advanced fault analysis attack on RC4 which requires 2^{10} faults was also introduced in the same paper.

Hojsík and Rudolf [20] presented an attack on another eSTREAM cipher, Trivium [7]. The attack recovers the secret internal state using 42 fault injections. The fault model used is the one in which the attacker is able to flip a random bit in the internal state of Trivium without being able to exactly control its location. This work was subsequently improved in [21], providing an attack that recovers Trivium inner state with only 3.2 fault injections on average. The authors used different cipher representation and were able to reduce high-degree equations to linear ones, concluding that a change in the way by which the cipher is represented may result in a better attack.

In this paper, we use the same model as the one used in fault analysis of Trivium [20,21]. The attacker is assumed to be able to flip a random bit in the internal state of the cipher without being able to exactly control its location. In other words, the exact location of induced fault is assumed to be unknown to the attacker.

The rest of the paper is organized as follows. The Rabbit specifications that are relevant to our attack are briefly reviewed in the next section. The main idea behind our attack is presented in section 4.1. The procedure used to determine the location of induced faults is described in section 4.2 and the complete attack is described in section 4.3. Finally, the attack success probability and its associated complexity are analyzed in section 5.

3   Specification of Rabbit Stream Cipher

Internal state of Rabbit consists of 513 bits. It includes: eight 32-bit values: $x_{0,t}$, $\cdots x_{7,t}$, eight 32-bit counters, $c_{0,t}, \ldots c_{7,t}$, and one additional bit $\phi_{7,t}$, used in the counter update. When the cipher steps from time t to time $t+1$, the counter is updated independently of x values, by adding known a_i values, corrected with carries ϕ as follows:

$$c_{0,t+1} = c_{0,t} + a_0 + \phi_{7,t}$$
$$c_{j,t+1} = c_{j,t} + a_j + \phi_{j-1,t+1}, 1 \leq j \leq 7$$

where

$$\phi_{j,t+1} = \begin{cases} 1 - 1_{\mathbb{Z}/2^{32}\mathbb{Z}}(c_{0,t} + a_0 + \phi_{7,t}) & \text{if } j = 0 \\ 1 - 1_{\mathbb{Z}/2^{32}\mathbb{Z}}(c_{j,t} + a_j + \phi_{j-1,t+1}) & \text{if } j > 0 \end{cases}$$

and $a_0 = a_3 = a_6 = 4D34D34D$, $a_1 = a_4 = a_7 = D34D34D3$, $a_2 = a_5 = 34D34D34D$. Function $1_{\mathbb{Z}/2^{32}\mathbb{Z}}$ is defined by

$$1_{\mathbb{Z}/2^{32}\mathbb{Z}}(x) = \begin{cases} 0 \text{ if } x \geq 2^{32} \\ 1 \text{ if } x < 2^{32} \end{cases}$$

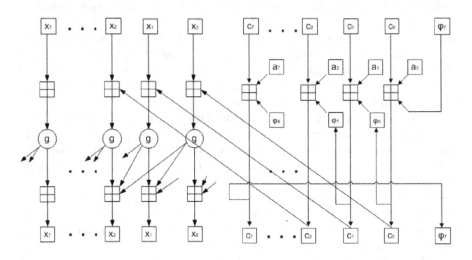

Fig. 1. Simplified view of the state update function of Rabbit, rotations omitted

The x values are updated by

$$
\begin{aligned}
x_{0,t+1} &= g_{0,t} + (g_{7,t} \lll 16) + (g_{6,t} \lll 16) \\
x_{1,t+1} &= g_{1,t} + (g_{0,t} \lll 8) + g_{7,t} \\
x_{2,t+1} &= g_{2,t} + (g_{1,t} \lll 16) + (g_{0,t} \lll 16) \\
x_{3,t+1} &= g_{3,t} + (g_{2,t} \lll 8) + g_{1,t} \\
x_{4,t+1} &= g_{4,t} + (g_{3,t} \lll 16) + (g_{2,t} \lll 16) \\
x_{5,t+1} &= g_{5,t} + (g_{4,t} \lll 8) + g_{3,t} \\
x_{6,t+1} &= g_{6,t} + (g_{5,t} \lll 16) + (g_{4,t} \lll 16) \\
x_{7,t+1} &= g_{7,t} + (g_{6,t} \lll 8) + g_{5,t}
\end{aligned}
\tag{1}
$$

where

$$
g_{j,t} = (x_{j,t} + c_{j,t+1})^2 \oplus [(x_{j,t} + c_{j,t+1})^2 \gg 32]
\tag{2}
$$

The 128-bit keystream output block $s_{t+1}^{[127..0]}$, is constructed as follows:

$$
\begin{aligned}
s_{t+1}^{[15..0]} &= x_{0,t+1}^{[15..0]} \oplus x_{5,t+1}^{[31..16]}, & s_{t+1}^{[31..16]} &= x_{0,t+1}^{[31..16]} \oplus x_{3,t+1}^{[15..0]} \\
s_{t+1}^{[47..32]} &= x_{2,t+1}^{[15..0]} \oplus x_{7,t+1}^{[31..16]}, & s_{t+1}^{[63..48]} &= x_{2,t+1}^{[31..16]} \oplus x_{5,t+1}^{[15..0]} \\
s_{t+1}^{[79..64]} &= x_{4,t+1}^{[15..0]} \oplus x_{1,t+1}^{[31..16]}, & s_{t+1}^{[95..80]} &= x_{4,t+1}^{[31..16]} \oplus x_{7,t+1}^{[15..0]} \\
s_{t+1}^{[111..96]} &= x_{6,t+1}^{[15..0]} \oplus x_{3,t+1}^{[31..16]}, & s_{t+1}^{[127..112]} &= x_{6,t+1}^{[31..16]} \oplus x_{1,t+1}^{[15..0]}
\end{aligned}
\tag{3}
$$

Figure 1 shows a simplified view of the Rabbit state update function. The description of the key setup scheme of Rabbit is omitted since it does not play a role in the attack outlined in this paper.

4   Differential Fault Analysis Attack

Throughout the rest of this paper, faulty words will be denoted same as non-faulty ones, except that a "'" sign will be added. This way, faulty Rabbit internal state words at time t will be denoted by $x'_{i,t}$, $c'_{j,t}$, $\phi'_{7,t}$. The whole Rabbit internal state at time t, consisting of $[(x_{i,t})_{i=0...7}, (c_{i,t})_{i=0...7}, \phi_{7,t}]$, will be denoted by S_t. Accordingly, its faulty counterpart will be denoted by S'_t. We will also use " $+$ " to denote addition mod 32, unless otherwise stated.

Faulty keystream output at step t will be denoted by s'_t. The i-th 16-bit segment of word s will be denoted by $s^{(i)}$. For example $s_t^{(1)}$ denotes $s_t^{[31..16]}$, i.e., bits 16 to 31 of word s_t.

According to our fault analysis model, the attacker has the power to flip a bit within the internal state of the cipher, that is $x_{i,t}, c_{i,t}, i = 0, \ldots 7$, $\phi_{7,t}$ but the attacker can not control or know the exact location of the induced fault (both at the bit and at the word level).

4.1 The Main Idea

Before stating the complete attack procedure, we provide a motivational example that illustrates the idea behind the attack. Let states of Rabbit at step t, S_t and S'_t, differ only in i-th bit of word $x_{0,t}$. Consequently, $x'_{0,t} + c'_{0,t+1} = x_{0,t} + c_{0,t+1} + \sigma 2^i$, for some unknown $\sigma \in \{-1, +1\}$ and $i \in \{0, \ldots 31\}$. Then, with high probability, $g'_{0,t} \neq g_{0,t}$ and $g'_{i,t} = g_{i,t}$ for $i = 1..7$. This implies that $x'_{i,t+1} \neq x_{i,t+1}$, for $i = 0, 1, 2$ and $x'_{i,t+1} = x_{i,t+1}$ for $i = 3..7$. In particular, since $x'^{[31..16]}_{5,t+1} = x^{[31..16]}_{5,t+1}$ and $x'^{[15..0]}_{3,t+1} = x^{[15..0]}_{3,t+1}$, then using the first line in Eq. (3), the following holds

$$s'^{[15..0]}_{t+1} = x'^{[15..0]}_{0,t+1} \oplus x^{[31..16]}_{5,t+1} \qquad s^{[15..0]}_{t+1} = x^{[15..0]}_{0,t+1} \oplus x^{[31..16]}_{5,t+1}$$

$$s'^{[31..16]}_{t+1} = x'^{[31..16]}_{0,t+1} \oplus x^{[15..0]}_{3,t+1} \qquad s^{[31..16]}_{t+1} = x^{[31..16]}_{0,t+1} \oplus x^{[15..0]}_{3,t+1}$$

Thus guessing $x^{[31..16]}_{5,t+1}$ and $x^{[15..0]}_{3,t+1}$ makes a candidate for values $x_{0,t+1}$ and $x'_{0,t+1}$ and consequently, using Eq. (1), a candidate for

$$x_{0,t+1} - x'_{0,t+1} =$$
$$(g_{0,t} + g_{7,t} \lll 16 + g_{6,t} \lll 16) - (g'_{0,t} + g'_{7,t} \lll 16 + g'_{6,t} \lll 16) = g_{0,t} - g'_{0,t}$$

Since inputs to $g_{0,t}$ and $g'_{0,t}$ differ by $\pm 2^i$ for some unknown $i = 0, \ldots 31$, this constraint can be described by a set of g function additive differentials $\{(\pm 2^i, \delta) | i = 0, \ldots 31\}$.

Suppose now the attacker obtains two more faulted keystream words s''_{t+1} and s'''_{t+1}, derived from states S''_t and S'''_t differing from S_t on bits j and k of word $x_{0,t}$, where $k \neq j, k \neq i, j \neq i$. Since in all three cases, values $x^{[31..16]}_{5,t+1}$ and $x^{[15..0]}_{3,t+1}$ do not change, using $s_{t+1}, s'_{t+1}, s''_{t+1}$ and s'''_{t+1} three sets of differentials $\{(\pm 2^i, \delta_1) | i = 0, \ldots 31\}$, $\{(\pm 2^i, \delta_2) | i = 0, \ldots 31\}$ and $\{(\pm 2^i, \delta_3) | i = 0, \ldots 31\}$ are obtained using the same guess in the way described above. As will be shown later, the probability that there exists an input x for the g function such that it satisfies all three sets of differentials at once, i.e., such that there exist mutually different i_1, i_2 and i_3 such that

$$g(x) - g(x \pm 2^{i_1}) = \delta_1,$$
$$g(x) - g(x \pm 2^{i_2}) = \delta_2,$$
$$g(x) - g(x \pm 2^{i_3}) = \delta_3$$

is small if the guess above is not correct. Thus, the attacker is able to discard wrong guesses for $x_{5,t+1}^{[31..16]}$ and $x_{3,t+1}^{[15..0]}$. Also, if the guess is a correct one, the attacker obtains candidates for g input value $x_{0,t} + c_{0,t+1}$. In the following we provide a full internal state recovery algorithm.

4.2   Determining the Position of the Fault

In the attack proposed in this paper, the first step after inducing a fault is to make restrictions on the position where the fault took place. The induced bit flipping can happen at one of the bits of words $x_{0,t}, \ldots x_{7,t}, c_{0,t}, \ldots c_{7,t}$ as well as at the 1-bit value $\phi_{7,t}$.

In the following we provide a tool for deducing important information on the location at which the fault occurred. Based on difference among faulty and non-faulty keystreams, information on the difference among internal states S_t and S_t' is deduced. More precisely, only keystream words s_t and s_t' will be used and according to the fault model, it will be assumed that internal states S_t and S_t' differ exactly on one bit.

To express these differences in a convenient way, we introduce the function d_{ST}, describing differences on the internal states and the function d_{KS}, describing differences among faulty and non-faulty keystream words. Let

$$d_{ST}(S, S') = \begin{cases} 0, & \text{if a fault occured either at } x_{0,t}, c_{0,t} \text{ or } \phi_{7,t} \\ 1 \leq i \leq 7, & \text{if a fault accured either at } x_{i,t} \text{ or } c_{i,t} \end{cases}$$

The function d_{ST} is defined for every pair of states (S, S') that differ exactly on one bit. If s and s' are two 128-bit keystream words at some step, then we define

$$d_{KS}(s, s') = \begin{cases} 0, & s^{(5)} = s'^{(5)}, s^{(6)} = s'^{(6)} \text{ and } s^{(i)} \neq s'^{(i)} \text{ for } i \neq 5, 6 \\ 1, & s^{(0)} = s'^{(0)}, s^{(5)} = s'^{(5)} \text{ and } s^{(i)} \neq s'^{(i)} \text{ for } i \neq 0, 5 \\ 2, & s^{(0)} = s'^{(0)}, s^{(7)} = s'^{(7)} \text{ and } s^{(i)} \neq s'^{(i)} \text{ for } i \neq 0, 7 \\ 3, & s^{(2)} = s'^{(2)}, s^{(7)} = s'^{(7)} \text{ and } s^{(i)} \neq s'^{(i)} \text{ for } i \neq 2, 7 \\ 4, & s^{(1)} = s'^{(1)}, s^{(2)} = s'^{(2)} \text{ and } s^{(i)} \neq s'^{(i)} \text{ for } i \neq 1, 2 \\ 5, & s^{(1)} = s'^{(1)}, s^{(4)} = s'^{(4)} \text{ and } s^{(i)} \neq s'^{(i)} \text{ for } i \neq 1, 4 \\ 6, & s^{(3)} = s'^{(3)}, s^{(4)} = s'^{(4)} \text{ and } s^{(i)} \neq s'^{(i)} \text{ for } i \neq 3, 4 \\ 7, & s^{(3)} = s'^{(3)}, s^{(6)} = s'^{(6)} \text{ and } s^{(i)} \neq s'^{(i)} \text{ for } i \neq 3, 6 \end{cases}$$

If a pair of 128 bit (s, s') words does not satisfy any of the conditions proposed by the right-hand side of the equation above, function $d_{KS}(s, s')$ is *undefined*.

To understand the motivation behind the above definition, assume that the injected fault affected the input to the function $g_{0,t}$. From Figure 1, it is clear that such fault *directly* affects the computation of $x_{0,t+1}$, $x_{1,t+1}$ and $x_{2,t+1}$. From Eq. (3), it follows that these three terms also *directly* affect the computation of all words on the output stream except $s'^{[95..80]}_{t+1} = s'^{(5)}$ and $s'^{[111..96]}_{t+1} = s'^{(6)}$ which explains the first line in the above definition. A similar argument applies to to rest of entries in the definition of $d_{KS}(s, s')$.

The criterion for determining the position of the fault $d_{ST}(S_t, S'_t)$ based on the first keystream word can now be simply stated as follows:

- If $d_{KS}(s_{t+1}, s'_{t+1})$ is defined, put $d_{ST}(S_t, S'_t) = d_{KS}(s_{t+1}, s'_{t+1})$
- Otherwise, leave $d_{ST}(S_t, S'_t)$ undefined

During the attack, when after a fault $d_{ST}(S_t, S'_t)$ value is undefined, the fault will be discarded and the attacker proceeds by inducing another fault.

The successfulness of this criterion can be measured by two types of errors, p_{incorr} and p_{undef}. Error p_{incorr} is defined as the probability that the criterion returns a wrong $d_{ST}(S, S')$ value, while error p_{undef} is defined as the probability that the criterion will leave $d_{ST}(S, S')$ undefined. The probability that the criterion returns correct $d_{ST}(S, S')$ value will be denoted as p_{corr}.

According to the theoretical estimate of these three probabilities (see Appendix A), $p_{corr} \approx 0.98635$, $p_{undef} \approx 0.013645$ and $p_{incorr} \approx 0$. To confirm these theoretical estimates, the following experiment was conducted. A Rabbit internal state was randomly initialized and a random fault was induced. The criterion was applied and it was noted which of the three options happened: correct position of the fault returned, incorrect position of the fault returned or position of the fault left undefined. After repeating the experiment for 10^6 times, the probabilities were obtained as $p_{corr} = 0.98408$, $p_{undef} = 0.015924$, and $p_{incorr} = 0$. Hence, given, say 100 faults, correct position will be determined for around 98 faults and 2 faults will be discarded. For no faults the incorrect position will be returned. Thus, it can be concluded that the proposed criterion represents reliable means for determining the position of faults.

4.3   The Complete Attack

Before stating the complete attack we introduce following definitions. Throughout the following three definitions, let $k \in \{0, 1, 2\}$ and $\sigma \in \{-1, +1\}$ be fixed values and let x be restricted to the set $\mathbb{Z}_{2^{32}}$. By $(\sigma 2^i, \delta)_k$ we denote a g function additive differential where the input difference is $\sigma 2^i$ and the output difference δ, taken after rotating g function output for $8 \times k$ bits.

Definition 1. *An x value will be considered to* satisfy *differential $(\sigma 2^i, \delta)_k$ if*

$$[g(x) \lll (8k)] - [g(x + \sigma 2^i) \lll (8k)] = \delta$$

Definition 2. *A set of differentials*

$$(\pm 2^i, \delta)_k|^{31}_{i=0} := \{(\sigma 2^i, \delta)_k | i = 0..31, \sigma = -1, +1\}$$

will be called a generalized differential. *An x value will be considered to* satisfy *generalized differential $(\pm 2^i, \delta)_k|^{31}_{i=0}$ if it satisfies any of the differentials contained in the set.*

Definition 3. *A set of generalized differentials*

$$\Delta = \{(\pm 2^i, \delta_1)_k|_{i=0}^{31}, (\pm 2^i, \delta_2)_k|_{i=0}^{31}, \ldots (\pm 2^i, \delta_n)_k|_{i=0}^{31})\}$$

will be considered satisfiable *if at least one* x *value satisfies them all, i.e., if there exists an* x *value as well as distinct values* $d_1, \ldots d_n$, *chosen from the set* $\{\pm 2^i | i = 0..31\}$, *such that*

$$[g(x) \lll 8k] - [g(x + d_j) \lll 8k] = \delta_j, j = 1, \ldots n$$

For such an x, *we shall say that it* satisfies set Δ.

The following procedure, flt_init(t) induces a sufficient number of faults at the internal state of the cipher at step t and arranges faulty keystream words to appropriate sets, using the mechanism described in Section 4.2.

- Let $FLTS_i = \emptyset$, $i = \ldots 7$.
- While $|FLTS_i| < 3$ for any $i = 0 \ldots 7$
  - Reinitialize the cipher, forward to step t and induce a fault. Obtain s'_{t+1}. If $d_{KS}(s'_{t+1}, s_{t+1})$ is defined, let $i = d_{KS}(s'_{t+1}, s_{t+1})$ and add s'_{t+1} to $FLTS_i$.

The procedure that follows, derive_inf(i,k), utilizes information in $FLTS_i$ to deduce the set of possible values for $x_{i,t} + c_{i,t+1}$. Parameter k can take values 0,1 and 2 and it determines the way $x_{i,t} + c_{i,t+1}$ value will be recovered. Namely, as will be seen from the algorithm, there are three different ways to derive candidates for this value and the logic of these three ways is encoded through values of $\alpha_{i,k}, \beta_{i,k}$, $k = 0, 1, 2$ in Table 1. The values for $\alpha_{i,k}, \beta_{i,k}$ have been derived utilizing Eq. (3). For example, running derive_inf(0,0), returns the set of candidates for $x_{0,t} + c_{0,t+1}$ by working on values $s_{t+1}^{(0)}$ and $s_{t+1}^{(1)}$, i.e., guessing values $x_{3,t+1}^{[15..0]}$ and $x_{5,t+1}^{[31..16]}$, creating the set of generalized differentials using $g_{0,t} - g'_{0,t} = x_{0,t+1} - x'_{0,t+1}$ and finally finding g-input values that satisfy it. On the other hand running derive_inf(0,1) aims to recover the same value $x_{0,t} + c_{0,t+1}$, but in a different way. Namely, in this case, the procedure operates on values $s_{t+1}^{(4)}$ and $s_{t+1}^{(7)}$, i.e., guesses $x_{4,t+1}^{[15..0]}$ and $x_{6,t+1}^{[31..16]}$, derives the generalized differential set by $g_{0,t} \lll 8 - g'_{0,t} \lll 8 = x_{1,t+1} - x'_{1,t+1}$ and then searches for g-input values that satisfy the set. The objective of obtaining the same value in three different ways is to take the intersection afterwards and hence minimize redundant candidates. Also, in the first case, a difference with no rotation was obtained and in the second, a difference after 8-bit rotations was found. The table is encoded so that whenever $k = 0$, $k = 1$ and $k = 2$, the number of rotations in the obtained difference will be 0, 8 and 16, respectively. This justifies the same value k present as index both for α, β and for generalized differentials themselves from $\Delta_i^k(A)$ sets in the procedure below.

The complete procedure derive_inf(i,k) follows:

- Let $Sat(\Delta_i^k) = \emptyset$
- For $A = 0, \ldots 2^{32} - 1$

Table 1. α and β index values used during the attack

i	0	1	2	3	4	5	6	7
$\alpha_{i,0}$	1	4	3	6	5	0	7	2
$\beta_{i,0}$	0	7	2	1	4	3	6	5
$\alpha_{i,1}$	4	3	6	5	0	7	2	1
$\beta_{i,1}$	7	2	1	4	3	6	5	0
$\alpha_{i,2}$	3	6	5	0	7	2	1	4
$\beta_{i,2}$	2	1	4	3	6	5	0	7

- Form the set of generalized differentials as follows:

$$\Delta_i^k(A) = \{\ (\pm 2^l, ([s^{(\alpha_{i,k})}||s^{(\beta_{i,k})}] \oplus A) - ([s'^{(\alpha_{i,k})}||s'^{(\beta_{i,k})}] \oplus A))_k|_{l=0}^{31}$$
$$|s' \in FLTS_i\}$$

- Let $Sat(\Delta_i^k) = Sat(\Delta_i^k) \cup Sat(\Delta_i^k(A))$, where $Sat(\Delta_i^k(A))$ is the set of x values that satisfy $\Delta_i^k(A)$

where $\alpha_{i,k}$, $\beta_{i,k}$, $i = 0 \ldots 7$, $k = 0, 1, 2$ are defined by Table 1. The Derivation of $Sat(\Delta_i^k(A))$ sets is done using precomputation, as explained in Section 5.2. To recover g input values at step t, i.e., values $x_{i,t} + c_{i,t+1}$, the procedure g_inp(t) can be invoked, as follows:

- flt_init(t)
- For $i = 0, \ldots 7$
  - Call derive_inf(i,0), derive_inf(i,1) and derive_inf(i,2) to find $Sat(\Delta_i^0)$, $Sat(\Delta_i^1)$ and $Sat(\Delta_i^2)$
  - $Cand(x_{i,t} + c_{i,t+1}) = Sat(\Delta_i^0) \cap Sat(\Delta_i^1) \cap Sat(\Delta_i^2)$

In the next section it will be shown that the probability that there will be more than one candidate for $x_{i,t} + c_{i,t+1}$, i.e., that there will be more than one element in the set $Cand(x_{i,t} + c_{i,t+1})$, is small.

Finally, the complete internal state at time $t = 1$ can be recovered by invoking the previous procedure for $t = 0$ which yields values $x_{i,0} + c_{i,1}$, $i = 0 \ldots 7$. This in turn yields $g_{i,0}$, $i = 0 \ldots 7$ values, which yield $x_{i,1}$, $i = 0 \ldots 7$, by Eq. 1. Invoking the previous procedure once again for $t = 1$ yields values $x_{i,1} + c_{i,2}$, $i = 0 \ldots 7$. Subtracting according values reveals $c_{i,2}$, $i = 0 \ldots 7$. Now $c_{i,1}$ values can be recovered by reversing the counter one step backward, according to the specification of counter update step. Whether $\phi_{7,1} = 0$ or $\phi_{7,1} = 1$ is found by mere trying both options and comparing the resulting keystream words.

5   Attack Success Probability and Complexity

5.1   Success Probability

In this section we show that the procedure from previous section determines the internal state uniquely. More precisely, it will be shown that $|Cand(x_{i,t} + $

$c_{i,t+1})| = 1$, for any $i = 0 \ldots 7$ and $t \geq 0$ with high probability. This will be done by modelling g as a random function and then showing that if differences $([s^{(\alpha_{i,k})}||s^{(\beta_{i,k})}] \oplus A)$ - $([s'^{(\alpha_{i,k})}||s'^{(\beta_{i,k})}] \oplus A)$ are chosen uniformly randomly, i.e., not corresponding to the actual values produced by the attack procedure, this set of candidates will have 0 elements with high probability. Then, this probability can be taken as probability that $|Cand(x_{i,t} + c_{i,t+1})| = 1$ since following the procedure with actual differences and using the real g function guarantees existence of one correct candidate for $x_{i,t} + c_{i,t+1}$. According to the way complete internal state is recovered from g input values at times $t = 0$ and $t = 1$, as described by the last paragraph of the previous section, it is clear that from uniqueness and correctness of g input values, uniqueness and correctness of the recovered internal state at step $t = 1$ follows.

Since during the algorithm, $Cand(x_{i,t} + c_{i,t+1})$ is derived as the intersection of $Sat(\Delta_i^0)$, $Sat(\Delta_i^1)$ and $Sat(\Delta_i^2)$, the probability distribution of the number of elements in these three sets is first examined. Assume g is a randomly chosen function and differences $(s^{(\alpha_{i,k})}|s^{(\beta_{i,k})}) \oplus A - (s'^{(\alpha_{i,k})}|s'^{(\beta_{i,k})}) \oplus A$ are chosen randomly uniformly. Sets $Sat(\Delta_i^k)$, $k = 0, 1, 2$ are formed as follows, as described by `derive_inf(i,k)`:

$$Sat(\Delta_i^k) = Sat(\Delta_i^k(0)) \cup \ldots \cup Sat(\Delta_i^k(2^{32} - 1))$$

For a given A, consider a generalized differential from $\Delta_i^k(A)$. The probability that random 32-bit x value will satisfy it is $63/2^{32}$. Since each set of generalized differentials $\Delta_i^k(A)$ contains at least three generalized differentials

$$P[\text{ x satisfies } \Delta_i^k(A)] \leq (63/2^{32})^3 = 2^{-78.068}$$

The probability that among 2^{32} possible x values there exists at least one that will satisfy $\Delta_i^k(A)$, i.e., the probability that $\Delta_i^k(A)$ is satisfiable, is

$$P[\text{ Some } x \in \{0, \ldots, 2^{32} - 1\} \text{ satisfies } \Delta_i^k(A)] \leq 1 - (1 - 2^{-78.068})^{2^{32}} \approx 2^{-46}$$

Finally, the probability that for at least one A there will exist an x that will satisfy $\Delta_i^k(A)$, i.e., the probability that $Sat(\Delta_i^k)$ is nonempty in a random model, is

$$P[\text{ } Sat(\Delta_i^k) \text{ is empty }] \geq (1 - 2^{-46})^{2^{32}} = 1 - 2^{-14} \tag{4}$$

The final set of candidates for $x_{i,t} + c_{i,t+1}$ in procedure `g_inp` is derived as an intersection of $Sat(\Delta_i^0)$, $Sat(\Delta_i^1)$ and $Sat(\Delta_i^2)$. The probability that, in a random model, the intersection of these three sets is non-empty is

$$P[\text{ Randomly modelled } Cand(x_{i,t} + c_{i,t+1}) \text{ nonempty }] \leq (2^{-14})^3 = 2^{-42} \tag{5}$$

This can finally be taken as an upper bound for the probability that there will be an element other than the correct one in $Cand(x_{i,t} + c_{i+1,t})$. Since $Cand(x_{i,t} + c_{i,t+1})$ is calculated for $i = 0, \ldots 7$ at times $t = 0, 1$ during the attack procedure, it can be concluded that there will be no redundant candidates for the internal state after the procedure is completed.

5.2 Attack Complexity

The attack complexity can be measured by the number of faults required, computational complexity as well as storage complexity. First, we examine the number of faults necessary to undertake an attack.

As described above, the input for the attack is a non-faulty keystream word s_{t+1} as well as certain number of faulty keystream words s'_{t+1}. Also, the set of faulty states from which s'_{t+1} values are produced needs to satisfy certain properties. More precisely, as specified by the flt_init procedure, at each of the following groups of bits

$$x_{0,t}, c_{0,t}, \phi_{7,t}$$
$$x_{1,t}, c_{1,t}$$
$$x_{2,t}, c_{2,t} \qquad\qquad\qquad (6)$$
$$\vdots$$
$$x_{7,t}, c_{7,t}$$

the attacker has to produce at least three different faults and obtain three corresponding s'_{t+1} values. It follows that the minimal number of required faults that will need to be induced is $3 \times 8 = 24$. However, since an attacker does not have the possibility to choose locations of faults he induces, the number of necessary faults will be higher.

Let n denote the overall number of induced faults. Let $p(n)$ denote the probability that that there will be at least 3 faults at each one of the 8 groups of bits above. Let A_i be the event that after inducing n random faults there will be at most 2 faults at $x_{i,t}, c_{i,t}$, or $x_{i,t}, c_{i,t}, \phi_{7,t}$ if $i = 0$. Then, $A_i = B_i^0 \cup B_i^1 \cup B_i^2$ where B_i^j, $j = 0, 1, 2$, $i = 0\ldots7$ is an event that at $x_{i,t}, c_{i,t}$ or $x_{i,t}, c_{i,t}, \phi_{7,t}$ if $i = 0$ there will be 0,1 or 2 different faults. Then, $p(n)$ can be approximated as follows:

$$p(n) =$$
$$1 - P[A_0 \cup \ldots \cup A_7] =$$
$$1 - P[(B_0^0 \cup B_0^1 \cup B_0^2) \cup \ldots \cup (B_7^0 \cup B_7^1 \cup B_7^2)] \approx$$
$$1 - (\sum_{i=0}^{7}\sum_{j=0}^{2} P[B_i^j]) - (\sum_{i_1=0}^{7}\sum_{j_1=0}^{2}\sum_{i_2=i_1+1}^{7}\sum_{j_2=0}^{2} P[B_{i_1}^{j_1} \cap B_{i_2}^{j_2}])$$

where the fact that $P[B_i^{j_1} \cap B_i^{j_2}] = 0$ for $j_1 \neq j_2$ has been used. For $i = 0\ldots7$

$$P[B_i^0] = (\frac{7}{8})^n, i = 0\ldots7$$

$$P[B_i^1] = \sum_{k_1+k_2=n-1} (\frac{7}{8})^{k_1}(\frac{1}{8})(\frac{7 \times 32 + 1}{8 \times 32})^{k_2}$$

$$P[B_i^2] = \sum_{k_1+k_2+k_3=n-2} (\frac{7}{8})^{k_1}(\frac{1}{8})(\frac{7 \times 32 + 1}{8 \times 32})^{k_2}(\frac{31}{8 \times 32})(\frac{7 \times 32 + 2}{8 \times 32})^{k_3}$$

and the second order probabilities are provided in Appendix B.

Substituting the according values of n yields $p(64) = 0.900$, $p(96) = 0.997$ and $p(128) = 0.999$. The quality of the approximation above has been verified by the following experiment. For 10^5 times, a data structure equivalent to Rabbit internal state was initialized with zeros and n faults were simulated by writing 1 to a uniformly random chosen bit location. After each iteration, if there was at least three 1-bits at each of the groups of bits in question, a counter was incremented. At the end of the experiment, the probability was obtained by dividing the counter by 10^5. Obtained ratios for $n = 64, 96, 128$ were 0.900, 0.996, 0.999 respectively. Consequently, throughout the rest of the paper, we assume that 64-128 faults are practically sufficient to guarantee that there will be at least 3 faults at each one of the 8 groups of bits defined in Eq. (6).

Since during the attack, as described in Section 4.3, procedure flt_init is called two times, the number of necessary faults is around $128 - 256$.

As for computational and storage complexity, the flt_init procedure can make use of precomputation. In particular, 32 tables $T_0^+, \ldots T_{31}^+$ can be created, such that cell $T_i^+[j]$ contains all the x values such that $j = g(x) - g(x + 2^i)$. Another 32 tables $T_0^-, \ldots T_{31}^-$ can be created, such that cell $T_i^-[j]$ contains all the x values such that $j = g(x) - g(x - 2^i)$. Analogous sets of tables can be created for $[g(x) \lll 8] - g(x \pm 2^i) \lll 8]$ and $[g(x) \lll 16] - g(x \pm 2^i) \lll < 16]$. Thus, the storage complexity is given by $3 \times 64 \times 2^{32} = 2^{39.6}$ words, i.e., $2^{41.6}$ bytes, and now the computational complexity for a query for x such that it satisfies a generalized differential is $O(1)$. Since around $2 \times 8 \times 3 \times 2^{32}$ such queries are made, the computational complexity of the attack is about 2^{38} steps.

To summarize, the proposed attack requires around $128 - 256$ faults, precomputed table of size $2^{41.6}$ bytes, and recovers the cipher internal state in about 2^{38} steps. Further refitment for the attack complexity and success probability is provided in Appendix C.

It should be noted that our proposed attack does not work if the induced faults have a Hamming weight > 1. Extending the ideas presented in this paper to the case where this assumption is relaxed seems to be a challenging research problem.

References

1. Anderson, R., Kuhn, M.: Low Cost Attacks on Tamper Resistant Devices. In: Christianson, B., Crispo, B., Lomas, M., Roe, M. (eds.) Security Protocols 1997. LNCS, vol. 1361, pp. 125–136. Springer, Heidelberg (1998)
2. Aumasson, J.P.: On a bias of Rabbit. In: Proc. of the State of the Art of Stream Ciphers, SASC (2007)
3. Biham, E., Shamir, A.: Differential Fault Analysis of Secret Key Cryptosystems. In: Kaliski Jr., B.S. (ed.) CRYPTO 1997. LNCS, vol. 1294, pp. 513–525. Springer, Heidelberg (1997)

4. Boesgaard, M., Vesterager, M., Pedersen, T., Christiansen, J., Scavenius, O.: Rabbit: A new high-performance Stream Cipher. In: Johansson, T. (ed.) FSE 2003. LNCS, vol. 2887, pp. 307–329. Springer, Heidelberg (2003)
5. Boneh, D., Demillo, R.A., Lipton, R.J.: On the importance of checking cryptographic protocols for faults. In: Fumy, W. (ed.) EUROCRYPT 1997. LNCS, vol. 1233, pp. 37–51. Springer, Heidelberg (1997)
6. Biham, E., Granboulan, L., Nguyen, P.Q.: Impossible Fault Analysis of RC4 and Differential Fault Analysis of RC4. In: Gilbert, H., Handschuh, H. (eds.) FSE 2005. LNCS, vol. 3557, pp. 359–367. Springer, Heidelberg (2005)
7. Cannière, C., Preneel, B.: TRIVIUM: A stream cipher construction inspired by block cipher design principles. In: Katsikas, S.K., López, J., Backes, M., Gritzalis, S., Preneel, B. (eds.) ISC 2006. LNCS, vol. 4176, pp. 171–186. Springer, Heidelberg (2006)
8. Cryptico A/S, Algebaric Analysis of Rabbit (2003), http://www.cryptico.com
9. Cryptico A/S, Analysis of the key setup function in Rabbit (2003), http://www.cryptico.com
10. Cryptico A/S, Hamming weights of the g-function (2003), http://www.cryptico.com
11. Cryptico A/S, Periodic properties of Rabbit (2003), http://www.cryptico.com
12. Cryptico A/S, Second degree approximations of the g-function (2003), http://www.cryptico.com
13. Cryptico A/S, Security Analysis of the IV-setup for Rabbit (2003), http://www.cryptico.com
14. Cryptico A/S, Mod n analysis of Rabbit (2003), http://www.cryptico.com
15. Dusart, P., Letourneux, G., Vivolo, O.: Differential fault analysis on AES. In: Zhou, J., Yung, M., Han, Y. (eds.) ACNS 2003. LNCS, vol. 2846, pp. 293–306. Springer, Heidelberg (2003)
16. Hoch, J., Shamir, A.: Fault Analysis of Stream Ciphers. In: Joye, M., Quisquater, J.-J. (eds.) CHES 2004. LNCS, vol. 3156, pp. 240–253. Springer, Heidelberg (2004)
17. Kocher, P., Jaffe, J., Jun, B.: Differential Power Analysis. In: Wiener, M. (ed.) CRYPTO 1999. LNCS, vol. 1666, pp. 388–397. Springer, Heidelberg (1999)
18. Kocher, P.: Timing Attacks on Implementations of Diffie-Hellman, RSA, DSS, and Other Systems. In: Koblitz, N. (ed.) CRYPTO 1996. LNCS, vol. 1109, pp. 104–113. Springer, Heidelberg (1996)
19. Lu, Y., Wang, H., Ling, S.: Cryptanalysis of Rabbit. In: Wu, T.-C., Lei, C.-L., Rijmen, V., Lee, D.-T. (eds.) ISC 2008. LNCS, vol. 5222, pp. 204–214. Springer, Heidelberg (2008)
20. Hojsík, M., Rudolf, B.: Differential fault analysis of Trivium. In: Nyberg, K. (ed.) FSE 2008. LNCS, vol. 5086, pp. 158–172. Springer, Heidelberg (2008)
21. Hojsík, M., Rudolf, B.: Floating fault analysis of Trivium. In: Chowdhury, D.R., Rijmen, V., Das, A. (eds.) INDOCRYPT 2008. LNCS, vol. 5365, pp. 239–250. Springer, Heidelberg (2008)
22. Maximov, A., Johansson, T.: Fast computation of large distributions and its cryptographic properties. In: Roy, B. (ed.) ASIACRYPT 2005. LNCS, vol. 3788, pp. 313–332. Springer, Heidelberg (2005)
23. Skorobogatov, S.P., Anderson, R.J.: Optical fault induction attacks. In: Kaliski Jr., B.S., Koç, Ç.K., Paar, C. (eds.) CHES 2002. LNCS, vol. 2523, pp. 2–12. Springer, Heidelberg (2003)

24. Zenner, E.: A Cache Timing Analysis of HC-256. In: Avanzi, R., Keliher, L., Sica, F. (eds.) SAC 2008. LNCS, vol. 5381. Springer, Heidelberg (2009)

A   Estimating p_{corr}, p_{incorr} and p_{undef}

In this section, we provide an analytical estimate for the probabilities that the criterion defined in Section 4.2 will return a correct, incorrect or undefined value.

Firstly, we estimate the probability that a fault in one of the $c_{i,t}$, $i = 0..6$ values propagates to $c_{i+1,t+1}$ and not only to $c_{i,t+1}$ during the update step, via carry transfer mechanism implemented by auxiliary $\phi_{i,t+1}$ value. Suppose the fault occurred at position $c_{i,t}$, the probability that $c_{i,t} + a_i + \phi_{i-1,t+1}$ will have a carry at 32-nd bit place is approximately equal to

$$p_{cr} \approx \frac{1}{32} \sum_{i=1}^{32} \frac{1}{2^i} = 0.03125.$$

This probability is given by the event that that addition of $\pm 2^i$, $i = 0..31$ to a random 32-bit number x changes value of $1_{\mathbb{Z}/2^{32}\mathbb{Z}}(x)$. While the a_i values in the actual cipher are fixed ($a_i \in \{4D34D34D, D34D34D3, 34D34D34D\}$), our experimental results confirmed the accuracy of the above approximation.

Then, values p_{corr}, p_{incorr}, p_{undef} are estimated in what follows. Suppose the a random fault was induced in the internal state of the cipher. Then, the position of the bit-flip can be at

- $x_{i,t}, i \in \{0, \ldots 7\}$ with probability $\frac{256}{513}$. In this case, our criterion will return a correct $d_{ST}(S, S')$ value if $g(x_{i,t} + c_{i,t+1}) \neq g(x'_{i,t} + c_{i,t+1})$, i.e., with probability $\frac{2^{32}-1}{2^{32}}$. In case that is not true, the criterion leaves $d_{ST}(S, S')$ undefined.
- $\phi_{7,t}$ with probability $\frac{1}{513}$. In this case, $c'_{0,t+1} \neq c_{0,t+1}$ with probability 1. Let $z \in \{0, \ldots 7\}$ such that $c'_{j,t+1} \neq c_{j,t+1}$ for $j = 0, ..z$ and $c'_{j,t+1} = c_{j,t+1}$ for $j = z + 1, ..7$. If
    - $z = 0$, which happens with probability $\frac{2^{32}-1}{2^{32}}$, and $g(x_{0,t} + c_{0,t+1}) \neq g(x'_{0,t} + c_{0,t+1})$, for which the probability is $\frac{2^{32}-1}{2^{32}}$, then our criterion returns a correct value. If, however, in this case $g(x_{0,t} + c_{0,t+1}) = g(x'_{0,t} + c_{0,t+1})$ which happens with probability $\frac{1}{2^{32}}$, the criterion leaves $d_{ST}(S, S')$ undefined.
    - $z = 1$ which happens with probability $\frac{1}{2^{32}}$. In this case, we consider only the case $g(x_{i,t} + c_{i,t+1}) \neq g(x'_{i,t} + c_{i,t+1})$, $i = 0, 1$, probability being $(\frac{2^{32}-1}{2^{32}})^3$ and in this case again our criterion leaves $d_{ST}(S, S')$ undefined. Other possibilities within the case $z = 1$ are highly improbable and hence do not have any practical implications on the success probability of our attack.
    - $z \geq 2$ occurs with probability $(\frac{1}{2^{32}})^2 \times \frac{2^{32}-1}{2^{32}}$. We do not go into further consideration since these events are highly improbable.

- $c_{i,t}, i \in \{0,\ldots 7\}$, with probability $\frac{256}{513}$. Again, let $z \in \{0,\ldots 7\}$ such that $c'_{j,t+1} \neq c_{j,t+1}$ for $j = i, \ldots i+z$ and $c'_{j,t+1} = c_{j,t+1}$ for $j = i+z+1, ..7$. If
  - $z = 0$, which happens with probability $1 - p_{cr}$ if $i \leq 6$ and with probability 1 if $i = 7$, the same analysis as with a fault on $x_{i,t}$ values applies. Namely, the correct $d_{ST}(S, S')$ value will be returned with probability $\frac{2^{32}-1}{2^{32}}$ and otherwise criterion value in question will be left undefined
  - $z = 1$, which happens with probability $p_{cr} \times \frac{2^{32}-1}{2^{32}}$ if $i \leq 5$, with probability p_{cr} if $i = 6$ and with probability 0 if $i = 7$, the following analysis applies. If values $g(x_{j,t} + c_{j,t+1})$ and $g(x_{j+1,t} + c_{j+1,t+1})$ are both equal to, or both different than $g(x'_{j,t} + c_{j,t+1})$ and $g(x'_{j+1,t} + c_{j+1,t+1})$, respectively, the criterion leaves $d_{ST}(S, S')$ undefined and the probability for this to happen is $(\frac{2^{32}-1}{2^{32}})^2 + (\frac{1}{2^{32}})^2$. However, if $g(x_{j,t} + c_{j,t+1}) = g(x_{j,t} + c'_{j,t+1})$ and $g(x_{j+1,t} + c_{j+1,t+1}) \neq g(x_{j+1,t} + c'_{j+1,t+1})$ the criterion returns the wrong value as an answer. The probability for this to happen is $\frac{1}{2^{32}} \times \frac{2^{32}-1}{2^{32}}$. In case $g(x_{j,t} + c_{j,t+1}) \neq g(x_{j,t} + c'_{j,t+1})$ and $g(x_{j+1,t} + c_{j+1,t+1}) = g(x_{j+1,t} + c'_{j+1,t+1})$, which occurs with the same probability, the criterion returns the right answer.
  - $z = 2$, which happens with probability $p_{cr} \times \frac{1}{2^{32}} \times \frac{2^{32}-1}{2^{32}}$ if $i \leq 4$, with probability $p_{cr} \times \frac{1}{2^{32}}$ if $i = 5$ and with probability 0 if $i \geq 6$, the following consideration applies. In the case where all three g values are changed, $d_{ST}(S, S')$ value is left undefined and the probability for this case is $(\frac{2^{32}-1}{2^{32}})^3$. Other cases are highly improbable and we do not consider them.
  - $z \geq 3$ occurs with probability $p_{cr} \times \frac{1}{2^{32}} \times \frac{2^{32}-1}{2^{32}}$. We do not go into consideration of further cases since their corresponding probabilities are negligible.

Using the probabilities from the discussion above, but ignoring parts that are less than $\frac{1}{2^{32}}$, provides a practically accurate estimate for the probability that the criterion will return a correct $d_{ST}(S, S')$ value as follows:

$$p_{corr} \approx \frac{256}{513} \frac{2^{32}-1}{2^{32}} + \frac{1}{513} (\frac{2^{32}-1}{2^{32}})^2 +$$
$$\frac{7 \times 32}{513}(1 - p_{cr})\frac{2^{32}-1}{2^{32}} + \frac{32}{513}(1 \times \frac{2^{32}-1}{2^{32}}) = 0.98635$$

Again, ignoring terms that are less than $\frac{1}{2^{32}}$, probability of $d_{ST}(S, S')$ being left undefined is given by

$$p_{undef} \approx \frac{6 \times 32}{513} \times p_{cr} \times \frac{2^{32}-1}{2^{32}}((\frac{2^{32}-1}{2^{32}})^2 + (\frac{1}{2^{32}})^2) +$$
$$\frac{32}{513} \times p_{cr} \times ((\frac{2^{32}-1}{2^{32}})^2 + (\frac{1}{2^{32}})^2) = 0.013645$$

Finally, ignoring terms less than $\frac{1}{2^{32}}$ yields probability for the criterion to return a false $d_{ST}(S, S')$ value is $p_{incorr} \approx 0$.

B   Second Order Terms in $p(n)$ Probability

As defined in section 5.2, $p(n)$ denotes the probability that there will be at least 3 faults at each one of the eight groups of bits defined in Eq. (6). In the following we provide formulas for second-order terms participating in the equation for $p(n)$:

$$P[B_{i_1}^0 \cap B_{i_2}^0] = (\frac{6}{8})^n$$

$$P[B_{i_1}^0 \cap B_{i_2}^1] = \sum_{k_1+k_2=n-1} (\frac{6}{8})^{k_1} \frac{1}{8} (\frac{6 \times 32 + 1}{8 \times 32})^{k_2}$$

$$P[B_{i_1}^0 \cap B_{i_2}^2] = \sum_{k_1+k_2+k_3=n-2} (\frac{6}{8})^{k_1} \frac{1}{8} (\frac{6 \times 32 + 1}{8 \times 32})^{k_2} \frac{31}{8 \times 32} (\frac{6 \times 32 + 2}{8 \times 32})^{k_3}$$

$$P[B_{i_1}^1 \cap B_{i_2}^2] =$$

$$3 \times \sum_{k_1+k_2+k_3+k_4=n-3} (\frac{6}{8})^{k_1} \frac{1}{8} (\frac{6 \times 32 + 1}{8 \times 32})^{k_2} \frac{1}{8} (\frac{6 \times 32 + 2}{8 \times 32})^{k_3} \frac{31}{8 \times 32} (\frac{6 \times 32 + 3}{8 \times 32})^{k_4}$$

$$P[B_{i_1}^2 \cap B_{i_2}^2] =$$

$$6 \times \sum_{k_1+k_2+k_3+k_4+k_5=n-4} (\frac{6}{8})^{k_1} \frac{1}{8} (\frac{6 \times 32 + 1}{8 \times 32})^{k_2} \frac{1}{8} (\frac{6 \times 32 + 2}{8 \times 32})^{k_3}$$

$$\frac{31}{8 \times 32} (\frac{6 \times 32 + 3}{8 \times 32})^{k_4} \frac{31}{8 \times 32} (\frac{6 \times 32 + 4}{8 \times 32})^{k_5}$$

C   On the Non-surjectiveness Property of the Function Used to Derive Differences

In this appendix, we examine the expression

$$([s^{(\alpha_{i,j})}||s^{(\beta_{i,j})}] \oplus A) - ([s'^{(\alpha_{i,j})}||s'^{(\beta_{i,j})}] \oplus A)$$

which is used to derive differences in derive_inf(i,j) procedure. One set of generalized differentials is created by fixing A and applying the expression to pairs of groups of bits of non-faulty and faulty keystream words (s, s'), (s, s''). If the function described by the expression above was $1 - 1$, the number of such different sets when A goes from $0, \ldots 2^{32} - 1$ would be 2^{32}. However, since the expression above is not a $1 - 1$ function, the question of what is the degree of repetition of generalized differential sets when A changes arises. In this section, we give an answer to this question and show how this can be used to provide some minor improvements in the computational complexity of the attack. Better lower bounds for the attack success probabilities are also provided.

Isolating the function given by the expression above yields

$$\Phi_{x,y}(A) = x \oplus A - y \oplus A$$

where x, y and A are 32-bit values. Clearly, the function is not $1-1$. For example, for A and A' such that $d(A, A') = \{31\}$, $\Phi_{x,y}(A) = \Phi_{x,y}(A')$ where $d(\cdot, \cdot)$ denotes the set of bit positions on which the enclosed bit strings differ, least significant bit being bit 0 (e.g., $d(0000, 1010) = \{0, 2\}$). Rephrasing the question from previous paragraph yields the problem of how many different sets

$$\{\Phi_{s,s'}(A), \Phi_{s,s''}(A), \ldots \Phi_{s,s'\ldots'}(A)\} \tag{7}$$

is expected to be constructed when A goes from 0 to $2^{32} - 1$.

To start, we are interested for which A, A' will $\Phi_{x,y}(A) = \Phi_{x,y}(A')$ hold. Let $z = \Phi_{x,y}(A)$ and $z' = \Phi_{x,y}(A')$. Then, $z_i = x_i \oplus A_i \oplus y_i \oplus A_i \oplus c_i$, where z_i denotes i-th bit of z. Cancelling out A_i and same reasoning for z' gives

$$z_i = x_i \oplus y_i \oplus c_i \tag{8}$$

$$z'_i = x_i \oplus y_i \oplus c'_i \tag{9}$$

where

$$c_i = \begin{cases} 0 \text{ if } (x_{i-1} \oplus A_{i-1}, y_{i-1} \oplus A_{i-1}, c_{i-1}) \in \\ \quad \{(1,0,0), (1,0,1), (1,1,0), (0,0,0)\} \\ 1 \text{ if } (x_{i-1} \oplus A_{i-1}, y_{i-1} \oplus A_{i-1}, c_{i-1}) \in \\ \quad \{(0,1,0), (0,1,1), (0,0,1), (1,1,1)\} \end{cases} \tag{10}$$

and c'_i is defined analogously, with A' instead of A.

As already noted, if $d(A, A') = \{31\}$, $\Phi_{x,y}(A) = \Phi_{x,y}(A')$ for every x and y. Let $S_1 = \{i | x_i = y_i, i \neq 31\}$.

Lemma 1. $\Phi_{x,y}(A) = \Phi_{x,y}(A') \Leftrightarrow d(A, A') \subseteq S_1 \cup \{31\}$

Proof. Let $z = \Phi_{x,y}(A)$ and $z' = \Phi_{x,y}(A')$.
($\Rightarrow$:) Assuming that the negation of the right side of equation is true, the goal is to prove that $z \neq z'$. According to (8) and (9), it is sufficient to show that $c_i \neq c'_i$ for some $i = 0, \ldots 30$. Let $i_0 = \min\{i | i \in d(A, A'), i \notin S_1\}$. According to (10), since if $c_{i+1} = 1$, then $c'_{i+1} = 0$ and vice versa. Thus, $z_{i_0+1} \neq z'_{i_0+1}$ and $z \neq z'$.
($\Leftarrow$:) It is sufficient to prove that $c_i = c'_i$ for $i = 0, \ldots 30$. We do this by induction. Let $i_0 = min(d(A, A'))$. Obviously, $c_i = c'_i$ for $i = 0, \ldots i_0 - 1$. According to (10) and using that $c_{i_0-1} = c'_{i_0-1}$ yields $c_{i_0} = c'_{i_0}$. Now let $i_n \in d(A, A')$. By induction hypothesis, $c_{i_n-1} = c'_{i_n-1}$. Again, using (10) yields $c_{i_n} = c_{i_n}$, which proves this part of the Lemma.

In other words, for any given A, changing bits from $S_1 \cup \{31\}$ will not change $\Phi_{x,y}(A)$. On the other hand, changing any other bits in A changes $\Phi_{x,y}(A)$. Hence

$$|\mathrm{Im}(\Phi_{x,y})| = 2^{32-(|S_1|+1)} \tag{11}$$

Since the expected value for $|S_1|$ is 15.5, the expected number for $E(|\mathrm{Im}(\Phi_{x,y})|) = 2^{32-15.5-1} = 2^{15.5}$. Analogous analysis can be used to estimate the expected number of different sets (7) when A goes from $0, \ldots 2^{32} - 1$. Denote s' by $s^{[1]}$, s'' by $s^{[2]}$, etc. Define the set S_n as follows

$$S_n = \{i | s_i = s_i^{[1]} = \ldots = s_i^{[n]}, i \neq 31\}$$

A generalization of the previous lemma can be used:

Lemma 2

$$(\Phi_{s,s^{[1]}}(A), \ldots, \Phi_{s,s^{[n]}}(A)) = (\Phi_{s,s^{[1]}}(A'), \ldots, \Phi_{s,s^{[n]}}(A'))$$
$$\Leftrightarrow d(A, A') \subseteq S_n \cup \{31\}$$

To calculate $E(|S_n|)$, let X_i, $i = 0, ..30$ denote a random variable which takes value 1 if $s_i = s_i^{[1]} = \ldots = s_i^{[n]}$ and value 0 otherwise. It is easy to see that $E(X_i) = \frac{1}{2^n}$, $i = 0, \ldots 30$. Then, for some particular instantiation of s values, the number of mutually equal bits is given by $X = X_0 + \ldots X_{30}$, the most significant bit being excluded. Hence $E(X) = E(X_0 + \ldots + X_{30}) = E(X_0) + \ldots + E(X_{30}) = 31 \times \frac{1}{2^n}$.

Now the question from the beginning of this appendix can be answered by

$$E_n = E(|\{\Phi_{s,s^{[1]}}(A), \Phi_{s,s^{[2]}}(A), \ldots \Phi_{s,s^{[n]}}(A)\}_{A=0}^{2^{32}-1}|) \approx$$
$$E(|(\Phi_{s,s^{[1]}}(A), \ldots, \Phi_{s,s^{[n]}}(A))_{A=0}^{2^{32}-1}|) = 2^{32-(|S_n|+1)} = 2^{32-((\frac{31}{2^n})+1)}$$

Table 2 shows the expected number of different $\Delta_i^j(A)$ sets when A varies from 0 to $2^{32} - 1$, for different $|FLTS_i| = n$ sizes.

Thus the complexity and success probability of the attack can now be further refined as follows:

- Computational complexity: in procedure `derive_inf(i,j)`, in the main loop, it is possible not to go through the whole set of A values, by fixing bit positions from determined by set S_n, i.e., bit positions on which $s, s', \ldots$, are mutually equal. If $|FLTS_i| = 3$, the loop will not have 2^{32} steps, but $2^{27.125}$ steps on average. Consequently, the overall attack complexity is now reduced to $2^{32.71}$ steps on average.
- Success probability: Eq. (4) now becomes

$$P[Sat(\Delta_i^j) \text{ is empty }] \geq (1 - 2^{-46})^k$$

where instead of $k = 2^{32}$, k takes values from the table, depending on $|FLTS_i| = n$. For $|FLTS_i| = 3$, the above value becomes equal to $1 - 2^{-18.9}$ instead of $1 - 2^{-14}$ and probability that the algorithm will return more than one element in set $Cand(x_{i,t} + c_{i,t+1})$ given by (5) will be 2^{-56} instead of 2^{-42}.

Table 2. Expected number of different $\Delta_i^j(A)$ sets

n	3	4	5	6	7	8
E_n	$2^{27.125}$	$2^{29.062}$	$2^{30.031}$	$2^{30.516}$	$2^{30.758}$	$2^{30.879}$

An Improved Recovery Algorithm for Decayed AES Key Schedule Images

Alex Tsow

The MITRE Corporation*
atsow@mitre.org

Abstract. A practical algorithm that recovers AES key schedules from decayed memory images is presented. Halderman et al. [1] established this recovery capability, dubbed the *cold-boot attack*, as a serious vulnerability for several widespread software-based encryption packages. Our algorithm recovers AES-128 key schedules tens of millions of times faster than the original proof-of-concept release. In practice, it enables reliable recovery of key schedules at 70% decay, well over twice the decay capacity of previous methods. The algorithm is generalized to AES-256 and is empirically shown to recover 256-bit key schedules that have suffered 65% decay. When solutions are unique, the algorithm efficiently validates this property and outputs the solution for memory images decayed up to 60%.

Keywords: anti-tamper, digital forensics, decayed memory, cold-boot attack, AES, key schedule.

1 Introduction

Cold-boot attacks are another troubling example of the increasingly sophisticated threats to security and privacy. In response to these threats we investigate defensive anti-tamper techniques in the hopes of better understanding the potential of specific attack vectors. In this paper we report on our investigation of cold boot attacks and demonstrate that the problem is more serious than previously thought. We present AES key recovery techniques that handle over twice the decay rate of prior methods at comparable computational effort.

The *cold-boot attack* [1] is a serious vulnerability for software-based encryption packages—including BitLocker, FileVault and the open-source project TrueCrypt—where one can recover secret keys from decayed memory images. Decryption with decayed AES keys does not produce original plaintexts. However, the redundancy of key material inherent in the AES key schedule can rectify these faults. When combined with *asymmetric decay*, where bits overwhelmingly decay to their ground state rather than their charged state, this redundancy enables reconstruction of the original key. Heninger and Halderman have developed a recovery algorithm for AES-128 that recovers keys from 30% decayed data in less than 20 minutes about half the time.

---

* Approved for Public Release; Distribution Unlimited; Tracking Number 09-1872.

M.J. Jacobson Jr., V. Rijmen, and R. Safavi-Naini (Eds.): SAC 2009, LNCS 5867, pp. 215–230, 2009.
© Springer-Verlag Berlin Heidelberg 2009

Our algorithm recovers keys up to several orders of magnitude faster than Heninger and Halderman's method. One case that took their algorithm more than 10 days to solve was solved by our improved method in 0.047 seconds. The speed increase enables key recovery from more severely degraded memory images. In an experimental evaluation, our algorithm recovered all keys from a 5,000 case test suite at 70% decay, with 4,927 instances recovered in less than 20 minutes—more than twice decay rate with almost double the success rate in 20 minutes. The speed increase also makes it feasible to enumerate *all* keys from which an image could have decayed, rather than halt on the first key that satisfies the decay and schedule constraints. In particular, the algorithm can determine that a solution is unique. Benchmarks demonstrate feasibility up to 60% decay where there is approximately a 2.5× slowdown compared with the halt-on-first-key search. The algorithm generalizes to 256-bit AES with only a moderate drop in the recovery capability. Empirically, the benchmarks show that AES-256 recovery begins to degrade around 65%; there are no other performance claims in the open literature about AES-256 recovery.

The AES key-schedule is the primary source of key redundancy. For the 128-bit version, the original key is bijectively mapped to 10 additional round-keys [2,3]. The mappings form a system of byte-level equations that constrain the space of likely key candidates.

The asymmetric decay property of DRAM provides a second set of constraints. When the refresh cycle of DRAM is interrupted, the data overwhelmingly decays to 0 (or 1 assuming the complementary encoding for the ground state) because the capacitance is lost over time. Occasionally bits invert to the charged state, although Halderman et al. bound these effects at 0.1%. The asymmetric decay property suggests a compatibility criterion for key candidates: if a candidate schedule subset differs from the decayed memory image only by inversions from the ground state, then it is compatible with the decayed memory image. The performance claims for our algorithm, and indeed those in [1] and [4], are based on the *perfect* asymmetry assumption, where no bit in the ground state ever inverts. The algorithm has also been adapted to accommodate inversions to the charged state by generalizing the compatibility criterion to allow a bounded number of such cases. No other logic changes are necessary.

Key reconstruction is possible because key candidates must satisfy both the asymmetric decay property and the system of equations defined by the AES key schedule. Our algorithm explores a tree of one-byte guesses. At each stage, or tree-depth, the new byte candidate and all bytes implied by the schedule equations are checked against the decayed image. The algorithm guesses bytes in an order such that guess n implies values for n schedule bytes for $n < 11$ in the 128-bit version; guesses 11-14 imply an additional 10 bytes each and the last guess implies the schedule's remaining 65 bytes.

Byte guessing proceeds in a depth-first manner. Each stage has 256 possibilities, but the schedule and decay-compatibility constraints quickly prune the possibilities, particularly in the later stages where a single byte guess implies several byte values. The selection and order of byte guessing is not unique.

Section 3.4 describes a path selection heuristic for byte guesses which improves recovery times by a factor of several hundred for decay rates over 50%, when compared with the static path implementation.

We make the following contributions: *1)* our recovery algorithm is several orders of magnitude faster than the best previously published method, *2)* the new method enables key recovery from images with significantly more decay, *3)* it enumerates all solutions to a decay image and *4)* we generalize the method to AES-256 with little loss of decay performance.

Organization: Section 2 reviews related works. Section 3 describes the algorithm in detail, including a heuristic to optimize the exploration path. Section 4 presents benchmarks for the algorithm with and without the path optimization heuristic. Benchmarks for the unique determination capability and AES-256 key recovery are also presented. Section 5 makes some observations about the benchmark results. Section 6 concludes the paper.

2   Related Work

Halderman et al. [1] established the cold-boot attack as a low-cost way to extract private key information from computers running software encryption. In particular, they extracted private keys for full-disk encryption packages such as BitLocker, FileVault, and the cross-platform open-source project TrueCrypt. Heninger and Halderman released proof-of-concept implementation that recovers 128-bit AES key-schedules.[1] It implements the algorithm from [1] which they have found to recover keys from 30% decayed memory within 20 minutes about half the time. Their archives also contain a recovery algorithm for the RSA cryptosystem.

Heninger and Shacham [4] vastly improved the ability to recover RSA private keys from decayed memory images. They improve recovery from 6% decay (running on the order of minutes) to 46% decay (running on the order of seconds) when p, q, d, d_p, and d_q are in the image. The paper further casts their recovery algorithm in terms of *known bits*, so that the bits may be randomly selected rather than simply the result of an asymmetric memory decay. We note that a perfect memory image maps to 50% known bits under the asymmetric decay assumption, since valid ground-state values theoretically could have decayed.

Nearly a decade before the cold boot attack was demonstrated, Handschuh, Paillier, and Stern modeled *probing attacks* [5] on the square-and-multiply algorithm for modular exponentiation, DES, and RC5. They reconstruct cryptographic secrets by tracing a few critical bits over the target operation's execution. Since cold-boot attacks capture a snapshot of the execution state, these techniques only apply if a trace has been preserved in memory.

Akavia, Goldwasser, and Vaikuntanathan [6] present a model of cold-boot memory attacks in terms of experiments with probabilistic polynomial time players. The recovery player chooses a sequence of probing functions that map private

---

[1] http://citp.princeton.edu/memory/code/

keys to bit vectors; this models key material leakage. They define *adaptive* and non-adaptive variants which may or may not alter the choice of probing function in response to the results of previous probes. They further show that the Regev public key cryptosystem [7] is secure under both definitions, but with different leak parameters.

Naor and Segev [8] revisit the above formalism for memory attacks and develop a schema for constructing public key cryptosystems that are resilient against key leakage. The schema relies on the assumptions of a *universal hash proof system* [9] enabling decisional Diffie-Hellman, quadratic residuosity, and Paillier's composite residuosity problem to instantiate the cryptosystem. Alwen, Dodis, and Wichs also examine leakage resilient public key cryptosystems, including identification schemes and authenticated key agreement protocols [10]. They extend their results to the bounded retrieval model, where they consider extremely large keys and an adversary can not learn more than a predetermined bound over a lifetime. Katz further constructs a leakage resistant signature scheme in the standard model [11].

Chari et al. propose the first theoretical model for power analysis [12]. Coron, Naccache, and Kocher develop a similar formalism for characterizing leakage immunity, and present several leakage detection tests [13]. Micali and Reyzin propose a general framework for security against side-channel analysis [14]. These models do not account for memory remanence or cold-boot attacks.

Countermeasures to cold-boot attacks remain scarce within the current technology paradigms. Migrating to hardware embedded encryption, such as that proposed by the Trusted Computing Group's Opal platform [15], will mitigate cold-boot attacks on full-disk encryption. Enck et al. [16] propose an encrypting memory controller that writes only encrypted data to main memory, but decrypts it on reads into the processor or cache. There have also been attempts to manipulate the Intel x86 cache-coherence model to ensure that keys and key-derived state (such as key schedules) remain in L2 caches, but not in main memory.[2] The feasibility of this approach has yet to be demonstrated with the current architectures, however vendor modification of the instruction set may indeed make this approach a reality.

Intel has developed specialized instructions for executing AES operations [17]. There are six kinds instructions (encrypt round, encrypt last round, decrypt round, decrypt last round, inverse mix columns, and key schedule assist) which use the 128-bit XMM registers to hold round keys and block data. In addition to improving execution speed, these instructions have been designed to eliminate vulnerabilities from *cache attacks* [18], an interprocess side-channel that exploits timing differences for operations dependent upon cached and uncached data. Intel does not claim that these instructions mitigate cold-boot attacks, however we speculate that the schedule derivation assistance may improve the performance of just-in-time round-key derivation, thereby reducing the number of round keys stored in memory.

---

[2] Jürgen Pabel. http://frozencache.blogspot.com/

3 Algorithmic Description

This primary exposition details the recovery algorithm for AES-128, although the concepts generalize to the 192-bit and 256-bit cases (Section 3.5). Their differing block and key sizes create more potential for confusion when referencing schedule elements. We have implemented the 128-bit and 256-bit cases; Section 4 presents performance results for both cases.

3.1 Preliminaries

A 128-bit AES key schedule expands a four by 32-bit-word key into a 44 word sequence. Schedule components are addressed with the following notation: Let sans-serif variables, S, refer to entire key schedule. Subscripting expresses a hierarchical view of schedule components. Let S_r refer to the four words of round r. Let $S_{r,w}$ refer to word w of round r. Let $S_{r,w,b}$ refer to byte b of word w of round r. It is also convenient to index the schedule in a flat manner. Let S^w_i refer to word i of the schedule. Let S^b_i refer to byte i of the schedule. The notation follows the least-significant-byte-first convention. Some additional function notation is necessary to express the key schedule. Let $sbox(S^b_i)$ apply the AES substitution box to the byte S^b_i. For convenience, let $sbox(S^w_i)$ apply the substitution box to each constituent byte when S^w_i is a word. Let $rot(S^w_i)$ rotate the word S^w_i by eight bit positions of increasing significance; e.g. $rot(S_{r,w,0}, S_{r,w,1}, S_{r,w,2}, S_{r,w,3}) = (S_{r,w,3}, S_{r,w,0}, S_{r,w,1}, S_{r,w,2})$, in the least-significant-byte-first representation. The round constants, denoted by $rcon[i]$, are the $(i-1)^{th}$ exponent of 2 in the field $GF(2^8)$ for the least significant byte and 0 for the other bytes.

For the 128-bit schedule, the first four words are the key itself. The subsequent words are prescribed by two equation schema.

$$\begin{aligned} S^w_i &= S^w_{i-1} \oplus sbox(rot(S^w_{i-3})) \oplus rcon[i/4], & \text{when } i \bmod 4 = 0 \\ S^w_i &= S^w_{i-1} \oplus S^w_{i-3}, & \text{when } i \bmod 4 \neq 0 \end{aligned} \tag{1}$$

The table below illustrates the indexing schema for the 128-bit key schedule. The column headings indicate the byte and word indices. Row labels indicate the round. This particular table shows how the flat byte-level references relate to the hierarchical tags. For instance, $S_{8,2,3}$ refers to the same byte as S^b_{139}.

w	0				1				2				3			
r \ b	0	1	2	3	0	1	2	3	0	1	2	3	0	1	2	3
0	S^b_0	S^b_1	S^b_2	S^b_3	S^b_4	S^b_5	S^b_6	S^b_7	S^b_8	S^b_9	S^b_{10}	S^b_{11}	S^b_{12}	S^b_{13}	S^b_{14}	S^b_{15}
1	S^b_{16}	S^b_{17}	S^b_{18}	S^b_{19}	S^b_{20}	S^b_{21}	S^b_{22}	S^b_{23}	S^b_{24}	S^b_{25}	S^b_{26}	S^b_{27}	S^b_{28}	S^b_{29}	S^b_{30}	S^b_{31}
2	S^b_{32}	S^b_{33}	S^b_{34}	S^b_{35}	S^b_{36}	S^b_{37}	S^b_{38}	S^b_{39}	S^b_{40}	S^b_{41}	S^b_{42}	S^b_{43}	S^b_{44}	S^b_{45}	S^b_{46}	S^b_{47}
3	S^b_{48}	S^b_{49}	S^b_{50}	S^b_{51}	S^b_{52}	S^b_{53}	S^b_{54}	S^b_{55}	S^b_{56}	S^b_{57}	S^b_{58}	S^b_{59}	S^b_{60}	S^b_{61}	S^b_{62}	S^b_{63}
4	S^b_{64}	S^b_{65}	S^b_{66}	S^b_{67}	S^b_{68}	S^b_{69}	S^b_{70}	S^b_{71}	S^b_{72}	S^b_{73}	S^b_{74}	S^b_{75}	S^b_{76}	S^b_{77}	S^b_{78}	S^b_{79}
5	S^b_{80}	S^b_{81}	S^b_{82}	S^b_{83}	S^b_{84}	S^b_{85}	S^b_{86}	S^b_{87}	S^b_{88}	S^b_{89}	S^b_{90}	S^b_{91}	S^b_{92}	S^b_{93}	S^b_{94}	S^b_{95}
6	S^b_{96}	S^b_{97}	S^b_{98}	S^b_{99}	S^b_{100}	S^b_{101}	S^b_{102}	S^b_{103}	S^b_{104}	S^b_{105}	S^b_{106}	S^b_{107}	S^b_{108}	S^b_{109}	S^b_{110}	S^b_{111}
7	S^b_{112}	S^b_{113}	S^b_{114}	S^b_{115}	S^b_{116}	S^b_{117}	S^b_{118}	S^b_{119}	S^b_{120}	S^b_{121}	S^b_{122}	S^b_{123}	S^b_{124}	S^b_{125}	S^b_{126}	S^b_{127}
8	S^b_{128}	S^b_{129}	S^b_{130}	S^b_{131}	S^b_{132}	S^b_{133}	S^b_{134}	S^b_{135}	S^b_{136}	S^b_{137}	S^b_{138}	S^b_{139}	S^b_{140}	S^b_{141}	S^b_{142}	S^b_{143}
9	S^b_{144}	S^b_{145}	S^b_{146}	S^b_{147}	S^b_{148}	S^b_{149}	S^b_{150}	S^b_{151}	S^b_{152}	S^b_{153}	S^b_{154}	S^b_{155}	S^b_{156}	S^b_{157}	S^b_{158}	S^b_{159}
10	S^b_{160}	S^b_{161}	S^b_{162}	S^b_{163}	S^b_{164}	S^b_{165}	S^b_{166}	S^b_{167}	S^b_{168}	S^b_{169}	S^b_{170}	S^b_{171}	S^b_{172}	S^b_{173}	S^b_{174}	S^b_{175}

A candidate key schedule byte, C^b_i, with ground states specified by M^b_i is *compatible with the decayed byte*, D^b_i, when D^b_i preserves all ground-state bits in C^b_i, or expressed equationally, when $(C^b_i \oplus D^b_i) \wedge (C^b_i \oplus M^b_i) = 0$.

3.2   Maximizing the Implied Schedule Bytes

For the first guess, the candidate byte, $C^b_{i_0}$, is only constrained by the known bits in corresponding decayed byte, $D^b_{i_0}$. Yet in the second stage, for a properly chosen i_1, $C^b_{i_1}$ is constrained by $D^b_{i_1}$ *and* a second byte D^b_j. A properly selected i_1 will instantiate a byte slice of one of the two schedule generating equations, (1). For example $C^b_4 \oplus C^b_{16} = C^b_{20}$ is the first byte-slice of the generating equation for $C^w_5 = (C^b_{20}, C^b_{21}, C^b_{22}, C^b_{23})$. If $i_0 = 4$ and $i_1 = 20$, then the implied byte is at index $j = 16$ and equals $C^b_{20} \oplus C^b_4$. Thus D^b_{16} constrains the implied value of $C^b_{20} \oplus C^b_4$.

This algorithm makes use of the following observations: *1)* Each schedule byte $S^b_i, 16 \leq i < 160$ is involved in three equations. *2)* There are 256 solutions to each equation when any one variable is fixed. There is a unique solution to each equation when any two variables are fixed. *3)* Guessing a single byte at stage n implies up to n other byte values for properly structured guessing orders.

Item 1 follows from the fact that every word is generated by its preceding word and the one 4 words ago; simply limit the scope to the byte-slice of the concerned byte. For example, consider $S_{1,0,0}=S^b_{16}$, the first byte of S^w_4.

$$
\begin{aligned}
S_{0,0,0} \oplus \mathsf{sbox}(S_{0,3,1}) \oplus \mathtt{0x01} &= S_{1,0,0} \\
S_{0,1,0} \oplus \phantom{\mathsf{sbox}()} S_{1,0,0} &= S_{1,1,0} \\
S_{1,0,0} \oplus \mathsf{sbox}(S_{1,3,1}) \oplus \mathtt{0x02} &= S_{2,0,0}
\end{aligned}
\tag{2}
$$

Item 2 is implied by the fact that $\oplus$ is the field addition operation for $GF(2^8)$ and that $\mathsf{sbox}()$ is a bijection in $GF(2^8)$.

To see item 3, consider the following candidate exploration order, C, for a decayed schedule D. First choose a candidate for $C_{0,0,0}$; it is only constrained by the known bits at that position. The next guess, $C_{1,0,0}$, is constrained by the known bits from $D_{1,0,0}$. Additionally, one can solve the equation $C_{0,0,0} \oplus$

Table 1. An order for byte-guesses and consequent values in AES-128; the end of Section 3.2 describes the scripting notation

w	0				1				2				3			
r\b	0	1	2	3	0	1	2	3	0	1	2	3	0	1	2	3
0	0_0	14_{10}			13_{10}				12_{10}				1_1	14_9		
1	1_0	13_9			12_9				2_2	14_8			2_1	13_8		
2	2_0	12_8			3_3	14_7			3_2	13_7			3_1	12_7		
3	3_0	4_4	14_6		4_3	13_6			4_2	12_6			4_1	5_5	14_5	
4	4_0	5_4	13_5		5_3	12_5			5_2	6_6	14_4		5_1	6_5	13_4	
5	5_0	6_4	12_4		6_3	7_7	14_3		6_2	7_6	13_3		6_1	7_5	12_3	
6	6_0	7_4	8_8	14_2	7_3	8_7	13_2		7_2	8_6	12_2	14_1	7_1	8_5	9_9	
7	7_0	8_4	9_8	13_1	8_3	9_7	12_1	14_0	8_2	9_6	10_{10}	13_0	8_1	9_5	10_9	
8	8_0	9_4	10_8	12_0 15_{10}	9_3	10_7	11_{10}	15_9	9_2	10_6	11_9	15_8	9_1	10_5	11_8	
9	9_0	10_4	11_7	15_7	10_3	11_6	15_6		10_2	11_5	15_5		10_1	11_4	15_4	
100	10_0	11_3	15_3		11_2	15_2			11_1	15_1			11_0	15_0		

$\mathsf{sbox}(C_{0,3,1}) = C_{1,0,0}$ for $C_{0,3,1}$, since $C_{0,0,0}$ and $C_{1,0,0}$ have candidate values. Thus the second guess is constrained by the known values at $D_{0,3,1}$ as well. By the same logic, $D_{2,0,0}$, $D_{1,3,1}$, and $D_{1,2,1}$ constrain the compatible guesses for $C_{2,0,0}$. Because $C_{r-1,3}$ rotates when computing $C_{r,0}$, continuing to propose bytes in the column $C_{r,0,0}$ causes the implied byte indices increment modulo 4 when the word index wraps around modulo 4.

Table 1 illustrates this behavior by enumerating the order of byte guesses and their consequent bytes. In the table entries, the full-sized number indicates the guessing stage and the subscript indicates the sequence of implied bytes. In particular, i_0 indicates a guess for stage i and i_1 is the first implied byte from this candidate. For example, $C_{5,0,0}$ is guess number 5. This value combined with the value for $C_{4,0,0}$ (chosen in step 4), allows one to solve for $C_{4,3,1}$. The following equations make the order of solution explicit.

$$\begin{aligned}
C_{4,3,1} &= \mathsf{sbox}^{-1}(C_{5,0,0} \oplus C_{4,0,0} \oplus 01) \\
C_{4,2,1} &= C_{4,3,1} \oplus C_{3,3,1} \\
C_{4,1,1} &= C_{4,2,1} \oplus C_{3,2,1} \\
C_{4,0,1} &= C_{4,1,1} \oplus C_{3,1,1} \\
C_{3,3,2} &= \mathsf{sbox}^{-1}(C_{4,0,1} \oplus C_{3,0,1} \oplus 01)
\end{aligned} \tag{3}$$

After selecting candidates for bytes 0-10, there are a number of ways to guess bytes 11-15. Table 1 illustrates a choice for these positions that implies values for an additional 10 bytes for each guess. The last guess implies 65 bytes because it causes round 8 to be fully specified; the entire schedule may be derived from any complete round.

3.3   The Recovery Algorithm

The algorithm, `recoverKeyRec`, explores the candidate space one byte at a time. It exploits the constraints on guesses and their consequent bytes to prune its exploration tree. For each guess, `recoverKeyRec` considers all 256 possible values. If the candidate satisfies all constraints imposed by the decayed image, D, then it guesses a value for the next step. Exploration proceeds in a depth-first manner, so that guess i is incremented to the next compatible candidate when all of descendant candidates have been ruled out.

A breadth-first search is also possible, however this strategy greatly increases the memory utilization. The advantage of the breadth-first method is that one could track the distance from decayed data for each candidate and explore the closest options first. The first implementations of this algorithm maintained the candidates on a binary heap indexed by their cost. In practice, there were many cases in the 60-70% decay range, where the process exceeded its 31-bit address space and halted before recovering the key. On the other hand, recovery speeds of depth-first search up to the 70% rate have proven fast enough in most cases and solvable in all attempted cases. The need to solve more cases with less memory dictated a transfer to depth-first search which consumes a fixed amount of memory, about 4.2 MB in the experimental implementation. Because depth-first search has a small memory footprint, it is also inexpensive to halt tree

```
recoverKeyRec(CandidateMatrix c, DecaySchedule d):
  if (c.length()==16):
    return c.key()
  for i=0 to 255:
    if(d.isCompatible(c.guess(i))):
      key = recoverKeyRec(c.guess(i),d)
      if (key != NULL)
        return key
  return NULL
```

Fig. 1. Recursive expression of key recovery; Appendix A details the core methods

exploration and resume later. The only data necessary to save is the candidate being examined at the halting time. Breadth-first search would require one to save the binary heap of candidates.

The following describes the semantics for tokens in `recoverKeyRec` (Figure 1). For simplicity of expression assume that operations do not mutate objects, but return newly constructed objects. Italicized tokens refer to classes and teletype tokens refer to fields and methods.

Let c be a *CandidateMatrix* that contains candidates for schedule bytes in the order indicated by Table 1. *CandidateMatrix* maintains a `count` of how many guesses have been made (0-16) and a flat schedule representation of 176 bytes to store byte candidates and their consequent bytes. The method `guess(`*Byte* b`)`, returns a new *CandidateMatrix* whose array has been updated to contain b at position $count_0$ and its consequent bytes at positions as specified by the path matrix (e.g., Table 1). In particular, guessing the 16th byte completes the entire schedule. The `key()` method returns the key once all 16 bytes have been guessed and validated. The `new CandidateMatrix()` constructor simply creates an empty array with `count` set to 0.

Let *DecaySchedule* contain the decayed key schedule and a predicate, `isCompatible(`*CandidateMatrix*`)`, that indicates whether or not a guess and its consequent bytes are compatible the decayed schedule. Compatibility is determined by checking that the *CandidateMatrix* contains all the known bits from the corresponding bytes of the decayed schedule.

The function `recoverKeyRec(`*CandidateMatrix,DecaySchedule*`)` returns a key whose schedule is decay-compatible and is the result of extending the incoming *CandidateMatrix*. It returns `NULL` when there is no compatible key schedule extension to the specified candidate prefix. Proper usage asserts that the starting *CandidateMatrix* and *DecaySchedule* are compatible. The initial call to `recoverKeyRec` begins with an empty *CandidateMatrix*. The recursive expression in Figure 1 makes the control logic explicit. Figure 1 halts on the first compatible key schedule, however one could modify it to halt after a full search of the key space; simply replace the `return` with a `print` on the third line. Section 4 benchmarks both variants.

3.4 Path Prioritization

The exploration path illustrated by Table 1 maximizes the number of implied schedule bytes with the goal of minimizing the number of compatible candidates at each stage. There are many ways to grow the selection path, and Table 1 illustrates just one of them. For instance, there are 3 symmetric alternatives obtained by rotating the bytes of each word. Different paths will encounter different constraints and therefore will result in varying recovery times. Within the stages 0-10, the selection order of $S_{x,0,0}$ may be altered and still obtain the same consequent bytes after 10 candidate stages. Growing the path with guesses adjacent to the body of previous guesses will preserve the set of inferred bytes; this claim has been verified experimentally. The following matrix shows that guesses may be grown from the middle of the schedule:

The exploration path in Table 2 starts at $S_{5,0,0}$. To maintain adjacency, the next choice may be $S_{4,0,0}$ or $S_{6,0,0}$. The path grows by extending either the top or bottom of previous choices in $S_{x,0,0}$, where $0 \leq x < 11$. This allows the number of inferred bytes to grow by one at each stage.

A first heuristic for choosing the best path might be to count up the known bits in the exploration path and its consequent bytes. However within a set of guesses and implied bytes, the selection order can also make a difference. Past a certain threshold, adding more constraints does not prune the exploration anymore because the byte is already uniquely determined (or its parent has been ruled out). Consider the case when the decayed data in the candidate position is 0xFF. If the last guess corresponds to this position, then all of the consequent bytes are wasted constraints because the byte is uniquely determined. On the other hand, that byte position would make an excellent initial position for stage 0 because the first guess never produces any consequent bytes. The final algorithm's path comparison heuristic estimates the number of branches at each stage. At a stage, the estimate simply counts the number of known bits in the initial and consequent positions. The intuition is that each known bit will reduce the number of valid branches by a factor of 2. Thus, the branch estimate for a stage i is $2^{min(8,k_i)}$ where k_i is the number of known bits in the ith stage's initial and consequent byte positions. This is clearly only an estimate, as some

Table 2. An exploration path starting in round 5; script notation parallels the example at the end of Section 3.2

w	0				1				2				3			
$r \backslash b$	0	1	2	3	0	1	2	3	0	1	2	3	0	1	2	3
0	10_0												10_1			
1	7_0								10_2				7_1			
2	6_0				10_3				7_2				6_1			
3	3_0	10_4			7_3				6_2				3_1	10_5		
4	2_0	7_4			6_3				3_2	10_6			2_1	7_5		
5	0_0	6_4			3_3	10_7			2_2	7_6			1_1	6_5		
6	1_0	4_4	10_8		4_3	7_7			4_2	6_6			4_1	5_5	10_9	
7	4_0	5_4	8_8		5_3	8_7			5_2	8_6	10_{10}		5_1	9_9		
8	5_0	8_4	9_8		8_3	9_7			8_2	9_6			8_1	9_5		
9	8_0	9_4			9_3				9_2				9_1			
10	9_0															

Table 3. AES-256 exploration path template; script notation parallels the example at the end of Section 3.2

	0	1	2	3	0	1	2	3
0_0	$15_7 23_7 31_7$	$14_7 22_7 30_7$	$13_7 21_7 29_7$	$12_7 20_7 28_7$	$10_6 18_6 26_6$	$9_6 17_6 25_6$	$8_6 16_6 24_6 31_6 1_1$	$15_6 23_6$
1_0	$14_6 22_6 30_6$	$13_6 21_6 29_6$	$12_6 20_6 28_6$	$10_5 18_5 26_5$	$9_5 17_5 25_5$	$8_5 16_5 24_5 31_5 2_2$	$15_5 23_5 30_5 2_1$	$14_5 22_5$
2_0	$13_5 21_5 29_5$	$12_5 20_5 28_5$	$10_4 18_4 26_4$	$9_4 17_4 25_4$	$8_4 16_4 24_4 31_4 3_3$	$15_4 23_4 30_4 3_2$	$14_4 22_4 29_4 3_1$	$13_4 21_4$
3_0	$12_4 20_4 28_4$	$10_3 18_3 26_3$	$9_3 17_3 25_3$	$8_3 16_3 24_3 31_3$	$4_4 15_3 23_3 30_3$	$4_3 14_3 22_3 29_3$	$4_2 13_3 21_3 28_3$	$4_1 12_3 20_3$
4_0	$10_2 18_2 26_2$	$9_2 17_2 25_2$	$8_2 16_2 24_2 31_2$	$5_5 15_2 23_2 30_2$	$5_4 14_2 22_2 29_2$	$5_3 13_2 21_2 28_2$	$5_2 12_2 20_2 26_1$	$5_1 10_1 18_1$
5_0	$9_1 17_1 25_1$	$8_1 16_1 24_1 31_1$	$6_6 15_1 23_1 30_1$	$6_5 14_1 22_1 29_1$	$6_4 13_1 21_1 28_1$	$6_3 12_1 20_1 26_0$	$6_2 10_0 18_0 25_0$	$6_1 10_1 7_0$
6_0	$8_0 16_0 24_0 31_0$	$7_7 15_0 23_0 30_0$	$7_6 14_0 22_0 29_0$	$7_5 13_0 21_0 28_0$	$7_4 12_0 20_0 27_6$	$7_3 11_6 19_6 27_5$	$7_2 11_5 19_5 27_4$	$7_1 11_4 19_4$
7_0	$11_3 19_3 27_3$	$11_2 19_2 27_2$	$11_1 19_1 27_1$	$11_0 19_0 27_0$				

bits may constrain portions of the guess that have already been determined. The total branching estimate is the product of the stage estimates. Since the stage estimates are all powers of two, it suffices to sum their exponents. Thus, the heuristic ranks paths by the scalar value $\sum_{i=0}^{15} min(8, k_i)$.

The algorithm considers variants of Table 1 along two axes. One axis is byte slice selection; Table 1 initiates on slice 0, although three other may be obtained by rotating all schedule words by the same amount. The other axis is byte guessing order within stages 0-10 as described above; choose a initial round and then extend the guesses by adding to the adjacent round on the top or the bottom. There are other paths not reached by these variables; e.g., growth may be rooted in words 1-3 of the round key rather than 0. Performance may well improve by selecting these paths from a larger space, however no additional path analysis is investigated in this work.

3.5   Generalizing to Other Instances of AES

Generalizing the algorithm to operate on 192-bit and 256-bit variants of AES is straightforward. One needs to construct a different path template and update the isCompatible() method to incorporate the schedule generating equations of the larger keys. Table 3 illustrates the path template used by the 256-bit implementation. The heuristic considers paths on the same axes as the 128-bit version. Since the key size is twice the length of the block size, the matrix is only 8 deep and therefore the heuristic considers fewer paths.

4   Benchmarks

Test cases are generated with OpenSSL's RAND_bytes() function. For a given decay rate d, the test generator derives a key schedule from randomly selected key bytes and then randomly zeroes $d\%$ of the bits. Tests assume a ground-state encoding of 0. Performance was evaluated on Dell Precision Workstation 7400 running a 3.4 Ghz quad-core Xeon processor with 4GB of RAM. The C99 reference implementation of the algorithm is compiled with the MinGW version of gcc-3.4.5 at the highest level of optimization, -O4. All computations cases were run to completion in a serial manner. Time was measured by entry and exit calls to clock() from the time.h library; clock resolution is 64 Hz. The original

Table 4. Run-time results for four versions of the algorithm; each decay rate test suite contains 10,000 cases

Case		Key size	Path selection	Halting condition
PathOpt-128		128 bits	Heuristically chosen (Section 3.4)	First match
PathOpt-256		256 bits	Heuristically chosen (Section 3.4)	First match
Basic-128		128 bits	Fixed to Table 1	First match
Exhaust-128		128 bits	Heuristically chosen (Section 3.4)	End of key space

PathOpt-128 — Run-time (seconds)

Decay		30%	40%	50%	60%
Total		90.120	93.559	142.322	1,736.321
Avg.		0.009	0.009	0.014	0.174
Med.		0.015	0.015	0.015	0.031
Max		0.015	0.015	0.078	2.094
Min		0.000	0.000	0.000	0.000
St.Dev		0.007	0.008	0.015	0.772

PathOpt-256 — Run-time (seconds)

Decay		30%	40%	50%	60%
Total		17.046	26.185	123.250	6,954.231
Avg.		0.002	0.003	0.012	0.695
Med.		0.000	0.000	0.000	0.062
Max		0.016	0.062	2.125	352.015
Min		0.000	0.000	0.000	0.000
St.Dev.		0.005	0.006	0.044	5.920

Basic-128 — Run-time (seconds)

Decay		30%	40%	50%	60%
Total		219.204	1,526.308	32,551.469	1,638,788.166
Avg.		0.022	0.153	3.255	163.879
Med.		0.015	0.015	0.078	1.968
Max		9.562	266.390	3,354.890	343,656.375
Min		0.000	0.000	0.000	0.000
St.Dev.		0.140	2.994	55.563	3,753.608

Exhaust-128 — Run-time (seconds)

Decay		30%	40%	50%	60%
Total		96.403	112.350	258.568	4,497.599
Avg.		0.010	0.011	0.026	0.450
Med.		0.015	0.015	0.015	0.110
Max		0.031	0.468	0.875	75.203
Min		0.000	0.000	0.000	0.000
St.Dev.		0.007	0.009	0.036	1.921

keys were found for all test cases, using the heuristically chosen path and halting on the first match.

Table 4 summarizes the benchmark results for four variants of the algorithm: PathOpt-128, PathOpt-256, Basic-128, and Exhaust-128. There are 10,000 cases for each of the four decay rates, 30%, 40%, 50%, and 60%. The 128-bit variants have been run on the same test cases, so their results are directly comparable. All times are measured in seconds. A time of 0.000 means that the computation finished in less than 1/64 second, or about 53 million processor cycles.

Additional testing (Table 5) was performed to estimate the maximum recoverable decay rates for PathOpt-128 and PathOpt-256. Only 5,000 cases were examined due to extended recovery times.

5   Analysis

PathOpt-128 solves all cases at 50% decay and less in under half a second. At 60% decay, PathOpt-128 recovered the worst case in 35.500 seconds while solving the average case in 0.174 seconds. At the extended decay rate of 70%, recovery time averages grew to just over 6 minutes with the median time at just under five seconds. Nearly half of the 17.4 day run was consumed by solving the worst case of the test suite; the second slowest case was over six times faster. 4927 cases were recovered in less than 20 minutes.

PathOpt-128 runs faster than Basic-128 across the board and the speedup quickly grows as the decay rate increases. The speedups for 30%, 40%, 50%, and

Table 5. Extended decay rate runs; each decay rate test suite has 5,000 cases

Case	Total	Avg.	Med.	Max	Min	St.Dev.
PathOpt-128 @ 70% decay	1,504,487.119 s	300.897 s	4.938 s	737,266.687 s	0.000 s	10,677.913 s
PathOpt-256 @ 65% decay	446,879.849 s	89.376 s	0.875 s	194,410.875 s	0.000 s	2,843.061 s

60% are 2.43×, 16.3×, 228×, and 943×, respectively. At 70%, only 10 cases had completed after a week when the experiment was terminated. The path selection heuristic makes 70% decay a feasibly solvable problem in the test environment. Even for the low decay rates, Basic-128 has a much higher standard deviation; their worst cases with Basic-128 are several orders of magnitude worse than their worst cases with PathOpt-128.

The profound impact of heuristic path selection at high decay rates suggests that a more thorough search for the best path could further extend the maximum feasible decay capacity. Only a small subset of possible paths are considered. The current analysis takes less than 1/64 second, as evidenced by the 0.000 timing results, so there is ample room for more startup analysis.

Full search of the key space appears to be a small factor slower than stopping at the first compatible key. It widens as the decay increases, but by 60% the Exhaust-128 only takes 2.590 times longer. We note that all 90,000 test cases had precisely one solution, so the exhaustive search seems unnecessary at the tested decay rates.

PathOpt-256 performs well up 60% decay rates, solving cases in an average of 0.695 seconds and in no more than 352.015 seconds. At 70%, no cases were solved in the test suite during a 1 week trial. At 65% the results are promising: the 5,000 cases have been solved in an average of 89.676 seconds. The longest case took 2.25 days to recover, while 99.4% of the cases have been recovered in less than 20 minutes.

Interestingly, the PathOpt-256 is slightly faster than PathOpt-128 on decay rates at 50% and below. The average solution times for these cases is within two units of the 64 Hz clock resolution. We conjecture that the heuristic path analysis takes less time with the 256-bit version since there are fewer paths to consider, due to the flatness of the path matrix (Section 3.5).

As a point of comparison, the original algorithm [1] was compiled in the same environment and run against the same test suite of 30% and 40% decayed AES-128 schedules. After three weeks of execution, only the first four cases at 30% had been solved and the first case at 40% had not yet finished.

6 Conclusion

We presented a new class of recovery capability that is several orders of magnitude faster than previous methods, particularly for higher decay rates. It more than doubles the decay rate recovery feasibility of prior work. The tree-pruning constraints enable efficient and exhaustive key space searches to determine solution uniqueness. We have generalized the implementation to AES-256 while maintaining excellent performance up to 65% decay.

Acknowledgements

I owe a special debt of gratitude to Dr. Adam L. Young for his insight and wisdom regarding writing, presentation, and the scientific community. I have additionally benefited from the thoughtful review of Kerry A. McKay. I am further indebted to Charles C. Howell, Steven M. Godin, and Eileen M. Boettcher for their strident advocacy—without which this paper would not be possible.

References

1. Halderman, J.A., Schoen, S.D., Heninger, N., Clarkson, W., Paul, W., Calandrino, J.A., Feldman, A.J., Appelbaum, J., Felten, E.W.: Lest we remember: cold boot attacks on encryption keys. In: USENIX Security Symposium, pp. 45–60. USENIX Association, Berkeley (2008)
2. Daemen, J., Rijmen, V.: The block ciptpher Rijndael. In: Schneier, B., Quisquater, J.-J. (eds.) CARDIS 1998. LNCS, vol. 1820, pp. 277–284. Springer, Heidelberg (2000)
3. Daemen, J., Rijmen, V.: The Design of Rijndael. Springer, Heidelberg (2002)
4. Heninger, N., Shacham, H.: Reconstructing RSA private keys from random key bits. In: Halevi, S. (ed.) CRYPTO 2009. LNCS, vol. 5677, pp. 1–17. Springer, Heidelberg (2009)
5. Handschuh, H., Paillier, P., Stern, J.: Probing attacks on tamper-resistant devices. In: Koç, Ç.K., Paar, C. (eds.) CHES 1999. LNCS, vol. 1717, pp. 303–315. Springer, Heidelberg (1999)
6. Akavia, A., Goldwasser, S., Vaikuntanathan, V.: Simultaneous hardcore bits and cryptography against memory attacks. In: Reingold, O. (ed.) TCC 2009. LNCS, vol. 5444, pp. 474–495. Springer, Heidelberg (2009)
7. Regev, O.: On lattices, learning with errors, random linear codes, and cryptography. In: STOC, pp. 84–93. ACM, New York (2005)
8. Naor, M., Segev, G.: Public-key cryptosystems resilient to key leakage. In: Halevi, S. (ed.) CRYPTO 2009. LNCS, vol. 5677, pp. 18–35. Springer, Heidelberg (2009)
9. Cramer, R., Shoup, V.: Universal hash proofs and a paradigm for adaptive chosen ciphertext secure public-key encryption. In: Knudsen, L.R. (ed.) EUROCRYPT 2002. LNCS, vol. 2332, pp. 45–64. Springer, Heidelberg (2002)
10. Alwen, J., Dodis, Y., Wichs, D.: Leakage-resilient public-key cryptography in the bounded-retrieval model. In: Halevi, S. (ed.) CRYPTO 2009. LNCS, vol. 5677, pp. 36–54. Springer, Heidelberg (2009)
11. Katz, J.: Signature schemes with bounded leakage resilience. Cryptology ePrint Archive: Report 2009/133 (March 22, 2009)
12. Chari, S., Jutla, C.S., Rao, J.R., Rohatgi, P.: Towards sound approaches to counteract power-analysis attacks. In: Wiener, M. (ed.) CRYPTO 1999. LNCS, vol. 1666, pp. 398–412. Springer, Heidelberg (1999)
13. Coron, J.S., Naccache, D., Kocher, P.: Statistics and secret leakage. ACM Trans. Embed. Comput. Syst. 3(3), 492–508 (2004)
14. Micali, S., Reyzin, L.: Physically observable cryptography. In: Naor, M. (ed.) TCC 2004. LNCS, vol. 2951, pp. 278–296. Springer, Heidelberg (2004)

15. TCG storage security subsystem class: Opal version 1.0,
    http://www.trustedcomputinggroup.org/ (January 27, 2009)
16. Enck, W., Butler, K., Richardson, T., McDaniel, P., Smith, A.: Defending against
    attacks on main memory persistence. In: ACSAC, Washington, DC, USA, pp. 65–
    74. IEEE Computer Society, Los Alamitos (2008)
17. Gueron, S.: Advanced encryption standard (AES) instructions set (April 27, 2009),
    http://www.intel.com/
18. Osvik, D.A., Shamir, A., Tromer, E.: Cache attacks and countermeasures: the
    case of AES. In: Pointcheval, D. (ed.) CT-RSA 2006. LNCS, vol. 3860, pp. 1–20.
    Springer, Heidelberg (2006)

A   Extended Pseudocode

```
path = ((  0),
        ( 16, 13),
        ( 32, 29, 25),
        ( 48, 45, 41, 37),
        ( 64, 61, 57, 53, 49),
        ( 80, 77, 73, 69, 65, 62),
        ( 96, 93, 89, 85, 81, 78, 74),
        (112,109,105,101, 97, 94, 90, 86),
        (128,125,121,117,113,110,106,102, 98),
        (144,141,137,133,129,126,122,118,114,111),
        (160,157,153,149,145,142,138,134,130,127,123),
        (173,169,165,161,158,154,150,146,143,139,135),
        (131,119,107, 95, 82, 70, 58, 46, 33, 21,  9),
        (124,115,103, 91, 79, 66, 54, 42, 30, 17,  5),
        (120,108, 99, 87, 75, 63, 50, 38, 26, 14,  1),
        (174,170,166,162,159,155,151,147,140,136,132))

class CandidateMatrix:
    Int count
    Byte m[176]

    def guess (Byte b):
        c = copy(self)
        c.count = count + 1
        c.m[path[count][0]] = b
        if count == 0:
            return c
        for i = 1 to len(path[count]):
            if defined(c.m[path[count][i]-16]):
                b4 = path[count][i-1]
                b3 = path[count][i]
                b0 = b4-16
                if inFirstWordOfRoundKey(b4):
                    c.m[b3] = unsbox(c.m[b4] XOR c.m[b0]
                              XOR rcon[getRound(b4)][getBytePos(b4)]
                else:
                    c.m[b3] = c.m[b0] XOR c.m[b4]
```

```
                else:
                    b0 = path[count][i]
                    b3 = path[count][i-1]
                    b4 = b0 + 16
                    if inFirstWordOfRoundKey(b4):
                        c.m[b0] = sbox(c.m[b3]) XOR c.m[b4]
                                  XOR rcon[getRound(b4)][getBytePos(b4)]
                    else:
                        c.m[b0] = c.m[b3] XOR c.m[b4]
            if c.count == 16:
                c = deriveFullScheduleFromRound8(c)
            return c

    def key ():
        return m[0:16]


class DecaySchedule:
    Byte decaySched[176]
    Byte gndEnc[176]
    def isCompatible (CandidateMatrix candidate):
        for i = 0 to 176:
            if defined(candidate.m[i]):
                if (candidate.m[i] XOR decaySched[i])
                   AND (candidate.m[i] XOR gndEnc[i]):
                    continue
                else:
                    return FALSE
        return TRUE


def recoverKeyRec(CandidateMatrix c, DecaySchedule d):
    if (c.length()==16):
        return c.key()
    for i=0 to 255:
        if(d.isCompatible(c.guess(i))):
            key = recoverKeyRec(c.guess(i),d)
            if (key != NULL):
                return key
    return NULL
```

The pseudocode follows a Python-like syntax, but with some additional explicit typing and field declaration. Variables may hold their declared types or undefined values—a property checked by defined().

The path variable encodes the guessing and inference order of Table 1 in the flat byte-level schedule view. The recursive exploration function, recoverKeyRec(), is the same as in Fig. 1.

The CandidateMatrix class is dominated by the guess method which guesses a value for the position determined by path[count][0] and infers the consequent bytes. The inference logic splits into two cases: when the unknown byte is in the first word, S^w_{i-4}, of the schedule generating equations (see (1)) or in the middle word, $S^w{}_{i-1}$. The variable names, b0, b3, and b4 reflect the relative position of their encapsulating words in the schedule; if b0 comes from an arbitrary word in the first 10 rounds, then b3 and b4 come from the words three and four words ahead of b0, respectively. The inference sequence in path has been chosen to account for the necessary rotations in byte slices when solving the s-box version of the generating equations. Upon completion of the 16 guesses the eighth round becomes fully specified, implying the remainder of the schedule. The key() method simply returns the first 16 bytes of completed key schedule.

DecaySchedule holds the observed decayed data and ground state encoding for each byte. Its one method isCompatible() checks that each defined bit of the candidate schedule equals the decayed data or the ground state.

Cryptanalysis of the Full MMB Block Cipher

Meiqin Wang[1], Jorge Nakahara Jr.[2], and Yue Sun[1]

[1] Key Laboratory of Cryptologic Technology and Information Security,
Ministry of Education, Shandong University, Jinan 250100, China
[2] EPFL, Lausanne, Switzerland
mqwang@sdu.edu.cn, jorge.nakahara@epfl.ch, yuesun@mail.sdu.edu.cn

Abstract. The block cipher MMB was designed by Daemen, Govaerts and Vandewalle, in 1993, as an alternative to the IDEA block cipher. We exploit and describe unusual properties of the modular multiplication in $\mathbb{Z}_{2^{32}-1}$, which lead to a differential attack on the full 6-round MMB cipher (both versions 1.0 and 2.0). Further **contributions** of this paper include detailed square and linear cryptanalysis of MMB. Concerning differential cryptanalysis (DC), we can break the full MMB with 2^{118} chosen plaintexts, $2^{95.91}$ 6-round MMB encryptions and 2^{64} counters, effectively bypassing the cipher's countermeasures against DC. For the square attack, we can recover the 128-bit user key for 4-round MMB with 2^{34} chosen plaintexts, $2^{126.32}$ 4-round encryptions and 2^{64} memory blocks. Concerning linear cryptanalysis, we present a key-recovery attack on 3-round MMB requiring $2^{114.56}$ known-plaintexts and 2^{126} encryptions. Moreover, we detail a ciphertext-only attack on 2-round MMB using $2^{93.6}$ ciphertexts and $2^{93.6}$ parity computations. These attacks do not depend on weak-key or weak-subkey assumptions, and are thus independent of the key schedule algorithm.

Keywords: MMB block cipher, differential cryptanalysis, square cryptanalysis, linear cryptanalysis, modular multiplication.

1 Introduction

The block cipher MMB (Modular Multiplication Based) block cipher [3] was designed by Daemen, Govaerts and Vandewalle in 1993, and its main innovation was the use of cyclic multiplication in the ring $\mathbb{Z}_{2^n-1}$, where n is the word size of the cipher. All internal operations of MMB are on n-bit words. The designers suggested $n = 32$, leading to the ring $\mathbb{Z}_{2^{32}-1}$. Note that $2^{32} - 1 = (2^{16}+1)\cdot(2^8+1)\cdot(2^4+1)\cdot(2^2+1)\cdot(2+1) = 65537\cdot257\cdot17\cdot5\cdot3 = 4294967295$, the product of all five known Fermat primes. MMB is an iterated cipher, composed of six rounds. MMB operates on 128-bit text blocks and uses 128-bit key. MMB was proposed as an alternative to the IDEA block cipher [7]. MMB has been designed particularly to resist differential cryptanalysis [5]. This paper presents differential, square and linear cryptanalysis of the MMB cipher. Previous cryptanalysis of MMB was a related-key attack and only applied to MMB version 1.0, according to [5]. In order to resist the related-key attack, MMB version 2.0 was proposed by revising

M.J. Jacobson Jr., V. Rijmen, and R. Safavi-Naini (Eds.): SAC 2009, LNCS 5867, pp. 231–248, 2009.
© Springer-Verlag Berlin Heidelberg 2009

only the key schedule algorithm. As far as we know, there is no previous attack on MMB version 2.0. However, our attacks are independent of the key schedule algorithm, so they can be applied to both versions 1.0 and 2.0.

In this paper, firstly we present differential cryptanalysis of the full 6-round MMB. Five-round differential characteristics have been identified, and we can recover the 128-bit user key with 2^{118} chosen plaintexts (CP), $2^{95.91}$ 6-round MMB encryptions and 2^{64} memory blocks. Secondly, we investigate the square attack on reduced-round MMB. We distinguished a new word type: X word, based on which we have found 2.75-round square distinguishers and applied the square attack to 4-round MMB. We can recover the 128-bit user key with 2^{34} CP, $2^{126.32}$ 4-round encryptions and 2^{64} memory blocks. Thirdly, we apply linear cryptanalysis to reduced-round MMB. We identified two linear approximations with bias $2^{-55.78}$ for 3-round MMB and recover one-bit subkey information for 3-round MMB with $2^{114.56}$ known plaintexts (KP) and equivalent parity computations; then recover 128-bit key for 3-round MMB with $2^{114.56}$ KP and 2^{126} 3-round MMB encryptions. Moreover, we can attack 2-round MMB with $2^{93.6}$ ciphertexts only (CO). From our attacks, particularly concerning differential cryptanalysis, we disprove the claims of the designers that MMB can resist DC.

The paper is organized as follows. Sec. 2 describes the MMB cipher. Sect. 3 presents the differential attack on the full MMB, and Sect. 4 details a square attack on a 4-round MMB. The linear attack on reduced-round MMB is provided in Sect. 5. Sect. 6 concludes the paper.

2   Description of the MMB Block Cipher

The MMB block cipher has a Substitution-Permutation Network (SPN) structure and operates on 128-bit text blocks, uses a 128-bit key, and iterates six rounds. One round of MMB consists of four transformations [5]:

- $\sigma[k^j]$: exclusive-or each data word with the j-th round subkey k^j. Formally,

$$\sigma[k^j](a_0, a_1, a_2, a_3) = (a_0 \oplus k_0^j, a_1 \oplus k_1^j, a_2 \oplus k_2^j, a_3 \oplus k_3^j),$$

  where $\oplus$ denotes bitwise exclusive-or, $a_i, k_i^j \in \mathbb{Z}_{2^{32}}$, for $0 \leq i \leq 3$. The $\sigma[k^j]$ operation is an involution, and is the only key-dependent operation in a round.

- γ: modular multiplication of each data word with fixed 32-bit constants G_i,

$$\gamma(a_0, a_1, a_2, a_3) = (a_0 \otimes G_0, a_1 \otimes G_1, a_2 \otimes G_2, a_3 \otimes G_3),$$

  where $a \otimes b = a * b \bmod (2^{32} - 1)$, $G_0 = 025F1CDB_x$, $G_1 = 2 \otimes G_0 = 04BE39B6_x$, $G_2 = 8 \otimes G_0 = 12F8E6D8_x$, and $G_3 = 128 \otimes G_0 = 2F8E6D81_x$ which can be efficiently computed since $(A * 2^x) \bmod (2^{32} - 1) = (A \lll x) \bmod (2^{32} - 1)$. There is a wrap-around effect in multiplication modulo $2^{32} - 1$, since $2^{32} \equiv 1 \bmod (2^{32} - 1)$, which means that the bits at the $(32 + i)$-th LSB position are shifted to the i-th LSB position. This effect is similar to the multiplication operation modulo $2^{16} + 1$ in IDEA. As cited in [5], the $\otimes$ operation can be expressed as:

$$a \otimes b = a * b \bmod (2^{32} - 1) = (a * b \bmod 2^{32} + \lfloor \frac{a * b}{2^{32}} \rfloor) \bmod (2^{32} - 1).$$

Notice that γ is invertible but is not an involution. Each 32-bit multiplication can be interpreted as a huge 32×32-bit S-box, since one of the operands in the multiplication is always fixed. There are two fixed points for any G_i: $0 \otimes G_i = 0$, and $(2^{32} - 1) \otimes G_i = 2^{32} - 1$.

- η: a data-dependent transformation operating on two out of the four input words (a_0, a_1, a_2, a_3):

$$\eta(a_0, a_1, a_2, a_3) = (a_0 \oplus (\mathrm{lsb}(a_0) * \delta), a_1, a_2, a_3 \oplus ((1 \oplus \mathrm{lsb}(a_3)) * \delta)),$$

where 'lsb' denotes the least significant bit, and $\delta = 2aaaaaaa_x$; η is an involution and a non-linear operation. η is used to resist the propagation of the differential characteristics with probability 1.

- θ: the only diffusion operation in MMB. Formally,

$$\theta(a_0, a_1, a_2, a_3) = (a_3 \oplus a_0 \oplus a_1, a_0 \oplus a_1 \oplus a_2, a_1 \oplus a_2 \oplus a_3, a_2 \oplus a_3 \oplus a_0),$$

where $a_i \in \mathbb{Z}_{2^{32}}$, with $0 \leq i \leq 3$. θ is an involution and has branch number four (see [10]).

There are two pairs of operations that can be interchanged: $(\theta, \sigma[k^j])$ and $(\eta, \sigma[k^j])$. In each case, the key k^j is transformed into an equivalent key $\theta(k^j)$ or $\eta(k^j)$, respectively.

The j-th (full) round transformation of MMB can be denoted:

$$\rho[k^j](X) = \theta \circ \eta \circ \gamma \circ \sigma[k^j](X) = \theta(\eta(\gamma(\sigma[k^j](X)))). \tag{1}$$

The full MMB encryption of a plaintext P can be denoted:

$$\mathrm{MMB}(P) = \sigma[k^6] \circ \rho[k^5] \circ \rho[k^4] \circ \rho[k^3] \circ \rho[k^2] \circ \rho[k^1] \circ \rho[k^0](P), \tag{2}$$

where $\sigma[k^6]$ is the output transformation or post-whitening operation.

In the original key schedule of MMB version 1.0, the first round subkey is simply the 128-bit user key $K = (k_0, k_1, k_2, k_3)$. Successive subkeys use K rotated by 32 bits to the left. So, for instance, (k_1, k_2, k_3, k_0), (k_2, k_3, k_0, k_1) and so forth. A redesigned key-schedule to avoid related-key attacks has led to a tweaked cipher called MMB version 2.0 [5] in which a constant value is xored to the leftmost 32-bit subkey word after each rotation.

3   Differential Cryptanalysis of the Full MMB

Differential cryptanalysis (DC) [2] exploits the propagation of particular differences of plaintext pairs across a cipher, to certain differences of the resultant ciphertext pairs. The designers of MMB claimed that an important design criterion was resistance against DC in [3], but we break the full MMB using DC.

3.1   Differential Characteristics for MMB

The main component in the round function of MMB responsible for the confusion property (according to C. Shannon) is γ. Thus, for the analyst it is very important to minimize the number of active multiplications in order to maximize the probability of the differential characteristics. The possible distributions of active modular multiplications are listed in Table 2. In the leftmost column, the input difference is said to cause (denoted with an arrow, $\xrightarrow{1r}$) the given output difference after one round. The second column shows the number of active multiplications. The rightmost column shows the restrictions on the output difference of active multiplications, which account for η. Due to θ, the output differences from the active multiplications in one round all have to be equal. For each row in Table 2, we denote the input difference as $\Delta_{ij}, (0 \le j \le 3)$ and the output difference as Δ_o.

In order to identify 2-round characteristics for MMB with the highest probability, we only consider two active multiplications per round. An important property for the modular multiplication operation γ has been described in [5]

$$R_p(\bar{0} \xrightarrow{\gamma} \bar{0}) = 1,$$

where $\bar{0} = 2^{32} - 1 = ffffffff_x$. This property means that the differential characteristic $\bar{0} \xrightarrow{\gamma} \bar{0}$ holds with probability 1, leading to the following 2-round characteristic with probability 1:

$$(0, \bar{0}, \bar{0}, 0) \xrightarrow{\sigma[k^0]} (0, \bar{0}, \bar{0}, 0) \xrightarrow{\gamma} (0, \bar{0}, \bar{0}, 0) \xrightarrow{\eta} (0, \bar{0}, \bar{0}, 0) \xrightarrow{\theta} (\bar{0}, 0, 0, \bar{0})$$

$$\xrightarrow{\sigma[k^1]} (\bar{0}, 0, 0, \bar{0}) \xrightarrow{\gamma} (\bar{0}, 0, 0, \bar{0}) \xrightarrow{\eta} (\bar{0} \oplus \delta, 0, 0, \bar{0} \oplus \delta) \xrightarrow{\theta} (0, \bar{0} \oplus \delta, \bar{0} \oplus \delta, 0).$$

where a single 0 denotes a 32-bit zero difference word. Then, we further extend the 2-round characteristic by two rounds above it and one round below it. For the lower round, the following differential characteristic needs to be determined:

$$(0, \bar{0} \oplus \delta, \bar{0} \oplus \delta, 0) \xrightarrow{\sigma[k^2]} (0, \bar{0} \oplus \delta, \bar{0} \oplus \delta, 0) \xrightarrow{\gamma} (0, \alpha_1, \alpha_2, 0).$$

We identified the characteristics $\bar{0} \oplus \delta \xrightarrow{G_1} fcfbdffff_x$ and $\bar{0} \oplus \delta \xrightarrow{G_2} f3ef7fff_x$, both of which have probability about 2^{-18}. With them, we construct a 3-round differential characteristic with probability 2^{-36} as follows:

$$(0, \bar{0}, \bar{0}, 0) \xrightarrow{1r} (\bar{0}, 0, 0, \bar{0}) \xrightarrow{1r} (0, \bar{0} \oplus \delta, \bar{0} \oplus \delta, 0) \xrightarrow{\sigma[k^i]} (0, \bar{0} \oplus \delta, \bar{0} \oplus \delta, 0)$$

$$\xrightarrow{\gamma} (0, fcfbdffff_x, f3ef7fff_x, 0) \xrightarrow{\eta} (0, fcfbdffff_x, f3ef7fff_x, 0)$$

$$\xrightarrow{\theta} (fcfbdffff_x, 0f14a000_x, 0f14a000_x, f3ef7fff_x).$$

For the upper round, the following differential characteristic needs to be determined:

$$(\beta_0, 0, 0, \beta_3) \xrightarrow{\sigma[k^i]} (\beta_0, 0, 0, \beta_3) \xrightarrow{\gamma} (\bar{0} \oplus \delta, 0, 0, \bar{0} \oplus \delta) \xrightarrow{\eta} (\bar{0}, 0, 0, \bar{0}) \xrightarrow{\theta}$$

$$(0, \bar{0}, \bar{0}, 0) \xrightarrow{1r} (\bar{0}, 0, 0, \bar{0}) \xrightarrow{1r} (0, \bar{0} \oplus \delta, \bar{0} \oplus \delta, 0) \xrightarrow{1r}$$

$$(fcfbdffff_x, 0f14a000_x, 0f14a000_x, f3ef7fff_x).$$

In order to further extend the above 4-round characteristic by one round above it, we only consider the cases $\beta_0 = \beta_3$. In this way, we identified the characteristics $a7cfdf7f_x \overset{G_0}{\to} \bar{0} \oplus \delta$ and $a7cfdf7f_x \overset{G_3}{\to} \bar{0} \oplus \delta$, both with probability about 2^{-21}. So, a 4-round characteristic with probability $2^{-42} \cdot 2^{-36} = 2^{-78}$ has been constructed. Then, we further extend the 4-round characteristic. The following characteristic needs to be determined,

$$(0, \xi_1, \xi_2, 0) \overset{\sigma[k^i]}{\to} (0, \xi_1, \xi_2, 0) \overset{\gamma}{\to} (0, a7cfdf7f_x, a7cfdf7f_x, 0)$$
$$\overset{\eta}{\to} (0, a7cfdf7f_x, a7cfdf7f_x, 0) \overset{\theta}{\to} (a7cfdf7f_x, 0, 0, a7cfdf7f_x)$$
$$\overset{1r}{\to} (0, \bar{0}, \bar{0}, 0) \overset{1r}{\to} (\bar{0}, 0, 0, \bar{0}) \overset{1r}{\to} (0, \bar{0} \oplus \delta, \bar{0} \oplus \delta, 0)$$
$$\overset{1r}{\to} (fcfbdfff_x, 0f14a000_x, 0f14a000_x, f3ef7fff_x).$$

We identified the characteristics $9bd3fdf7_x \overset{G_1}{\to} a7cfdf7f_x$ and $e6f4ff7d_x \overset{G_2}{\to} a7cfdf7f_x$, both with probability about 2^{-14}. With them, we construct a 5-round characteristic with probability $2^{-28} \cdot 2^{-78} = 2^{-106}$ as follows:

$$(0, 9bd3fdf7_x, e6f4ff7d_x, 0) \overset{1r}{\to} (a7cfdf7f_x, 0, 0, a7cfdf7f_x) \qquad (3)$$
$$\overset{1r}{\to} (0, \bar{0}, \bar{0}, 0) \overset{1r}{\to} (\bar{0}, 0, 0, \bar{0}) \overset{1r}{\to} (0, \bar{0} \oplus \delta, \bar{0} \oplus \delta, 0)$$
$$\overset{1r}{\to} (fcfbdfff_x, 0f14a000_x, 0f14a000_x, f3ef7fff_x).$$

With the 5-round characteristic in (3), we cannot attack the full 6-round MMB because the S/N is too small. In order to increase the S/N, we found another 5-round differential characteristic with probability 2^{-110} as follows:

$$(0, 9bd3fdf7_x, e6f4ff7d_x, 0) \overset{1r}{\to} (a7cfdf7f_x, 0, 0, a7cfdf7f_x) \overset{1r}{\to} (0, \bar{0}, \bar{0}, 0)$$
$$\overset{1r}{\to} (\bar{0}, 0, 0, \bar{0}) \overset{1r}{\to} (0, \bar{0} \oplus \delta, \bar{0} \oplus \delta, 0) \overset{1r}{\to} (40404040_x, 0, 0, 40404040_x), \qquad (4)$$

where the characteristics $\bar{0} \oplus \delta \overset{G_1}{\to} 40404040_x$ and $\bar{0} \oplus \delta \overset{G_2}{\to} 40404040_x$ have probability about 2^{-20}. Although the probability of (4) is lower than that of (3), the ratio of the counted to all pairs of ciphertext decreases prominently. Therefore we use (4) to attack the full 6-round MMB cipher.

The 6-round MMB encryption of a plaintext P is depicted in (2). In order to decrease the time complexity, we move $\sigma[k^6]$ to the front of θ in the 6th round; $\sigma[k^6]$ will be transformed to $\sigma[k^{6'}]$, where $k_0^{6'} = k_0^6 \oplus k_1^6 \oplus k_3^6$; $k_1^{6'} = k_0^6 \oplus k_1^6 \oplus k_2^6$; $k_2^{6'} = k_1^6 \oplus k_2^6 \oplus k_3^6$ and $k_3^{6'} = k_0^6 \oplus k_2^6 \oplus k_3^6$. Thus, we will recover the equivalent subkey $k^{6'}$.

3.2   Attack Algorithm

We choose 2^{54} structures of 2^{64} chosen plaintexts each. In each structure, the second and third words of the plaintext can together take 2^{64} possible values. There are 2^{63} plaintext pairs with the difference $(0, 9bd3fdf7_x, e6f4ff7d_x, 0)$ in each structure. So, the total number of pairs in 2^{54} structures is $2^{54} \cdot 2^{63} = 2^{117}$.

The differential characteristic has probability 2^{-110}, so the number of the right pairs is $2^{117} \cdot 2^{-110} = 2^7 = 128$. For each structure, there are about 2^{63} pairs of plaintexts to be considered in total.

Since the output difference of the 5th round for a right pair is $(40404040_x, 0, 0, 40404040_x)$, the difference of the ciphertext pairs should be $(\alpha \oplus \beta, \alpha, \beta, \alpha \oplus \beta)$, with $\alpha, \beta \in \mathbb{Z}_{2^{32}}$, so we can use this to discar wrong pairs. Thus, about $2^{63} \cdot 2^{-64} = 2^{-1}$ candidates for the right pairs remain from each structure.

The input difference of the 6th round is $(40404040_x, 0, 0, 40404040_x)$. We found that the numbers of possible output difference values given the input difference 40404040_x for the modular multiplication G_0 or G_3 is $6738641/2^{32} = 2^{-9.32}$, so about $2^{-1} \cdot 2^{-18.64} = 2^{-19.64}$ candidates for the right pairs remain for each structure. The total number of remaining pairs in all the 2^{54} structures is $2^{54} \cdot 2^{-19.64} = 2^{34.36}$.

For each remaining ciphertext pair (C_0, C_1, C_2, C_3) and (C_0', C_1', C_2', C_3'), we guess the equivalent subkey words $k_0^{6'}$ and $k_3^{6'}$, and the total number of guessed subkey bits is 64. Then, calculate $\xi_0 = (G_0^{-1} \otimes (\eta(C_0 \oplus C_1 \oplus C_3 \oplus k_0^{6'}))) \oplus (G_0^{-1} \otimes (\eta(C_0' \oplus C_1' \oplus C_3' \oplus k_0^{6'})))$ and $\xi_3 = (G_3^{-1} \otimes (\eta(C_0 \oplus C_2 \oplus C_3 \oplus k_3^{6'}))) \oplus (G_3^{-1} \otimes (\eta(C_0' \oplus C_2' \oplus C_3' \oplus k_3^{6'})))$. If both ξ_0 and ξ_3 are equal to 40404040_x, the counter corresponding to $(k_0^{6'}, k_3^{6'})$ will be incremented by one. For G_0 and G_3 with the inputxor 40404040_x and any given outputxor, there will be at most 2^{17} pairs, so the maximum count per counted pair of the subkey words will be $2^{17} \cdot 2^{17} = 2^{34}$.

In our attack, the signal-to-noise ratio is computed as follows:

$$S/N = \frac{p \cdot 2^k}{\alpha \cdot \beta} = \frac{2^{-110} \cdot 2^{64}}{2^{-64-18.64} \cdot 2^{34}} = 2^{2.64} = 6.23.$$

The success probability is computed as follows[11]:

$$Ps = \int_{-\frac{\sqrt{\mu S/N} - \Phi^{-1}(1-2^{-a})}{\sqrt{S/N+1}}}^{\infty} \Phi(x)dx = 0.99999999,$$

where $a = 64$ is the number of subkey bits involved in the decryption and μ is the number of right pairs which can be obtained $\mu = p \cdot N = 2^{-110} \cdot 2^{117} = 128$. With probability 0.99999999 the right key can be recovered.

The attack needs 2^{118} CP and $2^{35.36} \cdot 2^{64} \cdot 2 = 2^{100.36}$ modular multiplications, which is no more than $2^{100.36}/4 = 2^{98.36}$ 1-round MMB encryptions, equivalent to $2^{98.36}/6 = 2^{95.91}$ 6-round MMB encryptions. The memory requirements are about 2^{64} 64-bit counters. The remaining 64-bit equivalent subkey $k_1^{6'}$ and $k_2^{6'}$ can be recovered by exhaustive search with about 2^{64} 6-round MMB encryptions. Finally, the 128-bit user key can be derived. In all, the data complexity is 2^{118} CP, the time complexity is $2^{95.91}$ 6-round MMB encryptions and the memory requirements are 2^{64} 64-bit blocks.

4 Square Analysis of MMB

MMB is a word-oriented cipher. More precisely, it operates on neatly partitioned 32-bit words. This wordwise behavior motivates our square analysis. Our attacks use Λ-sets of 2^{32} CP. We use the terminology of [6].

4.1   Square Distinguisher

Due to the special property for modular multiplication, we discovered a new word type: X word, which can propagate across γ. The X word is very useful for us to identify four chains of Λ-sets, each of which represents a 2.75-round square distinguisher if we consider every round transformation σ, γ, η and θ as a fraction of 0.25 of a (full) round.

- $(A,C,C,C) \overset{1r}{\to} (A,A,C,A) \overset{1r}{\to} (X,B,B,E) \overset{\sigma[k^2]}{\to} (X,B,B,E) \overset{\gamma}{\to} (X,?,?,E)$
  $\overset{\eta}{\to} (B,?,?,E)$,
- $(C,A,C,C) \overset{1r}{\to} (A,A,A,C) \overset{1r}{\to} (B,B,E,B) \overset{\sigma[k^2]}{\to} (B,B,E,B) \overset{\gamma}{\to} (?,?,E,?)$
  $\overset{\eta}{\to} (?,?,E,?)$,
- $(C,C,A,C) \overset{1r}{\to} (C,A,A,A) \overset{1r}{\to} (B,E,B,B) \overset{\sigma[k^2]}{\to} (B,E,B,B) \overset{\gamma}{\to} (?,E,?,?)$
  $\overset{\eta}{\to} (?,E,?,?)$,
- $(C,C,C,A) \overset{1r}{\to} (A,C,A,A) \overset{1r}{\to} (E,B,B,X) \overset{\sigma[k^2]}{\to} (E,B,B,X) \overset{\gamma}{\to} (E,?,?,X)$
  $\overset{\eta}{\to} (E,?,?,B)$,

where 'A' indicates an active word; 'C' denotes a passive (or constant) word; 'B' denotes a balanced word, that is, the xor sum of whose contents gives zero; 'X' denotes another special balanced word in which any value x and $\neg x$ appear the same number of times; 'E' denotes a special balanced word in which each value appears an even number of times [1]; '?' indicates that the xor sum of the 32-bit in that word is an unpredictable value. The proofs of the propagation of Λ-sets can be found in Appendix A.

4.2   Square Attack on 4-Round MMB

With any of the above square distinguishers, the key-recovery attack on 4-round MMB can be applied.

Consider the square distinguisher $(A,C,C,C) \overset{1r}{\to} (A,A,C,A) \overset{1r}{\to} (X,B,B,E)$ $\overset{\sigma[k^2]}{\to} (X,B,B,E) \overset{\gamma}{\to} (X,?,?,E) \overset{\eta}{\to} (B,?,?,E)$. A full 4-round MMB consists of

$$\sigma[k^4] \circ \theta \circ \eta \circ \gamma \circ \sigma[k^3] \circ \theta \circ \eta \circ \gamma \circ \sigma[k^2] \circ \theta \circ \eta \circ \gamma \circ \sigma[k^1] \circ \theta \circ \eta \circ \gamma \circ \sigma[k^0].$$

We aim at recovering k^4 by partial decryption. We also move $\sigma[k^4]$ across θ. We denote the modified key as $k^{4'}$. Further, we can remove θ because it is invertible and key independent.

The attack procedure is as follows:

- Step 1: Choose 2^{32} plaintexts (x, c_1^0, c_2^0, c_3^0), $x \in \mathbb{Z}_{2^{32}}$, c_1^0, c_2^0 and c_3^0 are constants.
- Step 2: Guess the 32-bit words $k_0^{4'}$, $k_1^{4'}$ and $k_3^{4'}$ of $k^{4'}$. Apply the inverse of η and γ, xor the three words to obtain a new word. If the new word is balanced, save the subkey value. On average, 2^{64} subkey values are saved.
- Step 3: for $i := 1$ to 3 do

- Step 3.1: Choose a new group of 2^{32} plaintexts $(x, c_1{}^i, c_2{}^i, c_3{}^i), x \in \mathbb{Z}_{2^{32}}$, $c_1{}^i, c_2{}^i$ and $c_3{}^i$ are constants.
- Step 3.2: For each saved subkey value, apply the inverse of η and γ, xor the three words to obtain the new word. If the new word is not balanced, delete the subkey value.
- Step 4: The remaining subkey value should be the right subkey with high probability.

The total number of guessed subkey bits is 96, only one Λ-set cannot identify the right subkey; on average, 2^{64} wrong guesses also satisfy the balanced property. So, we choose different constant values of the later three words in plaintexts to construct four Λ-sets. We expect any wrong subkey value to satisfy the balanced property with probability 2^{-32}, but, the right subkey value must satisfy the balanced property always.

In total, the complexity is about $(2^{96} + 2^{64} + 2^{32} + 1) \cdot 2^{32} = 2^{128}$ 1.25-round decryptions; 2^{34} CP, and memory of 2^{64} 96-bit counters. For the third word of $k^{4'}$, we can recover it by exhaustive search. In total, the time complexity is $2^{128} \cdot 1.25 + 6 \cdot 2^{32} = 2^{128.32}$ 1-round decryptions, or equivalently, $2^{128.32}/4 = 2^{126.32}$ full 4-round MMB encryptions. The data complexity will be 2^{34} CP. The memory complexity is 2^{64} text blocks.

5   Linear Attacks on MMB

Linear cryptanalysis typically works in a known-plaintext or ciphertext-only setting (in the latter, assuming the plaintext is ASCII text), and its origin dates back to the works of Matsui on DES [8,9].

The main concept for this attack is the linear relation, which consists of a linear combination of text and key bits, holding with a high parity deviation from the uniform parity distribution. Initially, linear relations are obtained for each individual cipher component. Further, more extensive relations are obtained by the combination of smaller relations from consecutive cipher components, up to full round, and then for multiple rounds. The effectiveness of a linear relation is measured by a parameter called bias, which is the absolute value of the deviation of the parity of the linear relation from $1/2$ (the expected value for an unbiased relation). The higher the bias, the more attractive the linear relations, since they demand less plaintext-ciphertext pairs. These linear relations form the core of a linear distinguisher, namely, a tool that allows one to distinguish a given cipher from a random permutation, or to recover subkey bits.

5.1   Linear Approximations for MMB

In MMB, the main non-linear operation that limits the effectiveness of linear approximations, is the multiplication in $\mathbb{Z}_{2^{32}-1}$, namely γ. Let $M_i = (m_{i0}, m_{i1}, m_{i2}, m_{i3})$ and $M_o = (m_{o0}, m_{o1}, m_{o2}, m_{o3})$ denote the linear input mask and the linear output mask of γ, respectively. Any nonzero m_{ij} (and m_{oj}), for $0 \le j \le 3$, represents an active multiplication.

As in the differential cryptanalysis in Sect.3, the possible distributions for active multiplications in linear approximation are listed in Table 3. m_o and m_{ij}, $(0 \le j \le 3)$ represent the input mask and the output mask, respectively. Besides the γ component, η is also non-linear. To avoid the effect of η on linear approximations, it is necessary to guarantee that the output mask m_o for the active G_i satisfies $m_o \cdot \delta = 0$, where $\cdot$ is the dot product.

We recall the rotational invariant property [4] of multiplication modulo $2^{32} - 1$,

$$a \otimes (x << k) = (a \otimes x) << k.$$

The linear approximation for one multiplication can be used to obtain the linear approximation for the other three multiplications. The bias ϵ for $m_{i1} \stackrel{G_1}{\to} m_o$ and the bias ϵ' for $m_{i1} \lll 2 \stackrel{G_2}{\to} m_o$ will be equal. In particular, with the rotational property, for $m_i = \bar{0}$ and $m_o = \bar{0}$, the biases for the linear approximation of multiplication for all G_i are equal (to $2^{-12.0897}$). But, the mask $\bar{0}$ is not appropriate concerning η. The corresponding bias for a one-round linear approximation is zero because $\bar{0} \cdot \delta = 1$.

In order to construct multi-round linear approximations, the output masks for different active multiplications must be equal. From the experiments of different masks for G_i, we conjecture that the linear approximations for modular multiplication with maximum bias have the following forms:

$$
\begin{aligned}
mmmmmmmm_x &\stackrel{G_i}{\to} nnnnnnnn_x, \\
m_0 m_1 m_0 m_1 m_0 m_1 m_0 m_{1x} &\stackrel{G_i}{\to} n_0 n_1 n_0 n_1 n_0 n_1 n_0 n_{1x}, \\
m_0 m_1 m_2 m_3 m_4 m_5 m_6 m_{7x} &\stackrel{G_i}{\to} m_0 m_1 m_2 m_3 m_4 m_5 m_6 m_{7x},
\end{aligned}
\tag{5}
$$

where $m, n, m_i, n_i \in \mathbb{Z}_{2^4}$, $0 \le i \le 7$. The probability for the above linear relations with the maximum bias decreases gradually. We have only searched the first two linear approximations in (5). The last linear relation needs too large a test space, so we have not searched it.

Linear Approximations for Modulo Multiplication:

The best linear approximations we identified have bias $2^{-8.8}$ for each G_i, and some of them are

$$
\begin{aligned}
3c3c3c3c_x &\stackrel{G_0}{\to} 0f0f0f0f_x, & 3c3c3c3c_x &\stackrel{G_1}{\to} 1e1e1e1e_x, \\
3c3c3c3c_x &\stackrel{G_2}{\to} 78787878_x, & 3c3c3c3c_x &\stackrel{G_3}{\to} 87878787_x.
\end{aligned}
$$

Based on the above linear approximations for G_i, one-round linear approximations with only one active multiplication can be obtained with bias $2^{-8.8}$.

Two-Round Linear Approximations

Two-round linear approximations can be obtained with only two active modular multiplications in each round. For active G_0 and G_2, for instance

$$(m_{i0}, 0, m_{i2}, 0) \overset{1r}{\to} (m_2, 0, m_2, 0) \overset{1r}{\to} (m_3, 0, m_3, 0), \tag{6}$$

where m_{i0}, m_{i2}, m_2 and m_3 are independent 32-bit masks. The following local approximations are required: $m_{i0} \overset{G_0}{\to} m_2$, $m_{i2} \overset{G_2}{\to} m_2$ and $m_2 \overset{G_j}{\to} m_3$ for $j \in \{0, 2\}$ with nonzero bias. The maximum bias we identified was $2^{-36.82}$ for 2-round linear approximation:

$$\begin{aligned} &(1b1b1b1b_x, 0, 63636363_x, 0) \overset{1r}{\to} (6c6c6c6c_x, 0, 6c6c6c6c_x, 0) \\ &\overset{1r}{\to} (72727272_x, 0, 72727272_x, 0), \end{aligned} \tag{7}$$

where the linear approximations for G_0 and G_2 are

$$1b1b1b1b_x \overset{G_0}{\to} 6c6c6c6c_x, \epsilon = 2^{-9.40}; \quad 63636363_x \overset{G_2}{\to} 6c6c6c6c_x, \epsilon = 2^{-9.40};$$
$$6c6c6c6c_x \overset{G_0}{\to} 72727272_x, \epsilon = 2^{-9.78}; \quad 6c6c6c6c_x \overset{G_2}{\to} 72727272_x, \epsilon = 2^{-11.24}.$$

In addition, we identified 2-round linear approximation for active G_1 and G_3 as follows:

$$(0, m_{i1}, 0, m_{i3}) \overset{1r}{\to} (0, m_2, 0, m_2) \overset{1r}{\to} (0, m_3, 0, m_3). \tag{8}$$

We identified the maximum bias $2^{-36.95}$ for 2-round linear approximation as follows:

$$\begin{aligned} &(0, 99999999_x, 0, 66666666_x) \overset{1r}{\to} (0, 33333333_x, 0, 33333333_x) \\ &\overset{1r}{\to} (0, 66666666_x, 0, 66666666_x), \end{aligned} \tag{9}$$

where the linear approximations for G_1 and G_3 are

$$99999999_x \overset{G_1}{\to} 33333333_x, \epsilon = 2^{-9.56}; \quad 66666666_x \overset{G_3}{\to} 33333333_x, \epsilon = 2^{-9.56};$$
$$33333333_x \overset{G_1}{\to} 66666666_x, \epsilon = 2^{-9.56}; \quad 33333333_x \overset{G_3}{\to} 66666666_x, \epsilon = 2^{-11.27}.$$

Three-Round Linear Approximations

We found the 3-round linear approximation for active G_0 and G_2 as follows:

$$\begin{aligned} &(d8d8d8d8_x, 0, 1b1b1b1b_x, 0) \overset{1r}{\to} (63636363_x, 0, 63636363_x, 0) \\ &\overset{1r}{\to} (36363636_x, 0, 36363636_x, 0) \overset{1r}{\to} (63636363_x, 0, 63636363_x, 0), \end{aligned}$$

where the linear approximations for G_0 and G_2 are

$$d8d8d8d8_x \overset{G_0}{\to} 63636363_x, \epsilon = 2^{-9.40}; \quad 1b1b1b1b_x \overset{G_2}{\to} 63636363_x, \epsilon = 2^{-9.40};$$
$$63636363_x \overset{G_0}{\to} 36363636_x, \epsilon = 2^{-13.76}; \quad 63636363_x \overset{G_2}{\to} 36363636_x, \epsilon = 2^{-10.56};$$
$$36363636_x \overset{G_0}{\to} 63636363_x, \epsilon = 2^{-13.76}; \quad 36363636_x \overset{G_2}{\to} 63636363_x, \epsilon = 2^{-10.56}.$$

The bias for the 3-round linear approximation is $2^{-9.40\cdot2-13.76\cdot2-10.56\cdot2+5} = 2^{-62.44}$. Moreover, if G_1 and G_3 are active, we identified two linear approximations for 3-round MMB with the maximum bias $2^{-55.78}$ as follows:

$$(0, 99999999_x, 0, 66666666_x) \xrightarrow{1r} (0, 33333333_x, 0, 33333333_x)$$
$$\xrightarrow{1r} (0, 66666666_x, 0, 66666666_x) \xrightarrow{1r} (0, 33333333_x, 0, 33333333_x), \tag{10}$$

$$(0, 33333333_x, 0, cccccccc_x) \xrightarrow{1r} (0, 66666666_x, 0, 66666666_x)$$
$$\xrightarrow{1r} (0, 33333333_x, 0, 33333333_x) \xrightarrow{1r} (0, 66666666_x, 0, 66666666_x), \tag{11}$$

where the linear approximations for G_1 and G_3 are

$$99999999_x \xrightarrow{G_1} 33333333_x, \epsilon = 2^{-9.56}; \quad 66666666_x \xrightarrow{G_3} 33333333_x, \epsilon = 2^{-9.56};$$
$$33333333_x \xrightarrow{G_1} 66666666_x, \epsilon = 2^{-9.56}; \quad 33333333_x \xrightarrow{G_3} 66666666_x, \epsilon = 2^{-11.27};$$
$$66666666_x \xrightarrow{G_1} 33333333_x, \epsilon = 2^{-11.27}; \quad cccccccc_x \xrightarrow{G_3} 66666666_x, \epsilon = 2^{-9.56}.$$

The bias for the two 3-round linear approximations is $2^{-9.56\cdot2-9.56\cdot2-11.27\cdot2+5} = 2^{-55.78}$.

In Appendix B, we list a linear approximation for four rounds, but whose bias is too low for an effective attack.

5.2 Linear Attack on Reduced-Round MMB

Known Plaintext Linear Attack

With the 3-round linear approximation in (10), a linear relation involving some plaintext bits, ciphertext bits and subkey bits can be derived. Using Algorithm 1 in [8], we can deduce the XOR value for the subkey bits involved in the linear relation. So, we can recover one bit of key information from 3-round MMB using $8 \cdot (2^{-55.78})^{-2} = 2^{114.56}$ KP and equivalent parity computations. Further, we can use the 3-round linear approximation in (11) to recover another one bit of key information from 3-round MMB. In all, two bits of key information can be recovered. For this step, the time complexity is $2^{115.56}$ parity computations. The remaining 126-bit subkey can be obtained by exhaustive search with about 2^{126} 3-round encryptions. In all, we can recover 128-bit key for 3-round MMB with $2^{114.56}$ known plaintexts and 2^{126} 3-round encryptions.

Ciphertext-Only Linear Attack

If the plaintexts are ASCII, then particular bitmasks involving only the most significant bit of each plaintext byte may allow a ciphertext-only (CO) linear attack on MMB. This is a more attractive attack setting than the conventional known-plaintext (KP) setting, since an opponent only needs ciphertext blocks. For MMB, we have identified 2-round linear relations with bitmasks that involve only the most significant bits of bytes in plaintext blocks. The linear relation with active G_0 and G_2 is identified as follows:

$$(80808080_x, 0, 80808080_x, 0) \xrightarrow{1r} (65656565_x, 0, 65656565_x, 0)$$
$$\xrightarrow{1r} (1e1e1e1e_x, 0, 1e1e1e1e_x, 0), \tag{12}$$

where the linear approximations for G_0 and G_2 are $80808080_x \xrightarrow{G_0} 65656565_x$, with $\epsilon = 2^{-15.74}$; $80808080_x \xrightarrow{G_2} 65656565_x$, with $\epsilon = 2^{-8.85}$; $65656565_x \xrightarrow{G_0} 1e1e1e1e_x$, with $\epsilon = 2^{-15.57}$; $65656565_x \xrightarrow{G_2} 1e1e1e1e_x$, with $\epsilon = 2^{-11.54}$. The bias for relation (12) is $2^{-15.74-8.85-15.57-11.54+3} = 2^{-48.70}$. The linear relation with active G_1 and G_3 is identified as follows:

$$(0, 80808080_x, 0, 80808080_x) \xrightarrow{1r} (0, 59595959_x, 0, 59595959_x)$$
$$\xrightarrow{1r} (0, 74747474_x, 0, 74747474_x), \tag{13}$$

where the linear approximations for G_1 and G_3 are $80808080_x \xrightarrow{G_1} 59595959_x$, with $\epsilon = 2^{-8.85}$; $80808080_x \xrightarrow{G_3} 59595959_x$, with $\epsilon = 2^{-16.83}$; $59595959_x \xrightarrow{G_1} 74747474_x$, with $\epsilon = 2^{-11.85}$; $59595959_x \xrightarrow{G_3} 74747474_x$, with $\epsilon = 2^{-10.77}$. The bias for relation (13) is $2^{-8.85-16.83-11.85-10.77+3} = 2^{-45.30}$. This bias only leads to a distinguishing attack on 2-round MMB, with $8 \cdot (2^{-45.30})^{-2} = 2^{93.60}$ CO, and equivalent number of parity computations.

6   Conclusions

This paper described the first detailed differential, square and linear attacks on versions 1.0 and 2.0 of the MMB block cipher, a design by Daemen, Govaerts and Vandewalle, dated from 1993, as an alternative to the IDEA block cipher. For differential cryptanalysis, the characteristic $\bar{0} \xrightarrow{\gamma} \bar{0}$ with probability 1 is the key point towards successful attack of the full MMB cipher. For square attack, the identical property of $\bar{0} \xrightarrow{\gamma} \bar{0}$ leads us to identify a new word type, the X word, which is relevant to identify 2.75-round square distinguishers. Without it, only 2-round square distinguishers can be found. For linear cryptanalysis, although the designers did not claim resistance of the MMB cipher against linear cryptanalysis, it is interesting that we were able to find better differential attacks than linear ones.

A summary of our attacks is in Table 1. We have presented both distinguishing-from-random and key-recovery attacks on the full and reduced-round MMB cipher. Our attacks apply equally well to MMB version 2.0 [5], which only differs from the original MMB in the key schedule algorithm, designed to avoid related-key attacks.

An unusual property of the θ and γ layers of MMB under a square attack is described in Appendix C. This attack demonstrates the importance of the η layer in MMB, in order to resist square attacks.

Table 1. Summary of attacks on MMB

#Rounds	Time		Data		Memory	Type
2	$2^{93.6}$	PC	$2^{93.6}$	CO	—	LC, DR
3	$2^{114.56}$	PC	$2^{114.56}$	KP	—	LC, DR
3	2^{126}	EN	$2^{114.56}$	KP	—	LC, KR
4	$2^{126.32}$	EN	2^{34}	CP	2^{64}	SC, KR
6	$2^{95.91}$	EN	2^{118}	CP	2^{64}	DC, KR

PC: number of parity computations; EN: number of encryptions;
LC, DR: Linear Distinguishing Attack;
LC, KR: Key-recovery Attack with Linear Cryptanalysis;
DC, KR: Key-recovery Attack with Differential Cryptanalysis;
SC, KR: Key-recovery Attack with Square Cryptanalysis.

Acknowledgements

It is a pleasure to acknowledge Xiaoyun Wang for various discussions on this paper. We would like to thank Yinglong Wang and Jinshan Pan in Shandong Computer Science Center for their providing the cluster computers to finish our experiments. We would also like to thank the anonymous reviewers for their very important comments.

This research is supported by 973 Program of China (Grant No. 2007CB807902) and National Outstanding Young Scientist fund of China (Grant No. 60525201).

References

1. Biryukov, A., Shamir, A.: Structural Cryptanalysis of SASAS. In: Pfitzmann, B. (ed.) EUROCRYPT 2001. LNCS, vol. 2045, pp. 394–405. Springer, Heidelberg (2001)
2. Biham, E., Shamir, A.: Differential Cryptanalysis of DES-like Cryptosystems. Journal of Cryptology 4(1), 3–72 (1991)
3. Daemen, J., Govaerts, R., Vandewalle, J.: Block Ciphers Based on Modular Multiplication. In: Wolfowicz, W. (ed.) Proceedings of 3rd Symposium on State and Progress of Research in Cryptography, Fondazione Ugo Bordoni, pp. 80–89 (1993)
4. Daemen, J., Van Linden, L., Govaerts, R., Vandewalle, J.: Propagation Properties of Multiplication Modulo $2^n - 1$. In: Proceedings of the 13th Symposium on Information Theory in the Benelux, Werkgemeenschap voor informatie- en Communicatietheorie, Enschede, The Netherlands, pp. 111–118 (1992)
5. Daemen, J.: Cipher and Hash Function Design – Strategies based on Linear and Differential Cryptanalysis. PhD Thesis, Dept. Elektrotechniek, Katholieke Universiteit Leuven, Belgium (1995)
6. Daemen, J., Knudsen, L.R., Rijmen, V.: The Block Cipher SQUARE. In: Biham, E. (ed.) FSE 1997. LNCS, vol. 1267, pp. 149–165. Springer, Heidelberg (1997)
7. Lai, X.: On the Design and Security of Block Ciphers. In: Massey, J.L. (ed.) ETH Series in Information Processing, vol. 1. Hartung-Gorre Verlag, Konstanz (1995)

8. Matsui, M.: Linear Cryptanalysis Method for DES Cipher. In: Helleseth, T. (ed.) EUROCRYPT 1993. LNCS, vol. 765, pp. 386–397. Springer, Heidelberg (1994)
9. Matsui, M.: The First Experimental Cryptanalysis of the Data Encryption Standard. In: Desmedt, Y.G. (ed.) CRYPTO 1994. LNCS, vol. 839, pp. 1–11. Springer, Heidelberg (1994)
10. Rijmen, V., Daemen, J., Preneel, B., Bosselaers, A., De Win, E.: The Cipher SHARK. In: Gollmann, D. (ed.) FSE 1996. LNCS, vol. 1039, pp. 99–111. Springer, Heidelberg (1996)
11. Selçuk, A.A., Biçak, A.: On Probability of Success in Linear and Differential Cryptanalysis. In: Cimato, S., Galdi, C., Persiano, G. (eds.) SCN 2002. LNCS, vol. 2576, pp. 174–185. Springer, Heidelberg (2003)

Appendix

Table 2. One-round differential characteristics for MMB: $\Delta_{ij} (0 \leq j \leq 3)$, and Δ_o are nonzero 32-bit xor difference values

input difference $\xrightarrow{1r}$ output difference	# active multiplications	restriction on Δ_o
$(\Delta_{i0}, 0, 0, 0) \xrightarrow{1r} (\Delta_o, \Delta_o, 0, \Delta_o)$	1	lsb(Δ_o)=0
$(0, \Delta_{i1}, 0, 0) \xrightarrow{1r} (\Delta_o, \Delta_o, \Delta_o, 0)$	1	—
$(0, 0, \Delta_{i2}, 0) \xrightarrow{1r} (0, \Delta_o, \Delta_o, \Delta_o)$	1	—
$(0, 0, 0, \Delta_{i3}) \xrightarrow{1r} (\Delta_o, 0, \Delta_o, \Delta_o)$	1	lsb(Δ_o)=0
$(\Delta_{i0}, \Delta_{i1}, 0, 0) \xrightarrow{1r} (0, 0, \Delta_o, \Delta_o)$	2	lsb(Δ_o)=0
$(\Delta_{i0}, 0, \Delta_{i2}, 0) \xrightarrow{1r} (\Delta_o, 0, \Delta_o, 0)$	2	lsb(Δ_o)=0
$(\Delta_{i0}, 0, 0, \Delta_{i3}) \xrightarrow{1r} (0, \Delta_o, \Delta_o, 0)$	2	lsb(Δ_o)=0
$(0, \Delta_{i1}, \Delta_{i2}, 0) \xrightarrow{1r} (\Delta_o, 0, 0, \Delta_o)$	2	—
$(0, \Delta_{i1}, 0, \Delta_{i3}) \xrightarrow{1r} (0, \Delta_o, 0, \Delta_o)$	2	lsb(Δ_o)=0
$(0, 0, \Delta_{i2}, \Delta_{i3}) \xrightarrow{1r} (\Delta_o, \Delta_o, 0, 0)$	2	lsb(Δ_o)=0
$(\Delta_{i0}, \Delta_{i1}, \Delta_{i2}, 0) \xrightarrow{1r} (0, \Delta_o, 0, 0)$	3	lsb(Δ_o)=0
$(\Delta_{i0}, \Delta_{i1}, 0, \Delta_{i3}) \xrightarrow{1r} (\Delta_o, 0, 0, 0)$	3	lsb(Δ_o)=0
$(\Delta_{i0}, 0, \Delta_{i2}, \Delta_{i3}) \xrightarrow{1r} (0, 0, 0, \Delta_o)$	3	lsb(Δ_o)=0
$(0, \Delta_{i1}, \Delta_{i2}, \Delta_{i3}) \xrightarrow{1r} (0, 0, \Delta_o, 0)$	3	lsb(Δ_o)=0
$(\Delta_{i0}, \Delta_{i1}, \Delta_{i2}, \Delta_{i3}) \xrightarrow{1r} (\Delta_o, \Delta_o, \Delta_o, \Delta_o)$	4	lsb(Δ_o)=0

A   Proofs of Square Distinguishers

A.1 Proof of the First Distinguisher

For the first distinguisher, we denote each word after η in the first round as $S(v)$; S is the status for the word, such as A, B, E and X, and v represents the variable; the first distinguisher can be written as

$$(A,C,C,C) \overset{\sigma[k^0]}{\to} (A,C,C,C) \overset{\gamma}{\to} (A,C,C,C) \overset{\eta}{\to} (A(x),C(c_1),C(c_2),C(c_3))$$
$$\overset{\theta}{\to} (A(x \oplus c_1 \oplus c_3), A(x \oplus c_1 \oplus c_2), C(c_1 \oplus c_2 \oplus c_3), A(x \oplus c_2 \oplus c_3))$$
$$\overset{\sigma[k^1]}{\to} (A(x \oplus c_1 \oplus c_3 \oplus k_0^1), A(x \oplus c_1 \oplus c_2 \oplus k_1^1), C(c_1 \oplus c_2 \oplus c_3 \oplus k_2^1),$$
$$A(x \oplus c_2 \oplus c_3 \oplus k_3^1))$$
$$\overset{\gamma}{\to} (A(G_0 \otimes (x \oplus c_1 \oplus c_3 \oplus k_0^1)), A(G_1 \otimes (x \oplus c_1 \oplus c_2 \oplus k_1^1)),$$
$$C(G_2 \otimes (c_1 \oplus c_2 \oplus c_3 \oplus k_2^1)), A(G_3 \otimes (x \oplus c_2 \oplus c_3 \oplus k_3^1)))$$
$$\overset{\eta}{\to} (A(\eta(G_0 \otimes (x \oplus c_1 \oplus c_3 \oplus k_0^1))), A(G_1 \otimes (x \oplus c_1 \oplus c_2 \oplus k_1^1)),$$
$$C(G_2 \otimes (c_1 \oplus c_2 \oplus c_3 \oplus k_2^1)), A(\eta(G_3 \otimes (x \oplus c_2 \oplus c_3 \oplus k_3^1))))$$
$$\overset{\theta}{\to} (X(y_0), B(y_1), B(y_2), E(y_3))$$
$$\overset{\sigma[k^2]}{\to} (X(y_0 \oplus k_0^2), B(y_1 \oplus k_1^2), B(y_2 \oplus k_2^2), E(y_3 \oplus k_3^2))$$
$$\overset{\gamma}{\to} (X(z_0), ?(z_1), ?(z_2), E(z_3))$$
$$\overset{\eta}{\to} (B(u_0), ?(u_1), ?(u_2), E(u_3)).$$

In the above transitions, the first output word of η in the first round is an A word and the other three output words are constants, so we denote them as the variables x, c_1, c_2 and c_3, respectively. In addition, η only affects the output of the first and the last words. We denote the four status variables after the operation of θ in the second round as $y_i (0 \le i \le 3)$, the four words after γ in

Table 3. One-round linear relations for MMB: $m_{ij}(0 \le j \le 3)$, and m_o are nonzero 32-bit masks

input mask $\overset{1r}{\to}$ output mask	#active multiplications	restriction on m_o
$(m_{i0},0,0,0) \overset{1r}{\to} (m_o,m_o,0,m_o)$	1	$m_o \cdot \delta = 0$
$(0,m_i,0,0) \overset{1r}{\to} (m_o,m_o,m_o,0)$	1	—
$(0,0,m_i,0) \overset{1r}{\to} (0,m_o,m_o,m_o)$	1	—
$(0,0,0,m_i) \overset{1r}{\to} (m_o,0,m_o,m_o)$	1	$m_o \cdot \delta = 0$
$(m_{i0},m_{i1},0,0) \overset{1r}{\to} (0,0,m_o,m_o)$	2	$m_o \cdot \delta = 0$
$(m_{i0},0,m_{i2},0) \overset{1r}{\to} (m_o,0,m_o,m_o)$	2	$m_o \cdot \delta = 0$
$(m_{i0},0,0,m_{i3}) \overset{1r}{\to} (0,m_o,m_o,0)$	2	$m_o \cdot \delta = 0$
$(0,m_{i1},m_{i2},0) \overset{1r}{\to} (m_o,0,0,m_o)$	2	—
$(0,m_{i1},0,m_{i3}) \overset{1r}{\to} (0,m_o,0,m_o)$	2	$m_o \cdot \delta = 0$
$(0,0,m_{i2},m_{i3}) \overset{1r}{\to} (m_o,m_o,0,0)$	2	$m_o \cdot \delta = 0$
$(m_{i0},m_{i1},m_{i2},0) \overset{1r}{\to} (0,m_o,0,0)$	3	$m_o \cdot \delta = 0$
$(m_{i0},m_{i1},0,m_{i3}) \overset{1r}{\to} (m_o,0,0,0)$	3	$m_o \cdot \delta = 0$
$(m_{i0},0,m_{i2},m_{i3}) \overset{1r}{\to} (0,0,0,m_o)$	3	$m_o \cdot \delta = 0$
$(0,m_{i1},m_{i2},m_{i3}) \overset{1r}{\to} (0,0,m_o,0)$	3	$m_o \cdot \delta = 0$
$(m_{i0},m_{i1},m_{i2},m_{i3}) \overset{1r}{\to} (m_o,m_o,m_o,m_o)$	4	$m_o \cdot \delta = 0$

the third round as $z_i(0 \le i \le 3)$, and the four words after η in the third round as $u_i(0 \le i \le 3)$. The square distinguisher can be proved in three steps.

1. Prove that y_0 is X, y_3 is E and both y_1 and y_2 are B words.
2. Prove that z_0 is an X word and z_3 is an E word.
3. Prove that u_0 is a B word and u_3 is an E word.

Step 1: Prove That y_0 is X, y_3 is E and Both y_1 and y_2 are B Words

We extend y_i as follows:

$$y_0 = \eta(G_0 \otimes (x \oplus c_1 \oplus c_3 \oplus k_0^1)) \oplus (G_1 \otimes (x \oplus c_1 \oplus c_2 \oplus k_1^1)) \oplus$$
$$\eta(G_3 \otimes (x \oplus c_2 \oplus c_3 \oplus k_3^1)), \tag{14}$$

$$y_1 = \eta(G_0 \otimes (x \oplus c_1 \oplus c_3 \oplus k_0^1)) \oplus (G_1 \otimes (x \oplus c_1 \oplus c_2 \oplus k_1^1)) \oplus$$
$$(G_2 \otimes (c_1 \oplus c_2 \oplus c_3 \oplus k_2^1)), \tag{15}$$

$$y_2 = (G_1 \otimes (x \oplus c_1 \oplus c_2 \oplus k_1^1)) \oplus (G_2 \otimes (c_1 \oplus c_2 \oplus c_3 \oplus k_2^1)) \oplus$$
$$\eta(G_3 \otimes (x \oplus c_2 \oplus c_3 \oplus k_3^1)), \tag{16}$$

$$y_3 = \eta(G_0 \otimes (x \oplus c_1 \oplus c_3 \oplus k_0^1)) \oplus (G_2 \otimes (c_1 \oplus c_2 \oplus c_3 \oplus k_2^1)) \oplus$$
$$\eta(G_3 \otimes (x \oplus c_2 \oplus c_3 \oplus k_3^1)). \tag{17}$$

In (14)–(17), x is a variable of an A word, so the input x and $\neg x$ must appear once each. Then, we have

$$y_0(x) \oplus y_0(\neg x) = \eta(G_0 \otimes (x \oplus c_1 \oplus c_3 \oplus k_0^1)) \oplus \eta(G_0 \otimes (\neg x \oplus c_1 \oplus c_3 \oplus k_0^1))$$
$$\oplus(G_1 \otimes (x \oplus c_1 \oplus c_2 \oplus k_1^1)) \oplus (G_1 \otimes (\neg x \oplus c_1 \oplus c_2 \oplus k_1^1))$$
$$\oplus\eta(G_3 \otimes (x \oplus c_2 \oplus c_3 \oplus k_3^1)) \oplus \eta(G_3 \otimes (\neg x \oplus c_2 \oplus c_3 \oplus k_3^1)).$$

Due to

$$G_i \otimes (\neg x \oplus c) = G_i \otimes (\bar{0} \oplus x \oplus c) = G_i \otimes (\bar{0} \oplus (x \oplus c))$$
$$= G_i \otimes (\bar{0} - (x \oplus c)) = (G_i \otimes \bar{0}) - G_i \otimes (x \oplus c) \tag{18}$$
$$= \bar{0} - G_i \otimes (x \oplus c) = \bar{0} \oplus G_i \otimes (x \oplus c) = \neg(G_i \otimes (x \oplus c))$$

and

$$\eta(w) \oplus \eta(\neg w) = \delta \oplus \bar{0} = d5555555_x, \tag{19}$$

we obtain

$$y_0(x) \oplus y_0(\neg x) = \eta(G_0 \otimes (x \oplus c_1 \oplus c_3 \oplus k_0^1)) \oplus \eta(\neg(G_0 \otimes (x \oplus c_1 \oplus c_3 \oplus k_0^1)))$$
$$\oplus(G_1 \otimes (x \oplus c_1 \oplus c_2 \oplus k_1^1)) \oplus \neg(G_1 \otimes (x \oplus c_1 \oplus c_2 \oplus k_1^1))$$
$$\oplus\eta(G_3 \otimes (x \oplus c_2 \oplus c_3 \oplus k_3^1)) \oplus \eta(\neg(G_3 \otimes (x \oplus c_2 \oplus c_3 \oplus k_3^1)))$$
$$= d5555555_x \oplus \bar{0} \oplus d5555555_x = \bar{0}.$$

We derive $y_0(x) = \neg y_0(\neg x)$. As a variable of an A word, both x and $\neg x$ must appear once each. So, $y_0(x)$ and $y_0(\neg x) = \neg y_0(x)$ must appear just as often. There are 2^{31} pairs of $(x, \neg x)$, so there are 2^{31} pairs of $(y_0, \neg y_0)$, which means that the xor sum of 2^{32} y_0 is zero (equal to 2^{31} times of the xor sum of $\bar{0}$). Therefore, y_0 is an X word.

$$y_3(x) \oplus y_3(\neg x) = \eta(G_0 \otimes (x \oplus c_1 \oplus c_3 \oplus k_0^1)) \oplus \eta(G_0 \otimes (\neg x \oplus c_1 \oplus c_3 \oplus k_0^1))$$
$$\oplus (G_2 \otimes (c_1 \oplus c_2 \oplus c_3 \oplus k_2^1)) \oplus (G_2 \otimes (c_1 \oplus c_2 \oplus c_3 \oplus k_2^1))$$
$$\oplus \eta(G_3 \otimes (x \oplus c_2 \oplus c_3 \oplus k_3^1)) \oplus \eta(G_3 \otimes (\neg x \oplus c_2 \oplus c_3 \oplus k_3^1))$$
$$= d5555555_x \oplus d5555555_x = 0$$

We obtain $y_3(x) = y_3(\neg x)$, i.e. any value of y_3 will appear an even number of times, so y_3 is an E word.

$$y_1(x) \oplus y_1(\neg x) = \eta(G_0 \otimes (x \oplus c_1 \oplus c_3 \oplus k_0^1))) \oplus \eta(G_0 \otimes (\neg x \oplus c_1 \oplus c_3 \oplus k_0^1)))$$
$$\oplus (G_1 \otimes (x \oplus c_1 \oplus c_2 \oplus k_1^1)) \oplus (G_1 \otimes (\neg x \oplus c_1 \oplus c_2 \oplus k_1^1))$$
$$\oplus (G_2 \otimes (c_1 \oplus c_2 \oplus c_3 \oplus k_2^1)) \oplus (G_2 \otimes (c_1 \oplus c_2 \oplus c_3 \oplus k_2^1))$$
$$= d5555555_x \oplus \bar{0} = \delta$$

We cannot assure that y_1 and $\neg y_1$ appear at the same time, so y_1 is not an X word. But, 2^{31} pairs of $(x, \neg x)$ result in the xor sum of 2^{32} y_1 is zero (equal to 2^{31} times of the xor sum of δ). So, y_1 is a B word. In this way, we can prove y_2 is also a B word.

Step 2: Prove That z_0 is an X Word and z_3 is an E Word

We extend y_i as the following equations,

$$z_0 = G_0 \otimes (y_0 \oplus k_0^2), \, z_1 = G_1 \otimes (y_1 \oplus k_1^2),$$
$$z_2 = G_2 \otimes (y_2 \oplus k_2^2), \, z_3 = G_3 \otimes (y_3 \oplus k_3^2).$$

We have proved y_0 is an X word which means that y_0 and $\neg y_0$ must appear at the same time. From (18), $G_0 \otimes (y_0 \oplus k_0^2) = \neg G_0 \otimes (\neg y_0 \oplus k_0^2)$, we can obtain $z_0(y_0) = \neg z_0(\neg y_0)$, which means any value of z_0 and $\neg z_0$ will appear at the same time. So z_0 is also an X word.

In addition, y_3 is an E word which means that any value of y_3 will appear an even number of times and results any value of z_3 will appear an even number of times too. Thus, z_3 should be an E word, too. Because y_1 and y_2 are B words, and B words cannot usually cross γ, so the status for z_1 and z_2 cannot be decided.

Step 3: Prove That u_0 is a B Word and u_3 is an E Word

We extend u_i as the following $u_0 = \eta(z_0)$, $u_1 = \eta(z_1) = z_1$, $u_2 = \eta(z_2) = z_2$, $u_3 = \eta(z_3)$. Recall that z_0 is an X word, which means that any value of z_0 and $\neg z_0$ will appear at the same time. From (19), $u_0(z_0) \oplus u_0(\neg z_0) = d5555555_x$. There are 2^{31} pairs of $(u_0(z_0), u_0(\neg z_0))$. So, the xor sum of 2^{32} u_0 is zero (equal to 2^{31} times of the xor sum for $d5555555_x$). Therefore, u_0 is a B word but not an X word. Since z_3 is an E word, it follows that any value of $u_3(z_3)$ will appear even times. So, u_3 is an E word. After θ in the third round, the balanced property will be destroyed in all four words.

A.2 Proof of the Other Three Distinguishers

The proof of the other three distinguishers is similar to the above proof for the first distinguisher.

B   Four-Round Linear Approximation

We have identified a 4-round linear relation with bias $2^{3\cdot(-9.56-11.27)-9.56\cdot2+7} = 2^{-74.61}$ which is given as follows:

$$(0, 99999999_x, 0, 66666666_x) \xrightarrow{1r} (0, 33333333_x, 0, 33333333_x)$$
$$\xrightarrow{1r} (0, 66666666_x, 0, 66666666_x) \xrightarrow{1r} (0, 33333333_x, 0, 33333333_x)$$
$$\xrightarrow{1r} (0, 66666666_x, 0, 66666666_x).$$

C   A Note on the $\theta \circ \gamma \circ \sigma$ Layer

Consider a modified MMB cipher whose round structure does not include η (call it MMB$-\eta$, read "MMB minus η"), that is, a full round consists of only $\theta \circ \gamma \circ \sigma$.

We have verified very peculiar Λ-set propagations in MMB$-\eta$, such as $(A, C, C, C) \xrightarrow{1r} (A, A, C, A) \xrightarrow{1r} (X, E, E, E) \xrightarrow{1r} (X, X, E, X) \xrightarrow{1r} (X, E, E, E)$. After the fourth round, the patterns (X, X, E, X) and (X, E, E, E) alternate, that is, balanced Λ-sets propagate **indefinitely (for an arbitrary number of rounds)**. This unusual behavior can be explained similarly to that of other patterns in Sect. 4. We concluded that

- This property does not depend on the round subkeys, or on the user key or even on the key schedule;
- This property is independent of the particular permutation used in the initial A word, or the constants used in the C words;
- It highlights the importance of the η layer (a data-dependent, nonlinear operation) in the security of the original MMB against square attacks, since its presence destroys the propagation of balanced Λ-sets after 2.75 rounds;
- The above distinguish-from-random attack applies to an arbitrary number of rounds of MMB-η and costs only 2^{32} CP, an equivalent number of encryptions and negligible memory.

Weak Keys of Reduced-Round PRESENT for Linear Cryptanalysis

Kenji Ohkuma[1,2]

[1] Corporate R & D Center, Toshiba Corporation
[2] IT Security Center, Information-technology Promotion Agency, Japan

Abstract. The block cipher PRESENT designed as an ultra-light weight cipher has a 31-round SPN structure in which the S-box layer has 16-parallel 4-bit S-boxes and the diffusion layer is a bit permutation. The designers claimed that the maximum linear characteristic deviation is not more than 2^{-43} for 28 rounds and concluded that PRESENT is not vulnerable to linear cryptanalysis. But we have found that 32% of PRESENT keys are weak for linear cryptanalysis, and the linear deviation can be much larger than the linear characteristic value by the multi-path effect. And we discovered a 28-round path with a linear deviation of $2^{-39.3}$ for the weak keys. Furthermore, we found that linear cryptanalysis can be used to attack up to 24 rounds of PRESENT for the weak keys.

1 Introduction

The block cipher PRESENT designed as an ultra-light weight cipher has a 31-round SPN structure in which the S-box layer has 16-parallel 4-bit S-boxes and the diffusion layer is a bit permutation. The data randomizing part has a 31-round SPN structure of 64-bit block size, where each round consists of a key addition layer (addKeyLayer), an S-box layer (sBoxLayer) with 16 parallel 4-bit S-boxes, and a bit permutation layer (pLayer) as shown in Figure 1. Two key lengths, 80 bits and 128 bits, are supported. We consider 80-bit key in this paper.

PRESENT is similar to AES in structure, as the S-box layer consists of 16 S-boxes and each S-box is connected to 4 S-boxes in the next round. But there is a big difference in design philosophy between the two ciphers. The MDS matrices in AES's mixColumn operations are based on the wide-trail strategy, and there are 25 or more active S-boxes in 4 successive rounds for both linear and differential cryptanalyses. On the contrary, PRESENT's pLayer only permutes the order of bits and the trail can be very narrow. As a matter of fact, we can easily find a linear path with only one active S-box per round. We call such a path a single-bit path.

There are some security evaluations of PRESENT. The designers insist that the differential characteristic is upper-bounded by $2^{-100}(< 2^{-64})$ for 25 rounds, and that the absolute value of linear characteristic deviation are upper-bounded by $2^{-43}(< 2^{-32})$ for 28 rounds [1]. These evaluations are very loose, and do not show how many rounds are vulnerable to attack.

M.J. Jacobson Jr., V. Rijmen, and R. Safavi-Naini (Eds.): SAC 2009, LNCS 5867, pp. 249–265, 2009.
© Springer-Verlag Berlin Heidelberg 2009

M.R. Z'aba et al. applied an integral attack for bit-patterns, and showed that 7-round PRESENT can be attacked [7]. M. Wang demonstrates 16-round PRESENT can be attacked by differential cryptanalysis [6]. M. Albrecht and C. Cid applied the differential cryptanalysis strengthened by algebraic techniques, and found 16-round PRESENT can be attacked [8]. B. Collard and F.-X. Standaert applied a statistical saturation attack which works up to 24 rounds in theory [?].

The designers evaluated the resistance of PRESENT against linear cryptanalysis by using linear characteristic deviation [1]. But, the absolute value of linear deviation can be much larger than is that of linear characteristic deviation as T. Shimoyama et al. showed for RC6 in FSE 2002 [5]. In fact, we found that PRESENT has many linear single-bit paths with the same input/output masks for 4 or more rounds, and that the linear deviation can be very large for some portion of keys, which we call weak keys. We show that 24-round reduced round PRESENT can be cryptanalysed with $2^{63.5}$ known plaintexts for 32% of the 80-bit keys.

The construction of this paper is as follows. Section 2 describes the structure of PRESENT and some notations. In Section 3, some properties of linear single-bit paths are analyzed. In Section 4, reduced round PRESENTs are attacked with linear cryptanalysis. Section 5 is devoted to concluding remarks.

2   Description of PRESENT Encryption

Fig. 1 shows the structure of encryption for PRESENT. The data radomizing part consists of 31 iterations of key addition (addRoundKey), S-box layer (sBoxLayer), bit permutation (pLayer) followed by the final addRoundKey.

sBoxLayer consists of 16 parallel 4-bit S-boxes described in Table 1.

pLayer permutates the output bits of sBoxLayer. When the rightmost bit is 0-th and the leftmost bit is 63rd, the ℓ-th bit moves to $P(\ell)$-th where P is given by the following equation.

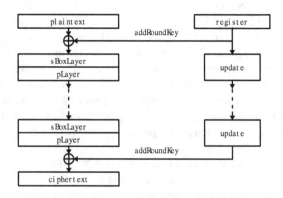

Fig. 1. Structure of PRESENT

Table 1. S-box

x	0	1	2	3	4	5	6	7	8	9	A	B	C	D	E	F
$S[x]$	C	5	6	B	9	0	A	D	3	E	F	8	4	7	1	2

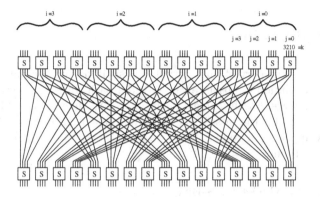

Fig. 2. Round structure

$$P(16 * i + 4 * j + k) = 16 * k + 4 * i + j, \quad 0 \leq i, j, k \leq 3 \quad (1)$$

Successive sBoxLayers connected by pLayer are shown in Fig. 2. Here, addRoundKeys just before sBoxLayer is not shown for simplicity.

As we focus on linear single-bit paths where only one bit is active in all masks, we represent the location of a bit using i, j, k as in Equation 1. The ℓ-th key bit in the r-th round[1] is denoted $k_{i,j,k}^{(r)}$, and the bit just before is denoted $x_{i,j,k}^{(r)}$. Therefore, the ℓ-th bit of the plaintext is $x_{i,j,k}^{(0)}$ and the ℓ-th bit of the ciphertext is $x_{i,j,k}^{(31)}$. A variable without the 3rd subscript k, such as $x_{i,j}^{(0)}$ means a 4-bit set for the $4 * i + j$-th S-box.

The key schedule of PRESENT for an 80-bit key is as follows. 80-bit key variables for the r-th round($0 \leq r \leq 31$) is denoted $\kappa_{79}^{(r)} \kappa_{78}^{(r)} \ldots \kappa_0^{(r)}$. The encryption key is used as $\kappa^{(0)} (r = 0)$, and its leftmost 64 bits are used as the 1st round key. The 80-bit key variable is updated 31 times according to the following steps, and its leftmost 64 bits are used as each round key.

$$k_{63}^{(r)} \mid k_{62}^{(r)} \mid \ldots \mid k_0^{(r)} = \kappa_{79}^{(r)} \mid \kappa_{78}^{(r)} \mid \ldots \mid \kappa_{16}^{(r)}$$

After a 19-bit right rotation, the leftmost 4 bits are transformed by an S-box, and 5 bits from the 15-th to the 19-th positions are XORed with a counter value.

$$\kappa_{79}^{'(r)} \kappa_{78}^{'(r)} \ldots \kappa_0^{'(r)}$$

$$\kappa_i^{'(r)} = \kappa_{i+19(\text{mod } 80)}^{(r)}$$

---

[1] $r = 0$ for the round key just after the plaintext.

$$\kappa_{79}^{(r+1)} \mid \kappa_{78}^{(r+1)} \mid \kappa_{77}^{(r+1)} \mid \kappa_{76}^{(r+1)} = S(\kappa_{79}^{'(r)} \mid \kappa_{78}^{'(r)} \mid \kappa_{77}^{'(r)} \mid \kappa_{76}^{'(r)})$$

$$\kappa_{19}^{(r+1)} \mid \kappa_{18}^{(r+1)} \mid \kappa_{17}^{(r+1)} \mid \kappa_{16}^{(r+1)} \mid \kappa_{15}^{(r+1)} = \kappa_{19}^{'(r)} \mid \kappa_{18}^{'(r)} \mid \kappa_{17}^{'(r)} \mid \kappa_{16}^{'(r)} \mid \kappa_{15}^{'(r)} \oplus (r+1)$$
$$\kappa_i^{(r+1)} = \kappa_i^{'(r)} \ (i \in \{0,\ldots,14,20,\ldots,75\})$$

3   Single-Bit Paths

The most important part of linear cryptanalysis is to find the paths with the largest linear deviation in absolute value. And a path with fewer active S-boxes tends to have a larger absolute linear deviation. As the linear layer of PRESENT pLayer only permutes the bits, it is easy to find linear paths in which only one active S-box appears per round. Thus, we focus on paths where only one bit is active in every round.

3.1   Single-Bit Masks for S-Box

Table 2 shows the linear deviation for the S-box of PRESENT.

There are 36 masks with the largest absolute linear deviation 2^{-2}, which do not include those with only one bit active in both input and output. Figure 3 shows 8 masks with one bit active in both input and output, which have non-zero absolute linear deviation(2^{-3}). Note that the 0-th bit is never active in the figure, but all combinations of the other 3 input and output bits appear, except for the combination in which the 3rd input bit and the 2nd output bit are active.

Table 2. Linear deviation of S-box(with sign)

	input masks														
	1_x	2_x	3_x	4_x	5_x	6_x	7_x	8_x	9_x	A_x	B_x	C_x	D_x	E_x	F_x
1_x	0	0	0	0	0	0	0	0	$\frac{1}{4}$	0	$-\frac{1}{4}$	0	$\frac{1}{4}$	0	$\frac{1}{4}$
2_x	0	$\frac{1}{8}$	$\frac{1}{8}$	$-\frac{1}{8}$	$-\frac{1}{8}$	0	0	$\frac{1}{8}$	$-\frac{1}{8}$	$\frac{1}{4}$	0	0	$\frac{1}{4}$	$\frac{1}{8}$	$-\frac{1}{8}$
3_x	0	$\frac{1}{8}$	$\frac{1}{8}$	$\frac{1}{8}$	$\frac{1}{8}$	$-\frac{1}{4}$	$\frac{1}{4}$	$-\frac{1}{8}$	$-\frac{1}{8}$	0	0	0	0	$\frac{1}{8}$	$\frac{1}{8}$
4_x	0	$-\frac{1}{8}$	$\frac{1}{8}$	$-\frac{1}{8}$	$\frac{1}{8}$	0	$\frac{1}{4}$	0	0	$\frac{1}{8}$	$-\frac{1}{8}$	$\frac{1}{8}$	$-\frac{1}{8}$	$-\frac{1}{4}$	0
5_x	$-\frac{1}{4}$	$-\frac{1}{8}$	$\frac{1}{8}$	$\frac{1}{8}$	$\frac{1}{8}$	0	0	0	0	$\frac{1}{8}$	$-\frac{1}{8}$	$\frac{1}{8}$	$-\frac{1}{8}$	$\frac{1}{4}$	0
6_x	0	0	$-\frac{1}{4}$	0	0	$-\frac{1}{4}$	0	$-\frac{1}{8}$	$\frac{1}{8}$	$\frac{1}{8}$	$\frac{1}{8}$	$-\frac{1}{8}$	$\frac{1}{8}$	$-\frac{1}{8}$	$-\frac{1}{8}$
7_x	$-\frac{1}{4}$	0	0	$\frac{1}{4}$	0	0	0	$\frac{1}{8}$	$-\frac{1}{8}$	$-\frac{1}{8}$	$-\frac{1}{8}$	$\frac{1}{8}$	$\frac{1}{8}$	$-\frac{1}{8}$	$-\frac{1}{8}$
8_x	0	$\frac{1}{8}$	$-\frac{1}{8}$	$-\frac{1}{8}$	$\frac{1}{8}$	0	0	$-\frac{1}{8}$	$-\frac{1}{8}$	0	$-\frac{1}{4}$	$\frac{1}{4}$	0	$-\frac{1}{8}$	$-\frac{1}{8}$
9_x	0	$-\frac{1}{8}$	$\frac{1}{8}$	$-\frac{1}{8}$	$\frac{1}{8}$	$-\frac{1}{4}$	$-\frac{1}{4}$	$\frac{1}{8}$	$-\frac{1}{8}$	0	0	0	0	$-\frac{1}{8}$	$\frac{1}{8}$
A_x	0	0	$-\frac{1}{4}$	0	$-\frac{1}{4}$	0	0	0	$-\frac{1}{4}$	0	0	0	0	0	$\frac{1}{4}$
B_x	0	$\frac{1}{4}$	0	$-\frac{1}{4}$	0	0	0	0	0	$-\frac{1}{4}$	0	$-\frac{1}{4}$	0	0	0
C_x	0	0	0	0	$\frac{1}{4}$	$\frac{1}{4}$	0	$-\frac{1}{8}$	$-\frac{1}{8}$	$\frac{1}{8}$	$\frac{1}{8}$	$-\frac{1}{8}$	$\frac{1}{8}$	$-\frac{1}{8}$	$\frac{1}{8}$
D_x	$-\frac{1}{4}$	$\frac{1}{4}$	0	0	0	0	0	$\frac{1}{8}$	$\frac{1}{8}$	$\frac{1}{8}$	$\frac{1}{8}$	$\frac{1}{8}$	$\frac{1}{8}$	$-\frac{1}{8}$	$\frac{1}{8}$
E_x	0	$-\frac{1}{8}$	$-\frac{1}{8}$	$\frac{1}{8}$	$\frac{1}{8}$	0	$\frac{1}{4}$	$\frac{1}{4}$	0	$-\frac{1}{8}$	$-\frac{1}{8}$	$\frac{1}{8}$	$\frac{1}{8}$	0	0
F_x	$\frac{1}{4}$	$\frac{1}{8}$	$-\frac{1}{8}$	$\frac{1}{8}$	$\frac{1}{8}$	0	0	$\frac{1}{4}$	0	$\frac{1}{8}$	$-\frac{1}{8}$	$-\frac{1}{8}$	$-\frac{1}{8}$	0	0

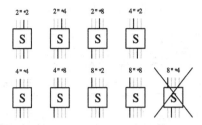

Fig. 3. Single-bit masks for S-box (absolute linear deviation: 2^{-3})

3.2 Continuable 1-Round Single-Path

As we saw in the previous subsection, S-box's single-bit masks do not contain the 0-th bit in both input and output. From this property it follows that the 1-round single-bit paths that can be included in single-bit paths with 3 or more rounds are limited to the 72 shown in Figure 4.

That is, single-bit paths with an arbitrary number of rounds can be constructed by connecting paths in Figure 4.

The above 72 masks have the property that none of i, j, k is 0 when the position of an active input or output bit is denoted (i, j, k). The reason that 0 does not appear for an active input bit can be explained as follows.

$i = 0$ *case.* Input bit $(0, j, k)$ proceeds to $(j', k', 0)$ after 2 rounds, and the single-bit path can not continue any more. That means the single-bit path can not continue to 3 or more rounds in the case of $i = 0$.

$j = 0$ *case.* Input bit $(i, 0, k)$ proceeds to $(k', i, 0)$ after 1 round, and the single-bit path can not continue any more. That means the single-bit path can not continue to 2 or more rounds in the case of $j = 0$.

$k = 0$ *case.* The active input bit is the rightmost for the S-box, and the single-bit path terminates here.

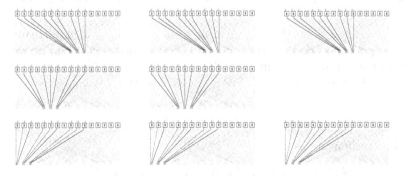

Fig. 4. Continuable 1-Round Single-bit Paths

Table 3. The number of single-bit paths for optimal input-output masks

# rounds	1	2	3	4	5	6	7	8	9	10	11	12	13	14	15
# paths	1	1	1	3	9	27	72	192	512	1,344	3,528	9,261	24,255	63,525	166,375

# rounds	16	17	18	19	20	21	22
# paths	435,600	1,140,480	2,985,984	7,817,472	20,466,576	53,582,633	140,281,323

# rounds	23	24	25	26	27
# paths	367,261,713	961,504,803	2,517,252,696	6,590,254,272	17,253,512,704

# rounds	28	29	30	31
# paths	45,170,283,840	118,257,341,400	309,601,747,125	810,547,899,975

3.3   Single-Bit Paths with the Same Input-Output Mask

In the previous subsection, we show that single-bit paths with 3 or more rounds can be made by connecting 1-round paths in Figure 4. For 4 or more rounds, there appear more than 1 single-bit paths with the same input-output mask. Figure 5 shows single-bit paths with the same input-output mask $\Gamma_{2,1,2}$-$\Gamma_{3,3,3}$. There are 3 paths for 4 rounds. The number of paths increases rapidly, as the number of rounds goes up (Table 3).

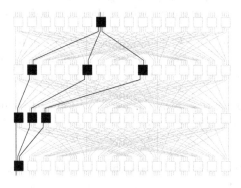

Fig. 5. 4-round single-bit paths with the same input-output mask ($\Gamma_{2,1,2}$-$\Gamma_{3,3,3}$)

The above case is for the input-output mask $\Gamma_{2,1,2}$-$\Gamma_{3,3,3}$, and we found this input-output mask has the maximum number of single-bit paths for all rounds. If we call this propery as optimal, there are 64 optimal input-output masks including the above one, which are given as arbitrary combinations of the 8 input masks

$$\Gamma_{1,1,1}, \Gamma_{1,1,2}, \Gamma_{1,2,1}, \Gamma_{1,2,2}, \Gamma_{2,1,1}, \Gamma_{2,1,2}, \Gamma_{2,2,1}, \Gamma_{2,2,2},$$

and the 8 output masks

$$\Gamma_{1,1,1}, \Gamma_{1,1,3}, \Gamma_{1,3,1}, \Gamma_{1,3,3}, \Gamma_{3,1,1}, \Gamma_{3,1,3}, \Gamma_{3,3,1}, \Gamma_{3,3,3}.$$

Table 3 shows the number of single-bit paths for R rounds $L(R)$ for the optimal input-output masks.

Table 4. Linear deviations with multiple-path effects for weak keys

R	16	17	18	19	20	21	22	23
$\log_2(\epsilon^{(R)})$	-23.634	-24.939	-26.245	-27.551	-28.857	-30.162	-31.468	-32.774

R	24	25	26	27	28	29	30	31
$\log_2(\epsilon^{(R)})$	-34.080	-35.385	-36.691	-37.997	-39.303	-40.608	-41.914	-43.220

3.4   Multiple-Path Effect and Weak Keys

As shown in the previous subsection, there are more than 1 single-bit paths for 4 or more rounds. For the case of R rounds, the absolute value of linear deviation for each path is evaluated as 2^{-2R-1} by using Matsui's Piling-up lemma [4]. It should be noted that the sign of each path depends on the S-box's single-bit approximation and the encryption key. Denote the number of positive paths by N^+, and the number of negative paths by N^-. Then, the linear deviation for the mask is approximated by $2^{-2R-1}(N^+ - N^-)$, which has been confirmed by computer simulations with random plaintexts. This is the multiple-path effect. Appendix A shows the theoretical analysis for the 3 single-bit path case, which can be regarded as a simplification of Figure 5.

Multiple-path effect depends on the extended key. If we assume the extended key distributes uniformly, the sign of each single-bit path follows the binary distribution of probability $1/2$. When the number of paths $L(R)$ is sufficiently large, the binary distribution is approximated by the normal distribution with the deviation $\sqrt{L(R)}$. Then the absolute deviation is considered to be not less than $\sqrt{L(R)}$ times larger than the deviation for a single-path for 32% of keys. In fact, by a computer simulation, we confirmed that the standard deviation of $(N^+ - N^-)$ and the rate where $\mid N^+ - N^- \mid$ is larger than the standard deviation are well fitted to the theoretical results.

We call the 32% of keys satisfying $\mid N^+ - N^- \mid >= \sqrt{L(R)}$ weak keys. The lower bound of absolute value of linear deviation for the weak keys is evaluated as follows.

$$\epsilon^{(R)} = 2^{-2R-1}\sqrt{L^{(R)}} \tag{2}$$

Table 4 shows their logarithms. For 28 rounds, linear deviation is about $2^{-39.3}$. This is larger than 2^{-43} which is the upper bound estimated by the designers using the linear characteristic deviations.

4   Key Recovery Attack

In the linear cryptanalysis with the linear deviation ϵ and the number of known plaintexts N, we assume the number of guessed keys is m-bits and they are independent. Then, the success rate of key guess p is denoted as follows (Appendix B).

Table 5. The number of known plaintexts needed for an $(R+1)$-round attack of upper 1-round elimination type

R	18	19	20	21	22
# plaintexts($\log_2$)	55.384	57.996	60.607	63.219	65.830

$$p = \left\{ \mathrm{erf}\left(\sqrt{N}\epsilon \right) \right\}^{2^m} \tag{3}$$

erf() is the Gaussian error function defined as follows here.

$$\mathrm{erf}(x) = \frac{1}{\sqrt{2\pi}} \int_{-\infty}^{x} e^{-y^2/2} dy \tag{4}$$

In reverse, the number of plaintexts N needed for the success rate p is evaluated as follows.

$$N = \left\{ \mathrm{erf}^{-1}\left(p^{1/2^m} \right) / \epsilon \right\}^2 \tag{5}$$

By taking the logarithm with base 2, the equation transforms as follows.

$$\log_2 N = 2\log_2 \left\{ \mathrm{erf}^{-1}\left(p^{1/2^m} \right) \right\} + 4R + 2 - \log_2 \left\{ L(R) \right\} \tag{6}$$

In the following, we apply linear cryptanalysis for 80-bit key PRESENT, with the linear approximation for an optimal single-bit path $\Gamma_{1,1,1} - \Gamma_{1,1,1}$. More specifically, 5 types of key guesses shown in Figures 7~11 of Appendix C have been applied.

4.1   Upper 1-Round Elimination Attack ($\Gamma_{1,1,1}$-$\Gamma_{1,1,1}$)

Figure 7 shows the $(R + 1)$-round attack for R-round optimal single-bit path $(\Gamma_{1,1,1}$-$\Gamma_{1,1,1})$ in which we guess key bits for 1 preceding round. We search for the 4-bit key which gives the largest absolute linear deviation.

$$\mathrm{Pr}\left(x_{1,1,1}^{(1)} \oplus x_{1,1,1}^{(R+1)} = 0 \right) - 1/2$$

Let γ_1 be a mask which takes 1st bit(2nd rightmost bit), then the above equation is transformed as follows.

$$\mathrm{Pr}\left(\gamma_1 \cdot S\left(x_{1,1}^{(0)} \oplus k_{1,1}^{(0)} \right) \oplus x_{1,1,1}^{(R+1)} = 0 \right) - 1/2 \tag{7}$$

Round key $k_{1,1}^{(0)}$ is selected such that the absolute value of the above equation is the largest. In this case, the number of guessed bits m is 4 for one S-box. Table 5 shows the logarithm of the number of plaintexts needed for 95% successful key guessing $2\log_2(\mathrm{erf}(0.95^{1/2^4})/\epsilon)$. The attack in Figure 7 is available

Table 6. The number of known plaintexts needed for an $(R+2)$-round attack for upper 1-round & lower 1-round elimination type

R	18	19	20	21	22
# plaintexts($\log_2$)	56.137	58.749	61.360	63.972	66.583

up to 22 rounds, and $2^{63.219}$ known texts are needed. The exhaustive search for the remaining key bits requires the calculation of 2^{76} encryptions.

4.2 Upper 1-Round & Lower 1-Round Elimination Attack $(\Gamma_{1,1,1}\text{-}\Gamma_{1,1,1})$

Figure 8 shows the $(R+2)$-round attack for R-round optimal single-bit path $(\Gamma_{1,1,1}\text{-}\Gamma_{1,1,1})$ in which we guess keys bits for 1 preceding and 1 following rounds. We search for the 8-bit key which gives the largest absolute linear deviation.

$$\Pr\left(x_{1,1,1}^{(1)} \oplus x_{1,1,1}^{(R+1)} = 0\right) - 1/2$$

Using γ_1, the above equation is transformed as follows.

$$\Pr\left(\gamma_1 \cdot S\left(x_{1,1}^{(0)} \oplus k_{1,1}^{(0)}\right) \oplus \gamma_1 \cdot S^{-1}\left(x_{1,1}^{(R+2)} \oplus k_{1,1}^{(R+2)}\right) = 0\right) - 1/2 \qquad (8)$$

Round key $\left(k_{1,1}^{(0)}, k_{1,1}^{(R+2)}\right)$ is selected such that the absolute value of the above equation is the largest. In this case, the number of guessed bits m is 8 for 2 S-boxes. Table 6 shows the logarithm of the number of plaintexts needed for 95% successful key guessing $2\log_2(\text{erf}(0.95^{1/2^8})/\epsilon)$. The attack in Figure 8 is available up to 23 rounds, and $2^{63.972}$ known texts are needed. The exhaustive search for the remaining key bits requires the calculation of 2^{72} encryptions.

4.3 Upper 2-Round Elimination Attack $(\Gamma_{1,1,1}\text{-}\Gamma_{1,1,1})$

Figure 9 shows the $(R+2)$-round attack for R-round optimal single-bit path $(\Gamma_{1,1,1}\text{-}\Gamma_{1,1,1})$ in which we guess keys bits for 2 preceding rounds. We search for the 20-bit key which gives the largest absolute linear deviation.

$$\Pr\left(x_{1,1,1}^{(2)} \oplus x_{1,1,1}^{(R+2)} = 0\right) - 1/2$$

Using γ_1, the above equation is transformed as follows.

$$\Pr\left(S\left(\gamma_1 \cdot S(x_{1,3}^{(0)} \oplus k_{1,3}^{(0)}) \mid \gamma_1 \cdot S(x_{1,2}^{(0)} \oplus k_{1,2}^{(0)}) \mid \gamma_1 \cdot S(x_{1,1}^{(0)} \oplus k_{1,1}^{(0)}) \mid \right.\right.$$

$$\left.\left. \gamma_1 \cdot S(x_{1,0}^{(0)} \oplus k_{1,0}^{(0)}) \oplus k_{1,1}^{(1)}\right) \oplus x_{1,1,1}^{(R+2)} = 0\right) - 1/2 \qquad (9)$$

Table 7. The number of known plaintexts needed for an $(R+2)$-round attack of upper 2-round elimination type

R	18	19	20	21	22
# plaintexts($\log_2$)	57.319	59.930	62.542	65.153	67.765

Table 8. The number of known plaintexts needed for an $(R+3)$-round attack for upper 2-round & lower 1-round elimination type

R	18	19	20	21	22
# plaintexts($\log_2$)	57.569	60.181	62.792	65.404	68.015

Round key $\left(k_{1,0}^{(0)}, k_{1,1}^{(0)}, k_{1,2}^{(0)}, k_{1,3}^{(0)}, k_{1,1}^{(1)}\right)$ is selected such that the absolute value of the above equation is the largest. In this case, the number of guessed bits m is 20 for 5 S-boxes. Table 7 shows the logarithm of the number of plaintexts needed for 95% successful key guessing $2\log_2(\mathrm{erf}(0.95^{1/2^{20}})/\epsilon)$. The attack in Figure 9 is available up to 22 rounds, and $2^{62.542}$ known texts are needed. The exhaustive search for the remaining key bits requires the calculation of 2^{60} encryptions.

4.4 Upper 2-Round & Lower 1-Round Elimination Attack $(\Gamma_{1,1,1}\text{-}\Gamma_{1,1,1})$

Figure 10 shows the $(R+3)$-round attack for R-round optimal single-bit path $(\Gamma_{1,1,1}\text{-}\Gamma_{1,1,1})$ in which we guess key bits for 2 preceding and 1 following round. We search for the 24-bit key which gives the largest absolute linear deviation.

$$\Pr\left(x_{1,1,1}^{(2)} \oplus x_{1,1,1}^{(R+2)} = 0\right) - 1/2$$

Using γ_1, the above equation is transformed as follows.

$$\Pr\left(\gamma_1 \cdot S\left(\gamma_1 \cdot S(x_{1,3}^{(0)} \oplus k_{1,3}^{(0)}) \mid \gamma_1 \cdot S(x_{1,2}^{(0)} \oplus k_{1,2}^{(0)}) \mid \gamma_1 \cdot S(x_{1,1}^{(0)} \oplus k_{1,1}^{(0)}) \mid \right.\right.$$

$$\left.\left. \gamma_1 \cdot S(x_{1,0}^{(0)} \oplus k_{1,0}^{(0)}) \oplus k_{1,1}^{(1)}\right) \oplus \gamma_1 \cdot S^{-1}\left(x_{1,1}^{(R+3)} \oplus k_{1,1}^{(R+3)}\right) = 0\right) - 1/2 \quad (10)$$

Round key $\left(k_{1,0}^{(0)}, k_{1,1}^{(0)}, k_{1,2}^{(0)}, k_{1,3}^{(0)}, k_{1,1}^{(1)}, k_{1,1}^{(R+3)}\right)$ is selected such that the absolute value of the above equation is the largest. In this case, the number of guessed bits m is 24 for 6 S-boxes. Table 8 shows the logarithm of the number of plaintexts needed for 95% successful key guessing $2\log_2(\mathrm{erf}(0.95^{1/2^{24}})/\epsilon)$. The attack in Figure 10 is available up to 23 rounds, and $2^{62.792}$ known texts are

Table 9. The number of known plaintexts needed for an $(R+4)$-round attack for upper 2-round & lower 2-round elimination type

R	18	19	20	21	22
# plaintexts($\log_2$)	58.285	60.897	63.508	66.120	68.731

needed. The exhaustive search for the remaining key bits requires the calculation of 2^{56} encryptions.

4.5 Upper 2-Round & Lower 2-Round Elimination Attack $(\Gamma_{1,1,1}\text{-}\Gamma_{1,1,1})$

Figure 11 shows the $(R + 4)$-round attack for R-round optimal single-bit path $(\Gamma_{1,1,1}\text{-}\Gamma_{1,1,1})$ with 2 preceeding and 2 following rounds for key guess. We search for the 40-bit key which gives the largest absolute linear deviation.

$$\Pr\left(x_{1,1,1}^{(2)} \oplus x_{1,1,1}^{(R+2)} = 0\right) - 1/2$$

Using γ_1, the above equation is transformed as follows.

$$\Pr\left(\gamma_1 \cdot S\left(\gamma_1 \cdot S(x_{1,3}^{(0)} \oplus k_{1,3}^{(0)}) \mid \gamma_1 \cdot S(x_{1,2}^{(0)} \oplus k_{1,2}^{(0)}) \mid \gamma_1 \cdot S(x_{1,1}^{(0)} \oplus k_{1,1}^{(0)}) \mid \right.\right.$$

$$\gamma_1 \cdot S(x_{1,0}^{(0)} \oplus k_{1,0}^{(0)}) \oplus k_{1,1}^{(1)}\Big)$$

$$\oplus \gamma_1 \cdot S^{-1}\left(\gamma_1 \cdot S^{-1}(x_{1,3}^{(R+4)} \oplus k_{1,3}^{(R+4)}) \mid \gamma_1 \cdot S^{-1}(x_{1,2}^{(R+4)} \oplus k_{1,2}^{(R+4)}) \mid \right.$$

$$\left.\left.\gamma_1 \cdot S^{-1}(x_{1,1}^{(R+4)} \oplus k_{1,1}^{(R+4)}) \mid \gamma_1 \cdot S^{-1}(x_{1,0}^{(R+4)} \oplus k_{1,0}^{(R+4)}) \oplus k_{1,1}^{(R+3)}\right) = 0\right) - 1/2 \quad (11)$$

Round key $\left(k_{1,0}^{(0)}, \; k_{1,1}^{(0)}, \; k_{1,2}^{(0)}, \; k_{1,3}^{(0)}, \; k_{1,1}^{(1)}, \; k_{1,1}^{(R+3)}, \; k_{1,0}^{(R+4)}, \; k_{1,1}^{(R+4)}, \; k_{1,2}^{(R+4)}, \right.$ $\left.k_{1,3}^{(R+4)}\right)$ is selected such that the absolute value of the above equation is the largest. In this case, the number of guessed bits m is 40 for 10 S-boxes. Table 9 shows the logarithm of the number of plaintexts needed for 95% successful key guessing $2\log_2(\mathrm{erf}(0.95^{1/2^{40}})/\epsilon)$. The attack in Figure 11 is available up to 24 rounds, and $2^{63.508}$ known texts are needed. The exhaustive search for the remaining key bits requires the calculation of 2^{40} encryptions.

5 Concluding Remarks

The block cipher PRESENT is designed so that the implementation is very small. As a bit permutation is used in the linear layer, the avalanche effect is very low, and it is easy to find single-bit paths with only one bit active in every round. We found that there are many such paths with the same input-output mask for 4 or more rounds.

Each single-bit path for the same input-output mask has its sign, and the absolute value of linear deviation is large when the portion of one sign is much larger than $1/2$. Let $L(R)$ be the number of single-bit paths for R rounds. Under the assumption that the signs follows the binary distribution, we determined that for 32% of all keys, which we call weak keys, the absolute linear deviation is not less than $\sqrt{L(R)}$ times of single path linear deviation. This phenomenon is called the multi-path effect.

By considering the multi-path effect, the linear deviation for 28 rounds is evaluated as $2^{-39.3}$ for weak keys, which is larger than 2^{-43} given by the designers with sigle-path evaluation. We applied 5 types of linear cryptanalysis to PRESENT, and 24-round reduced PRESENT is vulnerable with $2^{63.5}$ known plaintexts.

In this paper 32% key is weak keys for PRESENT, but the evaluation is for only one input-output mask. There are 64 masks with the same number of single-bit paths, and if a key is weak at least for one mask, it can be regarded as a weak key.

References

1. Bogdanov, A., Knudsen, L.R., Leander, G., Paar, C., Poschmann, A., Robshaw, M.J.B., Seurin, Y., Vikkelsoe, C.: PRESENT: An Ultra-Lightweight Block Cipher. In: Paillier, P., Verbauwhede, I. (eds.) CHES 2007. LNCS, vol. 4727, pp. 450–466. Springer, Heidelberg (2007)
2. Kaliski Jr., B.S., Robshaw, M.J.B.: Linear Cryptanalysis Using Multiple Approximations. In: Desmedt, Y.G. (ed.) CRYPTO 1994. LNCS, vol. 839, pp. 26–39. Springer, Heidelberg (1994)
3. Kaliski Jr., B.S., Robshaw, M.J.B.: Linear Cryptanalysis Using Multiple Approximations and FEAL. In: Preneel, B. (ed.) FSE 1994. LNCS, vol. 1008, pp. 249–264. Springer, Heidelberg (1995)
4. Matsui, M.: Linear Cryptanalysis Method for DES Cipher. In: Helleseth, T. (ed.) EUROCRYPT 1993. LNCS, vol. 765, pp. 386–397. Springer, Heidelberg (1994)
5. Shimoyama, T., Takenaka, M., Koshiba, T.: Multiple Linear Cryptanalysis of a Reduced Round RC6. In: Daemen, J., Rijmen, V. (eds.) FSE 2002. LNCS, vol. 2365, pp. 76–88. Springer, Heidelberg (2002)
6. Wang, M.: Differential Cryptanalysis of Reduced-Round PRESENT. In: Vaudenay, S. (ed.) AFRICACRYPT 2008. LNCS, vol. 5023, pp. 40–49. Springer, Heidelberg (2008)
7. Z'aba, M.R., Raddum, H., Henricksen, M., Dawson, E.: Bit-Pattern Based Integral Attack. In: Nyberg, K. (ed.) FSE 2008. LNCS, vol. 5086, pp. 363–381. Springer, Heidelberg (2008)
8. Collard, B., Standaert, F.-X.: A Statistical Saturation Attack against the Block Cipher PRESENT. In: Fischlin, M. (ed.) CT-RSA 2009. LNCS, vol. 5473, pp. 195–210. Springer, Heidelberg (2009)

A   Analysis of 3 Single-Bit Path Model

In this Appendix, multiple-path effect is analyzed for the simplest case.

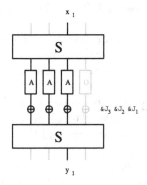

Fig. 6. 3 single-bit path model

Table 10. Linear deviation for 3-path model

$\kappa_1 + \kappa_2 + \kappa_3$	0	1	2	3
η	$3\epsilon - 4\epsilon^3$	$\epsilon + 4\epsilon^3$	$-\epsilon - 4\epsilon^3$	$-3\epsilon + 4\epsilon^3$

Figure 6 is a simplified model for Figure 5. 2 S's are the S-boxes of PRESENT, 3 A's are mutually independent paths with a positive linear deviation ϵ, and O is a path with 0 deviation. The deviation of A can take both positive and negative signs. But, without loss of generality, we can assume the positive sign, as the sign can be absorbed in the key bits κ's.

$$\Pr(x \oplus A(x) = 0) = 1/2 + \epsilon$$

$$\Pr(x \oplus O(x) = 0) = 1/2$$

Table 10 shows the deviation η for an equation $x_1 \oplus y_1$.

$$\Pr(x_1 \oplus y_1 = 0) = 1/2 + \eta$$

$\kappa_1 + \kappa_2 + \kappa_3$ means the number of paths with negative correlation for $x_1 \oplus y_1$. When $\kappa_1 + \kappa_2 + \kappa_3 = 0$, all 3 paths induce positive correlations. To the contrary, when $\kappa_1 + \kappa_2 + \kappa_3 = 3$, all 3 paths induce negative correlations. When $\kappa_1 + \kappa_2 + \kappa_3 = 1$ or $\kappa_1 + \kappa_2 + \kappa_3 = 2$, one positive and one negative are canceled out, one positive or negative correlation remains.

The 1st order terms of ϵ in Table 10 shows an effect which agrees with the above consideration. The 3rd order terms of ϵ is regarded as shifts from the simple sum of 3 correlations, which can be negligible for a sufficiently small ϵ. Let the number of $\kappa_i = 0$ as N^+ and the number of $\kappa_i = 1$ as N^- in Figure 6, then, the linear deviation η is approximated as follows.

$$\eta \simeq (N^+ - N^-)\epsilon. \tag{12}$$

B   Evaluation of Successful Key Recovery Rate

Let ϵ be the absolute value of linear deviation. Then, the probability that a linear approximation is satisfied is considered to be $1/2 \pm \epsilon$ for the correct key guess, and $1/2$ for the wrong key guess. Without loss of generality, we assume the sign is negative.

When known plaintexts are assumed to be uniformly distributed, both valid linear approximation rates are considered to follow the binary distribution. When the number of plaintexts is N, the average and the variance for both rates are evaluated as follows

$$correctkey\ average : \mu_T = 1/2 - \epsilon$$
$$standarddeviation : \sigma_T = \sqrt{(1/4 - \epsilon^2)/N}$$
$$wrongkeys\ average : \mu_F = 1/2$$
$$standarddeviation : \sigma_F = 2/\sqrt{N}$$

When N is sufficiently large, the 2 distributions can be approximated by the normal distributions. Let x_0 be the value for the cross point. When we compare the correct key and one wrong key, and choose the key with the probability which is more distant from $1/2$, the probability to choose the correct key p_s is given as follows.

$$p_s = \int_{-\infty}^{x_0 - \mu_T} \frac{1}{\sqrt{2\pi}\sigma_T} e^{-(x-\mu_T)^2/2\sigma_T^2} dx \tag{13}$$

When ϵ is sufficiently small, terms of order ϵ^2 are negligible, and the next approximated equations are given.

$$\sigma_T = 1/2\sqrt{N}$$

$$x_0 = 1/2 - \epsilon/2$$

From the above equations and the next transform

$$y = (x - \mu_T)/\sigma_T,$$

p_s is evaluated as follows.

$$p_s = \int_{-\infty}^{(x_0 - \mu_T)/\sigma_T} \frac{1}{\sqrt{2\pi}\sigma_T} e^{-y^2/2} \sigma_T\ dy$$

$$= \int_{-\infty}^{\sqrt{N}\epsilon} \frac{1}{\sqrt{2\pi}} e^{-y^2/2} dy = \mathrm{erf}\left(\sqrt{N}\epsilon\right)$$

When m-bit key is guessed, $2^m - 1$ keys are wrong, and the probability of correct key selection p is $(2^m - 1)$-th power of p_s

$$p = p_s^{2^m - 1} = \left\{ \mathrm{erf}\left(\sqrt{N}\epsilon\right) \right\}^{2^m - 1}$$

When m is sufficiently large, deleting -1 is a good approximation for this equation.

C   Figures of Key Recovery Attacks

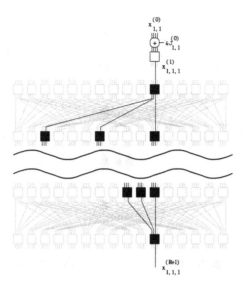

Fig. 7. Upper 1-round elimination attack

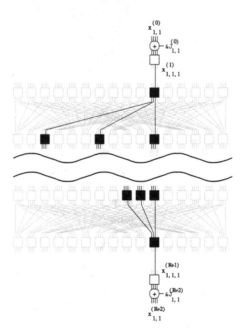

Fig. 8. Upper 1-round & lower 1-round elimination attack

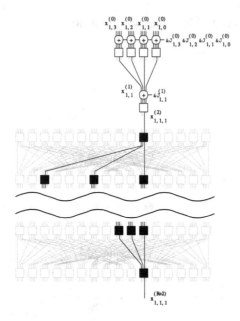

Fig. 9. Upper 2-round elimination attack

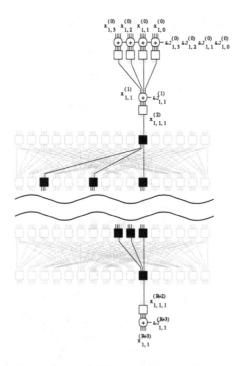

Fig. 10. Upper 2-round & lower 1-round elimination attack

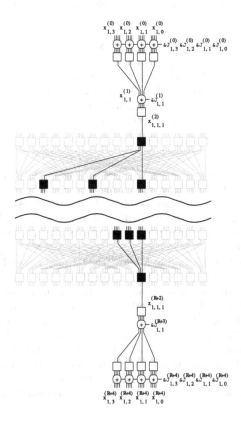

Fig. 11. Upper 2-round & lower 2-round elimination attack

Improved Integral Attacks on MISTY1[*]

Xiaorui Sun and Xuejia Lai

Department of Computer Science
Shanghai Jiao Tong University
Shanghai, 200240, China
sunsirius@sjtu.edu.cn, lai-xj@cs.sjtu.edu.cn

Abstract. We present several integral attacks on MISTY1 using the FO Relation. The FO Relation is a more precise form of the Sakurai-Zheng Property such that the functions in the FO Relation depend on 16-bit inputs instead of 32-bit inputs used in previous attacks, and that the functions do not change for different keys while previous works used different functions.

We use the FO Relation to improve the 5-round integral attack. The data complexity of our attack, 2^{34} chosen plaintexts, is the same as previous attack, but the running time is reduced from 2^{48} encryptions to $2^{29.58}$ encryptions. The attack is then extended by one more round with data complexity of 2^{34} chosen plaintexts and time complexity of $2^{107.26}$ encryptions. By exploring the key schedule weakness of the cipher, we also present a chosen ciphertext attack on 6-round MISTY1 with all the FL layers with data complexity of 2^{32} chosen ciphertexts and time complexity of $2^{126.09}$ encryptions. Compared with other attacks on 6-round MISTY1 with all the FL layers, our attack has the least data complexity.

1 Introduction

The MISTY1 algorithm is a block cipher with a 64-bit block size and a 128-bit key size proposed by Matsui [8]. It was recommended by the European NESSIE project and the CRYPTREC project, and became an ISO standard in 2005. The cipher generally uses an 8-round Feistel structure with a round function FO. Before each odd round and after the last round, there is an additional FL layer.

Many cryptanalysis results on MISTY1 have been published [1, 2, 3, 4, 5, 6, 7, 10, 11]. The integral attack on 5 rounds with all but the last FL layers [4] requires 2^{34} chosen plaintexts, and has a time complexity of 2^{48} encryptions. The impossible differential attack on 6 rounds with all the FL layers [2] requires 2^{51} chosen plaintexts, and has a time complexity of $2^{123.4}$ encryptions. With all the FL functions absent, the impossible differential attack [2] could break the 7 rounds with data complexity of $2^{50.2}$ known plaintexts and time complexity of $2^{114.1}$ encryptions.

---

[*] This work was supported by NSFC Grant No.60573032, 60773092 and 11th PRP of Shanghai Jiao Tong University.

M.J. Jacobson Jr., V. Rijmen, and R. Safavi-Naini (Eds.): SAC 2009, LNCS 5867, pp. 266–280, 2009.

In this paper, we present several integral attacks using a more precise form of the variant Sakurai-Zheng Property for the round function FO. We call this new property the FO Relation. Sakurai-Zheng Property was founded by Sakurai and Zheng in [9]. Knudsen and Wagner used a variation of this property for the FO function to attack the 5-round MISTY1 [4]. Compared with the variant Sakurai-Zheng property, there are two merits of the FO Relation: the inputs of the functions in the FO Relation are shortened from 32 bits to 16 bits, and these functions do not change for different keys while the previous property used different functions for different keys.

We use this new relation to improve the integral attack [4] on 5 rounds with all but the last FL layers. The data complexity of our improved attack is 2^{34} chosen plaintexts, and the time complexity of the attack is $2^{29.58}$ encryptions. Compared with the 5-round integral attack [4], the time complexity of our attack is reduced from 2^{48} encryptions to $2^{29.58}$ encryptions with the same data complexity.

Next, we extend the 5-round attack by one more round. Using the equivalent description of the FO function [5,11] and the FO Relation, we modify the partial decryption process of computing the required intermediate values to reduce the key bits needed. The data complexity of this 6-round attack is 2^{34} chosen plaintexts, and the time complexity of the attack is $2^{107.26}$ encryptions.

Table 1. Attacks on MISTY1

Rounds	Attack	FL functions	Data	Time	Ref.
5	Higher-Order Differential	None	$2^{10.5}$ CP	2^{17}	[1]
6	Impossible Differential	None	2^{54} CP	2^{61}	[5]
6	Impossible Differential	None	2^{39} CP	2^{106}	[5]
6	Impossible Differential	None	2^{39} CP	2^{85}	[7]
7	Impossible Differential	None	$2^{50.2}$ KP	$2^{114.1}$	[2]
4	Impossible Differential	Most	2^{23} CP	$2^{90.4}$	[5]
4	Impossible Differential	Most	2^{38} CP	2^{62}	[5]
4	Collision Search	Most	2^{20} CP	2^{89}	[5]
4	Collision Search	Most	2^{28} CP	2^{76}	[5]
4†	Slicing	All	$2^{22.25}$ CP	2^{45}	[6]
4	Slicing	All	$2^{27.2}$ CP	$2^{81.6}$	[6]
4	Impossible Differential	All	$2^{27.5}$ CP	2^{116}	[6]
5*	Integral	Most	2^{34} CP	2^{48}	[4]
5†	Impossible Differential	All	2^{38} CP	$2^{46.45}$	[2]
6	Impossible Differential	All	2^{51} CP	$2^{123.4}$	[2]
5*	Integral	Most	2^{34} CP	$2^{29.58}$	Section 4
6	Integral	Most	2^{34} CP	$2^{107.26}$	Section 5
6	Integral	All	2^{32} CC	$2^{126.09}$	Section 6

KP - Known Plaintext CP - Chosen Plaintext CC - Chosen Ciphertext
None - the version of MISTY1 without all the FL layers
Most - the version of MISTY1 without the final FL layer
All - the version of MISTY1 with all the FL layers
† - the attack retrieves 41.36 bits of information about the key.
* - the attack retrieves 50 bits of information about the key.

We also provide an attack on 6 rounds with all the FL layers. The attack is a chosen ciphertext attack starting from the FL_3, FL_4 layer to the end of the cipher. We explore the key schedule weakness to speed up the computation of the required intermediate values. The data complexity of the attack is 2^{32} chosen ciphertexts, and the time complexity of the attack is $2^{126.09}$ encryptions. Compared with other attacks on 6 rounds with all the FL layers, our attack has the least data complexity. The summarization of our attacks and previous attacks is listed in Table 1, where the data complexity is measured by the number of plaintexts/ciphertexts and the time complexity is measured by the number of encryptions needed in the attack.

The paper is organized as follows: In Section 2 we give a brief description of MISTY1 block cipher. We present the FO Relation in Section 3, and then use this new property to improve the integral attack on 5-round MISTY1 with all but the last FL layers in Section 4. Section 5 extends the 5-round attack to 6 rounds with all but the last FL layers. In Section 6 we present the attack on 6 rounds with all the FL layers. Section 7 concludes this paper.

2   The MISTY1 Block Cipher

MISTY1 is a block cipher with a 64-bit block size and a 128-bit key size. Let P and C denote the 64 bit plaintext and ciphertext, respectively. We use the superscript without brackets to distinguish the values corresponding to different plaintexts, e.g. C^1 and C^2 denote the ciphertexts for P^1 and P^2 respectively. The superscript with brackets denotes the bits of the words, e.g. $C^{1(1...7)}$ denote the left 7 bits of C^1. The subscript(without brackets) is used to distinguish the intermediate values for different rounds.

MISTY1 has a recursive structure. As shown in Figure 1(a), the cipher generally uses an 8-round Feistel structure with a round function FO. In each round, the left 32-bit part is functioned with the FO function, and then is XORed with the right 32 bits. This new 32-bit value is the left 32-bit input of the next round and the right 32-bit input of the next round is the original 32 left bits. Before each odd round and after the last round, there is an additional FL layer. We also use subscripted FO(or FL) to distinguish FO(or FL) functions with different keys, e.g, FO_3 denote FO function keyed with AKO_3 and AKI_3.

In the original specification of MISTY1 [8], the round function FO uses a 112-bit key. Several equivalent descriptions of the FO function [5,11] have been proposed, which use less key bits. Here we use the equivalent FO description similar to [5] as Figure 1(b) and (c). The round function FO_i itself has a 3-round Feistel-like structure. The 32-bit input of FO_i is divided into two blocks of 16 bits, denoted as IL_i and IR_i, respectively. In each round, the left 16-bit part is XORed with a subkey $AKO_{i,1}$ and then functioned with FI using a 9-bit key $AKI_{i,1}$. The output of the FI is XORed with the right 16 bits. The left 16 bits and the right 16 bits are then swaped. The same procedure is repeated three times, the left 16 bits and the right 16 bits after the third round are denoted as ML_i and MR_i. OL_i is the XOR of ML_i and the subkey $AKO_{i,4}$, and OR_i

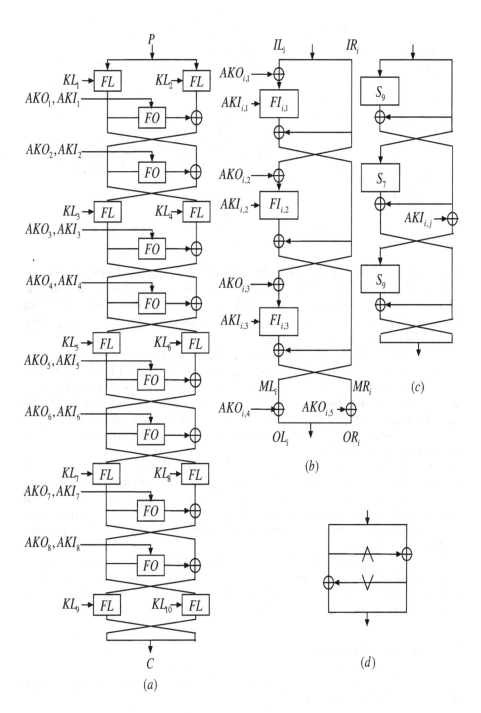

Fig. 1. (a) MISTY1 general structure (b) FO function (c) FI function (d) FL function

Table 2. The Key Schedule for MISTY1

Subkey	Correspondence
$KL_{i,1}$	$K_{\frac{i+1}{2}}$(odd i) or $K'_{\frac{i}{2}+2}$(even i)
$KL_{i,2}$	$K'_{\frac{i+1}{2}+6}$(odd i) or $K_{\frac{i}{2}+4}$(even i)
$AKO_{i,1}$	K_i
$AKO_{i,2}$	K_{i+2}
$AKO_{i,3}$	$K_{i+7} \oplus K'^{(1...7)}_{i+5}\|\|00\|\|K'^{(1...7)}_{i+5}$
$AKO_{i,4}$	$K_{i+4} \oplus K'^{(1...7)}_{i+5}\|\|00\|\|K'^{(1...7)}_{i+5} \oplus K'^{(1...7)}_{i+1}\|\|00\|\|K'^{(1...7)}_{i+1}$
$AKO_{i,5}$	$K'^{(1...7)}_{i+5}\|\|00\|\|K'^{(1...7)}_{i+5} \oplus K'^{(1...7)}_{i+1}\|\|00\|\|K'^{(1...7)}_{i+1} \oplus K'^{(1...7)}_{i+3}\|\|00\|\|K'^{(1...7)}_{i+3}$
$AKI_{i,1}$	$K'^{(8...16)}_{i+5}$
$AKI_{i,2}$	$K'^{(8...16)}_{i+1}$
$AKI_{i,3}$	$K'^{(8...16)}_{i+3}$

is the XOR of MR_i and $AKO_{i,5}$. The output of FO_i is $OL_i\|\|OR_i$ ($\|\|$ denotes concatenation). All the $AKO_{i,k}(1 \le k \le 5)$ are 16-bit subkeys, and all the $AKI_{i,k}(1 \le k \le 3)$ are 9-bit subkeys, hence the FO_i function uses a 107-bit key. Since $AKO^{(8...9)}_{i,5}$ is zero and $AKO^{(1...7)}_{i,5}$ is equal to $AKO^{(10...16)}_{i,5}$, the FO_i function actually takes a 98-bit key.

The FI function also has a 3-round Feistel-like structure. In the first round, the left 9-bit input enters a S-box $S9$, and then is XORed with the right 7-bit input(padded two zero bits left to the 7 bits). Swap the left and the right parts. In the second round, the left 7-bit part enters a S-box $S7$ and then is XORed with the right 9 bits(truncated the left 2 bits). The right 9-bit part is XORed with $AKI_{i,1}$ and then is swaped with the left 7 bits. The third round of FI is the same as the first round.

In the FL layer, the left 32 bits and the right 32 bits are put into the FL functions. In each FL function, the 32-bit input is divided into two blocks of 16 bits. The left 16-bit part is ANDed with the subkey $KL_{i,1}$, and then XORed with the right 16 bits to produce the right 16-bit output. This right 16-bit output is ORed with the subkey $KL_{i,2}$ and then XORed with left 16 bits to produce the left 16-bit output.

The key schedule of MISTY1 divides the 128-bit key into eight blocks of 16-bit words K_1, K_2, ..., K_8. Another eight 16-bit words are computed by $K'_i = FI_{K_{i+1}}(K_i)$. The correspondence of these 16 bit words and the subkeys used in the encryption is listed in Table 2.

3 The FO Relation

The following proposition on the FO function, which is a variant for Sakurai-Zheng Property [9], is presented in [4].

Proposition 1 ([4]). [1] *For the FO function of round i, the following equation holds*

$$OL_i^{(1...7)} = f_{AKO_{i,1}}(IL_i \| IR_i) \oplus g_{AKO_{i,2}}(IL_i \| IR_i) \oplus k \qquad (1)$$

where $f_{AKO_{i,1}}$ and $g_{AKO_{i,2}}$ are functions related to the subkeys $AKO_{i,1}$ and $AKO_{i,2}$, respectively, and k is a constant related to the key used in this FO_i function.

As shown in Figure 2, ML_i is the XOR of the values corresponding to the point α and β, hence $OL_i^{(1...7)} = \alpha \oplus \beta \oplus AKO_{i,4}^{(1...7)}$. Since the left 7 bits of the values at α and β are not related to $AKI_{i,1}$ and $AKI_{i,2}$, Equation (1) holds by letting $f_{AKO_{i,1}}(IL_i \| IR_i)$ correspond to the left 7 bits of the value at α, $g_{AKO_{i,2}}(IL_i \| IR_i)$ correspond to the left 7 bits of the value at β and the key-related constant k correspond to $AKO_{i,4}^{(1...7)}$. Expanding $f_{AKO_{i,1}}$ and $g_{AKO_{i,2}}$, Equation (1) is

$$OL_i^{(1...7)} = [FI^{(1...7)}(IL_i \oplus AKO_{i,1}) \oplus IR_i^{(1...7)}] \oplus [FI^{(1...7)}(IR_i \oplus AKO_{i,2})] \oplus AKO_{i,4}^{(1...7)} \qquad (2)$$

where $FI^{(1...7)}$ denotes the partial FI function which inputs 16-bit input of FI, and outputs the left 7-bit output of FI. By identical transformation, Equation (2) can be rewritten as follows:

$$OL_i^{(1...7)} = [FI^{(1...7)}(IL_i \oplus AKO_{i,1})] \oplus [FI^{(1...7)}(IR_i \oplus AKO_{i,2}) \oplus (IR_i \oplus AKO_{i,2})^{(1...7)}]$$
$$\oplus [AKO_{i,4}^{(1...7)} \oplus AKO_{i,2}^{(1...7)}] \qquad (3)$$

Let $f_{AKO_{i,1}}(IL_i \| IR_i)$ be $FI^{(1...7)}(IL_i \oplus AKO_{i,1})$, $g_{AKO_{i,2}}(IL_i \| IR_i)$ be $FI^{(1...7)}(IR_i \oplus AKO_{i,2}) \oplus (IR_i \oplus AKO_{i,2})^{(1...7)}$ and the key-related constant k be $AKO_{i,4}^{(1...7)} \oplus AKO_{i,2}^{(1...7)}$, Equation (1) still holds, but $f_{AKO_{i,1}}$ is not related to IR_i and $g_{AKO_{i,2}}$ is not related to IL_i. Hence Proposition 1 can be refined as follows:

Lemma 1 (the FO Relation). *For the FO function of round i, the following equation holds*

$$OL_i^{(1...7)} = f(IL_i \oplus AKO_{i,1}) \oplus g(IR_i \oplus AKO_{i,2}) \oplus k \qquad (4)$$

where f and g are two fixed functions, and k is a constant related to the key used in this FO_i function.

The FO Relation can be viewed as an improvement of Proposition 1. There are two folds of the improvement:

---

[1] The description of this proposition in [4] uses the subkeys $KO_{i,1}$ and $KO_{i,2}$ corresponding to the original form of the FO function described in [8]. Here we use the subkeys $AKO_{i,1}$ and $AKO_{i,2}$ as described in Section 2. The proposition does not change because $AKO_{i,1} = KO_{i,1}$ and $AKO_{i,2} = KO_{i,2}$.

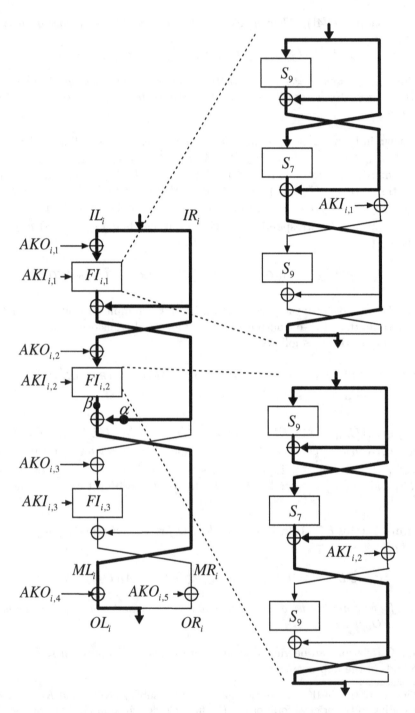

Fig. 2. The FO Relation. The thick lines denote the pathes related to the calculation of $OL_i^{(1...7)}$.

1. The functions f and g used in Lemma 1 rely only on the 16-bit partial input of the FO function instead of the whole 32-bit input used in Proposition 1 (the original Sakurai-Zheng Property [9] is similar to this form, however, Proposition 1 for the FO function proposed in [4] does not have this property).
2. The functions f and g are not related to the subkeys $AKO_{i,1}$ and $AKO_{i,2}$. The Subkeys $AKO_{i,1}$ and $AKO_{i,2}$ are moved into the inputs of the functions f and g.

These two merits will benefit our attack.

Based on the FO Relation, the following theorem can be obtained:

Theorem 1. *Let* $IL^1\|IR^1$, $IL^2\|IR^2$, $\ldots,IL^{2n}\|IL^{2n}$ *denote* $2n$ *inputs of the* FO *function of round* i *for some even number* $2n$, *the following equation holds:*

$$\bigoplus_{j=1}^{2n} OL_i^{j(1\ldots7)} = \bigoplus_{j=1}^{2n} f(IL_i^j \oplus AKO_{i,1}) \oplus \bigoplus_{j=1}^{2n} g(IR_i^j \oplus AKO_{i,2}) \qquad (5)$$

This theorem indicates that to obtain the value of $\bigoplus_{j=1}^{2n} OL_i^{j(1\ldots7)}$, we can treat the left 16 bits and the right 16 bits separately to compute the value of $\bigoplus_{j=1}^{2n} f(IL_i^j \oplus AKO_{i,1})$ and $\bigoplus_{j=1}^{2n} g(IR_i^j \oplus AKO_{i,2})$. Based on this theorem, we are ready to present our attacks.

4   Improved Integral Attack on 5-Round MISTY1

The integral attack on 5-round MISTY1 with all but the last FL layers, which was proposed in [4], uses the following four-round integral:

Proposition 2 ([4]). *Consider a structure (named integral structure) of* 2^{32} *plaintexts where the left 32 bits are held constant and the right 32 bits take on all possible values. The four round integral after* FL_6 *(the XOR of all the* 2^{32} *32-bit corresponding intermediate values of the structure after* FL_6*) is equal to zero.*

The main idea of the previous attack is to partially decrypt the encryptions of the structure and check whether Proposition 2 holds. Proposition 1 shown in Section 3 is used for fast checking whether the left seven bits of the integral are equal to zero predicated by Proposition 2.

We improve the above attack by using the FO Relation, which provides a more efficient method for checking the left seven bits of the integral than Proposition 1. The improved attack is as follows:

1. Ask for the encryptions of four different integral structures. Each structure includes all plaintexts that have the same left 32 bits and all possible right 32 bits.
2. For encryptions of each integral structure:
   (a) For every possible $AKO_{5,1}$, compute the value of $\bigoplus_{j=1}^{2^{32}} f(IL_5^j \oplus AKO_{5,1})$.

(b) For every possible $AKO_{5,2}$, compute the value of $\bigoplus_{j=1}^{2^{32}} g(IR_5^j \oplus AKO_{5,2})$.

(c) Discard all the $AKO_{5,1}, AKO_{5,2}$ pairs such that $\bigoplus_{j=1}^{2^{32}} f(IL_5^j \oplus AKO_{5,1})$
$\oplus \bigoplus_{j=1}^{2^{32}} g(IR_5^j \oplus AKO_{5,2})$ does not equal to $\bigoplus_{j=1}^{2^{32}} C^{j(1...7)}$.

3. For the remaining $AKO_{5,1}$, $AKO_{5,2}$ pairs, guess all the possible values of $AKI_{5,1}, AKI_{5,2}$ to get full 16 bit $\bigoplus_{j=1}^{2^{32}} OL^j$, discard all guesses such that Proposition 2 is not satisfied.

For each integral structure in Step 2(a), directly computing $\bigoplus_{j=1}^{2^{32}} f(IL_5^j \oplus AKO_{5,1})$ for each possible $AKO_{5,1}$ takes about 2^{48} encryptions. We develop one one technique when implementing this step to reduce the time needed from 2^{48} encryptions to $2^{26.58}$ encryptions for each integral structure as follows.

There are only 2^{16} possible different IL_5 values. For one 16-bit value that occurs an even number of times in all IL_5^j ($1 \leq j \leq 2^{32}$), the XOR of all the corresponding $f(IL_5^j \oplus AKO_{5,1})$ is zero. Hence, in Step 2(a) the attack first counts the occurrences of each 16-bit value in all IL_5^j. Then for each guessed $AKO_{5,1}$, using the 16-bit values that occur odd times in all IL_5^j to compute $\bigoplus_{j=1}^{2^{32}} f(IL_5^j \oplus AKO_{5,1})$.

For an integral structure in Step 2(a), counting the occurrences of every possible 16 bits among all IL_5^j can be accomplished by 2^{32} simple instructions. Since each simple instruction takes about 2^{-6} encryptions, the workload of the counting is 2^{26} encryptions. There are expected 2^{15} different 16 bit values which occur odd times in all IL_5^j. So, for one fixed $AKO_{5,1}$ we could compute $\bigoplus_{j=1}^{2^{32}} f(IL_5^j \oplus AKO_{5,1})$ by computing 2^{15} times function f. If we precompute all the possible value of function f(the time for this preprocess is neglectable compared with the total time complexity), it is possible to use table look-up to speed up. Since one table look-up takes no more than 2^{-6} encryptions, the running time for one integral is no more than $2^{26} + 2^{15} \cdot 2^{-6} \cdot 2^{16} = 2^{26.58}$ encryptions for an integral structure. Hence, Step 2(a) needs about $2^{26.58} \cdot 4 = 2^{28.58}$ encryptions for all the four integral structures. By using similar technique, Step 2(b) also needs about $2^{28.58}$ encryptions.

In Step 2(c), each guess of $AKO_{5,1}$, $AKO_{5,2}$ pair has a probability of 2^{-7} passing the check of an integral structure. For each integral structure, if we generate all the values could pass this check and then check whether they have already been discarded, there are at most $2^{32} \cdot 2^{-7} = 2^{25}$ candidates need to check. Since checking one pair needs only one table look-up, this step needs about $2^{25} \cdot 2^{-6} \cdot 4 = 2^{21}$ encryptions, which is neglectable compared with the time used in Step 2(a) and 2(b).

After checking of four integral structures, the probability of one $AKO_{5,1}$, $AKO_{5,2}$ pair not being discarded is 2^{-28}, thus there are about 2^4 such pairs entering Step 3. Using similar technique of Step 2, it is possible to finish this step within $2^4 \cdot 2^{19} = 2^{23}$ encryptions. After Step 3, only the correct guess remains and the wrong guesses are all discarded with high probability.

As shown above, the total time needed is dominated by Step 2(a) and Step 2(b). Hence, the time complexity of this attack is about $2^{28.58} + 2^{28.58} = 2^{29.58}$ encryptions.

5   Attack on 6-Round MISTY1 without the Last FL Layer

In this section, we extend the improved 5-round integral attack to 6-round without the last FL layer. To apply the method of the 5-round integral attack, the 6-round attack needs to recover the actual value of the input of FO_5, which means the attack needs to partially decrypt the sixth round. However, directly guessing 98 key bits used in FO_6 will make the attack slower than exhaustive key search. To reduce the time needed, we start from the following observation.

As shown in Figure 3, the input of FO_5, $IL_5 || IR_5$, can be written as $C^{(1\ldots 16)} \oplus ML_6 \oplus AKO_{6,4} || C^{(17\ldots 32)} \oplus MR_6 \oplus AKO_{6,5}$. The corresponding form of Equation (5) is then

$$\bigoplus_{j=1}^{2n} OL_5^{j(1\ldots 7)} = \bigoplus_{j=1}^{2n} f(C^{j(1\ldots 16)} \oplus ML_6^j \oplus AKO_{6,4} \oplus AKO_{5,1}) \oplus \bigoplus_{j=1}^{2n} g(C^{j(17\ldots 32)} \oplus MR_6^j \oplus AKO_{6,5} \oplus AKO_{5,2})$$
(6)

where $AKO_{5,1}$ and $AKO_{5,2}$ are then replaced by $AKO_{5,1} \oplus AKO_{6,4}$ and $AKO_{5,2} \oplus AKO_{6,5}$, respectively, and the input of FO_5 is then replaced by $C^{(1\ldots 16)} \oplus ML_6 || C^{(17\ldots 32)} \oplus MR_6$, because the subkeys $AKO_{6,4}$, $AKO_{6,5}$, $AKO_{5,1}$ and $AKO_{5,2}$ are not related to compute ML_6 and MR_6.

To compute the intermediate values $C^{(1\ldots 16)} \oplus ML_6$ and $C^{(17\ldots 32)} \oplus MR_6$, $AKO_{6,1}$, $AKO_{6,2}$, $AKO_{6,3}$, $AKI_{6,1}$, $AKI_{6,2}$ and $AKI_{6,3}$ are required. These six subkeys, which are corresponding to K_6, K_8, $K_5 \oplus K_3^{\prime(1\ldots 7)} || 00 || K_3^{\prime(1\ldots 7)}$, $K_3^{\prime(8\ldots 16)}$, $K_7^{\prime(8\ldots 16)}$, $K_1^{\prime(8\ldots 16)}$, only take 75 key bits. The attack can be described as follows:

1. Ask for the encryptions of four different integral structures. Each structure includes all plaintexts that have the same left 32 bits and all possible right 32 bits.

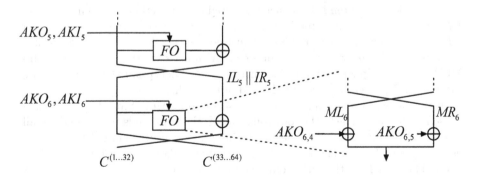

Fig. 3. Partial decryption in the attack on 6-round MISTY1 with all but the last FL layers

2. Guess 75 key bits, and partially decrypt all the 2^{34} encryptions to obtain the value of $C^{(1...16)} \oplus ML_6$ and $C^{(17...32)} \oplus MR_6$.

3. For each integral structure:

   (a) For every possible $AKO_{6,4} \oplus AKO_{5,1}$, compute the value of $\bigoplus_{j=1}^{2^{32}}$ $f(C^{j(1...16)} \oplus ML_6^j \oplus AKO_{6,4} \oplus AKO_{5,1})$.

   (b) For every possible $AKO_{6,5} \oplus AKO_{5,2}$, compute the value of $\bigoplus_{j=1}^{2^{32}}$ $f(C^{j(17...32)} \oplus MR_6^j \oplus AKO_{6,5} \oplus AKO_{5,2})$.

   (c) Discard all $AKO_{6,4} \oplus AKO_{5,1}$, $AKO_{6,5} \oplus AKO_{5,2}$ pairs such that $\bigoplus_{j=1}^{2^{32}}$ $f(C^{j(1...16)} \oplus ML_6^j \oplus AKO_{6,4} \oplus AKO_{5,1}) \oplus \bigoplus_{j=1}^{2^{32}} g(C^{j(17...32)} \oplus MR_6^j \oplus$ $AKO_{6,5} \oplus AKO_{5,2})$ does not equal to $\bigoplus_{i=1}^{2^{32}} C^{j(33...39)}$

4. For the guessed keys that are not discarded, exhaustively search for remaining key bits.

For each guessed 75-bit key, Step 2 partially decrypts 2^{34} encryptions. Each partial decryption takes no more than $1/4$ encryption. So this step needs about 2^{32} encryptions for each guessed 75-bit key.

As shown in Section 4, Both Step 3(a) and 3(b) need $2^{28.58}$ encryptions for the four integral structures, and Step 3(c) needs neglectable time compared with Step 3(a) and 3(b).

There are expected 2^4 out of 2^{32} possible $AKO_{5,1} \oplus AKO_{6,4}$, $AKO_{5,2} \oplus AKO_{6,5}$ pairs entering Step 4 for each guessed 75 bits in Step 2. For each guess entering Step 4, the attack exhaustively searches the 2^{21} possible remaining key bits. So the running time of this step is about 2^{25} encryptions for each guessed 75-bit key in Step 2.

As a result, the total time complexity of this attack is $2^{75} \cdot (2^{32} + 2^{28.58} + 2^{28.58} + 2^{25}) = 2^{107.26}$ encryptions.

6   Attack on 6-Round MISTY1 with All FL Layers

In this section, we extend the attack presented in last section to 6-round with all FL layers. If we simply extend the MISTY1 used in last section with FL_9 and FL_{10}, the attack will slower than exhaustive key search. By exploring the key schedule algorithm, we can perform a chosen ciphertext attack on last 6 round MISTY1 block cipher with all FL functions. The encryption then starts before the FL_3, FL_4 layer and ends at the end of the cipher.

In this attack, we also make use of the 4-round integral. Since the attack is a chosen ciphertext attack, the four round integral corresponds to the XOR of all the 2^{32} 32-bit intermediate values of the integral structure before FL_5. We also use Theorem 1 for fast checking the integral. As shown in Figure 4, the Equation (5) of Theorem 1 for the forth round can be rewritten as:

$$\bigoplus_{j=1}^{2n} OL_4^{j(1...7)} = \bigoplus_{j=1}^{2n} f(D^{j(33...48)} \oplus ML_3^j \oplus AKO_{3,4} \oplus AKO_{4,1}) \oplus \bigoplus_{j=1}^{2n} g(D^{j(49...64)} \oplus MR_3^j \oplus AKO_{3,5} \oplus AKO_{4,2})$$

$$(7)$$

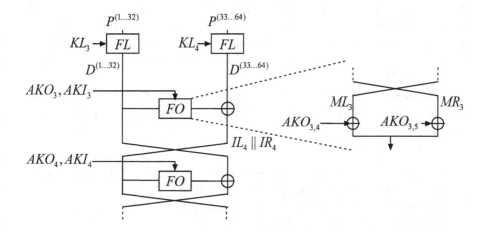

Fig. 4. Partial decryption in attack on 6-round MISTY1 with all layers

where D denotes the result of the plaintext P passing through the first FL_3 and FL_4 layer as shown in Figure 4.

We need to obtain the values of $D^{(33...48)} \oplus ML_3$ and $D^{(49...64)} \oplus MR_3$ from decryptions. For one decryption, if the attack partially decrypts for these values directly, it needs to guess at least total 105 key bits. Such a guess together with 2^{32} partial decryption will make the attack slower than exhaustive key search. However, to check Equation (7), we could obtain all the 2^{32} $D^{j(33...48)} \oplus ML_3^j$ to compute $\bigoplus_{j=1}^{2^{32}} f(D^{j(33...48)} \oplus ML_3^j \oplus AKO_{3,4} \oplus AKO_{4,1})$ and obtain all the 2^{32} $D^{j(49...64)} \oplus MR_3^j$ to compute $\bigoplus_{j=1}^{2^{32}} g(D^{j(49...64)} \oplus MR_3^j \oplus AKO_{3,5} \oplus AKO_{4,2})$ separately.

By exploring the key schedule weakness, we notice that none of the two processes needs all the 105 key bits. To obtain $D^{(33...48)}$, the attack needs $KL_{4,1}$ and $KL_{4,2}$, which are K_4' and K_6. To obtain ML_3 from the plaintext P, the attack needs only the subkeys $KL_{3,1}$, $KL_{3,2}$, $AKO_{3,1}$, $AKO_{3,2}$, $AKI_{3,1}$ and $AKI_{3,2}$, which correspond to K_2, K_8', K_3, K_5, $K_8'^{(8...16)}$, $K_4'^{(8...16)}$. Thus, only K_4', K_2, K_8', K_3, K_5 and K_6 are required.

Proposition 3. *For one decryption, computing $D^{(33...48)} \oplus ML_3$ from plaintext needs only 96 key bits.*

Consider the process of computing $D^{(49...64)} \oplus MR_3$. To obtain $D^{(49...64)}$, the attack only needs to know $KL_{4,1}(K_4')$ but not $KL_{4,2}(K_6)$. To obtain MR_3, the attack needs only $KL_{3,1}$, $KL_{3,2}$, $AKO_{3,1}$, $AKO_{3,2}$, $AKO_{3,3}$, $AKI_{3,1}$, $AKI_{3,2}$ and $AKI_{3,3}$, which correspond to K_2, K_8', K_3, K_5, $K_2 \oplus K_8'^{(1...7)}||00||K_8'^{(1...7)}$, $K_8'^{(8...16)}$, $K_4'^{(8...16)}$, $K_6'^{(8...16)}$, as Table 2.

Proposition 4. *For one decryption, computing $D^{(49...64)} \oplus MR_3$ from plaintext needs only 89 key bits.*

When computing the value of $\bigoplus_{j=1}^{2^{32}} f(D^{j(33...48)} \oplus ML_3^j \oplus AKO_{3,4} \oplus AKO_{4,1})$ for an integral structure, the $AKO_{4,1}$ and $AKO_{3,4}$ used correspond to K_4 and $K_7 \oplus K_8'^{(1...7)}||00||K_8'^{(1...7)} \oplus K_4'^{(1...7)}||00||K_4'^{(1...7)}$. The subkey K_7 is not included in the guessed 96 key bits. Hence K_7 should be guessed after obtaining all the $D^{(33...48)} \oplus ML_3^j$.

To obtain $\bigoplus_{j=1}^{2^{32}} g(D^{j(49...64)} \oplus MR_3^j \oplus AKO_{3,5} \oplus AKO_{4,2})$, the attack still needs to guess $K_6 \oplus K6'^{(1...7)}||00||K6'^{(1...7)}$. This guess can be done after obtaining all the $D^{(49...64)} \oplus MR_3^j$.

The attack is as follows:

1. Ask for decryptions of one integral structure in which all ciphertexts have the same left 32 bits and all possible right 32 bits.
2. For each guess of the 80-bit K_2, K_8', K_3, K_5, K_4':
   (a) Compute the value of $\bigoplus_{j=1}^{2^{32}} D^{j(1...7)}$.
   (b) Guess 16-bit K_6 and obtain all the 2^{32} $D^{j(33...48)} \oplus ML_3^j$. Continue to guess 16 bit words K_7 and compute $\bigoplus_{j=1}^{2^{32}} f(D^{j(33...48)} \oplus ML_3^j \oplus AKO_{3,4} \oplus AKO_{4,1})$.
   (c) Guess 9-bit $K_6'^{(8...16)}$ and obtain all the 2^{32} $D^{j(49...64)} \oplus MR_3^j$. Continue to guess 16 bit words $K_6 \oplus K_6'^{(1...7)}||00||K_6'^{(1...7)}$ and obtain $\bigoplus_{j=1}^{2^{32}} g(D^{j(49...64)} \oplus MR_3^j \oplus AKO_{3,5} \oplus AKO_{4,2})$.
   (d) Discard all guesses of K_6, K_7, $K_6'^{(8...16)}$ and $K_6 \oplus K_6'^{(1...7)}||00||K_6'^{(1...7)}$ such that Equation (7) does not hold or the guesses that result in conflict (K_6 and K_7 do not produce K_6' corresponding to $K_6'^{(8...16)}$ and $K_6 \oplus K_6'^{(1...7)}||00||K_6'^{(1...7)}$).
3. For the guesses not discarded, exhaustively search for the remaining key bits.

In Step 2(a), the computation of $\bigoplus_{j=1}^{2^{32}} D^{j(1...7)}$ takes no more than $2^{32} \cdot 1/4 = 2^{30}$ encryptions for each guessed 80 key bits.

In Step 2(b), for each guessed 16-bit K_6, the calculation of 2^{32} $D^{j(33...48)} \oplus ML_3^j$ takes no more than $2^{32} \cdot 1/4 = 2^{30}$ encryptions. For each K_7, the calculation of $\bigoplus_{j=1}^{2^{32}} f(D^{j(33...48)} \oplus ML_3^j \oplus AKO_{3,4} \oplus AKO_{4,1})$ takes about 2^{15} table lookups, which is equivalent to about $2^{15} \cdot 2^{-6} = 2^9$ encryptions. Hence, the running time of Step 2(b) is no more than $2^{16} \cdot (2^{30} + 2^{16} \cdot 2^9) = 2^{46.04}$ encryptions for each guessed 80 key bits. Similarly, Step 2(c) needs about $2^{39.04}$ encryptions for each guessed 80 key bits.

In Step 2(d), for each K_6 and K_7, we calculate the value of $K_6'^{(8...16)}$ and $K_6 \oplus K_6'^{(1...7)}||00||K_6'^{(1...7)}$, and then check whether Equation (7) holds. Hence Step 2(d) checks 2^{32} guesses, and the time needed is also neglectable to Step 2(b). For each guess of the 32 bit K_6 and K_7, the probability of satisfying Equation (7) is 2^{-7}. Hence, for each guessed 80 bits in Step 2, there are expected 2^{25} guesses out of 2^{32} possible K_6, K_7 entering Step 3. We notice that for each guess entering Step 3, the attack still needs to exhaustively search for the remaining 16 key bits. Therefore, Step 3 takes 2^{41} encryptions for each guess in Step 2.

The running time of the whole attack is dominated by Step 2(b). The time complexity of the attack is $(2^{30} + 2^{46.04} + 2^{39.04} + 2^{41}) \cdot 2^{80} = 2^{126.09}$ encryptions.

7 Conclusion

In this paper, we presented several integral attacks on reduced MISTY1 block cipher. Our attack improved the 5-round integral attack presented in [4] with the use of the *FO* Relation. We also extended the attack to 6-round with all *FL* layers by exploring the key schedule algorithm.

The existence of the *FO* Relation stems from the structure of the *FO* function and the fact that the key is XORed in the *FO* function; the resulting diffusion effect is too weak to defeat popular cryptanalysis techniques, such as differential cryptanalysis and integral cryptanalysis.

Our attack also indicates that the correspondence between subkeys used and the 128-bit key might be simple. Further exploration of this weakness of the key schedule is still worthy studying.

Acknowledgement

The authors would like to acknowledge Zheng Gong and Ruoyao Shi for their helpful advices.

References

1. Babbage, S., Frisch, L.: On MISTY1 Higher Order Differential Cryptanalysis. In: Won, D. (ed.) ICISC 2000. LNCS, vol. 2015, pp. 22–36. Springer, Heidelberg (2001)
2. Dunkelman, O., Keller, N.: An Improved Impossible Differential Attack on MISTY1. In: Pieprzyk, J. (ed.) ASIACRYPT 2008. LNCS, vol. 5350, pp. 441–454. Springer, Heidelberg (2008)
3. Hatano, Y., Tanaka, H., Kaneko, T.: An Optimized Algebraic Method for Higher Order Differential Attack. In: Fossorier, M.P.C., Høholdt, T., Poli, A. (eds.) AAECC 2003. LNCS, vol. 2643, pp. 61–70. Springer, Heidelberg (2003)
4. Knudsen, L.R., Wagner, D.: Integral Cryptanalysis. In: Daemen, J., Rijmen, V. (eds.) FSE 2002. LNCS, vol. 2365, pp. 112–127. Springer, Heidelberg (2002)
5. Kühn, U.: Cryptanalysis of Reduced-Round MISTY. In: Pfitzmann, B. (ed.) EUROCRYPT 2001. LNCS, vol. 2045, pp. 325–339. Springer, Heidelberg (2001)
6. Kühn, U.: Improved Cryptanalysis of MISTY1. In: Daemen, J., Rijmen, V. (eds.) FSE 2002. LNCS, vol. 2365, pp. 61–75. Springer, Heidelberg (2002)
7. Lu, J., Kim, J., Keller, N., Dunkelman, O.: Improving the Efficiency of Impossible Differential Cryptanalysis of Reduced Camellia and MISTY1. In: Malkin, T.G. (ed.) CT-RSA 2008. LNCS, vol. 4964, pp. 370–386. Springer, Heidelberg (2008)
8. Matsui, M.: New Block Encryption Algorithm MISTY. In: Biham, E. (ed.) FSE 1997. LNCS, vol. 1267, pp. 54–68. Springer, Heidelberg (1997)
9. Sakurai, K., Zheng, Y.: On Non-Pseudorandomness from Block Ciphers with Provable immunity Against Linear Cryptanalysis. In: Fossorier, M.P.C., Imai, H., Lin, S., Poli, A. (eds.) IEICE Trans. Fund., vol. E80-A(1), pp. 19–24 (1997)

10. Tanaka, H., Hatano, Y., Sugio, N., Kaneko, T.: Security Analysis of MISTY1. In: Kim, S., Yung, M., Lee, H.-W. (eds.) WISA 2007. LNCS, vol. 4867, pp. 215–226. Springer, Heidelberg (2008)
11. Tanaka, H., Hisamatsu, K., Kaneko, T.: Strength of MISTY1 without FL Function for Higher Order Differential Attack. In: Fossorier, M.P.C., Imai, H., Lin, S., Poli, A. (eds.) AAECC 1999. LNCS, vol. 1719, pp. 221–230. Springer, Heidelberg (1999)

New Results on Impossible Differential Cryptanalysis of Reduced–Round Camellia–128

Hamid Mala[1], Mohsen Shakiba[1], Mohammad Dakhilalian[1], and Ghadamali Bagherikaram[2]

[1] Cryptography & System Security Research Laboratory, Department of Electrical and Computer Engineering, Isfahan University of Technology, Isfahan, Iran
{hamid_mala@ec, m.shakiba@ec, mdalian@cc}.iut.ac.ir
[2] Department of Electrical and Computer Engineering, University of Waterloo, Waterloo, Ontario, Canada
gbagheri@cst.uwaterloo.ca

Abstract. Camellia, a 128–bit block cipher which has been accepted by ISO/IEC as an international standard, is increasingly being used in many cryptographic applications. In this paper, using the redundancy in the key schedule and accelerating the filtration of wrong pairs, we present a new impossible differential attack to reduced–round Camellia. By this attack 12–round Camellia–128 without FL/FL^{-1} functions and whitening is breakable with a total complexity of about $2^{116.6}$ encryptions and $2^{116.3}$ chosen plaintexts. In terms of the numbers of the attacked rounds, our attack is better than any previously known attack on Camellia–128.

1 Introduction

Camellia [1] is a 128–bit block cipher that supports several key lengths. For the sake of simplicity, Camellia with n–bit keys is denoted by Camellia–n, n=128, 192, 256. Camellia was jointly proposed in 2000 by NTT and Mitsubishi and then was submitted to several standardization and evaluation projects. It was selected as a winner of CRYPTREC e-government recommended ciphers in 2002 [5], NESSIE block cipher portfolio in 2003 [17] as well as the standardization activities at IETF [18]. Finally Camellia was selected as an international standard by ISO/IEC in 2005 [9]. As one of the most widely used block ciphers, Camellia has received a significant amount of cryptanalytic attention. The most efficient cryptanalytic results on Camellia include linear and differential attacks [19], truncated differential attack [5,10,13,20], higher order differential attack [7,11], collision attack [14,21], square attack [8,14,24], a square like attack [6] and impossible differential attack [15,20,22,23].

Impossible differential cryptanalysis, an extension of the differential attack [4], is one of the most powerful methods used for block cipher cryptanalysis. This method was first introduced by Biham [3] and Knudsen [12] independently. Impossible differential attacks use differentials that hold with probability zero (impossible differentials) to eliminate the wrong keys and leave the right key.

M.J. Jacobson Jr., V. Rijmen, and R. Safavi-Naini (Eds.): SAC 2009, LNCS 5867, pp. 281–294, 2009.
© Springer-Verlag Berlin Heidelberg 2009

The most efficient impossible differential attacks, recently proposed to reduced variants of Camellia, are as follows. The initial analysis of the security of Camellia to impossible differential cryptanalysis was given in [20]. They presented some 7–round impossible differentials for Camellia. In [23] Wu et al. introduced a nontrivial 8–round impossible differential that lead to an impossible differential attack on Camellia–192 and Camellia–256 without the FL/FL^{-1} functions with complexity of about 2^{118} chosen plaintexts and a time complexity of about 2^{126} memory accesses. Introducing the early abort technique, Lu et al. improved the impossible differential attack on Camellia in [16]. Later in [22] Wu et al. found a flaw in [16] and presented an impossible differential attack on 12–round Camellia–128 and claimed that their attack has a data complexity of 2^{65} chosen plaintexts and a time complexity of about $2^{111.5}$ encryptions. In this paper, we point out a flaw in their attack and show that its time complexity is more than exhaustive key search. However, their work is the first impossible differential attack on Camellia that considers the weakness in its key schedule.

In this paper, using the same 8–round impossible differential of [23], considering the weakness in the key schedule of Camellia–128, and also exploiting a hash table to simplify the selection of proper pairs, we present the first successful 12–round attack on Camellia–128. The proposed attack requires $2^{116.3}$ chosen plaintexts and has a total time complexity equivalent to about $2^{116.6}$ encryptions. We summarize our results along with previously known results on Camellia–128 in Table 1. The results of [16] in Table 1 come from its early version reported in Lu's PhD thesis [15], so we mark them with "†". In this table, time complexity is measured in encryption units unless MA is mentioned for memory accesses.

The rest of this paper is organized as follows: Section 2 provides a brief description of Camellia. We propose our new impossible differential attack on 12–round Camellia–128 in Section 3. Section 3 includes the previously known 8–round impossible differential (in Subsection 3.1), some observations on the key schedule of Camellia–128 (in Subsection 3.2), the proposed attack procedure on 12–round

Table 1. Summary of previous attacks and our new attack on Camellia–128

#Rounds	FL/FL^{-1}	Data	Time	Attack type	Source
8	no	$2^{83.6}$	$2^{55.6}$	Truncated Diff.	[13]
8	no	2^{20}	2^{120}	Higher Order Diff.	[7]
9	no	2^{92}	2^{111}	Higher Order Diff.	[7]
9	yes	2^{48}	2^{122}	Square.	[14]
9	no	$2^{113.6}$	2^{121}	Collision.	[21]
9	no	2^{88}	2^{90}	Square.	[14]
9	no	2^{105}	2^{105}	Differential.	[19]
9	no	2^{66}	$2^{84.8}$	Square like.	[6]
10	no	2^{120}	2^{121}	Linear.	[19]
11	no	2^{118}	2^{126}MA&2^{118}	Impossible Diff.	[16]†
11	no	2^{118}	2^{126}MA	Impossible Diff.	[16]†
12	no	$2^{116.3}$	$2^{116.6}$	Impossible Diff.	This work

Camellia–128 (in Subsection 3.3), and the analysis of the attack complexity (in Subsection 3.4). Finally, we conclude the paper in Section 4.

2   Preliminaries

2.1   Notations

In this paper, we will use the following notations:

L^{r-1}	: the left 64–bit half of the r–th round input,
R^{r-1}	: the right 64–bit half of the r–th round input,
k^r	: the subkey used in r–th round,
k_l^r	: the l–th byte of a subkey k^r,
$k_l^r[i-j]$	: the i–th to the j–th bits of $k_l^r, i, j = 1, 2, ..., 8, i \leq j$,
$x\|y$	: bit string concatenation of x and y,
$\oplus$	: bit-wise exclusive or operation,
$x <<<_l$	: the rotation of x by l bits to the left.

2.2   Description of Camellia

The 128–bit block cipher Camellia [1] has an 18–round (for 128 bit keys) or 24–round (for 192/256–bit keys) Feistel structure. The FL/FL^{-1} functions layer is inserted every 6 rounds. Before the first round and after the last round, there are pre– and post–whitening layers. In this paper we will consider a reduced variant of Camellia without FL/FL^{-1} functions and whitening layers. The Feistel structure of the r–th round is

$$L^r = R^{r-1} \oplus F(L^{r-1}, k^r), \ R^r = L^{r-1},$$

where function F consists of a key–addition layer, a substitution transformation S and a diffusion layer P. The S transformation contains 4 types of 8×8 S–boxes s_1, s_2, s_3 and s_4 as follows:

$$S(x_1|x_2|x_3|x_4|x_5|x_6|x_7|x_8)$$
$$= s_1(x_1)|s_2(x_2)|s_3(x_3)|s_4(x_4)|s_2(x_5)|s_3(x_6)|s_4(x_7)|s_1(x_8).$$

The transformation $P : (\{0,1\}^8)^8 \rightarrow (\{0,1\}^8)^8$ maps $(z_1, ..., z_8)$ to $(z_1', ..., z_8')$. This transformation and its inverse, P^{-1}, are defined as:

$$
\begin{aligned}
z_1' &= z_1 \oplus z_3 \oplus z_4 \oplus z_6 \oplus z_7 \oplus z_8 & z_1 &= z_2' \oplus z_3' \oplus z_4' \oplus z_6' \oplus z_7' \oplus z_8' \\
z_2' &= z_1 \oplus z_2 \oplus z_4 \oplus z_5 \oplus z_7 \oplus z_8 & z_2 &= z_1' \oplus z_3' \oplus z_4' \oplus z_5' \oplus z_7' \oplus z_8' \\
z_3' &= z_1 \oplus z_2 \oplus z_3 \oplus z_5 \oplus z_6 \oplus z_8 & z_3 &= z_1' \oplus z_2' \oplus z_4' \oplus z_5' \oplus z_6' \oplus z_8' \\
z_4' &= z_2 \oplus z_3 \oplus z_4 \oplus z_5 \oplus z_6 \oplus z_7 & z_4 &= z_1' \oplus z_2' \oplus z_3' \oplus z_5' \oplus z_6' \oplus z_8' \\
z_5' &= z_1 \oplus z_2 \oplus z_6 \oplus z_7 \oplus z_8 & z_5 &= z_1' \oplus z_2' \oplus z_5' \oplus z_7' \oplus z_8' \\
z_6' &= z_2 \oplus z_3 \oplus z_5 \oplus z_7 \oplus z_8 & z_6 &= z_2' \oplus z_3' \oplus z_5' \oplus z_6' \oplus z_8' \\
z_7' &= z_3 \oplus z_4 \oplus z_5 \oplus z_6 \oplus z_8 & z_7 &= z_3' \oplus z_4' \oplus z_5' \oplus z_6' \oplus z_7' \\
z_8' &= z_1 \oplus z_4 \oplus z_5 \oplus z_6 \oplus z_7 & z_8 &= z_1' \oplus z_4' \oplus z_6' \oplus z_7' \oplus z_8'
\end{aligned}
$$

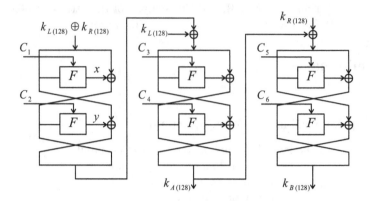

Fig. 1. Key schedule of Camellia

Table 2. The first 12 round keys of Camellia–128

Round	Subkey	Value	Round	Subkey	Value
1	k^1	$(k_A <<<_0)_L$	7	k^7	$(k_L <<<_{45})_L$
2	k^2	$(k_A <<<_0)_R$	8	k^8	$(k_L <<<_{45})_R$
3	k^3	$(k_L <<<_{15})_L$	9	k^9	$(k_A <<<_{45})_L$
4	k^4	$(k_L <<<_{15})_R$	10	k^{10}	$(k_L <<<_{60})_R$
5	k^5	$(k_A <<<_{15})_L$	11	k^{11}	$(k_A <<<_{60})_L$
6	k^6	$(k_A <<<_{15})_R$	12	k^{12}	$(k_A <<<_{60})_R$

Fig. 1 shows the key schedule of Camellia. For Camellia–128, two 128–bit variables k_L and k_R are defined as follows. The 128–bit user key is used as k_L, and k_R is a 128–bit string of 0 bits. Two 128–bit variables k_A and k_B are generated from k_L and k_R as shown in Fig. 1, in which $C_i, i = 1, ..., 6$ are constants used as the keys of the Feistel round function. The round keys of Camellia are rotations of variables k_A, k_B, k_L and k_R. Note that k_B is used only if the length of the user key is 192 or 256 bits. Here, we only give the first 12 round keys for Camellia–128 in Table 2.

2.3   Analysis of Wu et al.'s Attack on Camellia–128

In step (3.c.iii) of Section 4.1 in [22], the authors write: "Furthermore, the probability that a subkey guess may remain after this test is about $(1 - 2^{-8})$." At the first look, it seems to be true, but we show that this statement and thus the resulted complexity are not true. We show that the correct value for this probability is $(1 - 2^{-68})$, and also we calculate the dominant part of the time complexity of the attack on Camellia–128 proposed in [22].

At the end of step (3.b) there remain 2^{5+m} pairs. Below, we specify the list and the number of subkeys that are determined for each of these pairs:

1. Only one value for subkey bytes $(k_1^1, k_2^1, k_3^1, k_5^1, k_8^1)$ and $(k_1^{12}, k_2^{12}, k_3^{12}, k_5^{12}, k_8^{12})$ that satisfy the required differences in Round 1 and 12, and the 28–bit condition suggested by Property 1-1,
2. 2^{16} guesses of the 16 unknown bits $(k_4^1[1-4], k_6^1, k_7^1[1-4])$,
3. according to Property 1-3, only one value for the $(k_4^{12}, k_6^{12}, k_7^{12})$,
4. only one value for k_1^2 which is obtained from the difference distribution table of S–boxes, and
5. 2^{-8} value for k_1^{11}, because the only value obtained for $(k_1^{11}$ in step (c.ii) must also satisfy the 8–bit condition $k_1^{11} = (k_8^1[5-8]|k_1^2[1-4])$.

Thus, the number of 76–bit target subkeys that satisfy the impossible differential for each of the 2^{5+m} remaining pairs is $1 \times 1 \times 2^{16} \times 1 \times 1 \times 2^{-8} = 2^8$. Thus, the probability that a 76–bit target subkey guess be discarded by each of these 2^{5+m} pairs is $\frac{2^8}{2^{76}} = 2^{-68}$. Hence the number of 76–bit wrong subkeys remained at the end of the attack procedure is $(2^{76} - 1).(1 - 2^{-68})^{2^{m+5}}$. If we choose $m = 9$, as [22] proposes, the number of remaining wrong subkeys becomes:

$$(2^{76} - 1).(1 - 2^{-68})^{2^{14}} \approx 2^{76}.e^{-2^{-54}} \approx 2^{76}$$

If we accept that only one wrong subkey remains, m can be obtained as below:

$$(2^{76} - 1).(1 - 2^{-68})^{2^{m+5}} = 1 \Rightarrow m \approx -5 + 68 + \log_2(\frac{76}{\log_2 e}) \approx 68.7$$

Thus, the number of the required chosen plaintexts is about $2^{m+56} = 2^{124.7}$. Also the dominant part of time complexity which is related to step 2, will be about $2^{50} \times 2^{124.7} = 2^{174.7}$ memory accesses. Hence, this attack is infeasible. It seems that there is a similar mistake in computing the complexity of the attack on Camellia-256 proposed in [22].

3 Impossible Differential Cryptanalysis of Reduced Camellia–128

In this section, we first present the 8–round impossible differential of Camellia introduced in [23], then we propose an impossible differential attack on 12–round Camellia–128 without the FL/FL^{-1} functions. Finally, we analyze the complexity of our attack in Section 3.4.

3.1 8–Round Impossible Differentials of Camellia

In 2007, Wu et al. [23] found the following 8–round impossible differentials of Camellia: $(0|0|0|0|0|0|0|0, a|0|0|0|0|0|0|0) \to_8 (h|0|0|0|0|0|0|0, 0|0|0|0|0|0|0|0)$, where a and h are any two non–zero bytes. Fig. 2 illustrates more details. A detailed explanation of these 8–round impossible differentials is given in [23].

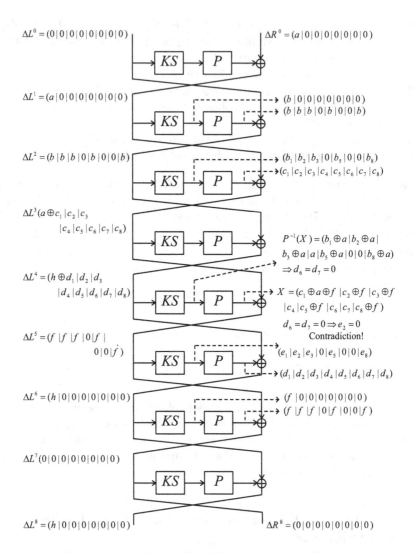

Fig. 2. 8–round impossible differentials of Camellia

3.2 Some Observations on the Key Schedule of Camellia–128

Redundancy in the Key Schedule: We first consider the relation between the target subkeys in our attack. The 18–byte target subkeys include the 8 bytes of k^1, the byte k_1^2, the byte k_1^{11} and 8 bytes of the last round key, k^{12}. Considering the key schedule of Camellia–128, we immediately observe that these 144 target bits are not distinct. From Table 2 we know that the four additional round keys k^1, k^2, k^{11}, k^{12} are rotations of the intermediate value k_A, as below:

$$k^1 = (k_A <<<_0)_L \quad , \quad k^2 = (k_A <<<_0)_R,$$
$$k^{11} = (k_A <<<_{60})_L \quad \text{and} \quad k^{12} = (k_A <<<_{60})_R$$

Table 3. Target subkeys in the attack represented in Fig. 4

Target Byte	Equivalent 8 bits of k_A	Target Byte	Equivalent 8 bits of k_A
k_1^1	$k_1\|k_2...\|k_8$	k_1^{11}	$k_{61}\|k_{62}...\|k_{68}$
k_2^1	$k_9\|k_{10}...\|k_{16}$	k_1^{12}	$k_{125}\|...\|k_{128}\|k_1\|...\|k_4$
k_3^1	$k_{17}\|k_{18}...\|k_{24}$	k_2^{12}	$k_5\|k_6...\|k_{12}$
k_4^1	$k_{25}\|k_{26}...\|k_{32}$	k_3^{12}	$k_{13}\|k_{14}...\|k_{20}$
k_5^1	$k_{33}\|k_{34}...\|k_{40}$	k_4^{12}	$k_{21}\|k_{22}...\|k_{28}$
k_6^1	$k_{41}\|k_{42}...\|k_{48}$	k_5^{12}	$k_{29}\|k_{30}...\|k_{36}$
k_7^1	$k_{49}\|k_{50}...\|k_{56}$	k_6^{12}	$k_{37}\|k_{38}...\|k_{44}$
k_8^1	$k_{57}\|k_{58}...\|k_{64}$	k_7^{12}	$k_{45}\|k_{46}...\|k_{52}$
k_1^2	$k_{65}\|k_{66}...\|k_{72}$	k_8^{12}	$k_{53}\|k_{54}...\|k_{60}$

Let us denote the intermediate value k_A by its bits as $k_A = k_1|k_2|...|k_{128}$. Then we can distinguish 18 target subkey bytes in bit strings of k_A in Table 3. It is obvious that the 18 target bytes are composed of only 76 distinct bits. This fact will help us to reduce the complexity of our attack. These distinct 76 bits include $k_1|k_2|...|k_{72}$ in Rounds 1, 2 and the four bits $k_{125}|...|k_{128}$ in the last round. This fact has previously been considered in [22].

Relation between k_L and k_A: Since in our attack some bits of k_A are recovered, here we investigate the relation between the master key of Camellia–128, $k_L = k_L^L|k_L^R$ and the intermediate key value $k_A = k_A^L|k_A^R$. In other words, we will show that k_L can be extracted from k_A. According to the key schedule of Camellia–128, k_R is zero. Let the outputs of round functions F in first and second rounds of the key schedule be denoted by x and y, respectively. According to Fig. 1, we can obtain x and y as functions of only k_A as below:

$$y = (k_A^L \oplus F_{C_4}(k_A^R) \oplus k_L^L) \oplus k_L^L = k_A^L \oplus F_{C_4}(k_A^R)$$
$$x = (k_A^R \oplus F_{C_3}(k_A^L \oplus F_{C_4}(k_A^R)) \oplus k_L^R) \oplus k_L^R = k_A^R \oplus F_{C_3}(k_A^L \oplus F_{C_4}(k_A^R))$$
$$= k_A^R \oplus F_{C_3}(y)$$

In a same way k_L can be represented in terms of x and y as below:

$$k_L^R = F_{C_2}^{-1}(y) \oplus x \quad k_L^L = F_{C_1}^{-1}(x)$$

So according to above equations we can obtain the master key of Camellia–128, $k_L = k_L^L|k_L^R$ in terms of k_A as below:

$$k_L^R = F_{C_2}^{-1}(k_A^L \oplus F_{C_4}(k_A^R)) \oplus k_A^R \oplus F_{C_3}(k_A^L \oplus F_{C_4}(k_A^R))$$
$$k_L^L = F_{C_1}^{-1}(k_A^R \oplus F_{C_3}(k_A^L \oplus F_{C_4}(k_A^R)))$$

Hence, the complexity of obtaining k_L from k_A is about four 1–round Camellia encryptions.

3.3 Impossible Differential Attack on 12–Round Camellia–128

In this section, we present the first successful impossible differential attack on 12 rounds of Camellia–128 without the FL/FL^{-1} functions and whitening. We

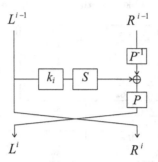

Fig. 3. Equivalent structure for one round of Camellia

attack Rounds 1 to 12, and use the 8–round impossible differential in Rounds 3 to 10. The attack is illustrated in Fig. 4. For the sake of simplicity, in Fig. 4 we use the equivalent round functions of Camellia in the Rounds 1, 2 and 12. The equivalent round function, which is shown in Fig. 3, is obtained by moving the P function to the output of the XOR operation and applying a transformation P^{-1} to the data line entering the XOR operation. According to Fig. 3, the equivalence of this modified structure to the original version can be verified easily as below:

$$R^i = L^{i-1},$$
$$L^i = P(S(k^i \oplus L^{i-1}) \oplus P^{-1}(R^{i-1}))$$
$$= P(S(k^i \oplus L^{i-1})) \oplus R^{i-1}$$
$$= F(L^{i-1}, k^i) \oplus R^{i-1}$$

In a traditional impossible differential attack where there exist additional rounds on both sides of the impossible differential, the attacker first checks a series of conditions in one side and choose pairs (or keys) that satisfy these conditions. She moves to the other side when she finishes checking all the conditions in the first side. When analyzing the Camellia, we observed that its structure allows us to change the side before finishing the investigation of all the conditions of one side. Thus we can check the condition that filters a greater number of pairs (or keys) before the other conditions. This strategy reduces the time complexity without any effect on the data complexity. So in the proposed attack, we first check some conditions in Round 1, then we conduct the attack in Rounds 12 and 11, and then we return to Rounds 1 and 2.

The attack procedure is as follows:

1. Take 2^n structures of plaintexts such that each structure contains 2^{56} plaintexts $P_i = L_i^0 | R_i^0$ with:

$$L_i^0 = (a'|a'|a'|\alpha_4|a'|\alpha_6|\alpha_7|a'),$$
$$R_i^0 = P(y_1'|y_2'|y_3'|\beta_4|y_5'|\beta_6|\beta_7|y_8') \oplus (y'|\gamma_2|\gamma_3|\gamma_4|\gamma_5|\gamma_6|\gamma_7|\gamma_8)$$

where the 7 bytes $(a', y', y_1', y_2', y_3', y_5', y_8')$ take all the possible values, and the bytes with the forms $\alpha_\times, \beta_\times$ and $\gamma_\times$ are fixed values in each structure.

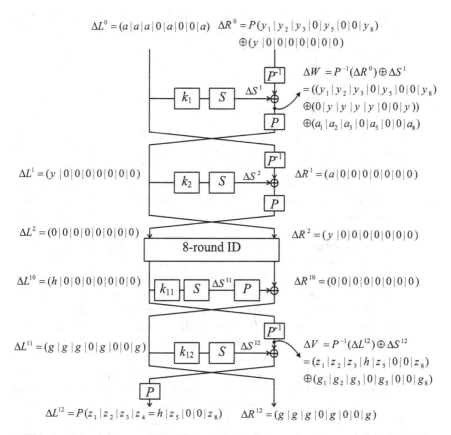

$\Delta L^0 = (a|a|a|0|a|0|0|a)$ $\Delta R^0 = P(y_1|y_2|y_3|0|y_5|0|0|y_8)$
$\oplus (y|0|0|0|0|0|0|0)$

$\Delta W = P^{-1}(\Delta R^0) \oplus \Delta S^1$
$= ((y_1|y_2|y_3|0|y_5|0|0|y_8)$
$\oplus (0|y|y|y|y|0|0|y))$
$\oplus (a_1|a_2|a_3|0|a_5|0|0|a_8)$

$\Delta L^1 = (y|0|0|0|0|0|0|0)$ $\Delta R^1 = (a|0|0|0|0|0|0|0)$

$\Delta L^2 = (0|0|0|0|0|0|0|0)$ $\Delta R^2 = (y|0|0|0|0|0|0|0)$

8-round ID

$\Delta L^{10} = (h|0|0|0|0|0|0|0)$ $\Delta R^{10} = (0|0|0|0|0|0|0|0)$

$\Delta L^{11} = (g|g|g|0|g|0|0|g)$ $\Delta V = P^{-1}(\Delta L^{12}) \oplus \Delta S^{12}$
$= (z_1|z_2|z_3|h|z_5|0|0|z_8)$
$\oplus (g_1|g_2|g_3|0|g_5|0|0|g_8)$

$\Delta L^{12} = P(z_1|z_2|z_3|z_4 = h|z_5|0|0|z_8)$ $\Delta R^{12} = (g|g|g|0|g|0|0|g)$

Fig. 4. 12–round impossible differential attack on reduced-round Camellia–128

It is obvious that each structure proposes about 2^{56} plaintexts, and 2^{111} plaintext pairs can be obtained from each structure. Totally, we can collect about 2^{n+56} plaintexts and 2^{n+111} plaintext pairs with the difference $\Delta L^0 = (a|a|a|0|a|0|0|a)$ and $\Delta R^0 = P(y_1|y_2|y_3|0|y_5|0|0|y_8) \oplus (y|0|0|0|0|0|0|0)$.

2. Obtain the ciphertexts of each structure and keep only the pairs that satisfy the following ciphertext difference:

$$\Delta L^{12} = P(z_1|z_2|z_3|h|z_5|0|0|z_8) \text{ and } \Delta R^{12} = (g|g|g|0|g|0|0|g)$$

where h, g and $z_\times$ are any non–zero byte values. The probability of this condition is $2^{-16} \times 2^{-56} = 2^{-72}$.

Thus the expected number of the remaining pairs is $2^{n+111} \times 2^{-72} = 2^{n+39}$.

3. Perform the following substeps:

(a) Guess the 8–bit value of k_1^1 and partially encrypt every remaining plaintext pair to get ΔW_1 in the output of the XOR of Round 1 (see Fig. 4). Keep only the pairs whose ΔW_1 is zero. The probability of this event is 2^{-8}, thus we expect about $2^{n+39} \times 2^{-8} = 2^{n+31}$ pairs remain.

(b) For $l = 2, 3, 5, 8$ guess the 8–bit value of k_l^1 and partially encrypt every remaining plaintext pair to get ΔW_l. Keep only the pairs whose ΔW_l is equal to y (consider that y is already determined by ΔR^0 for each plaintext pair). The probability of this event for each l is 2^{-8}, thus the expected number of remaining pairs is $2^{n+31} \times 2^{-8 \times 4} = 2^{n-1}$.

4. In this step consider the corresponding ciphertext pairs (C, C^*) of the remaining pairs then perform the following substeps:

(a) Guess the 8-bit value of k_1^{12}. Notice that according to Table 3, four bits of k_1^{12} is already fixed by k_1^1 previously guessed in step 3.a. Partially decrypt every remaining ciphertext pair (C, C^*) to get the first byte of the intermediate value ΔV in the output of the XOR of Round 12 (see Fig. 4). Keep only the pairs whose ΔV_1 is equal to zero. The probability of this condition is 2^{-8}, thus we expect about $2^{n-1} \times 2^{-8} = 2^{n-9}$ pairs remain.

(b) For $l = 2, 3$ obtain the 8–bit value of k_l^{12}. Notice that according to Table 3, all bits of k_l^{12} is already fixed by $k_{1,2,3}^1$ previously guessed in step 3. Partially decrypt every remaining ciphertext pair (C, C^*) to get the l–th byte of the intermediate value ΔV. Keep only the pairs whose ΔV_l is equal to h (consider that h is already determined by ΔL^{12} for each remaining ciphertext pair). The probability of this event for each l is 2^{-8}, thus the expected number of remaining pairs is $2^{n-9} \times 2^{-8 \times 2} = 2^{n-25}$.

(c) For $l = 5, 8$ guess the 8–bit value of k_l^{12}. Notice that according to Table 3, four bits of k_5^{12} and four bits of k_8^{12} are already fixed by previously guessed k_5^1 and k_8^1, respectively. Partially decrypt every remaining ciphertext pair (C, C^*) to get the l–th byte of the intermediate value ΔV. Keep only the pairs whose ΔV_l is equal to h (consider that h is already determined by ΔL^{12} for each of the remaining ciphertext pairs). The probability of this event for each l is 2^{-8}, thus the expected number of remaining pairs is $2^{n-25} \times 2^{-8 \times 2} = 2^{n-41}$.

(d) Guess the 24-bit value of $k_{4,6,7}^{12}$. Notice that according to Table 3, four bits of k_4^{12} and four bits of k_6^{12} are already fixed by previously guessed k_3^1 and k_5^1, respectively. Partially decrypt every remaining ciphertext pair (C, C^*) to get the exact value of intermediate pairs (L_1^{10}, L_1^{*10}). Consider that with probability 1, $\Delta V_{4,6,7} = h|0|0$. So this step does not affect the number of the remaining pairs.

5. Guess the 8-bit value of k_1^{11}. Notice that according to Table 3, four bits of k_1^{11} is already fixed by k_8^1 previously guessed in step 3. For every remaining pair, partially decrypt the (L_1^{10}, L_1^{*10}) through the first s-box of Round 11 to obtain ΔS_1^{11} and check if ΔS_1^{11} is equal to g, where g is already determined by ΔR^{12} for each ciphertext pair (see Fig. 4). The probability of this event is 2^{-8}, thus the expected number of remaining pairs is $2^{n-41} \times 2^{-8} = 2^{n-49}$.

In this stage of the attack, for every 72–bit guess of the subkeys $k_{1,2,3,5,8}^1$, k_7^{12}, 4 bits of $k_{1,4,5,6,8}^{12}$, and 4 bits of k_1^{11} we expect to obtain about 2^{n-49} pairs that satisfy the output difference of the 8–round impossible differential and also satisfy the difference $\Delta L_1 = \Delta R_2 = (y|0|0|0|0|0|0|0)$.

6. In this step, consider the corresponding plaintext pairs (P, P^*) of the remaining pairs then obtain the 24–bit value of $k^1_{4,6,7}$ (Notice that according to Table 3, all these 24 bits are already fixed by $k^{12}_{4,5,6,7,8}$). Now all bytes of k^1 are known, so partially encrypt every remaining pair to get the exact value of intermediate pairs (L^1_1, L^{*1}_1). Consider that with probability 1, $\Delta W_{4,6,7} = y|0|0$. So this step does not affect the number of the remaining pairs.

7. Guess the 8–bit value of k^2_1. Notice that according to Table 3, four bits of k^2_1 is already fixed by k^{11}_1 previously guessed in step 5. Then partially encrypt the (L^1_1, L^{*1}_1) through the first s–box of Round 2 to obtain ΔS^2_1 and check if ΔS^2_1 is equal to a, where a is already determined by ΔL^0 (see Fig. 4). If there exists a pair that passes this test, i.e. a pair that meets the input difference of the 8–round impossible differential, then discard the 76–bit subkey guess, and try another; otherwise for every 76-bit subkey guess, exhaustively search for the remaining 52 bits to recover the whole of k_A. Considering the relation between k_L and k_A, described in Section 3.2, this will lead to recovering the master key k_L.

3.4   Complexity of the Attack

In step 7, the probability that the difference ΔS^2_1 is equal to a fixed value a, is about 2^{-8}. So we expect only about $\epsilon = 2^{76}(1 - 2^{-8})^{2^{n-49}}$ guesses for 76–bit target subkey remain. If we accept the ϵ be equal to 1, then n will be 62.7. Thus the attack requires $2^{n+56} = 2^{118.7}$ chosen plaintexts.

In step 2, to get the qualified pairs, we first store the ciphertexts of each structure in a hash table indexed by the 4–th, 6–th and 7–th bytes of R^{12}, the XOR of the 1–st and 2–nd bytes of R^{12}, the XOR of the 1–st and 3–rd bytes of R^{12}, the XOR of the 1–st and 5–th bytes of R^{12}, the XOR of the 1–st and 8–th bytes of R^{12}, the 6–th and 7–th bytes of $P^{-1}(L^{12})$. Thus, every 2 ciphertexts with the same index in this table have the proper difference:

$$\Delta C = \Delta L^{12}|\Delta R^{12} = P(z_1|z_2|z_3|h|z_5|0|0|z_8)|(g|g|g|0|g|0|0|g).$$

Computing the 6–th and 7–th bytes of $P^{-1}(L^{12})$, requires 8 XOR operations, while each round of Camellia requires 24 XOR operations and 8 substitutions [2]. Thus, the total time complexity of computing the 6–th and 7–th bytes of $P^{-1}(L^{12})$ is less than about $8 \times 2^{118.7} \times \frac{1}{24} \times \frac{1}{12} \approx 2^{113.5}$ encryptions. Considering the complexity of obtaining the ciphertexts, step 2 requires about $2^{118.7} + 2^{113.5} \approx 2^{118.7}$ encryptions. At the end of this step, we expect about $2^{n+39} = 2^{101.7}$ proper pairs to be accessible.

According to procedure described in section 3.3 the time complexity (in terms of encryption units) of steps 3–7 for recovering 76 bits of k_A is as follows:

$$\text{Step 3(a)} : 2 \times \frac{1}{8} \times \frac{1}{12} \times 2^{n+39} \times 2^8 = \frac{1}{12} \times 2^{n+45}$$

$$\text{Step 3(b)} : 2 \times \frac{1}{8} \times \frac{1}{12} \times \sum_{i=0}^{3}(2^{n+31-8i} \times 2^{8+8\times(i+1)}) = \frac{1}{12} \times 2^{n+47}$$

Step 4(a) : $2 \times \dfrac{1}{8} \times \dfrac{1}{12} \times 2^{n-1} \times 2^{40+4} = \dfrac{1}{12} \times 2^{n+41}$

Step 4(b) : $2 \times \dfrac{1}{8} \times \dfrac{1}{12} \times \displaystyle\sum_{i=0}^{1}(2^{n-9-8i} \times 2^{44}) \approx \dfrac{1}{12} \times 2^{n+33}$

Step 4(c) : $2 \times \dfrac{1}{8} \times \dfrac{1}{12} \times \displaystyle\sum_{i=0}^{1}(2^{n-25-8i} \times 2^{44+4\times(i+1)}) = \dfrac{1}{12} \times (2^{n+21} + 2^{n+19})$

Step 4(d) : $2 \times \dfrac{3}{8} \times \dfrac{1}{12} \times 2^{n-41} \times 2^{52+4+4+8} = 2^{n+23}$

Step 5 : $2 \times \dfrac{1}{8} \times \dfrac{1}{12} \times 2^{n-41} \times 2^{68+4} = \dfrac{1}{12} \times 2^{n+29}$

Step 6 : $2 \times \dfrac{3}{8} \times \dfrac{1}{12} \times 2^{n-49} \times 2^{72+0} = 2^{n+19}$

Step 7 : $2 \times \dfrac{1}{8} \times \dfrac{1}{12} \times 2^{72+4} \times \displaystyle\sum_{i=0}^{2^{n-49}-1}(1-2^{-8})^i \approx \dfrac{1}{12} \times 2^{82} \times (1 - e^{-2^{n-57}})$

Thus the dominant part of time complexity to recover 76 bits of k_A is related to steps 2, 3(a) and 3(b) which is about $\frac{1}{12} \times (2^{n+45} + 2^{n+47}) + 2^{118.7} \approx 2^{118.7}$ encryptions. In order to recover the whole of master key (k_L), for each of the 76-bit candidates (outputs of the procedure described in Section 3.3 which is expected to be about ϵ) we have to exhaustively search the remaining 52 bits of k_A. Then using the second result of Section 3.2, we can obtain k_L for each of these 52-bit guesses. As we described in Section 3.2, this operation requires about $\epsilon \times 4 \times 2^{52} \times \frac{1}{12}$ encryptions. Also one additional encryption is required to check the key with a plaintext/ciphertext pair. Finally, the overall time complexity to recover the master key is about $2^{118.7} + \epsilon \times 2^{52} \times (\frac{4}{12} + 1)$. For $\epsilon = 1$, the complexity will be about $2^{118.7}$ encryptions.

If we let the ϵ be about 2^{62}, then n will be equal to 60.3. Thus, data complexity of the proposed attack reduces to $2^{n+56} = 2^{116.3}$ and the dominant time complexity is composed of the time complexity of steps 2, 3(a), 3(b) and the exhaustive search in step 7, as below

$$2^{n+56} + \dfrac{1}{12} \times (2^{n+45} + 2^{n+47}) + \epsilon \times 2^{52} \times (\dfrac{4}{12} + 1) \approx 2^{116.6}.$$

4   Conclusion

In this paper, we proposed a new impossible differential attack on 12–round Camellia–128 without the FL/FL^{-1} functions. The attack uses a previously known 8–round impossible differential to retrieve the whole of the master key. The proposed attack exploits the redundancy in the key schedule of Camellia–128 to reduce the complexity. In this attack also we use the strategy of moving between the additional rounds in a zigzag path to accelerate the filtration of wrong pairs for each key guesses. Using these techniques along with a hash table to simplify the selection of proper pairs, the proposed attack requires about $2^{116.3}$

plaintexts, and has a time complexity equivalent to about $2^{116.6}$ encryptions. Our attack is the first successful impossible differential attack on 12 rounds of Camellia–128.

References

1. Aoki, K., Ichikawa, T., Kanda, M., Matsui, M., Moriai, S., Nakajima, J., Tokita, T.: Camellia: a 128-bit Block Cipher Suitable for Multiple Platforms-Design and Analysis. In: Stinson, D.R., Tavares, S. (eds.) SAC 2000. LNCS, vol. 2012, pp. 39–56. Springer, Heidelberg (2001)
2. Aoki, K., Ichikawa, T., Kanda, M., Matsui, M., Moriai, S., Nakajima, J., Tokita, T.: Specification of Camellia – a 128-bit Block Cipher. version 2.0 (2001), http://info.isl.ntt.co.jp/crypt/eng/camellia/specifications.html
3. Biham, E., Biryukov, A., Shamir, A.: Cryptanalysis of Skipjack Reduced to 31 Rounds Using Impossible Differentials. In: Stern, J. (ed.) EUROCRYPT 1999. LNCS, vol. 1592, pp. 12–23. Springer, Heidelberg (1999)
4. Biham, E., Shamir, A.: Differential Cryptanalysis of the Data Encryption Standard. Springer, Heidelberg (1993)
5. CRYPTREC – Cryptography Research and Evaluation Committees, report, Archive (2002), http://www.ipa.go.jp/security/enc/CRYPTREC/index-e.html
6. Duo, L., Li, C., Feng, K.: Square Like Attack on Camellia. In: Qing, S., Imai, H., Wang, G. (eds.) ICICS 2007. LNCS, vol. 4861, pp. 269–283. Springer, Heidelberg (2007)
7. Hatano, Y., Sekine, H., Kaneko, T.: Higher Order Differential Attack of Camellia (II). In: Nyberg, K., Heys, H.M. (eds.) SAC 2002. LNCS, vol. 2595, pp. 129–146. Springer, Heidelberg (2003)
8. He, Y., Qing, S.: Square Attack on Reduced Camellia Cipher. In: Qing, S., Okamoto, T., Zhou, J. (eds.) ICICS 2001. LNCS, vol. 2229, pp. 238–245. Springer, Heidelberg (2001)
9. International Standardization of Organization (ISO), International Standard - ISO/IEC 18033-3, Information technology - Security techniques - Encryption algorithms - Part 3: Block ciphers (July 2005)
10. Kanda, M., Matsumoto, T.: Security of Camellia against Truncated Differential Cryptanalysis. In: Matsui, M. (ed.) FSE 2001. LNCS, vol. 2355, pp. 119–137. Springer, Heidelberg (2002)
11. Kawabata, T., Kaneko, T.: A Study on Higher Order Differential Attack of Camellia. In: The 2nd open NESSIE workshop (2001)
12. Knudsen, L.R.: DEAL – a 128-bit Block Cipher. Technical report, Department of Informatics, University of Bergen, Norway (1998)
13. Lee, S., Hong, S., Lee, S., Lim, J., Yoon, S.: Truncated Differential Cryptanalysis of Camellia. In: Kim, K.-c. (ed.) ICISC 2001. LNCS, vol. 2288, pp. 32–38. Springer, Heidelberg (2002)
14. Lei, D., Chao, L., Feng, K.: New Observation on Camellia. In: Preneel, B., Tavares, S. (eds.) SAC 2005. LNCS, vol. 3897, pp. 51–64. Springer, Heidelberg (2006)
15. Lu, J.: Cryptanalysis of Block Ciphers. PhD Thesis, Department of Mathematics, Royal Holloway, University of London, England (2008)
16. Lu, J., Kim, J., Keller, N., Dunkelman, O.: Improving the Efficiency of Impossible Differential Cryptanalysis of Reduced Camellia and MISTY1. In: Malkin, T.G. (ed.) CT-RSA 2008. LNCS, vol. 4964, pp. 370–386. Springer, Heidelberg (2008)

17. NESSIE – New European Schemes for Signatures, Integrity, and Encryption, final report of European project IST-1999-12324. Archive (1999), https://www.cosic.esat.kuleuven.be/nessie/Bookv015.pdf
18. NTT Information Sharing Platform Laboratories: Internationally Standardized Encryption Algorithm from Japan"Camellia", http://info.isl.ntt.co.jp/crypt/camellia/dl/Camellia20061108v4_eng.pdf
19. Shirai, T.: Differential, Linear, Boomerang and Rectangle Cryptanalysis of Reduced-Round Camellia. In: Proceedings of 3rd NESSIE workshop (November 2002)
20. Sugita, M., Kobara, K., Imai, H.: Security of Reduced Version of the Block Cipher Camellia against Truncated and Impossible Differential Cryptanalysis. In: Boyd, C. (ed.) ASIACRYPT 2001. LNCS, vol. 2248, pp. 193–207. Springer, Heidelberg (2001)
21. Wu, W., Feng, D., Chen, H.: Collision Attack and Pseudorandomness of Reduced-Round Camellia. In: Handschuh, H., Hasan, M.A. (eds.) SAC 2004. LNCS, vol. 3357, pp. 252–266. Springer, Heidelberg (2004)
22. Wu, W., Zhang, L., Zhang, W.: Improved Impossible Differential Cryptanalysis of Reduced-Round Camellia. In: Avanzi, R., Keliher, L., Sica, F. (eds.) SAC 2008. LNCS, vol. 5381, pp. 442–456. Springer, Heidelberg (2009)
23. Wu, W., Zhang, W., Feng, D.: Impossible Differential Cryptanalysis of Reduced-Round ARIA and Camellia. Journal of Computer Science and Technology 22(3), 449–456 (2007)
24. Yeom, Y., Park, S., Kim, I.: On the security of Camellia against the Square attack. In: Daemen, J., Rijmen, V. (eds.) FSE 2002. LNCS, vol. 2365, pp. 89–99. Springer, Heidelberg (2002)

Format-Preserving Encryption

Mihir Bellare[1], Thomas Ristenpart[1], Phillip Rogaway[2], and Till Stegers[2]

[1] Dept. of Computer Science & Engineering, UC San Diego, La Jolla, CA 92093, USA
[2] Dept. of Computer Science, UC Davis, Davis, CA 95616, USA

Abstract. Format-preserving encryption (FPE) encrypts a plaintext of some specified format into a ciphertext of identical format—for example, encrypting a valid credit-card number into a valid credit-card number. The problem has been known for some time, but it has lacked a fully general and rigorous treatment. We provide one, starting off by formally defining FPE and security goals for it. We investigate the natural approach for achieving FPE on complex domains, the "rank-then-encipher" approach, and explore what it can and cannot do. We describe two flavors of unbalanced Feistel networks that can be used for achieving FPE, and we prove new security results for each. We revisit the cycle-walking approach for enciphering on a non-sparse subset of an encipherable domain, showing that the timing information that may be divulged by cycle walking is not a damaging thing to leak.

1 Introduction

BACKGROUND. During the last few years, *format-preserving encryption* (FPE) has emerged as a useful tool in applied cryptography. The goal is this: under the control of a symmetric key K, deterministically encrypt a plaintext X into a ciphertext Y that has the same *format* as X. Examples include encryption of US social security numbers (SSNs), credit card numbers (CCNs) of a given length, 512-byte disk sectors, postal addresses of some particular country, and jpeg files of some given length. In our formalization of FPE, the format of a plaintext X will be a name N describing a finite set $\mathcal{X}_N$ over which the encryption function induces a permutation. For example, with SSNs this is the set of all nine-decimal-digit numbers.

The FPE goal is actually quite old. For one thing, a blockcipher itself can be seen as one kind of FPE: each N-bit string, where N is the block size, is mapped to some N-bit string. But what makes FPE an interesting and powerful idea is that the notion reaches far beyond blockciphers, which normally encipher strings of some one, convenient length.

SOME PRIOR WORK. In FIPS 74 (1981) [27], a DES-based approach is described to encipher strings over some fixed alphabet, say the decimal digits $\mathsf{D} = \{0, 1, \ldots, 9\}$. Each plaintext $X \in \mathsf{D}^N$ would be mapped to a ciphertext $Y \in \mathsf{D}^N$. Here each plaintext $X \in \mathsf{D}^*$ has a unique format $N = |X|$ and we must encipher X relative to the set $\mathcal{X}_N = \mathsf{D}^N$.

M.J. Jacobson Jr., V. Rijmen, and R. Safavi-Naini (Eds.): SAC 2009, LNCS 5867, pp. 295–312, 2009.
© Springer-Verlag Berlin Heidelberg 2009

Brightwell and Smith (1997) [6] considered a more general scenario, identifying what they termed *datatype-preserving encryption*. They wanted to encrypt database entries of some particular datatype without disrupting that datatype. A field containing an SSN (a nine-digit decimal string) should get mapped to another SSN. The authors colorfully explain the difficulty of doing this, saying that, with conventional encryption schemes, a "Ciphertext ... bears roughly the same resemblance to plaintext ... as a hamburger does to a T-bone steak. A social security number, encrypted using the DES encryption algorithm, not only does not resemble a social security number but will likely not contain any numbers at all" [6, p. 142]. The authors provide a proposed solution, though, as with FIPS 74, definitions or proofs for it are not likely or claimed.

Black and Rogaway [4] provided a provable-security investigation of a special case of FPE, asking how to make a cipher $E \colon \mathcal{K} \times \mathcal{X} \to \mathcal{X}$ with an arbitrary domain $\mathcal{X}$. Their solutions focused on $\mathcal{X} = \mathbb{Z}_N$, the integers $\{0, 1, \ldots, N-1\}$. The authors offer no general definition for FPE but they clearly intend that ciphers with domains of $\mathbb{Z}_N$ be used to construct schemes with other domains, like the set of valid CCNs of a given length.

The term *format-preserving encryption* is due to Terence Spies, Voltage Security's CTO [40]. Voltage, Semtek and other companies have been active in productizing FPE and explaining its utility [39]. FPE can enable a simpler migration path when encryption is added to legacy systems and databases, as required, for example, by the payment-card industry's data security standard (PCI DSS) [34]. Use of FPE enables upgrading database security in a way transparent to many applications, and minimally invasive to others. Spies has gone on to submit to NIST a proposed mechanism, FFSEM, that combines cycle walking and an AES-based balanced Feistel network [41].

SYNTAX. The current paper aims to help cryptographic theory "catch up" with cryptographic practice in this FPE domain. We initiate a general treatment of the problem, doing this within the provable-security tradition of modern cryptography.

We begin with a very general definition for FPE. Unlike a conventional cipher, an FPE scheme has associated to it a *collection* of domains, $\{\mathcal{X}_N\}_{N \in \mathcal{N}}$. We call each $\mathcal{X}_N$ a *slice* (the overall domain is their union, $\mathcal{X} = \bigcup_N \mathcal{X}_N$). The set $\mathcal{N}$ is the *format space*. For every key K, format N, and tweak T the FPE scheme E names a permutation $E_K^{N,T}$ on $\mathcal{X}_N$. We are careful to make FPEs tweakable [20] because, in this context, use of a tweak can significantly enhance security.

Returning to the CCN example, suppose we want to do FPE of CCNs with a zero Luhn-checksum [18]. Let's assume that the map should be length-preserving and that the possible lengths range from 12 to 19 decimal digits. Then we could let $\mathcal{N} = \{12, \ldots, 19\}$ and let $\mathcal{X}_N$ be the set of all N-digit numbers X such that LuhnOK(X) is true. Now an FPE scheme E with slices $\{\mathcal{X}_N\}_{N \in \mathcal{N}}$ does the job. You encrypt CCN X with key K and tweak T by letting $Y = E_K^{N,T}(X)$, where $N = \mathsf{len}(X)$.

SECURITY NOTIONS. We define multiple notions of security for FPE schemes. Our strongest adapts the traditional PRP notion to capture the idea that FPE is a good approximation for a family of uniform permutations on the slices. Our weaker notions are denoted SPI, MP, and MR. SPI (single-point indistinguishability) is a variant of the PRP notion in which there is a only a single challenge point. MP (message privacy) lifts semantic security to the FPE setting by adapting earlier notions of deterministic encryption [3,5]. MR (message recovery) formalizes an adversary's inability to recover a challenge message, in its entirety, from the message's ciphertext. All of these notions can be made with respect to an adaptive or nonadaptive adversary, and can also be strengthened to allow chosen-ciphertext attacks (for PRP, this would result in what is called a *strong* PRP).

Why bother with SPI, MP, and MR when they are implied by PRP? SPI is useful because it is easy to work with and implies MP and MR with a tight bound. MP and MR are interesting because they, even in their nonadaptive form, are what an application will most typically need. An attack against the PRP notion may be no threat in practice, and achieving good PRP security may be overkill. Good concrete security bounds become particularly a focus when slices are small: a bound permitting $q \approx 2^{n/4}$ queries provides limited assurance when $n = 20$ bits.

CONSTRUCTIONS. We next investigate the construction of FPE schemes. Suppose we wish to build an FPE scheme $\mathcal{E}$ with a complex *specification*—the slices $\{\mathcal{X}_N\}$ on which it should encipher. A natural approach is to arbitrarily number the points in each $\mathcal{X}_N$, say $\mathcal{X}_N = \{X_0, X_1, \ldots, X_{n-1}\}$ where $n = |\mathcal{X}_N|$. Then, to encipher $X \in \mathcal{X}_N$, find its index i in the enumeration, encipher i to j in $\mathbb{Z}_n$, and then return X_j as the encryption of X. We call this strategy the *rank-then-encipher* approach. It's the obvious, one could say folklore, approach. To implement it, we need an *integer* FPE that can encipher on $\mathbb{Z}_n$ for any needed n, as well a *ranking function*, rank, that maps each (N, X) with $X \in \mathcal{X}_N$ to a point in $\mathbb{Z}_n$ with $\mathrm{rank}(N, \cdot) \colon \mathcal{X}_N \to \mathbb{Z}_n$ a bijection for all $N \in \mathcal{N}$.

We will show how to build ranking functions for any FPE problem whose domain is a regular language (the slices being strings of each possible length). This includes many practical problems. This can be extended to domains that are context-free languages having unambiguous grammars.

Our starting point for building integer FPEs is the construction of Black and Rogaway [4], which combines a generalization of an unbalanced Feistel network (the left and right hand side are numbers in $\mathbb{Z}_a$ and $\mathbb{Z}_b$ rather than strings) and a technique the authors call *cycle walking*, a method apparently going back to the rotor machines of the early 1900's [37]. We extend their work to handle multiple slices with the same key, and to incorporate tweaks.

The type of unbalanced Feistel network that was extended in [4] is the type due to Lucks [22]. It is not the only kind of unbalanced Feistel network. An equally natural possibility is the unbalanced Feistel design of Schneier and Kelsey [36]. Extended to $\mathbb{Z}_N$ where $N = ab$, we call this a *type-1* Feistel, as opposed to the *type-2* unbalanced Feistel network of [4,22]. Our FPE schemes FE1 and FE2, based on type-1 and type-2 unbalanced Feistel networks, comprise a flexible, efficient, and customizable means for enciphering domains $\mathbb{Z}_N$ where $N = ab$ is the product of integers greater than one. Its round function can be based, for example, on AES. Combining FE1 or FE2 with the rank-then-encipher approach lets one achieve FPE in a wide variety of contexts.

SECURITY. Ideally, we would like to prove good bounds on the strong-PRP security for FE1 and FE2, assuming the round function to be a good PRF. But we run into a limitation, namely that the proven strength of Feistel ciphers [4,21,24,26,28,29,30,31,32,33], in terms of quality of bounds, falls short of what is wanted, and what appears to be the actual strength of the techniques. We address this in a couple of ways.

First, proofs have always targeted PRP. Instead, we target MP and MR, thereby getting better bounds more easily. We prove that FE2 with only *three* rounds hides all partial information with respect to a nonadaptive chosen-plaintext attack: one achieves nonadaptive SPI, MP, and MR security with reasonable bounds. Even then, we feel that being guided purely by what can be proved would lead to an overly quite pessimistic security estimate. The most realistic picture may be obtained by also assessing resistance to attacks. We consider known attacks and discuss their implications for our parameter choices (principally the number of rounds). We also provide a novel attack against (heavily) unbalanced type-2 Feistel networks, one that achieves message recovery with success probability exponentially small in the number of rounds. The attack is damaging if the number of rounds is too small.

Finally, reaching beyond PRP/SPI/MP/MR security, we consider a particular kind of side-channel attack. The use of cycle-walking in the rank-then-encipher approach raises the fear of timing attacks: might the number of times one has to apply the underlying cipher leak adversarially valuable information? We prove that cycle-walking will *not*, on its own, give rise to timing attacks. This is because the correct distribution on the number of iterations of the cipher on any input can be computed by a simulator that does not attend to the inputs. Due to space constraints, we present this result in the full version only [1].

THE FUTURE. We expect FPE to be increasingly deployed. The complex systems that process financial transactions impose a powerful legacy constraint. Using classical blockcipher-based modes would require far larger changes to these systems, which is costly and error-prone. FPE can be realized by simple, AES-based modes of operation, avoiding the need to design and review any fundamentally new primitive. Besides the enciphering of database fields, FPE may prove useful in networking applications, allowing datagrams to have their fields protected without changing their format. What one might lose in security when employing a *deterministic* encryption scheme can often be erased by sensibly tweaking the FPE scheme [20]. Moreover, such loss of security may be entirely overshadowed by the reduced need for random bits and disruption in infrastructure, protocols, and code.

2   FPE Syntax

SYNTAX. A scheme for *format-preserving encryption* (FPE) is a function $E\colon \mathcal{K} \times \mathcal{N} \times \mathcal{T} \times \mathcal{X} \to \mathcal{X} \cup \{\bot\}$ where the sets $\mathcal{K}$, $\mathcal{N}$, $\mathcal{T}$, and $\mathcal{X}$ are called the *key space*, *format space*, *tweak space*, and *domain*, respectively. All of these sets are nonempty and $\bot \notin \mathcal{X}$. We write $E_K^{N,T}(X) = E(K,N,T,X)$ for the encryption of X with respect to key K, format N, and tweak T. We require that whether or not $E_K^{N,T}(X) = \bot$ depends only on N, X and not on K, T, and let

$$\mathcal{X}_N = \{X \in \mathcal{X}\colon E_K^{N,T}(X) \in \mathcal{X} \text{ for all } (K,T) \in \mathcal{K} \times \mathcal{T}\}$$

be the N-indexed *slice* of the domain. We demand that a point $X \in \mathcal{X}$ live in at least one slice, $X \in \mathcal{X}_N$ for some N (if X is in no slice it should not be included in E's domain). We demand that there be finitely many points in each slice, meaning $\mathcal{X}_N$ is finite for all $N \in \mathcal{N}$. We require that $E_K^{N,T}(\cdot)$ be a permutation on $\mathcal{X}_N$ for any $(K,T) \in \mathcal{K} \times \mathcal{T}$. Its inverse $D\colon \mathcal{K} \times \mathcal{N} \times \mathcal{T} \times \mathcal{X} \to \mathcal{X} \cup \{\bot\}$ is defined by $D_K^{N,T}(Y) = D(K,N,T,Y) = X$ if $E_K^{N,T}(X) = Y$. In summary, an FPE enciphers the points within each of the (finite) slices that collectively comprise its domain.

A practical FPE scheme $E\colon \mathcal{K} \times \mathcal{N} \times \mathcal{T} \times \mathcal{X} \to \mathcal{X} \cup \{\bot\}$ must be realizable by efficient algorithms: an algorithm E to encrypt, an algorithm D to decrypt, and an algorithm to sample uniformly from the key space $\mathcal{K}$. Thus $\mathcal{K}$, $\mathcal{N}$, $\mathcal{T}$, and $\mathcal{X}$ should consist of strings or points easily encoded as strings, and E and D should return $\bot$ when presented a point outside of $\mathcal{K} \times \mathcal{N} \times \mathcal{T} \times \mathcal{X}$. We will not draw any distinction between an integer element of $\mathcal{X}$, say, and a string that encodes such a point.

THE FORMAT OF A POINT. Let $E\colon \mathcal{K} \times \mathcal{N} \times \mathcal{T} \times \mathcal{X} \to \mathcal{X} \cup \{\bot\}$ be an FPE scheme. Then we can speak of $X \in \mathcal{X}$ as having *format N* if $X \in \mathcal{X}_N$. One could associate to E a *format function* $\varphi\colon \mathcal{X} \to \mathcal{P}(\mathcal{N}) \setminus \{\emptyset\}$ that maps each $X \in \mathcal{X}$ to its possible formats; formally, $\varphi(X) = \{N \in \mathcal{N}\colon X \in \mathcal{X}_N\}$.

Note that, under our definitions, a point may have multiple formats. But often this will not be the case: each $X \in \mathcal{X}$ will belong to exactly one $\mathcal{X}_N$. In that case we can regard the format function as mapping $\varphi\colon \mathcal{X} \to \mathcal{N}$ and interpret $\varphi(X)$ as *the* format of X. FPE is somewhat simpler to understand for such *unique-format* FPEs: you can examine an X and know from it the slice $\mathcal{X}_{\varphi(X)}$ on which you mean to encipher it. For a unique format FPE one can write $E_K^T(X)$ rather than $E_K^{N,T}(X)$ since N is determined by X.

SPECIFICATIONS. An FPE problem, as needed by some application, will specify the desired collection of slices, $\{\mathcal{X}_N\}_{N \in \mathcal{N}}$. It will also specify the desired tweak space $\mathcal{T}$. Typically it is easy to support whatever tweak space one wants, but it may be quite hard to support a given collection of slices $\{\mathcal{X}_N\}_{N \in \mathcal{N}}$ (indeed it may be hard to accommodate a single slice, depending on what it is). We therefore call the collection of slices $\{\mathcal{X}_N\}_{N \in \mathcal{N}}$ the *specification* for an FPE scheme. We will write $\mathcal{X} = \{\mathcal{X}_N\}_{N \in \mathcal{N}}$ for a specification, only slightly abusing notation because the domain $\mathcal{X}$ *is* the union of slices in $\{\mathcal{X}_N\}_{N \in \mathcal{N}}$. The question confronting the cryptographer is *how to design an FPE scheme with a given specification.* We now provide some example possibilities.

EXAMPLES. **(1)** AES-128 can be regarded as an FPE with a single slice, $\{0,1\}^{128}$. The key space is $\mathcal{K} = \{0,1\}^{128}$ and the format space and tweak space are trivial (have size one). **(2)** To encipher 16-digit decimal numbers, take $\mathcal{X} = \{0,1,\ldots,9\}^{16}$ and just the one slice. **(3)** To encipher 512-byte disk sectors using an 8-byte sector index as the tweak, let $\mathcal{X} = \{0,1\}^{4096}$, $\mathcal{T} = \{0,1\}^{64}$, and just the one slice. **(4)** Suppose you want to encipher CCNs of 12–19 digits with a proper Luhn checksum, the ciphertext having the same length as the plaintext. Then the specification could be $\mathcal{X} = \{\mathcal{X}_N\}_{N \in \mathcal{N}}$ where $\mathcal{N} = \{12, 13, \ldots, 18, 19\}$ and $\mathcal{X}_N$ is the set of all strings $X \in \{0,1,\ldots,9\}^N$ satisfying the predicate $\mathsf{LuhnOK}(X)$. Here $|\mathcal{X}_N| = 10^{N-1}$. **(5)** One nice FPE has slices that are $\{0,1\}^N$ for each $N \geq 0$. It allows length-preserving encryption of any binary string. **(6)** One can FPE rather unusual spaces. For example, slice $\mathcal{X}_N$ could encode all N-vertex graphs. Or $\mathcal{X}_N$ could be all valid C-programs on N bytes. Designing an efficient FPE with this specification might be impossible. All of the examples just given are unique-format FPEs. The following example is not.

INTEGER FPES. The specification for a particularly handy kind of FPE is the following. The slices are $\mathcal{X}_N = \mathbb{Z}_N$, for $N \in \mathcal{N} \subseteq \mathbb{N}$. This allows enciphering natural numbers with respect to any permitted modulus N. Assuming the tweak space is similarly rich, say $\mathcal{T} = \{0,1\}^*$, we call such scheme an *integer FPE*. When used within the rank-then-encipher paradigm, integer FPEs enable the construction of FPEs with quite complex specifications.

3   FPE Security Notions

GAMES. Our definitions and proofs use *code-based games* [2], so we first review that material. A game has an **Initialize** procedure, an optional **Finalize** procedure, and any number of additional procedures. A game G is executed with an adversary $\mathcal{A}$ as follows. First, **Initialize** executes, possibly returning an output s, and then $\mathcal{A}(\mathsf{run}, s)$ is run ($s = \varepsilon$ if **Initialize** returns no string). As $\mathcal{A}$ executes it may call any procedure G (but not **Initialize** or **Finalize**) provided by G. If there is no **Finalize** procedure, the output of $\mathcal{A}$ is the output of the game. If the game does specify a **Finalize**, then, when $\mathcal{A}$ terminates, $\mathcal{A}$'s output is **Finalize**'s input and the game's output is that of **Finalize**. Game procedures may call $\mathcal{A}(\mathsf{identifier}[, x])$, which invokes an instance of the caller with distinct coins for each distinct identifier. Conceptually, then, each identifier thus names a separate adversarial algorithm. State is not shared among them. Let $\mathrm{G}^{\mathcal{A}} \Rightarrow y$ denote the event that the game outputs y. We write $S \xleftarrow{\cup} x$ as shorthand for $S \leftarrow S \cup \{x\}$. Later we write $c \xleftarrow{+} d$ for $c \leftarrow c + d$.

Boolean variables, including *bad*, are silently initialized to FALSE, set variables to $\emptyset$, integer variables to 0. Games G and H are said to be identical-until-*bad* if their code differs only in the sequel of statements that first set *bad* to true. We say that "$\mathrm{G}^{\mathcal{A}}$ sets *bad*" for the event that game G, when executed with adversary $\mathcal{A}$, sets *bad* to true. If G, H are identical-until-*bad* and $\mathcal{A}$ is an adversary then $\Pr\left[\,G^{\mathcal{A}} \text{ sets } bad\,\right] = \Pr\left[\,H^{\mathcal{A}} \text{ sets } bad\,\right]$. It is also standard ("the fundamental lemma") that if G, H are identical-until-*bad* then $\Pr\left[\,G^{\mathcal{A}} \Rightarrow y\,\right] - \Pr\left[\,H^{\mathcal{A}} \Rightarrow y\,\right] \leq \Pr\left[\,G^{\mathcal{A}} \text{ sets } bad\,\right]$.

SECURITY NOTIONS. We will extend the standard PRP notion to our setting, but we will also describe notions weaker than it, because they can be achieved with better proven concrete security for the same efficiency and, at the same time, they suffice for typical applications. Coming at it from the latter perspective, the most basic and often sufficient requirement is security against message recovery (MR), under either an adaptive or nonadaptive attack. We define this as well as a stronger notion of message privacy (MP) that requires that partial information about the message is not leaked by the ciphertext. We also consider a weakening of the PRP notion that we call SPI. The reason for considering this notion is that it is simpler than MP and MR to work with yet implies them; at the same time, it can be achieved with better concrete security bounds than we currently know how to get for the ordinary PRP notion.

In the following let $E \colon \mathcal{K} \times \mathcal{N} \times \mathcal{T} \times \mathcal{X} \to \mathcal{X} \cup \{\bot\}$ be an FPE scheme. We consider the games in Figure 1. It is assumed that any query of the form (N, T, X) satisfies $N \in \mathcal{N}$, $X \in \mathcal{X}_N$, and $T \in \mathcal{T}$.

PRP security. The standard notion of PRP security is extended to FPE schemes via game PRP_E and the corresponding adversary advantage is

$$\mathbf{Adv}_E^{\mathrm{prp}}(\mathcal{A}) = 2 \cdot \Pr\left[\,\mathrm{PRP}_E^{\mathcal{A}} \Rightarrow \mathsf{true}\,\right] - 1\,.$$

In the game $\mathrm{Perm}(\mathcal{X}_N)$ is the set of all permutations on $\mathcal{X}_N$.

SPI *security.* Single-point indistinguishability (SPI) requires that the adversary be unable to distinguish between the encryption of a single chosen message or a random range point, even when given adaptive access to a true encryption oracle. The formalization is based on game SPI_E. An adversary $\mathcal{A}$ is allowed to make only a single **Test** query, and this must be its first oracle query. Its associated advantage is

$$\mathbf{Adv}_E^{\mathrm{spi}}(\mathcal{A}) = 2 \cdot \Pr\left[\,\mathrm{SPI}_E^{\mathcal{A}} \Rightarrow \mathsf{true}\,\right] - 1\,.$$

The SPI notion is closely related to (and inspired by) a definition originally from [12], variants of which were also considered in [9,25]. It is easy to see that PRP security implies SPI security, but there is an additive loss of q/M in the advantage bound, where q is the number of queries by the adversary and M is the minimum size of $\mathcal{X}_N$ over all $N \in \mathcal{N}$. This is perhaps unfortunate, but SPI is only used as a tool anyway. A hybrid argument following [9,12] shows that SPI security likewise implies PRP security. Here, $\mathbf{Adv}_E^{\mathrm{spi}}(\mathcal{A}) \leq q \cdot \mathbf{Adv}_E^{\mathrm{prp}}(B) + q^2/M$ where q is the number of **Enc** queries of starting prp adversary $\mathcal{A}$, and constructed spi adversary B makes $q - 1$ **Enc** queries.

Message recovery. An FPE scheme secure against message recovery is one for which an adversary is unable to recover plaintexts from ciphertexts, even given an encryption oracle and a favorable distribution of plaintexts, formats, and tweaks.

Initialize	**Enc**(N,T,X)	**Finalize**(b') // Game PRP_E
$b \xleftarrow{\$} \{0,1\}$; $K \xleftarrow{\$} \mathcal{K}$	if $b=1$ then ret $E_K^{N,T}(X)$	ret $(b=b')$
for $(N,T) \in \mathcal{N} \times \mathcal{T}$	if $b=0$ then ret $\pi_{N,T}(X)$	
do $\pi_{N,T} \xleftarrow{\$} \mathrm{Perm}(\mathcal{X}_N)$		

Initialize	**Test**(N^*,T^*,X^*)	**Finalize**(b') // Game SPI_E
$b \xleftarrow{\$} \{0,1\}$; $K \xleftarrow{\$} \mathcal{K}$	if $(N^*,T^*,X^*) \in S$ then	ret $(b=b')$
	ret $\perp$	
Enc(N,T,X)	$S \xleftarrow{\cup} (N^*,T^*,X^*)$	
if $(N,T,X) \in S$ then	if $b=1$ then	
ret $\perp$	$Y^* \leftarrow E_K^{N^*,T^*}(X^*)$	
$S \xleftarrow{\cup} (N,T,X)$	else $Y^* \xleftarrow{\$} \mathcal{X}_{N^*}$	
ret $E_K^{N,T}(X)$	ret Y^*	

Initialize	**Enc**(N,T,X)	**Test** // Game MP_E
$K \xleftarrow{\$} \mathcal{K}$	ret $E_K^{N,T}(X)$	ret Y^*
$(N^*,T^*,X^*) \xleftarrow{\$} \mathcal{A}(\mathrm{dist})$		
$Y^* \leftarrow E_K^{N^*,T^*}(X^*)$	**Eq**(X)	**Finalize**(Z)
ret (N^*,T^*)	ret $(X=X^*)$	ret $(Z = \mathcal{A}(\mathrm{func}, X^*))$

Initialize	**Enc**(N,T,X)	**Test** // Game MR_E
$K \xleftarrow{\$} \mathcal{K}$	ret $E_K^{N,T}(X)$	ret Y^*
$(N^*,T^*,X^*) \xleftarrow{\$} \mathcal{A}(\mathrm{dist})$		
$Y^* \leftarrow E_K^{N^*,T^*}(X^*)$	**Eq**(X)	**Finalize**(X)
ret (N^*,T^*)	ret $(X=X^*)$	ret $(X=X^*)$

Fig. 1. Games used for defining FPE security notions SPRP, PRP, SPI, MP, and MR. Procedure $\mathcal{A}$, invoked by games MP and MR, denotes the caller of the game.

If the encryption were randomized we would require that the target ciphertext Y^* and encryption oracle E_K be of *no* use in recovering the plaintext, but this is too much to ask for with a deterministic encryption scheme, as an adversary can always encrypt candidate messages $X_1, \ldots, X_q$ to ciphertexts $Y_1, \ldots, Y_q$ and, if $Y_i = Y^*$ for some i, it will know that the target plaintext is $X^* = X_i$. Our security definition will formalize that this attack is (up to the adversary's advantage) the best one possible.

The idea is formalized as game MR_E in Figure 1. An MR-adversary $\mathcal{A}$ must begin with a **Test** query and have $Q_{\mathrm{Test}}(\mathcal{A}) = 1$ and $Q_{\mathrm{Eq}}(\mathcal{A}) = 0$, while a simulator $\mathcal{S}$ for $\mathcal{A}$ is an adversary that has $\mathcal{S}(\mathrm{dist}) = \mathcal{A}(\mathrm{dist})$, $Q_{\mathrm{Test}}(\mathcal{S}) = Q_{\mathrm{Enc}}(\mathcal{S}) = 0$ and $Q_{\mathrm{Eq}}(\mathcal{S}) = Q_{\mathrm{Enc}}(\mathcal{A})$. Here $Q_{\mathrm{Proc}}(\mathcal{C})$ is the maximum number of calls that adversary $\mathcal{C}$ might make to procedure **Proc**, the maximum over all coins of $\mathcal{C}$ and all possible oracle responses. The MR-advantage of adversary $\mathcal{A}$ is then defined as

$$\mathbf{Adv}_E^{\mathrm{mr}}(\mathcal{A}) = \Pr\left[\mathrm{MR}_E^{\mathcal{A}} \Rightarrow \mathrm{true}\right] - p_{\mathcal{A}}$$

where $p_{\mathcal{A}} = \max_{\mathcal{S}} \Pr\left[\mathrm{MR}_E^{\mathcal{S}} \Rightarrow \mathrm{true}\right]$ with the maximum over all simulators $\mathcal{S}$ for $\mathcal{A}$. Translating our formalism into English, an adversary making a **Test** query and some number of **Enc**-queries could do just as well forgoing its **Test** query and trading its **Enc** queries for **Eq** queries.

In our experiment defining $p_\mathcal{A}$ it is easy to see *what* strategy an optimal $\mathcal{S}$ should use: it makes q **Eq**-queries, $X_1, \ldots, X_q$, where X_1 is a most likely point output by $\mathcal{A}(\mathsf{dist})$ for the known (N^*, T^*); X_2 is a second most likely point $(X_2 \neq X_1)$; X_3 is a third most likely point $(X_3 \notin \{X_1, X_2\})$; and so on. If the **Eq**-oracle returns true for some X_i then $\mathcal{S}$ calls **Finalize**(X_i); otherwise, it calls **Finalize**(X_{q+1}) where $X_{q+1} \notin \{X_1, \ldots, X_q\}$ is the next most likely point after X_q. In this way $\mathcal{S}$ will win with probability $p_\mathcal{A} = \sum_{i=1}^{q+1} p_i$ where $p_i = \Pr[\mathcal{A}(\mathsf{dist}) \Rightarrow (N, T, X_i) \mid (N, T) = (N^*, T^*)]$.

Message privacy. In message privacy we are trying to measure the ability of an adversary with an encryption oracle to compute some function of a challenge plaintext X^* from its encryption C^*. If the encryption is randomized we would require that the challenge ciphertext C^* is of no use in such an attack. The formalization of this is semantic security [13]. For deterministic encryption, the intuition we aim to capture is that the adversary should do no better than it could if the encryption were ideal. In this case, the encryption oracle provides no more than the capability of testing whether a message of the adversary's choice equals the challenge message.

Our formalization closely resembles that for MR. A difference is that $\mathcal{A}$ is asked not only to come up with the distribution on plaintexts, but also the function on which it hopes to do well. See game MP in Figure 1. An MP-adversary $\mathcal{A}$ must begin with a **Test** query and have $Q_{\mathrm{Test}}(\mathcal{A}) = 1$ and $Q_{\mathrm{Eq}}(\mathcal{A}) = 0$, while a simulator $\mathcal{S}$ for $\mathcal{A}$ is an adversary that has $\mathcal{S}(\mathsf{dist}) = \mathcal{A}(\mathsf{dist})$, $Q_{\mathrm{Test}}(\mathcal{S}) = Q_{\mathrm{Enc}}(\mathcal{S}) = 0$, $Q_{\mathrm{Eq}}(\mathcal{S}) = Q_{\mathrm{Enc}}(\mathcal{A})$ and $\mathcal{S}(\mathsf{func}) = \mathcal{A}(\mathsf{func})$. The advantage of $\mathcal{A}$ is defined as

$$\mathbf{Adv}_E^{\mathrm{mp}}(\mathcal{A}) = \Pr\left[\mathrm{MP}_E^{\mathcal{A}} \Rightarrow \mathsf{true}\right] - p_\mathcal{A}$$

where $p_\mathcal{A} = \max_\mathcal{S} \Pr\left[\mathrm{MP}_E^{\mathcal{S}} \Rightarrow \mathsf{true}\right]$ with the maximum over all simulators $\mathcal{S}$ for $\mathcal{A}$. Translating our formalism into English, an adversary making a **Test** query and some number of **Enc**-queries could do just as well in guessing $Z = \mathcal{A}(\mathsf{func}, X^*)$ forgoing its **Test** query and trading its **Enc** queries for **Eq** queries. Note that MR-security amounts to a special case of MP-security where the function $\mathcal{A}(\mathsf{func}, \cdot)$ is the identity function.

RELATIONS BETWEEN NOTIONS. One can pictorially describe the relationships between our four security notions like this:

$$\mathrm{PRP} \rightleftarrows \mathrm{SPI} \rightleftarrows \mathrm{MP} \rightleftarrows \mathrm{MR}$$

The solid arrows indicate tight implications and the broken arrows indicate lossy ones. We already noted the implications between PRP and SPI above. These can be shown to be the best possible, with the counter-example in the first case being a perfect FPE scheme and in the second case following [9]. We also noted that MP tightly implies MR. The non-obvious implication is that SPI tightly implies MP, and is proved below. Finally, MP does not imply SPI, and MR does not imply MP. For the former separation, consider an FPE scheme that has a fixed point for all keys; for the latter separation, consider an FPE that always leaks a single bit of the plaintext. The proof of the following is given in the full version [1].

Proposition 1. [SPI ⇒ MP] *Let* $E\colon \mathcal{K} \times \mathcal{N} \times \mathcal{T} \times \mathcal{X} \to \mathcal{X} \cup \{\perp\}$ *be an FPE scheme and let* $\mathcal{A}$ *be an MP adversary. Then there is an SPI adversary* $\mathcal{B}$ *such that* $\mathbf{Adv}_E^{\mathrm{mp}}(\mathcal{A}) \le \mathbf{Adv}_E^{\mathrm{spi}}(\mathcal{B})$. *In addition, adversary* $\mathcal{B}$ *runs in time that of* $\mathcal{A}$ *and* $Q_{\mathrm{Enc}}(\mathcal{B}) = Q_{\mathrm{Enc}}(\mathcal{A})$. $\qquad\square$

NONADAPTIVE SECURITY, STRONG SECURITY. We expect that nonadaptive adversaries (the "static" security setting) are sufficient for many applications of FPE—the constructed scheme is not so much a tool as an end. We consider the class of static adversaries $\mathcal{S}$. An adversary $\mathcal{A} \in \mathcal{S}$, on input run, decides at the beginning of its execution the sequence of queries it will ask, their number and their kind being fixed. The relations between the non-adaptive notions of security remain the same as for their adaptive counterparts as described above.

In the other direction, the notions can be strengthened to require CCA-security. This is done by adding to the games a decryption procedure. In the PRP case, procedure $\mathbf{Dec}(N, T, Y)$ would return $D_K^{N,T}(Y)$ if $b = 1$ and $\pi_{N,T}^{-1}(Y)$ otherwise, where $D = E^{-1}$ denotes the inverse of E, as defined earlier. The resulting notion is the FPE analog of what is sometimes called strong-PRP (SPRP). In the games for SPI, MP and MR, $\mathbf{Dec}(N, T, Y)$ would return $D_K^{N,T}(Y)$. The adversary is not allowed to call it on inputs N^*, T^*, Y^* and the simulator is not allowed to call it at all.

ASYMPTOTIC NOTIONS. We can adapt our definitions to the asymptotic setting. We illustrate this for PRP-security. Recall first that, in speaking of complexity, we assume that $\mathcal{K}$, E, and D are all given by algorithms. Also, algorithm $\mathcal{K}$ took no input. We must slightly adjust the syntax of our FPE schemes. In particular, we provide $\mathcal{K}$ an input of the form 1^k. The algorithm must run in probabilistic polynomial time. Algorithm E and its inverse D must run in deterministic polynomial time in the sum of their input lengths. We then say that E is *PRP-secure* if, for any PPT adversary $\mathcal{A}$, the function $\varepsilon(k) = \mathbf{Adv}_E^{\mathrm{prp}}(\mathcal{A}(1^k))$ is negligible, meaning $\varepsilon(k) \in k^{-\omega(1)}$. We emphasize that it is the key K output by $\mathcal{K}$ that, presumably, grows with the security parameter k; the specification $\mathcal{X} = \{\mathcal{X}_N\}$ does not grow with or otherwise depend on the security parameter.

4   The Rank-then-Encipher Approach

THE IDEA. Suppose we want to build an FPE scheme $\mathcal{E}$ the slices of which may be quite complex. As an example, we might want to do length-preserving encryption of credit cards of various lengths, the CCNs of each length having a particular checksum and satisfying specified constraints on allowable substrings. It would be undesirable to design an encryption schemes whose internal workings were tailored to the specialized task in hand. Instead, what one can do is this. First, arbitrarily order and then number the points in each slice, $\mathcal{X}_N = \{X_0, X_1, \ldots, X_{n-1}\}$ where $n = |\mathcal{X}_N|$. Then, to encipher $X \in \mathcal{X}_N$, find its index i in the enumeration, encipher i to j in $\mathbb{Z}_n$ using an integer FPE scheme, and then return X_j as the encryption of X. We call this strategy the *rank-then-encipher* approach. The method will be efficient if there is an efficient way to map each point X to its index i, to encipher i to j, and to map j back to the corresponding point X_j. Details now follow, attending more closely to formats and tweaks, and also allowing the enumeration used for mapping j to X_j to differ from that used for ranking.

DEFINITIONS. To formalize RtE encryption, we first define a *ranking* and an *unranking* function for a specification $\mathcal{X} = \{\mathcal{X}_N\}$. A ranking function is a map *rank*: $\mathcal{N} \times \mathcal{X} \to \mathbb{N} \cup \{\perp\}$ for which $rank_N(\cdot) = rank(N, \cdot)$ is a bijection from $\mathcal{X}_N$ to $\mathbb{Z}_{|\mathcal{X}_N|}$. In addition, $rank_N(X) = \perp$ if $N \notin \mathcal{N}$ or $X \notin \mathcal{X}_N$. An unranking function is a map *unrank*: $\mathcal{N} \times \mathbb{N} \to \mathcal{X} \cup \{\perp\}$ for which $unrank_N(\cdot) = unrank(N, \cdot)$ is a bijection from $\mathbb{Z}_{|\mathcal{X}_N|}$ to $\mathcal{X}_N$. In addition, $unrank_N(i) = \perp$ if $i \notin \mathbb{Z}_{|\mathcal{X}_N|}$.

For the asymptotic tradition, we say that a specification $\mathcal{X} = \{\mathcal{X}_N\}$ can be *efficiently ranked* if there are (deterministic) polynomial-time computable ranking and unranking functions for $\mathcal{X} = \{\mathcal{X}_N\}$. Polynomiality is in the sum of the input lengths. Note that the security parameter is not an input to the ranking or unranking functions, but it is already built in that larger slices may take more time to rank and unrank, as the input to these functions includes the format N.

THE SCHEME. Suppose one aims to create an FPE scheme $\mathcal{E}$ with specification $\mathcal{X} = \{\mathcal{X}_N\}_{N \in \mathcal{N}}$. Let the desired tweak space for $\mathcal{E}$ be the set $\mathcal{T}$. Let $\mathbb{N}_0 = \{|\mathcal{X}_N| : N \in \mathcal{N}\} \subseteq \mathbb{N}$ be the sizes of the different slices. Then we can construct our desired FPE scheme $\mathcal{E}$ if we have in hand: (1) an integer FPE scheme $E : \mathcal{K} \times \mathbb{N}_0 \times \{0,1\}^* \to \mathbb{N}$ (it enciphers points in $\mathbb{Z}_n$ for each $n \in \mathbb{N}_0$), and (2) a ranking function *rank* and an unranking function *unrank* for $\mathcal{X} = \{\mathcal{X}_N\}_{N \in \mathcal{N}}$. Given such objects, define $\mathcal{E} = \text{RtE}[E, rank, unrank]$ as the map $\mathcal{E} : \mathcal{K} \times \mathcal{N} \times \mathcal{T} \times \mathcal{X} \to \mathcal{X} \cup \{\perp\}$ with

$$\mathcal{E}_K^{N,T}(X) = unrank_N(E_K^{|\mathcal{X}_N|, \langle N, T \rangle}(rank_N(X)))$$

when $X \in \mathcal{X}_N$, and $\mathcal{E}_K^{N,T}(X) = \perp$ otherwise. We call this *rank-then-encipher* approach. In words: convert the N-formatted string X to its corresponding number i; encipher $i \in \mathbb{Z}_{|\mathcal{X}_N|}$ to some $j \in \mathbb{Z}_{|\mathcal{X}_N|}$, employing a tweak that encodes both the format N of X and the tweak of $\mathcal{E}$; finally, convert j back to a domain point in $Y \in \mathcal{X}_N$ using a possibly unrelated enumeration of points.

We will omit formalizing and proving the rather obvious statements that, if E is secure with respect to the strong-PRP, PRP, SPI, MP, or MR notion of security, then so too will be the FPE scheme $\mathcal{E} = \text{RtE}[E, rank, unrank]$, the reduction being tight and having time complexity that is approximately the sum of the times to perform the ranking and unranking.

By way of the rank-then-encipher approach, one can take an integer FPE (based, e.g., on the techniques described in [4]) and create from it an FPE with a quite intricate specification $X = \{X_N\}_{N \in \mathbb{N}}$.

For many specifications the needed ranking and unranking functions are simple to design and fast to compute: an *ad hoc* approach will work fine. But what can one say in general about the power of the rank-then-encipher FPE approach? We now turn our attention to this.

5   FPE for Arbitrary Regular Languages

THE PROBLEM. Let Σ be a (finite) alphabet and let $L \subseteq \Sigma^*$ be a language over it. We say that an FPE scheme $E \colon \mathcal{K} \times \mathcal{N} \times \mathcal{T} \times X \to X \cup \{\bot\}$ is an FPE scheme *for* L if $X = L$, $\mathcal{N} = \mathbb{N}$, and the slices are $X_n = L_n = L \cap \Sigma^n$ for all $n \in \mathbb{N}$. In this section we show how to build an FPE for an arbitrary regular language L by describing how to compute a corresponding ranking and unranking function.

Why attend to regular languages? Many FPE specifications can be cast as asking for an FPE for a regular language. This is trivially true when the domain is finite. Some important domains are finite and without an easily summarized structure; a domain like "a valid postal address" is likely to be defined by a database such as the US Address Information System (AIS) and, given such a database, ranking is easy. Other finite domains are large but have a concise description as a regular language, either in terms of a regular expression or a DFA. For example, a US social security number is a string in the regular language $(0 \cup 1 \cup \cdots \cup 9)^9$. Alternatively, one may subtract from this any set of numbers that have not been assigned, such as those starting with an 8 or 9, having 0000 as the last four digits, or having 00 as the preceding two digits, but the resulting set will again have a concise description. For credit card numbers, a simple 20-state DFA M recognizes the language $Luhn^R$ of strings that are the reversals of numbers with a valid Luhn checksum [18]. Namely, the DFA is $M = (Q, \Sigma, \delta, q_0, F)$ with states $Q = \mathbb{Z}_{10} \times \mathbb{Z}_2$, final states $F = \{0\} \times \mathbb{Z}_2$, start state $q_0 = (0,0)$, and transition rule $\delta((a,b),d) = (a + 2d + a\lceil d/5 \rceil \bmod 10, 1-b)$. We will continue to use the $M = (Q, \Sigma, \delta, q_0, F)$ syntax below, following the convention of Sipser's book [38].

RANK COMPUTATION FOR REGULAR LANGUAGES. We will describe efficient ranking and unranking functions for the specification $X = \{X_M\}$ where M is a DFA and $X_M = L(M)$ is its language. First impose a total order $a_1 \prec \cdots \prec a_{|\Sigma|}$ on the elements of the alphabet $\Sigma = \{a_1, \ldots, a_{|\Sigma|}\}$ and extend this to the lexicographic order $\prec$ on each Σ^n. For $a \in \Sigma$ let $\mathrm{ord}(a)$ be the index i such that $a = a_i$ and for every $n \in \mathbb{N}$ let the ranking function be given by $rank_L(X) = |\{Y \in L : |X| = |Y| = n \text{ and } Y \prec X\}|$. We omit the argument $n = |X|$ because it is determined by X. Assume we have an integer FPE scheme E. Provided that we can efficiently compute each $rank_L(\cdot)$ and its inverse $unrank_L(\cdot)$, applying the RtE paradigm gives a practical FPE $\mathcal{E} = \mathrm{RtE}[E, rank_L, unrank_L]$ with $\mathcal{E} \colon \mathcal{K} \times \mathcal{N} \times \mathcal{T} \times L \to L \cup \{\bot\}$.

```
algorithm BuildTable(n)
for q ∈ Q do
    if q ∈ F then T[q, 0] ← 1
for i ← 1, ..., n do
    for q ∈ Q do
        for a ∈ Σ do
            T[q, i] ←± T[δ(q, a), i − 1]
```

algorithm rank(X)	algorithm unrank(c)

$q \leftarrow q_0$; $c \leftarrow 0$; $n \leftarrow |X|$
 for $i \leftarrow 1, \ldots, n$ do
   for $j \leftarrow 1, \ldots, \mathrm{ord}(X[i]) - 1$ do
     $c \leftarrow^{\pm} T[\delta(q, a_j), n - i]$
   $q \leftarrow \delta(q, X[i])$
 ret c
	$X \leftarrow \varepsilon$; $q \leftarrow q_0$; $j \leftarrow 1$
	 for $i \leftarrow 1, \ldots, n$ do
	   while $c \geq T[\delta(q, a_j), n - i]$ do
	     $c \leftarrow c - T[\delta(q, a_j), n - i]$; $j \leftarrow^{\pm} 1$
	   $X[i] \leftarrow a_j$; $q \leftarrow \delta(q, x[i])$; $j \leftarrow 1$
	 ret X

Fig. 2. Bottom left: Algorithm for computing the rank of a word in the regular language L of a DFA $M = (Q, \Sigma, \delta, q_0, F)$. **Top:** Initializing the table T. Each $T[\cdot, \cdot]$ starts at zero. **Bottom right:** How to compute the inverse of the ranking function.

Let $M = (Q, \Sigma, \delta, q_0, F)$ be a DFA recognizing the regular language $L \subseteq \Sigma^*$. Let $X[i]$ denote the i-th character of $X \in \Sigma^*$ (numbering from the left and starting at 1). Extend δ to $Q \times \Sigma^*$ so that $\delta(q, X)$ is the state we end up in by starting from q and following $X \in \Sigma^*$. Formally, set $\delta(q, \epsilon) = q$ for all $q \in Q$ and recursively define $\delta(q, x) = \delta(\delta(q, X[1]) \cdots X[n-1]), X[n])$ for all $q \in Q$ and all $X \in \Sigma^*$ with $n = |X| \geq 1$.

We compute the ranking function for M by dynamic programming, following [11]. Let $T[q, n]$ be the number of strings $X \in \Sigma^n$ such that $\delta(q, X) \in F$. The first algorithm of Figure 2, on input n, uses dynamic programming to compute, for all $q \in Q$ and $j \in [1..n]$, the number $T[q, j]$ of accepting paths of length j that start at q. The rank of a word in L can be computed based on T as shown by the second algorithm in Figure 2. The third algorithm in the figure computes the inverse, deriving a word in L by its rank. In the unit-cost model of computation, where arbitrary integer multiplications and additions are performed in unit time, $rank_M$ and $unrank_M$ can be computed in $O(|\Sigma| \cdot n)$ time, while the preprocessing step BuildTable(n) takes time $O(|Q| \cdot |\Sigma| \cdot n)$ time.

We comment that ranking can be further sped up to require about n sums instead of $n|\Sigma|$ by precomputing the needed partial sums, adding a third coordinate to T. The unranking function would need a binary search, or some other method, to map a number into the corrected (precomputed) interval $[0..\beta_1)$, $[\beta_1..\beta_2), \ldots, [\beta_{\sigma-1}, \beta_\sigma)$ that contains it, where $\sigma = |\Sigma|$. Regardless, ranking and unranking are linear-time for any regular language L, with modest constants in terms of the DFA representation of L.

ON THE IMPORTANCE OF REPRESENTATIONS. It is important that we represented our regular language in terms of a DFA; had L been represented in terms of an NFA or a regular expression, we could not have efficiently computed the ranking and unranking functions. In particular, remember that it is NP-hard (even PSPACE-hard) to decide if the language of an NFA M (or a regular expression α) is Σ^* [10, #AL1], [14]. Consequently, if P $\neq$ NP, we can't compute $unrank(2^n - 1)$ efficiently for all n, as such functionality would provide immediate means to decide if $L(M) = \Sigma^*$. Formally, if P$\neq$NP then $\mathfrak{X}_M$ can't be efficiently ranked, where $\mathfrak{X}_M = L(M)$ is the language of the NFA M. Note, however that this does not imply an inability to make an efficient FPE scheme for this specification—it only means that such a scheme could not use the RtE approach.

RANKING NON-REGULAR LANGUAGES. Beyond regular languages, we can also apply the RtE approach with Mäkinen's ranking algorithm for the language generated by an unambiguous context-free grammar [23]. Efficient ranking algorithms exist for various other classes of combinatorial objects. For example, if we wish to encrypt the domain $\mathfrak{X}_{n!}$ consisting of the set of permutations on n elements, the Lucas-Lehmer encoding [16] provides an efficient ranking. Other examples are spanning trees of a graph [7], B-trees [19], and Dyck languages [17]. Efficient rankings have also been studied in coding theory, starting with [8].

Given the ease of ranking regular languages and beyond, it is natural to ask if *every* language for which there is an efficient FPE scheme admits an RtE-style one. In the full version [1] we show that the answer is *no*. More specifically, we exhibit a specification $\mathfrak{X} = \{\mathfrak{X}_N\}_{N\in\mathcal{N}}$ where efficient FPE is possible but efficient ranking is not.

6   Feistel-Based Integer FPEs

We present two Feistel-based constructions of integer FPE schemes $E\colon \mathcal{K} \times \mathcal{N} \times \mathcal{T} \times \mathcal{X} \to \mathcal{X} \cup \{\bot\}$ with format space $\mathcal{N} = \mathbb{N} \times \mathbb{N}$ and $\mathcal{X}$ such that $\mathfrak{X}_N = \mathbb{Z}_{ab}$ for $N = (a, b)$ with $a \le b$. Both are parameterized by the following: (1) a round function $F\colon \mathcal{K} \times \mathcal{N} \times \mathcal{T} \times \mathbb{N} \times \mathbb{N} \to \mathbb{N}$; and (2) a function $r\colon \mathcal{N} \to \mathbb{N}$ specifying the number of rounds.

Figure 3 defines encryption and decryption for the two integer FPE schemes FE1 and FE2. We refer to Feistel networks, such as FE1, that utilize the same kind of round function every round as type-1. Type-1 Feistel networks were previously treated in [26,36] for the case of bit strings. We refer to Feistel networks, such as FE2, that alternate the kind of round function as type-2. Type-2 Feistel networks for the case of bit strings are due to Lucks [22]. Type-2 Feistel networks with modular arithmetic were first used in [4].

ROUND FUNCTIONS. The round functions should be PRFs. It is not clear what this means when the range is the infinite set $\mathbb{N}$. To specify a round function, we will first specify a range function $w\colon \mathcal{N} \to \mathbb{N}$ such that for all $N \in \mathcal{N}$ we have $w(N) \ge b$ where $N = (a, b)$. The PRF advantage of an adversary $\mathcal{A}$ is then defined by

$$\mathbf{Adv}_F^{\mathrm{prf}}(\mathcal{A}) = \Pr\left[\mathcal{A}^{F(K,\cdot,\cdot,\cdot,\cdot)} \Rightarrow 1\right] - \Pr\left[\mathcal{A}^{\$(\cdot,\cdot,\cdot,\cdot)} \Rightarrow 1\right]$$

algorithm $\mathbf{FE1}_K^{N,T}(X)$

$(a,b) \leftarrow N$; $X_0 \leftarrow X$
for $i = 1,\ldots,r(N)$ do
   $L_{i-1} \leftarrow X_{i-1} \operatorname{div} b$
   $R_{i-1} \leftarrow X_{i-1} \bmod b$
   $W_i \leftarrow L_{i-1} + F_K(N,T,i,R_{i-1}) \bmod a$
   $X_i \leftarrow aR_{i-1} + W_i$
ret $X_{r(N)}$

algorithm $\mathbf{FD1}^{N,T}(Y)$

$(a,b) \leftarrow N$; $Y_{r(N)} \leftarrow Y$
for $i = r(N),\ldots,1$ do
   $W_i \leftarrow Y_i \bmod a$
   $R_{i-1} \leftarrow Y_i \operatorname{div} b$
   $L_{i-1} \leftarrow W_i - F_K(N,T,i,R_{i-1}) \bmod a$
   $Y_{i-1} \leftarrow aR_{i-1} + Z_i$
ret Y_0

algorithm $\mathbf{FE2}_K^{N,T}(X)$

$(a,b) \leftarrow N$
$L_0 \leftarrow X \bmod a$; $R_0 \leftarrow X \operatorname{div} a$
for $i = 1,\ldots,r(N)$ do
   If $i \bmod 2 = 1$ then $s \leftarrow a$ else $s \leftarrow b$
   $L_i \leftarrow R_{i-1}$
   $R_i \leftarrow L_{i-1} + F_K(N,T,i,R_{i-1}) \bmod s$
ret $sL_{r(N)} + R_{r(N)}$

algorithm $\mathbf{FD2}_K^{N,T}(Y)$

$(a,b) \leftarrow N$
If $r(N) \bmod 2 = 1$ then $s \leftarrow a$ else $s \leftarrow b$
$R_{r(N)} \leftarrow Y \bmod s$; $L_0 \leftarrow Y \operatorname{div} s$
for $i = r(N),\ldots,1$ do
   If $i \bmod 2 = 1$ then $s \leftarrow a$ else $s \leftarrow b$
   $R_{i-1} \leftarrow L_i$
   $L_{i-1} \leftarrow R_i - F_K(N,T,i,R_{i-1}) \bmod s$
ret $sR_0 + L_0$

Fig. 3. Top: Encryption and decryption algorithms for the integer FPE scheme FE1 where $K \in \mathcal{K}$, $T \in \mathcal{T}$, $F \in \mathcal{N}$, and $X,Y \in \mathcal{X}_F$. Here $x \operatorname{div} y$ is short-hand for $\lfloor x/y \rfloor$. **Bottom:** Encryption and decryption algorithms for the integer FPE scheme FE2.

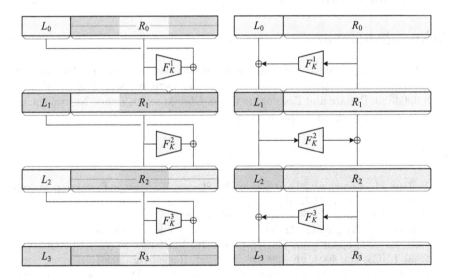

Fig. 4. Diagrams of three rounds of FE1 (**left**) and FE2 (**right**) for format $N = (a,b) = (2^{n_0}, 2^{n_1})$ and input $X \in \mathbb{Z}_{ab}$. For both mechanisms, $L_0, L_1, L_2, L_3 \in \mathbb{Z}_a$ and $R_0, R_1, R_2, R_3 \in \mathbb{Z}_b$.

where $\mathcal{A}$'s oracle in the second case returns a random point in $\mathbb{Z}_{w(N)}$ in response to a query F, T, i, X. Adversary $\mathcal{A}$ is not allowed to repeat an oracle query.

In cases of practical interest, we can build suitable round functions based on block ciphers (e.g. 3DES or AES) or cryptographic hash functions (e.g. SHA-256). In the full version [1] we detail example instantiations. We also discuss there the use of precomputation for speed improvements (deriving from the fact that several of the inputs to F are the same across all rounds).

DISCUSSION. The round function takes as input the format and tweak, which effectively provides "separate" instances of the cipher for each format, tweak pair. To ensure independence between rounds, the round number is also input into the PRF.

FE1 and FE2 support domains of the form $\mathbb{Z}_{ab}$ and only provide security when $a > 1$. To handle arbitrary $\mathbb{Z}_n$ one can choose $N = (a, b)$ so that $ab > N$ and then utilize the cycle walking technique with FE1 or FE2 (see [4] for a treatment). Alternatively, one might utilize the off-by-one construction (see [4]) to avoid cycle-walking. But for typical applications like the encryption of credit card numbers, the requisite domains will be $\mathbb{Z}_n$ for which $n = ab$ for a and b that are almost balanced.

SECURITY OF FE1, FE2. In the full version [1] we discuss in detail the security of FE1 and FE2 in terms of best known attacks and proven security bounds. Beyond prior results, we give a novel MR attack that breaks FE2 when it is used with very unbalanced (a, b) and a relatively small number of rounds. We also give novel provable SPI security bounds for both schemes, which by Proposition 1 establishes MP and MR security.

Acknowledgments

Rogaway thanks Terence Spies for many useful discussions, and for sparking his interest in this topic. Rogaway and Stegers were supported by NSF grant CNS 0904380. Bellare and Ristenpart thank Clay Mueller, Lance Nakamura and Semtek for useful discussions and support.

References

1. Bellare, M., Ristenpart, T., Rogaway, P., Stegers, T.: Format-Preserving Encryption. In: Jacobson, M.J., Rijmen, V., Safavi-Naini, R. (eds.) SAC 2009. LNCS, vol. 5867, pp. 295–312. Springer, Heidelberg (2009)
2. Bellare, M., Rogaway, P.: The security of triple encryption and a framework for code-based game-playing proofs. In: Vaudenay, S. (ed.) EUROCRYPT 2006. LNCS, vol. 4004, pp. 409–426. Springer, Heidelberg (2006)
3. Bellare, M., Fischlin, M., O'Neill, A., Ristenpart, T.: Deterministic encryption: definitional equivalences and constructions without random oracles. In: Wagner, D. (ed.) CRYPTO 2008. LNCS, vol. 5157, pp. 360–378. Springer, Heidelberg (2008)
4. Black, J., Rogaway, P.: Ciphers with arbitrary finite domains. In: Preneel, B. (ed.) CT-RSA 2002. LNCS, vol. 2271, pp. 114–130. Springer, Heidelberg (2002)

5. Boldyreva, A., Fehr, S., O'Neill, A.: On notions of security for deterministic encryption, and efficient constructions without random oracles. In: Wagner, D. (ed.) CRYPTO 2008. LNCS, vol. 5157, pp. 335–359. Springer, Heidelberg (2008)

6. Brightwell, M., Smith, H.: Using datatype-preserving encryption to enhance data warehouse security. In: 20th NISSC Proceedings, pp. 141–149 (1997), http://www.csrc.nist.gov/nissc/1997

7. Colbourn, C., Day, R., Nel, L.: Unranking and ranking spanning trees of a graph. Journal of Algorithms 10(2), 271–286 (1989)

8. Cover, T.: Enumerative source encoding. IEEE Transactions on Information Theory 19(1), 73–77 (1977)

9. Desai, A., Miner, S.: Concrete security characterizations of pRFs and pRPs: Reductions and applications. In: Okamoto, T. (ed.) ASIACRYPT 2000. LNCS, vol. 1976, pp. 503–516. Springer, Heidelberg (2000)

10. Garey, M., Johnson, D.: Computers and Intractability: A Guide to the Theory of NP-Completeness. W.H. Freeman and Company, New York (1979)

11. Goldberg, A., Sipser, M.: Compression and Ranking. In: 17th Annual ACM Symposium on the Theory of Computing (STOC 1985), pp. 440–448. ACM Press, New York (1985)

12. Goldreich, O., Goldwasser, S., Micali, S.: How to construct random functions. Journal of the ACM 33(4), 792–807 (1986)

13. Goldwasser, S., Micali, S.: Probabilistic encryption. J. Comput. Syst. Sci. 28(2), 270–299 (1984)

14. Hopcroft, J., Ullman, J.: Formal Languages and their Relation to Automata. Addison-Wesley, Reading (1969)

15. Jerrum, M.: A very simple algorithm for estimating the number of k-colorings of a low-degree graph. Random Structures and Algorithms 7(2), 157–165 (1995)

16. Knuth, D.: The Art of Computer Programming, 3rd edn. Seminumerical Algorithms, vol. 2. Addison-Wesley, Reading (1997)

17. Liebehenschel, J.: Ranking and unranking of a generalized Dyck language and the application to the generation of random trees. Séminaire Lotharingien de Combinatoire 43 (2000)

18. ISO/IEC 7812-1:2006. Identification cards – Identification of issuers – Part 1: Numbering system

19. Kelsen, P.: Ranking and unranking trees using regular reductions. In: Puech, C., Reischuk, R. (eds.) STACS 1996. LNCS, vol. 1046, pp. 581–592. Springer, Heidelberg (1996)

20. Liskov, M., Rivest, R., Wagner, D.: Tweakable block ciphers. In: Yung, M. (ed.) CRYPTO 2002. LNCS, vol. 2442, pp. 31–46. Springer, Heidelberg (2002)

21. Luby, M., Rackoff, C.: How to construct pseudorandom permutations from pseudorandom functions. SIAM Journal of Computing 17(2), 373–386 (1988)

22. Lucks, S.: Faster Luby-Rackoff ciphers. In: Gollmann, D. (ed.) FSE 1996. LNCS, vol. 1039, pp. 189–203. Springer, Heidelberg (1996)

23. Mäkinen, E.: Ranking and unranking left Szilard languages. Report A-1997-2, Department of Computer Science, University of Tampere (1997)

24. Maurer, U., Pietrzak, K.: The security of many-round Luby-Rackoff pseudo-random permutations. In: Biham, E. (ed.) EUROCRYPT 2003. LNCS, vol. 2656, pp. 544–561. Springer, Heidelberg (2003)

25. Morris, B., Rogaway, P., Stegers, T.: How to encipher messages on a small domain: deterministic encryption and the Thorp shuffle. In: Halevi, S. (ed.) CRYPTO 2009. LNCS, vol. 5677, pp. 286–302. Springer, Heidelberg (2009)

26. Naor, M., Reingold, O.: On the construction of pseudorandom permutations: Luby-Rackoff revisited. Journal of Cryptology 12(1), 29–66 (1999)
27. National Bureau of Standards. FIPS PUB 74. Guidelines for Implementing and Using the NBS Data Encryption Standard (April 1, 1981)
28. Patarin, J.: New results on pseudorandom permutation generators based on the DES Scheme. In: Feigenbaum, J. (ed.) CRYPTO 1991. LNCS, vol. 576, pp. 301–312. Springer, Heidelberg (1992)
29. Patarin, J.: Generic attacks on Feistel schemes. In: Boyd, C. (ed.) ASIACRYPT 2001. LNCS, vol. 2248, pp. 222–238. Springer, Heidelberg (2001)
30. Patarin, J.: Luby-Rackoff: 7 rounds are enough for $2^{n(1-\epsilon)}$ security. In: Boneh, D. (ed.) CRYPTO 2003. LNCS, vol. 2729, pp. 513–529. Springer, Heidelberg (2003)
31. Patarin, J.: Security of random Feistel schemes with 5 or more rounds. In: Franklin, M. (ed.) CRYPTO 2004. LNCS, vol. 3152, pp. 106–122. Springer, Heidelberg (2004)
32. Patarin, J., Nachef, V., Berbain, C.: Generic attacks on unbalanced Feistel schemes with contracting functions. In: Lai, X., Chen, K. (eds.) ASIACRYPT 2006. LNCS, vol. 4284, pp. 396–411. Springer, Heidelberg (2006)
33. Patel, S., Ramzan, Z., Sundaram, G.: Efficient constructions of variable-input-length block ciphers. In: Handschuh, H., Hasan, M.A. (eds.) SAC 2004. LNCS, vol. 3357, pp. 326–340. Springer, Heidelberg (2004)
34. PCI Security Standards Council. Payment Card Industry Data Security Standard Version 1.2,
https://www.pcisecuritystandards.org/security_standards/pci_dss.shtml
35. Petrank, E., Rackoff, C.: CBC MAC for real-time data sources. J. of Cryptology 13(3), 315–338 (2000)
36. Schneier, B., Kelsey, J.: Unbalanced Feistel networks and block cipher design. In: Gollmann, D. (ed.) FSE 1996. LNCS, vol. 1039, pp. 121–144. Springer, Heidelberg (1996)
37. Schroeppel, R.: Personal communication, approximately (2001)
38. Sipser, M.: Introduction to the Theory of Computation, 2nd edn. Thomson Press (2006)
39. Spies, T.: Format preserving encryption. Unpublished white paper, www.voltage.com Database and Network Journal (December 2008), Format preserving encryption: www.voltage.com
40. Spies, T.: Personal communications (February 2009)
41. Spies, T.: Feistel finite set encryption mode. Manuscript, posted on NIST's website on (February 6, 2008),
http://www.csrc.nist.gov/groups/ST/toolkit/BCM/documents/proposedmodes/ffsem/ffsem-spec.pdf
42. Valiant, L.: The complexity of computing the permanent. Theoretical Computer Science 8, 189–201 (1979)

BTM: A Single-Key, Inverse-Cipher-Free Mode for Deterministic Authenticated Encryption

Tetsu Iwata[1] and Kan Yasuda[2]

[1] Dept. of Computational Science and Engineering, Nagoya University, Japan
iwata@cse.nagoya-u.ac.jp
[2] NTT Information Sharing Platform Laboratories, NTT Corporation, Japan
yasuda.kan@lab.ntt.co.jp

Abstract. We present a new blockcipher mode of operation named BTM, which stands for Bivariate Tag Mixing. BTM falls into the category of Deterministic Authenticated Encryption, which we call DAE for short. BTM makes all-around improvements over the previous two DAE constructions, SIV (Eurocrypt 2006) and HBS (FSE 2009). Specifically, our BTM requires just one blockcipher key, whereas SIV requires two. Our BTM does not require the decryption algorithm of the underlying blockcipher, whereas HBS does. The BTM mode utilizes *bivariate* polynomial hashing for authentication, which enables us to handle vectorial inputs of dynamic dimensions. BTM then generates an initial value for its counter mode of encryption by *mixing* the resulting *tag* with one of the two variables (hash keys), which avoids the need for an implementation of the inverse cipher.

Keywords: Bivariate, universal hash function, counter mode, random-until-bad game, systematic proof.

1 Introduction

The modes of operation for blockciphers can be divided into the following three groups: encryption modes, message authentication codes, and authenticated encryption modes. The first group achieves privacy, the second ensures integrity, and the third does both at the same time. On one hand, the third group is attractive to users for its providing both kinds of security concurrently. On the other hand, the third group tends to employ rather complex mechanisms, since authenticated encryption modes are essentially a combination of an encryption mode and a message authentication code [2].

The complexity of authenticated encryption adds variety to the constructions, such as CCM [16], GCM [10] and OCB [14]. These constructions, however, have one thing in common—their security is based on the so-called nonce assumption. The assumption requires that nonce values be never repeated. Otherwise, the security of the overlying scheme is seriously compromised, which is a difficult aspect of the nonce-based constructions.

The problem of the nonce assumption was settled by the notion of Deterministic Authenticated Encryption (DAE), which Rogaway and Shrimpton introduced

M.J. Jacobson Jr., V. Rijmen, and R. Safavi-Naini (Eds.): SAC 2009, LNCS 5867, pp. 313–330, 2009.

at Eurocrypt 2006 [15]. DAE constructions are more robust than nonce-based ones. Namely, a DAE construction can be used as a nonce-based one by embedding a nonce value into part of its input, in which case the DAE construction achieves the same security level as a nonce-based one. Furthermore, a DAE construction maintains a certain level of security even when no nonce value is combined with. In such a scenario, an adversary can detect a repetition of exactly the same inputs being encrypted, but nothing more.

Though more robust, DAE modes are also more difficult to construct than nonce-based ones. To date, there have been two concrete DAE constructions, SIV [15] and HBS [7]. The SIV construction utilized blockcipher iteration both for its encryption algorithm and for authentication. SIV had a number of attractive features but had one major disadvantage that it required two blockcipher keys. This disadvantage was removed by the more recent single-key HBS construction. HBS also accelerated the speed by employing polynomial hashing, rather than blockcipher iteration, for its authentication algorithm.

Nevertheless, at the same time, the HBS construction sacrificed many of SIV's advantages in the interest of single-key usage and of polynomial-hashing design. In the following we point out those disadvantages which HBS suffered.

1. INVERSE-CIPHER REQUIREMENT. The decryption process of the HBS mode required the decryption algorithm of the underlying blockcipher. This requirement involved numerous drawbacks. First, HBS increased the size of its footprint (e.g., the number of gates or slices). Second, the security proof of HBS relied on the stronger SPRP (Strong Pseudo-Random Permutation) assumption about the underlying blockcipher. Third, the tag size of HBS was fixed to the full n-bit, disabling any kind of tag truncation for saving the bandwidth. These problems did not exist within the SIV construction, which worked without the inverse cipher.

2. WORSE SECURITY BOUND. The security bound of HBS was of the form $\ell^2 q^2/2^n$, where ℓ denotes the maximum length of each query and q the total number of queries. This should be contrasted with the security bound of SIV which was of the form $\sigma^2/2^n$, where σ denotes the maximum query complexity (i.e., the total length of all queries). The former becomes markedly worse than the latter when queries are of varying lengths.

3. INFLEXIBLE VECTOR DIMENSIONS. HBS needed to fix in advance the dimension of vectorial inputs (i.e., the maximum number of headers). Note that handling flexible vector dimensions is an important advantage, because some applications may be unable to set a limit reflecting the unpredictable nature of the dimension, or some applications may want to update and increase the limit on the dimension while maintaining backward compatibility. HBS suggested using the square-root operation $\sqrt{\ }$ as a remedy for dynamic changes in the dimension, but such a technique was more complex and less elegant than the vectorized-CMAC [6,13] solution offered by SIV.

4. ENLARGED COUNTER REGISTER. The HBS mode specified an unusual increment method $S \oplus \underline{1}_n, S \oplus \underline{2}_n, \ldots$, where S denotes an initial value, $\oplus$ the bitwise xor operation, and $\underline{a}_n$ an n-bit binary representation of an integer a.

Table 1. Comparison between SIV, HBS and BTM. The figures show the number of computations for a single header H and a message M, where $h = \lceil |H|/n \rceil$ and $m = \lceil |M|/n \rceil$.

	SIV	HBS	BTM
# of blockcipher keys	2	1	1
Inverse-cipher-free	yes	no	yes
Blockcipher assumption	PRP	SPRP	PRP
Tag truncation	possible	impossible	possible
Security bound	$O(\sigma^2/2^n)$	$O(\ell^2 q^2/2^n)$	$O(\sigma^2/2^n)$
Vector dimension	dynamic	static	dynamic
Counter register size	n-bit	$1.5n$-bit	n-bit
Total # of computations	$h + 2m + 2$	$h + 2m + 4$	$h + 2m + 2$
# of blockcipher calls	$h + 2m + 2$	$m + 2$	$m + 3$
# of multiplications	0	$h + m + 2$	$h + m - 1$

This method required a $1.5n$-bit register in order to maintain the current counter value, namely n bits for keeping the value S and $0.5n$ bits for incrementing a. On the other hand, SIV worked with almost any increment method, such as $S+1$, $S+2$, ... (arithmetic addition modulo 2^n) and $\underline{2}_n \cdot S$, $\underline{2}_n^2 \cdot S$, ... (doubling in the finite field of 2^n elements). These usual methods require only an n-bit register and maintain smaller sizes of footprints.

5. INCREASED NUMBER OF COMPUTATIONS. Although HBS accelerated the speed for large data by employing polynomial hashing, the total number of computations, which is the sum of the number of blockcipher calls plus that of finite-field multiplications, increased by 2. This caused the degradation in performance for short messages, depending on implementations.

The goal of the current work is to overstep this line of tradeoff and to construct a new single-key, polynomial-hashing DAE mode of operation which does not counterbalance any of the advantages that SIV had. For this, we propose a new mode of operation called BTM, which stands for Bivariate Tag Mixing. See Table 1 for summary. The BTM construction achieves our goal by utilizing the following two techniques.

1. BIVARIATE POLYNOMIAL HASHING. Our polynomial hashing utilizes two variables (keys) L and U, which are derived from a single blockcipher key K as $L := E_K(\underline{0}_n)$ and $U := E_K(\underline{1}_n)$. The bivariate hashing is capable of handling vectorial inputs of dynamic dimensions. The bivariate hashing also contributes to reducing the number of finite-field multiplications.

2. TAG MIXING. The tag value T is mixed with the hash key U via a special type of arithmetic addition, for which we write $T \boxplus U$. The mixed value is used as an initial value for the counter mode of encryption. In this way we avoid the use of the inverse cipher. In addition, this mixing allows us to use the increment $\boxplus \underline{1}_n$, saving the register size of the counter.

2   Preliminaries

Notation. We have already introduced the symbols $X \oplus Y$, a_n and $X \cdot Y$. The symbol $X \| Y$ denotes the concatenation of two strings X and Y. Given a string X, we write $|X|$ to represent its length in bits. If X is a finite set, then $|X|$ denotes its cardinality. We write $x \xleftarrow{\$} X$ for sampling an element from the set X uniformly at random and assigning its value to the variable x. Given a positive integer a and a string X such that $a \leq |X|$, $\mathrm{msb}(a, X)$ represents the leftmost a bits of the string X. The set $\{0,1\}^n$ of n-bit strings is regarded in multiple ways. It corresponds to the set $\{0, 1, \ldots, 2^n - 1\}$ of non-negative integers less than 2^n. It is also treated as the finite field $GF(2^n)$ of 2^n elements (with respect to some irreducible polynomial).

Blocks and Vectors. Throughout the paper we fix the *block size* n. Typical values of n are 64 and 128. The *block decomposition* $X[0] \cdots X[x-1]$ of a string $X \in \{0,1\}^*$ is computed as follows. If $X = \emptyset$ (the null string), then we set $x \leftarrow 1$ and $X[0] \leftarrow \emptyset$. Otherwise (i.e., when $|X| \geq 1$), we set $x \leftarrow \lceil |X|/n \rceil$, and blocks $X[0], \ldots, X[x-1]$ are defined as the unique set of strings satisfying the conditions $X[0] \| \cdots \| X[x-1] = X$, $|X[0]| = \cdots = |X[x-2]| = n$, and $1 \leq |X[x-1]| \leq n$. We call x the *block length* of the string X.

Given a string $X \in \{0,1\}^*$ such that $|X| \leq n$, we define

$$\pi(X) := \begin{cases} X \| 1 \| 0^{n-1-|X|} & \text{if } 0 \leq |X| \leq n-1, \\ X & \text{if } |X| = n, \end{cases}$$

so that we always have $|\pi(X)| = n$. Similarly, we define

$$\delta(X) := \begin{cases} \underline{1}_n & \text{if } 0 \leq |X| \leq n-1, \\ \underline{2}_n & \text{if } |X| = n. \end{cases}$$

We consider d-dimensional vectors $\overrightarrow{X} = (X_0, \ldots, X_{d-2}, X_{d-1})$ of strings $X_i \in \{0,1\}^*$ for $i = 0, \ldots, d-2, d-1$.

Headers and Messages. A plaintext (i.e., data which is input to the encryption algorithm) consists of header information $\overrightarrow{H} = (H_0, \ldots, H_{d-2})$ and a message M, being a d-dimensional vector $(\overrightarrow{H}, M) = (H_0, \ldots, H_{d-2}, M)$ of strings. In the paper it is understood that when $d = 1$ the notation $(\overrightarrow{H}, M)$ represents the 1-dimensional vector (M) (i.e., the case of no header information). Note that the message part M of a plaintext gets both authenticated and encrypted, while header information $\overrightarrow{H}$ gets authenticated but remains unencrypted.

Blockciphers and DAEs. A blockcipher is a family of permutations. We often write $E : \{0,1\}^k \times \{0,1\}^n \to \{0,1\}^n$ for a blockcipher, where $K \in \{0,1\}^k$ is a key and $E_K := E(K, \cdot) : \{0,1\}^n \to \{0,1\}^n$ is the permutation specified by the key K. An adversary A is an oracle machine that outputs a bit. The goal of a PRP(Pseudo-Random Permutation)-adversary A is to distinguish the blockcipher E_K (with a random key K) from a truly random permutation P : $\{0,1\}^n \to \{0,1\}^n$ [9,1]. The success probability of A is measured by

$$\mathbf{Adv}_E^{\mathrm{prp}}(A) := \Pr[A^{E_K(\cdot)} = 1] - \Pr[A^{P(\cdot)} = 1],$$

where in the first game A has access to the E_K oracle and in the second the P oracle. We fix a model of computation and a choice of encoding. We write $\mathbf{Adv}_E^{\mathrm{prp}}(t,\sigma) := \max_A \mathbf{Adv}_E^{\mathrm{prp}}(A)$, where max runs over adversaries A whose time complexity is at most t and whose query complexity is at most σ. The time complexity is the running time plus the code size. The query complexity is the total length in blocks of the queries made to the oracles.

A DAE scheme is a pair of algorithms $\mathcal{E}_K$ and $\mathcal{D}_K$. The encryption algorithm $\mathcal{E}_K$ takes a vectorial input $(\vec{H}, M)$ and outputs a pair of a tag T and a ciphertext C, where $|T| = n$ and $|C| = |M|$. The decryption $\mathcal{D}_K$ takes an input $(\vec{H}, T, C)$ and outputs either the corresponding plaintext M or a special symbol $\perp$. The goal of a DAE-adversary A is to distinguish between the pair $(\mathcal{E}_K, \mathcal{D}_K)$ and the pair $(\mathcal{R}, \perp)$, where $\mathcal{R}$ is an oracle that returns, upon a query $(\vec{H}, M)$, random strings of $n + |M|$ bits, and $\perp$ is an oracle that always returns the $\perp$ symbol. We define

$$\mathbf{Adv}_{\mathcal{E},\mathcal{D}}^{\mathrm{dae}}(A) := \Pr[A^{\mathcal{E}_K(\cdots),\mathcal{D}_K(\cdots)} = 1] - \Pr[A^{\mathcal{R}(\cdots),\perp(\cdots)} = 1],$$

where trivial queries are excluded. As before we also define $\mathbf{Adv}_{\mathcal{E},\mathcal{D}}^{\mathrm{dae}}(t,\sigma)$.

3 Bivariate Polynomials and L-Polynomials

A *bivariate polynomial* is a polynomial in two variables L and U over the field of 2^n elements, i.e., an element of $GF(2^n)[L,U]$. A function G of two arguments L and U is said to be an L-*polynomial* if the following conditions are satisfied.

1. $G(L,U)$ is a polynomial in the variable L. Let x be the degree of $G(L,U)$ as a polynomial in L.
2. We then have $x \geq 1$.
3. The coefficient of the leading term L^x is a polynomial function of U. Let y be the degree of this coefficient function as a polynomial in U.

We define $\deg_L G := x$ and $\deg_L^U G := y$. Observe that any non-constant bivariate polynomial is either an L-polynomial or a U-polynomial (or both). Also note that if G is a bivariate polynomial, then we have $\deg_L^U G \leq \deg_U G$. Now the following lemma is a basic result, and its proof can be found in Appendix A.

Lemma 1. *Let G be an L-polynomial. We have*

$$\Pr[G(L,U) = \underline{0}_n \mid L \xleftarrow{\$} \{0,1\}^n, U \xleftarrow{\$} \{0,1\}^n] \leq \frac{\deg_L^U G + \deg_L G}{2^n}.$$

4   Specification of BTM

In this section we give the specification of the BTM algorithms. First, we define the bivariate polynomial hashing $F_{L,U}$. Second, we describe the way of mixing the tag value with one of the hash keys. Finally, we give the definition of the BTM encryption and decryption algorithms.

4.1   Bivariate Hashing $F_{L,U}$

We begin with defining a polynomial f_L in one variable L. Given a string $X \in \{0,1\}^*$, define

$$f_L(X) := \delta\big(X[x-1]\big) \cdot \big(L^x \oplus L^{x-1} \cdot X[0] \oplus \cdots \oplus L \cdot X[x-2] \oplus \pi(X[x-1])\big),$$

where the addition and the multiplication are done in the finite field $GF(2^n)$ of 2^n elements, so that $f_L(X)$ is an element of $GF(2^n)[L]$. Recall that the polynomial $f_L(X)$ can be computed recursively, using the relation

$$f_L(X) = \delta\big(X[x-1]\big) \cdot \big((\cdots((L \oplus X[0]) \cdot L \oplus X[1])\cdots) \cdot L \oplus \pi(X[x-1])\big).$$

Note that the polynomial $f_L(X)$ always has a degree x due to the leading term L^x.

Now we define the polynomial $F_{L,U}$ in two variables L and U as

$$F_{L,U}\big(\vec{H}, M\big) := U^{d-1} \cdot f_L(H_0) \oplus \cdots \oplus U \cdot f_L(H_{d-2}) \oplus f_L(M),$$

which is an element of $GF(2^n)[L,U]$. Roughly speaking, we first hash each of $H_0, H_1, \ldots, H_{d-2}, M$ in terms of the variable L, which results in d-many hash values $f_L(H_0), \ldots, f_L(M)$, and then we hash these values in terms of the variable U. See Fig. 1 for an illustration.

Observe that $f_L(X)$ requires $x-1$ multiplications by L. Neglecting the xor operations and the last multiplication by $\underline{1}_n/\underline{2}_n$, the computation of $f_L(X)$ can

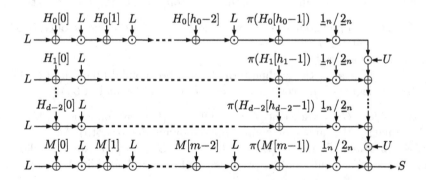

Fig. 1. Illustration of the bivariate hashing $F_{L,U}(H_0, H_1, \ldots, H_{d-2}, M)$. The symbol $X \odot Y$ denotes the multiplication $X \cdot Y$ in the field of 2^n elements. The multiplication by $\underline{1}_n/\underline{2}_n$ corresponds to the function $\delta\big(\pi(X[x-1])\big)$.

Algorithm $\text{BTM.Enc}_K(\vec{H}, M)$	**Subroutine** $\text{MAC}_K^{L,U}(\vec{H}, M)$
1. $L \leftarrow E_K(\underline{0}_n), U \leftarrow E_K(\underline{1}_n)$	1. $S \leftarrow F_{L,U}(\vec{H}, M)$
2. $T \leftarrow \text{MAC}_K^{L,U}(\vec{H}, M)$	2. $T \leftarrow E_K(S)$
3. $C \leftarrow \text{CTR}_K^{T \boxplus U}(M)$	3. **return** T
4. **return** (T, C)	

Algorithm $\text{BTM.Dec}_K(\vec{H}, (T, C))$	**Subroutine** $\text{CTR}_K^N(X)$
1. $L \leftarrow E_K(\underline{0}_n), U \leftarrow E_K(\underline{1}_n)$	1. $x \leftarrow \lceil |X|/n \rceil$
2. $M \leftarrow \text{CTR}_K^{T \boxplus U}(C)$	2. **for** $i \leftarrow 0$ **to** $x - 1$ **do**
3. $T' \leftarrow \text{MAC}_K^{L,U}(\vec{H}, M)$	3. $R[i] \leftarrow E_K(N \boxplus \underline{i}_n)$
4. **if** $T \neq T'$ **then**	4. **end for**
5. $M \leftarrow \bot$	5. $R \leftarrow R[0] \, \| \, \cdots \, \| \, R[x-1]$
6. **end if**	6. $Y \leftarrow X \oplus \text{msb}(|X|, R)$
7. **return** M	7. **return** Y

Fig. 2. Pseudocode of the BTM encryption and decryption algorithms. The subroutines MAC and CTR are extracted from the algorithms and shown on the right-hand side.

be done in about $x - 1$ finite-field multiplications. This means that for a two-dimensional vector (H, M) the computation of $F_{L,U}(H, M)$ can be done in about $(h - 1) + (m - 1) + 1 = h + m - 1$ finite-field multiplications (The last "+1" comes from the multiplication by U). This explains the figures in Table 1. More generally, for a d-dimensional vector $(\vec{H}, M)$, it takes about

$$(h_0 - 1) + (h_1 - 1) + \cdots + (h_{d-2} - 1) + (m - 1) + d - 1 = h_0 + h_1 + \cdots + h_{d-2} + m - 1$$

finite-field multiplications to compute the bivariate hashing $F_{L,U}(\vec{H}, M)$ (the value h_i being the block length of H_i and m that of M).

It can be directly verified that if $(\vec{H}, M) \neq (\vec{H}', M')$ are two distinct inputs, then we have an inequality of polynomials $F_{L,U}(\vec{H}, M) \neq F_{L,U}(\vec{H}', M')$. This fact plays an important role in our security analysis.

4.2   Tag Mixing $T \boxplus U$

The operation $\boxplus$ is defined as follows. For two strings $X, Y \in \{0,1\}^n$, divide them into two equal-length parts as $X = X_1 \| X_2$ and $Y = Y_1 \| Y_2$, so that $X_1, X_2, Y_1, Y_2 \in \{0,1\}^{n/2}$.[1] Then define

$$X \boxplus Y := (X_1 + Y_1) \| (X_2 + Y_2),$$

where the addition $+$ is done modulo $2^{n/2}$. We use $\boxplus$ rather than $+$ modulo 2^n, because $\boxplus$ is less costly and is sufficient for security up to the birthday bound.

---

[1] We assume that the block size n is an even number.

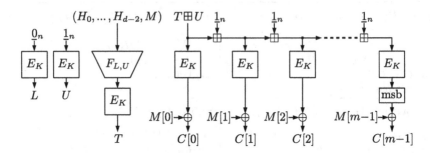

Fig. 3. Illustration of the BTM encryption algorithm

4.3 BTM Encryption and Decryption Algorithms

We are now ready to describe our BTM mode of operation. The encryption algorithm BTM.Enc and the decryption BTM.Dec are described in Fig. 2. See also Fig. 3 for a diagrammatic representation of the BTM encryption algorithm.

Note that when $|M| = 0$ (i.e., $M = \emptyset$ the null string), the algorithm BTM.Enc returns only the tag T. Also note that the one-dimensional input (M) (without a header) and the two-dimensional input (H, M) with $H = \emptyset$ generally result in different outputs for the same value of M.

5   Security Analysis of BTM

We prove the security of our BTM mode as a DAE construction. We first introduce a simple tool which makes our analysis easy. We then prove the privacy and the integrity of BTM. Thanks to the tool, our proofs are quite systematic, consisting of counting "bad" events and computing their probabilities.

5.1   A Simple Tool: Random-Until-Bad Games

We consider a special type of game called a "random-until-**bad**" game. This type of game can be systematically analyzed, making our security proofs simple and easy.

In a random-until-bad game, the adversary's goal is to set a **bad** flag written somewhere in the description of the overlying game. There may be multiple **bad** flags, as **bad**[0], **bad**[1], etc. The adversary wins the game as soon as one of the **bad** flags gets set. The only way for the adversary to set a **bad** flag is by making a query to its oracle. A query is processed according to the description of the game. If a query sets a **bad** flag, then the game terminates immediately. Otherwise, the oracle returns a truly random string of a specified length[2] to the adversary.

In a random-until-bad game, we only need to consider non-adaptive adversaries. Recall that oracles only return random strings to an adversary until the

---

[2] We consider the special symbol $\perp$ as a random string of length zero.

game terminates. This means that any adaptive adversary can be transformed into a non-adaptive one without changing its winning probability, by feeding a random tape to the adversary. More precisely, let A be an adaptive adversary which makes (exactly) q queries. Using A, we can construct a non-adaptive adversary B which has about the same running time as A and exactly the same winning probability, as follows: We let B run A by simulating A's oracles via B's internal random coins. The adversary B records A's queries $x_1, x_2, \ldots, x_q$. Then B outputs the sequence of queries $x_1, x_2, \ldots, x_q$. Note that at this point B has made no queries to B's oracles, and the values of the queries $x_1, x_2, \ldots, x_q$ have been already fixed. Hence we see that B is non-adaptive. It is also easy to see that B's winning probability is exactly the same as that of A.

Furthermore, we only need to consider deterministic adversaries. For this, we show that for any non-adaptive, probabilistic adversary there exists a non-adaptive, deterministic adversary having the same or better winning probability. So let A be a non-adaptive, probabilistic adversary making q queries to its oracles. For each sequence of queries $x_1, x_2, \ldots, x_q$ (the total length being no more than σ) the adversary A outputs this sequence with some probability. The winning probability of A is the weighted average (arithmetic mean) of the winning probabilities over all sequences $x_1, x_2, \ldots, x_q$. Then there exists a sequence $x_1^*, x_2^*, \ldots, x_q^*$ having the maximum winning probability. So let A^* be the adversary that always outputs the sequence $x_1^*, x_2^*, \ldots, x_q^*$. Then we see that the winning probability of A^* is no less than that of A.

In a random-until-bad game, we can systematically compute each probability that a **bad** flag gets set and then sum up the probabilities. This gives us the bound of the adversaries' winning probability.

5.2   From Computational to Information-Theoretic

The first step is to replace the blockcipher E_K (using a random key K) with a random permutation $P : \{0,1\}^n \rightarrow \{0,1\}^n$. We write BTM[$P$] for such a DAE scheme. Let A be a DAE-adversary whose time and query complexities are at most t and σ, respectively. We can directly construct an adversary B that uses A and tries to distinguish between the blockcipher E_K and the random permutation P. The simulation requires two calls to E for computing L and U, q calls to E for computing tags (together with necessary polynomial hashing), and σ calls to E for encrypting the messages. Any difference between $\mathbf{Adv}^{\mathrm{dae}}_{\mathrm{BTM}[E]}(A)$ and $\mathbf{Adv}^{\mathrm{dae}}_{\mathrm{BTM}[P]}(A)$ contributes to B's advantage $\mathbf{Adv}^{\mathrm{prp}}_{E}(B)$, so we have

$$\mathbf{Adv}^{\mathrm{dae}}_{\mathrm{BTM}[E]}(t, \sigma) \leq \mathbf{Adv}^{\mathrm{prp}}_{E}(t', 2 + q + \sigma) + \mathbf{Adv}^{\mathrm{dae}}_{\mathrm{BTM}[P]}(\sigma),$$

where the running time t' is about t plus the complexity to compute $2 + q + \sigma$ times the blockcipher E (and the complexity to perform corresponding polynomial hashing). Note that we have omitted the time complexity from the notation, since it becomes irrelevant to the context of BTM[P].

We then replace P with a random *function* $F : \{0,1\}^n \to \{0,1\}^n$ (not to be confused with $F_{L,U}$). Using the PRP/PRF switching lemma [3], we obtain

$$\mathbf{Adv}_{\text{BTM}[P]}^{\text{dae}}(\sigma) \le \binom{2+q+\sigma}{2} \cdot \frac{1}{2^n} + \mathbf{Adv}_{\text{BTM}[F]}^{\text{dae}}(\sigma).$$

Therefore, it amounts to evaluating the security of the scheme BTM$[F]$.

Now let $\mathcal{E}_F$ and $\mathcal{D}_F$ denote the encryption and decryption algorithms of the BTM$[F]$ scheme, respectively. For an adversary A we have

$$
\begin{aligned}
\mathbf{Adv}_{\text{BTM}[F]}^{\text{dae}}(A) &= \Pr\big[A^{\mathcal{E}_F(\cdots),\mathcal{D}_F(\cdots)} = 1\big] - \Pr\big[A^{\mathcal{R}(\cdots),\perp(\cdots)} = 1\big] \quad {}^{\bullet} \\
&= \Pr\big[A^{\mathcal{E}_F(\cdots),\mathcal{D}_F(\cdots)} = 1\big] - \Pr\big[A^{\mathcal{E}_F(\cdots),\perp(\cdots)} = 1\big] \quad \text{(integrity)} \\
&\quad + \Pr\big[A^{\mathcal{E}_F(\cdots),\perp(\cdots)} = 1\big] - \Pr\big[A^{\mathcal{R}(\cdots),\perp(\cdots)} = 1\big]. \quad \text{(privacy)}
\end{aligned}
$$

We shall evaluate the privacy first and then integrity.

5.3   Privacy Proof of BTM$[F]$

Theorem 1. *Let A be an adversary whose total query complexity is at most σ blocks. Then we have*

$$\Pr\big[A^{\mathcal{E}_F(\cdots),\perp(\cdots)} = 1\big] - \Pr\big[A^{\mathcal{R}(\cdots),\perp(\cdots)} = 1\big] \le \frac{8\sigma^2}{2^n}.$$

Proof. Consider the eight **bad** events listed in Table 2. These **bad** flags are placed in the description of the $\mathcal{E}_F$ oracle. Observe that these **bad** events cause the $\mathcal{E}_F$ oracle to return some non-random values by invoking the function F on some "old" inputs. In other words, as long as none of these **bad** flags gets set, the $\mathcal{E}_F$ oracle behaves exactly the same as the ideal $\mathcal{R}$ oracle, since the values returned by the $\mathcal{E}_F$ oracle are then outputs of the random function F on some fresh inputs.

Therefore, by the fundamental lemma of game playing [3], we get

$$\Pr\big[A^{\mathcal{E}_F(\cdots),\perp(\cdots)} = 1\big] - \Pr\big[A^{\mathcal{R}(\cdots),\perp(\cdots)} = 1\big] \le \Pr\big[A^{\mathcal{E}_F(\cdots),\perp(\cdots)} \text{ sets } \mathbf{bad}\big],$$

and now the game under consideration is random-until-**bad**. Hence, we can systematically compute the winning probability of A. We simply sum up the probabilities in Table 2 as

$$\Pr\big[A^{\mathcal{E}_F(\cdots),\perp(\cdots)} \text{ sets } \mathbf{bad}\big] \le \frac{\sigma + \cdots + 2(q-1)(\sigma-q)}{2^n} \le \frac{8\sigma^2}{2^n},$$

which gives us the desired bound. $\qquad\square$

5.4   Integrity Proof of BTM$[F]$

Theorem 2. *Let A be an adversary whose total query complexity is at most σ blocks. Then we have*

$$\Pr\big[A^{\mathcal{E}_F(\cdots),\mathcal{D}_F(\cdots)} = 1\big] - \Pr\big[A^{\mathcal{E}_F(\cdots),\perp(\cdots)} = 1\big] \le \frac{11\sigma^2}{2^n}.$$

Table 2. The **bad** events in the privacy game. The superscript (i) means that the variable comes from the i-th query. The adversary makes q queries. The indices run over $1 \leq i \leq q$, $1 \leq j \leq i-1$, $0 \leq \alpha \leq m^{(i)} - 1$ and $0 \leq \beta \leq m^{(j)} - 1$, where $m^{(i)}$ is the length in blocks of the queried message $M^{(i)}$. The computations of the probabilities can be found in Appendix B.

flag	event	type	probability
bad[0]	$S^{(i)} = \underline{0}_n$		$\sigma/2^n$
bad[1]	$S^{(i)} = \underline{1}_n$		$\sigma/2^n$
bad[2]	$S^{(i)} = S^{(j)}$	bivariate polynomial	$(q-1)\sigma/2^n$
bad[3]	$S^{(i)} = T^{(j)} \boxplus U \boxplus \underline{\beta}_n$	L-polynomial	$\sigma^2/2^n$
bad[4]	$T^{(i)} \boxplus U \boxplus \underline{\alpha}_n = \underline{0}_n$		$\sigma/2^n$
bad[5]	$T^{(i)} \boxplus U \boxplus \underline{\alpha}_n = \underline{1}_n$		$\sigma/2^n$
bad[6]	$T^{(i)} \boxplus U \boxplus \underline{\alpha}_n = S^{(j)}$	L-polynomial	$\sigma^2/2^n$
bad[7]	$T^{(i)} \boxplus U \boxplus \underline{\alpha}_n = T^{(j)} \boxplus U \boxplus \underline{\beta}_n$		$2(q-1)(\sigma-q)/2^n$

Proof. The two games are identical unless the $\mathcal{D}_F$ oracle returns something other than $\perp$. Therefore, by the fundamental lemma of game playing, we have

$$\Pr\big[A^{\mathcal{E}_F(\cdots),\mathcal{D}_F(\cdots)} = 1\big] - \Pr\big[A^{\mathcal{E}_F(\cdots),\perp(\cdots)} = 1\big] \leq \Pr\big[A^{\mathcal{E}_F(\cdots),\mathcal{D}_F(\cdots)} \text{ forges}\big].$$

Using the adversary A, we shall construct a new adversary B that finds a forgery of the message authentication code $\mathcal{G}_F : (\vec{H}, M) \mapsto T$. We let B gain access to an auxiliary oracle $\mathcal{O}_F$, which returns the value $F(U \boxplus W)$ upon a query $W \in \{0,1\}^n$. Here recall that $U := F(\underline{0}_n)$. We let B simulate oracles for A in the natural way. This yields

$$\Pr\big[A^{\mathcal{E}_F(\cdots),\mathcal{D}_F(\cdots)} \text{ forges}\big] \leq \Pr\big[B^{\mathcal{G}_F(\cdots),\mathcal{V}_F(\cdots),\mathcal{O}_F(\cdot)} \text{ forges}\big],$$

where $\mathcal{V}_F$ is the verification oracle of the message authentication code $\mathcal{G}_F$. We note that B makes at most $\sigma - q$ queries to the $\mathcal{O}_F$ oracle.

Now we introduce a random oracle $\mathcal{R}_n^n : \{0,1\}^n \to \{0,1\}^n$ (independent of F) and replace $\mathcal{O}_F$ with the ideal $\mathcal{R}_n^n$. Then we have

$$\Pr\big[B^{\mathcal{G}_F(\cdots),\mathcal{V}_F(\cdots),\mathcal{O}_F(\cdot)} \text{ forges}\big]$$
$$\leq \Pr\big[B^{\mathcal{G}_F(\cdots),\mathcal{V}_F(\cdots),\mathcal{O}_F(\cdot)} \text{ forges}\big] - \Pr\big[B^{\mathcal{G}_F(\cdots),\mathcal{V}_F(\cdots),\mathcal{R}_n^n(\cdot)} \text{ forges}\big] \quad (1)$$
$$+ \Pr\big[\tilde{B}^{\mathcal{G}_F(\cdots),\mathcal{V}_F(\cdots)} \text{ forges}\big], \quad (2)$$

where $\tilde{B}$ is the adversary that runs B by simulating the $\mathcal{R}_n^n$ oracle using $\tilde{B}$'s internal random coins.

First we evaluate the quantity (1). The two games proceed exactly the same as long as the oracle $\mathcal{O}_F$ returns only random strings. Consider the **bad** flags listed in Table 3. The flags **bad**[8 − 11] are placed in the description of the $\mathcal{O}_F$ oracle, with the hash values $S^{(i)}$ being recorded upon queries to the $\mathcal{G}_F$ and $\mathcal{V}_F$

Table 3. The **bad** events in the integrity game. The index runs over $1 \le r \le \sigma - q$. The computations of the probabilities can be found in Appendix C.

flag	event	type	probability
bad[8]	$W^{(r)} \boxplus U = \underline{0}_n$		$(\sigma - q)/2^n$
bad[9]	$W^{(r)} \boxplus U = \underline{1}_n$		$(\sigma - q)/2^n$
bad[10]	$W^{(r)} \boxplus U = S^{(j)}$	L-polynomial	$(\sigma - q)(\sigma + q)/2^n$
bad[11]	$S^{(i)} = W^{(r)} \boxplus U$	L-polynomial	$(\sigma - q)(\sigma + q)/2^n$
bad[12]	$S^{(i)} = \underline{0}_n$		$\sigma/2^n$
bad[13]	$S^{(i)} = \underline{1}_n$		$\sigma/2^n$
bad[14]	$S^{(i)} = S^{(j)}$	bivariate polynomial	$(q-1)\sigma/2^n$
bad[15]	$\mathcal{V}_F(\cdots) \ne \perp$		$q/2^n$

oracles. The $\mathcal{O}_F$ oracle behaves just like the ideal $\mathcal{R}_n^n$ unless one of these **bad** flags gets set. Therefore, by the fundamental lemma of game playing, we get

$$(1) \le \Pr\left[B^{\mathcal{G}_F(\cdots), \mathcal{V}_F(\cdots), \mathcal{O}_F(\cdot)} \text{ sets } \mathbf{bad}[8-11]\right].$$

Using the adversary B, we shall construct a new adversary C that has access only to the $\mathcal{G}_F$ and $\mathcal{O}_F$ oracles. The adversary C simulates the $\mathcal{V}_F$ oracle in the natural way using its $\mathcal{G}_F$ oracle. We add more flags **bad**[12 − 14], listed in Table 3, to the description of the $\mathcal{G}_F$ oracle. Clearly we have

$$\Pr\left[B^{\mathcal{G}_F(\cdots), \mathcal{V}_F(\cdots), \mathcal{O}_F(\cdot)} \text{ sets } \mathbf{bad}[8-11]\right] \le \Pr\left[C^{\mathcal{G}_F(\cdots), \mathcal{O}_F(\cdot)} \text{ sets } \mathbf{bad}[8-14]\right],$$

and we see that C plays a random-until-**bad** game. Therefore, we have

$$\Pr\left[C^{\mathcal{G}_F(\cdots), \mathcal{O}_F(\cdot)} \text{ sets } \mathbf{bad}[8-14]\right] \le \frac{\sigma + \cdots + \sigma}{2^n} \le \frac{6\sigma^2}{2^n},$$

rather than $7\sigma^2/2^n$, since one of the σ's gets cancelled out by $(q-1)\sigma = q\sigma - \sigma$ in **bad**[14].

It remains to evaluate the quantity (2). For this, we introduce a random oracle $\mathcal{R}_n$ which returns an n-bit random string upon a query $(\vec{H}, M)$. We replace the $\mathcal{G}_F$ oracle with the ideal $\mathcal{R}_n$, as

$$\Pr\left[\tilde{B}^{\mathcal{G}_F(\cdots), \mathcal{V}_F(\cdots)} \text{ forges}\right]$$
$$= \Pr\left[\tilde{B}^{\mathcal{G}_F(\cdots), \mathcal{V}_F(\cdots)} \text{ forges}\right] - \Pr\left[\tilde{B}^{\mathcal{R}_n(\cdots), \mathcal{V}_F(\cdots)} \text{ forges}\right] \quad (3)$$
$$+ \Pr\left[\tilde{B}^{\mathcal{R}_n(\cdots), \mathcal{V}_F(\cdots)} \text{ forges}\right]. \quad (4)$$

By the fundamental lemma of game playing, we see that

$$(3) \le \Pr\left[\tilde{B}^{\mathcal{G}_F(\cdots), \mathcal{V}_F(\cdots)} \text{ sets } \mathbf{bad}[12-14]\right],$$

where the **bad** flags are placed across the two oracles $\mathcal{G}_F$ and $\mathcal{V}_F$. We obtain the following random-until-**bad** games.

$$(3) \leq \Pr\left[\tilde{B}^{\mathcal{G}_F(\cdots),\mathcal{V}_F(\cdots)} \text{ sets } \mathbf{bad}[12-15]\right] \leq \frac{4\sigma^2}{2^n}, \text{ and}$$

$$(4) = \Pr\left[\tilde{B}^{\mathcal{G}_F(\cdots),\mathcal{V}_F(\cdots)} \text{ sets } \mathbf{bad}[15]\right] \leq \frac{\sigma^2}{2^n},$$

which gives us $(6+4+1)\sigma^2/2^n = 11\sigma^2/2^n$ as desired. $\qquad\qquad\square$

6   Alternative Way of Tag Mixing

Here we mention a variant of BTM. We could use $T \oplus U$ in place of $T \boxplus U$. Then the counter increment is done via the multiplication by $\underline{2}_n$ in the finite field of 2^n elements. This variant has both advantages and disadvantages. After careful consideration, we have decided to choose $T \boxplus U$.

The $T \oplus U$ method does not require the arithmetic addition, which somewhat reduces the size of hardware footprint. Moreover, there exist quite efficient hardware implementations of the multiplication by $\underline{2}_n$. On the other hand, however, the software implementations of the multiplication by $\underline{2}_n$ become a bit costly, depending on the block size n and on the available word size(s) of the platform.

We have observed that the software inefficiency of the $T \oplus U$ method appears to be a little high price to pay for the hardware efficiency. There also exist fairly efficient hardware implementations of the $\boxplus$ operation, and the $\boxplus \underline{1}_n$ increment gains much better software performance on most of the platforms.

7   Improving Security via Tweakable Blockciphers

BTM gives excellent performance but provides security only up to the standard birthday bound. Here we consider the problem of constructing DAE whose security is *beyond* the birthday bound (BBB). The problem was addressed in [7], and BBB constructions are of particular interest if we consider *key wrap* [12], an important application of DAE. With a key-wrap algorithm, one encrypts and authenticates specialized data such as cryptographic keys, where one might desire to ensure the highest security possible. BBB constructions can be used also when one prefers to use a 64-bit blockcipher as the underlying primitive and at the same time ensure security better than the $O(2^{32})$ birthday bound.

The BBB construction described in [7] requires about twelve blockcipher calls to encrypt two blocks of a message, which is hopelessly inefficient and impractical. Here we present a BBB construction that uses a *tweakable* blockcipher [8] as the underlying primitive, instead of using an ordinary blockcipher. Our construction is somewhat more efficient than the one in [7] in a situation where one can start with such a tweakable blockcipher.

Let $\tilde{E}_K : \mathcal{T} \times \{0,1\}^n \to \{0,1\}^n$ be a tweakable blockcipher, where $\mathcal{T} = \{0,1\}^n$ is a tweak space. First construct a $2n$-to-$2n$-bit blockcipher E'_{K_1,K_2,K_3} :

$\{0, 1\}^{2n} \to \{0, 1\}^{2n}$, by using the sENR (simplified Extended Naor-Reingold) construction [11], where K_1 and K_2 are the keys for the underlying tweakable blockcipher, and $K_3 \in \{0, 1\}^n$. Recall that one call to E' requires one multiplication over $GF(2^n)$ and two calls to $\tilde{E}$. Then construct the BTM mode having a block size of $2n$ bits, using E' as its underlying blockcipher. This construction ensures security beyond the $O(2^{n/2})$ bound. The construction is not too inefficient; to encrypt two blocks of a message, it requires about one multiplication over $GF(2^{2n})$, one multiplication over $GF(2^n)$, and two calls to $\tilde{E}$.

Unfortunately, the construction still has the following problems.

1. The key length is more than n bits; the key space of E' is rather large.
2. The tag size is $2n$ bits instead of n bits; the ciphertext is somewhat long.

It remains open to provide a BBB construction (based on a tweakable blockcipher) which resolves the two problems. Also, our unorthodox method involves using a tweakable blockcipher as the underlying primitive, which itself must have BBB security. This implies that the standard construction of a tweakable blockcipher in [8] is not suitable for our purpose. Although the constructions in [5,4,11] stand as potential candidates for our $\tilde{E}$, there is no known construction that completely fulfills our requirements. The basic problem of designing an efficient tweakable blockcipher with BBB security, possibly from scratch, still remains to be solved.

Acknowledgments

The authors would like to express their thanks to the anonymous reviewers of SAC 2009 for their helpful comments.

References

1. Bellare, M., Kilian, J., Rogaway, P.: The security of the cipher block chaining message authentication code. J. Comput. Syst. Sci. 61(3), 362–399 (2000)
2. Bellare, M., Namprempre, C.: Authenticated encryption: Relations among notions and analysis of the generic composition paradigm. In: Okamoto, T. (ed.) ASIACRYPT 2000. LNCS, vol. 1976, pp. 531–545. Springer, Heidelberg (2000)
3. Bellare, M., Rogaway, P.: The security of triple encryption and a framework for code-based game-playing proofs. In: Vaudenay, S. (ed.) EUROCRYPT 2006. LNCS, vol. 4004, pp. 409–426. Springer, Heidelberg (2006)
4. Ferguson, N., Lucks, S., Schneier, B., Whiting, D., Bellare, M., Kohno, T., Callas, J., Walker, J.: The Skein hash function family. Submission to NIST (2008), http://csrc.nist.gov/groups/ST/hash/sha-3/Round1/submissions_rnd1.html
5. Goldenberg, D., Hohenberger, S., Liskov, M., Schwartz, E.C., Seyalioglu, H.: On tweaking Luby-Rackoff blockciphers. In: Kurosawa, K. (ed.) ASIACRYPT 2007. LNCS, vol. 4833, pp. 342–356. Springer, Heidelberg (2007)
6. Iwata, T., Kurosawa, K.: OMAC: One-key CBC MAC. In: Johansson, T. (ed.) FSE 2003. LNCS, vol. 2887, pp. 129–153. Springer, Heidelberg (2003)

7. Iwata, T., Yasuda, K.: HBS: A single-key mode of operation for deterministic authenticated encryption. In: Dunkelman, O. (ed.) FSE 2009. LNCS, vol. 5665, pp. 394–415. Springer, Heidelberg (2009)
8. Liskov, M., Rivest, R.L., Wagner, D.: Tweakable block ciphers. In: Yung, M. (ed.) CRYPTO 2002. LNCS, vol. 2442, pp. 31–46. Springer, Heidelberg (2002)
9. Luby, M., Rackoff, C.: How to construct pseudorandom permutations from pseudorandom functions. SIAM J. Comput. 17(2), 373–386 (1988)
10. McGrew, D.A., Viega, J.: The security and performance of the Galois/counter mode (GCM) of operation. In: Canteaut, A., Viswanathan, K. (eds.) INDOCRYPT 2004. LNCS, vol. 3348, pp. 343–355. Springer, Heidelberg (2004)
11. Minematsu, K.: Beyond-birthday-bound security based on tweakable block cipher. In: Dunkelman, O. (ed.) FSE 2009. LNCS, vol. 5665, pp. 308–326. Springer, Heidelberg (2009)
12. NIST: AES key wrap specification (2001)
13. NIST: Recommendation for block cipher modes of operation: The CMAC mode for authentication (2005)
14. Rogaway, P., Bellare, M., Black, J., Krovetz, T.: OCB: A block-cipher mode of operation for efficient authenticated encryption. In: ACM CCS, pp. 196–205. ACM Press, New York (2001)
15. Rogaway, P., Shrimpton, T.: A provable-security treatment of the key-wrap problem. In: Vaudenay, S. (ed.) EUROCRYPT 2006. LNCS, vol. 4004, pp. 373–390. Springer, Heidelberg (2006)
16. Whiting, D., Housley, R., Ferguson, N.: Counter with CBC-MAC (CCM). Submission to NIST (2002), http://csrc.nist.gov/groups/ST/toolkit/BCM/index.html

A   Proof of Lemma 1

We have

$$\Pr\big[G(L,U) = \underline{0}_n \mid L, U \xleftarrow{\$} \{0,1\}^n\big]$$
$$= \sum_{U_0 \in \{0,1\}^n} \Pr\big[U = U_0 \wedge G(L,U) = \underline{0}_n \mid L, U \xleftarrow{\$} \{0,1\}^n\big]$$
$$= \sum_{U_0 \in \{0,1\}^n} \Pr\big[U = U_0 \mid U \xleftarrow{\$} \{0,1\}^n\big] \cdot \Pr\big[G(L,U_0) = \underline{0}_n \mid L \xleftarrow{\$} \{0,1\}^n\big].$$

Now put $x := \deg_L G$ and let $Z \subset \{0,1\}^n$ be the set of U_0 which makes the coefficient of L^x zero. Note that we have $|Z| \leq \deg_L^U G$. We get

$$\sum_{U_0 \in Z} \Pr\big[U = U_0 \mid U \xleftarrow{\$} \{0,1\}^n\big] \cdot \Pr\big[G(L,U_0) = \underline{0}_n \mid L \xleftarrow{\$} \{0,1\}^n\big]$$
$$+ \sum_{U_0 \notin Z} \Pr\big[U = U_0 \mid U \xleftarrow{\$} \{0,1\}^n\big] \cdot \Pr\big[G(L,U_0) = \underline{0}_n \mid L \xleftarrow{\$} \{0,1\}^n\big]$$
$$\leq |Z| \cdot \frac{1}{2^n} \cdot 1 + \sum_{U_0 \notin Z} \frac{1}{2^n} \cdot \frac{\deg_L G}{2^n}$$
$$\leq \frac{\deg_L^U G + \deg_L G}{2^n},$$

as desired.

B   Computing the Probabilities in Table 2

In the following computations we introduce a variable

$$\mu := \max\{h_0, \ldots, h_{d-2}, m\}.$$

In other words, μ denotes the maximum length of a component in the vector $(\overrightarrow{H}, M)$. We also introduce a new variable

$$\lambda := h_0 + \cdots h_{d-2} + m.$$

The superscript (i) denotes the fact that the variable comes from the i-th query.

We compute the probabilities one by one. We start with **bad**[0]. This event corresponds to outputting the hash key L. From Lemma 1 we have

$$\Pr[\mathbf{bad}[0]] \leq \sum_{i=1}^{q} \Pr\left[F_{L,U}\left(H_0^{(i)}, \ldots, H_{d^{(i)}-2}^{(i)}, M^{(i)}\right) = \underline{0}_n\right]$$

$$\leq \sum_{i=1}^{q} \frac{d^{(i)} - 1 + \mu^{(i)}}{2^n} \leq \sum_{i=1}^{q} \frac{\lambda^{(i)}}{2^n},$$

which is then bounded by $\sigma/2^n$.

The probability of **bad**[1] can be done in the exactly same manner, as this corresponds to outputting the other hash key U. We have $\Pr[\mathbf{bad}[1]] = \Pr[\mathbf{bad}[0]] \leq \sigma/2^n$.

The event **bad**[2] means a collision among the hash values. Using Lemma 1, we compute

$$\Pr[\mathbf{bad}[2]] \leq \sum_{i=2}^{q} \sum_{j=1}^{i-1} \Pr\left[F_{L,U}\left(\overrightarrow{H}^{(i)}, M^{(i)}\right) = F_{L,U}\left(\overrightarrow{H}^{(j)}, M^{(j)}\right)\right]$$

$$\leq \sum_{i=2}^{q} \sum_{j=1}^{i-1} \frac{\max\{d^{(i)} - 1, d^{(j)} - 1\} + \max\{\mu^{(i)}, \mu^{(j)}\}}{2^n}$$

$$\leq (q-1) \sum_{i=1}^{q} \frac{d^{(i)} - 1 + \mu^{(i)}}{2^n} \leq (q-1) \sum_{i=1}^{q} \frac{\lambda^{(i)}}{2^n},$$

which is then bounded by $(q-1)\sigma/2^n$.

We proceed to **bad**[3]. We use Lemma 1 to get

$$\Pr[\mathbf{bad}[3]] \leq \sum_{i=2}^{q} \sum_{j=1}^{i-1} \sum_{\beta=0}^{m^{(j)}-1} \Pr\left[F_{L,U}\left(H_0^{(i)}, \ldots, H_{d^{(i)}-2}^{(i)}, M^{(i)}\right) = T^{(j)} \boxplus U \boxplus \underline{\beta}_n\right]$$

$$\leq \sum_{i=2}^{q} \sum_{j=1}^{i-1} \sum_{\beta=0}^{m^{(j)}-1} \frac{d^{(i)} - 1 + \mu^{(i)}}{2^n}$$

$$\leq \sum_{i=2}^{q} \sum_{j=1}^{i-1} \sum_{\beta=0}^{m^{(j)}-1} \frac{\lambda^{(i)}}{2^n} = \sum_{i=2}^{q} \frac{\lambda^{(i)}}{2^n} \sum_{j=1}^{i-1} \sum_{\beta=0}^{m^{(j)}-1} 1 = \sum_{i=2}^{q} \frac{\lambda^{(i)}}{2^n} \sum_{j=1}^{i-1} m^{(j)},$$

which we can bound as $\sigma/2^n \cdot \sigma = \sigma^2/2^n$.

The quantity $\Pr[\mathbf{bad}[4]]$ can be evaluated relatively easily. We get

$$\Pr[\mathbf{bad}[4]] \leq \sum_{i=2}^{q} \sum_{\alpha=0}^{m^{(i)}-1} \Pr[T^{(i)} \boxplus U \boxplus \underline{\alpha}_n = \underline{0}_n] \leq \sum_{i=2}^{q} \frac{m^{(i)}}{2^n},$$

which must be less than $\sigma/2^n$.

The event $\Pr[\mathbf{bad}[5]]$ can be treated in a way similar to $\Pr[\mathbf{bad}[4]]$. We have $\Pr[\mathbf{bad}[5]] = \Pr[\mathbf{bad}[4]] \leq \sigma/2^n$.

Also, the quantity $\Pr[\mathbf{bad}[6]]$ is exactly the same as $\Pr[\mathbf{bad}[3]]$. We have $\Pr[\mathbf{bad}[6]] = \Pr[\mathbf{bad}[3]] \leq \sigma^2/2^n$.

Lastly, we go on to treat the event $\mathbf{bad}[7]$. We have

$$\Pr[\mathbf{bad}[7]] \leq \sum_{i=2}^{q} \sum_{j=1}^{i-1} \Pr\left[\bigvee_{\alpha=0}^{m^{(i)}-1} \bigvee_{\beta=0}^{m^{(j)}-1} (T^{(i)} \boxplus U \boxplus \underline{\alpha}_n = T^{(j)} \boxplus U \boxplus \underline{\beta}_n) \right]$$

$$\leq \sum_{i=2}^{q} \sum_{j=1}^{i-1} \Pr\left[\bigvee_{\gamma=-m^{(i)}+1}^{m^{(j)}-1} (T^{(i)} = T^{(j)} \boxplus \underline{\gamma}_n) \right]$$

$$\leq \sum_{i=2}^{q} \sum_{j=1}^{i-1} \frac{m^{(i)} + m^{(j)} - 2}{2^n}$$

$$\leq (q-1) \sum_{i=1}^{q} \frac{2m^{(i)} - 2}{2^n},$$

which is less than $2(q-1)(\sigma-q)/2^n$. This concludes the computation of the probabilities in Table 2.

C   Computing the Probabilities in Table 3

Again we use the variable

$$\mu := \max\{h_0, \ldots, h_{d-2}, m\}.$$

As usual, the superscript (i) denotes the fact that the variable comes from the i-th query that the adversary makes.

We begin with $\Pr[\mathbf{bad}[8]]$. We get

$$\Pr[\mathbf{bad}[8]] \leq \sum_{r=1}^{\sigma-q} \Pr[W^{(r)} \boxplus U = \underline{0}_n] \leq \sum_{r=1}^{\sigma-q} \frac{1}{2^n},$$

which must be bounded by $(\sigma - q)/2^n$.

The event $\mathbf{bad}[9]$ can be treated in a similar way. We obtain $\Pr[\mathbf{bad}[9]] = \Pr[\mathbf{bad}[8]] \leq (\sigma - q)/2^n$.

Next we evaluate the probability $\Pr[\mathbf{bad}[10]]$. Using Lemma 1, we compute as

$$\Pr[\mathbf{bad}[10]] \leq \sum_{r=1}^{\sigma-q} \sum_{j=1}^{q} \Pr\left[W^{(r)} \boxplus U = F_{L,U}\left(H_0^{(j)}, \ldots, H_{d^{(j)}-2}^{(j)}, M^{(j)}\right)\right]$$

$$\leq \sum_{r=1}^{\sigma-q} \sum_{j=1}^{q} \frac{\max\{d^{(j)}-1, 1\} + \mu^{(j)}}{2^n},$$

which can be bounded by $(\sigma - q)(\sigma + q)/2^n$.

The probability $\Pr[\mathbf{bad}[11]]$ is exactly the same as $\Pr[\mathbf{bad}[10]]$. We have $\Pr[\mathbf{bad}[11]] = \Pr[\mathbf{bad}[10]] \leq (\sigma - q)(\sigma + q)/2^n$.

The events $\mathbf{bad}[12]$, $\mathbf{bad}[13]$ and $\mathbf{bad}[14]$ can be treated in ways similar to $\mathbf{bad}[0]$, $\mathbf{bad}[1]$ and $\mathbf{bad}[2]$, respectively, whose probabilities have been already computed in Appendix B. We get

$$\Pr[\mathbf{bad}[12]] = \Pr[\mathbf{bad}[0]] \leq \frac{\sigma}{2^n},$$

$$\Pr[\mathbf{bad}[13]] = \Pr[\mathbf{bad}[1]] \leq \frac{\sigma}{2^n},$$

$$\Pr[\mathbf{bad}[14]] = \Pr[\mathbf{bad}[2]] \leq \frac{(q-1)\sigma}{2^n},$$

as expected.

Lastly, we handle $\mathbf{bad}[15]$. Observe that this event is nothing but a forgery by making random guesses. So we obtain

$$\Pr[\mathbf{bad}[15]] \leq \sum_{i=1}^{q} \Pr[\mathcal{V}_F(\cdots) \neq \bot] \leq \frac{q}{2^n}.$$

Thus we have completed computing the probabilities in Table 3.

On Repeated Squarings in Binary Fields

Kimmo U. Järvinen

Helsinki University of Technology (TKK)
Department of Information and Computer Science
P.O. Box 5400, FI-02015 TKK, Finland
kimmo.jarvinen@tkk.fi

Abstract. In this paper, we discuss the problem of computing repeated squarings (exponentiations to a power of 2) in finite fields with polynomial basis. Repeated squarings have importance, especially, in elliptic curve cryptography where they are used in computing inversions in the field and scalar multiplications on Koblitz curves. We explore the problem specifically from the perspective of efficient implementation using field-programmable gate arrays (FPGAs) where the look-up table (LUT) structure helps to reduce both area and delay overheads. In fact, we show that the optimum construction depends on the size of the LUTs. We propose several repeated squarer architectures and demonstrate their practicability for FPGA-based implementations. Finally, we show that the proposed repeated squarers can offer significant speedups and even improve resistivity against side-channel attacks.

1 Introduction

Squaring is the operation where an element is multiplied by itself. We explore the problem of computing *repeated squarings*, i.e. several successive squarings, in finite binary fields, $\mathbb{F}_{2^m}$, with polynomial basis and present hardware solutions for it. In this paper, we consider repeated squarings mainly in the context of elliptic curve cryptography [1,2], but generalizations to other application domains are straightforward.

Repeated squarings have two important applications in elliptic curve cryptography:

1. *Itoh-Tsujii inversion* [3] is a method based on Fermat's Little Theorem that finds the multiplicative inverse of $a \in \mathbb{F}_{2^m}$ by efficiently computing $a^{2^m - 2}$. In this exponentiation the number of squarings is considerably higher than the number of multiplications and, hence, it includes many repeated squarings. Originally, Itoh-Tsujii inversion was proposed for binary fields over normal basis where (repeated) squarings are trivial, but it is a viable solution also for polynomial basis [4].

2. *Koblitz curves* [5] are a class of elliptic curves over $\mathbb{F}_{2^m}$, where fast Frobenius maps can be used instead of computationally more demanding point doublings resulting in considerably faster computations. Frobenius is computed by squaring the coordinates of a point on the curve; thus, successive Frobenius maps require a repeated squaring for each coordinate.

M.J. Jacobson Jr., V. Rijmen, and R. Safavi-Naini (Eds.): SAC 2009, LNCS 5867, pp. 331–349, 2009.
© Springer-Verlag Berlin Heidelberg 2009

Recent studies show that when the above cases are implemented in hardware, a considerable portion of the total computation time is consumed in squarings [6,7]; hence, they motivate the research on efficient computation of repeated squarings.

Field-programmable gate arrays (FPGAs) are popular implementation platforms for cryptographic algorithms because their combination of speed and programmability offers many advantages over general-purpose processors and application specific circuits [8]. The basic building blocks of an FPGA are n-to-1 bit *look-up tables* (n-LUTs). The most typical LUT size is $n = 4$, but larger n are used in some contemporary FPGA architectures. Despite the vast amount of papers describing FPGA-based cryptographic implementations, the effects of LUT size have been studied surprisingly little. Papers introducing finite field arithmetic units, e.g. multipliers, typically consider only 2-to-1 bit gates. In this paper, we demonstrate how optimizations for a specific LUT structure can considerably improve finite field arithmetic units.

There are only few studies on hardware implementation of repeated squarings. Lutz and Hasan [9] showed how repeated squarings can be accelerated by attaching a register directly after a squarer and feeding its content back to the input of the squarer. This removed the need of storing intermediate values outside the squarer and, as a result, computing e repeated squarings required only e clock cycles [9]. Similar repeated squarers were later used by Järvinen and Skyttä [6]. Recently, Rebeiro and Mukhopadhyay [7] observed that quading, i.e. a^4, is only slightly more expensive than squaring in FPGAs. They utilized this observation by building a repeated squarer using a chain of quadings and used it for accelerating Itoh-Tsujii inversions.

This paper contributes in the following ways:

- We provide simple tools for analyzing repeated squarings in binary fields defined by arbitrary irreducible polynomials and use them for analyzing repeated squarings in fields defined by NIST in [10];
- We generalize the observation of [7] and demonstrate how LUTs can be exploited in designing efficient circuitries for repeated squarings;
- We develop efficient repeated squarers with (a) fixed exponents and (b) varying exponents and provide implementation results on Xilinx FPGAs (Spartan-3A 3S1400-5 and Virtex-5 5VLX50-3) demonstrating the practicability of the repeated squarers; and
- We show how repeated squarings can improve speed of Itoh-Tsujii inversions (or, more generally, exponentiations in binary fields) and elliptic curve cryptography on Koblitz curves and even increase resistivity against side-channel attacks.

We begin by introducing the preliminaries of repeated squarings in Sec. 2. In Sec. 3, we analyze repeated squarings with a focus on fields defined by NIST in [10]. Based on the results of the analysis, we design repeated squarers with both a fixed exponent and a varying exponent in Sec. 4 and present results on Xilinx FPGAs in Sec 5. We end with conclusions and possible directions for future research in Sec. 6.

2 Squaring in Binary Fields

A *binary field*, $\mathbb{F}_{2^m}$, is generated from the ring of polynomials over $\mathbb{F}_2$, $\mathbb{F}_2[x]$, with an *irreducible polynomial*, $p(x)$, with a degree m by setting $\mathbb{F}_{2^m} : \mathbb{F}_2[x]/p(x)$. In *polynomial basis*, an element of $\mathbb{F}_{2^m}$ is represented as a binary polynomial with a degree at most $m - 1$, i.e.,

$$a(x) = \sum_{i=0}^{m-1} a_i x^i. \tag{1}$$

Operations in polynomial basis are computed modulo $p(x)$. Addition, $a(x)+b(x)$, is simply a bitwise exclusive-or (xor). Multiplication, $a(x) \times b(x)$, is more complicated and divides into two steps: (1) multiplication in $\mathbb{F}_2[x]$ and (2) reduction modulo $p(x)$. *Squaring* is a special case of multiplication where $a(x) = b(x)$. For squaring the first step of multiplication is performed simply by inserting zeros between each bit in the bitvector representing $a(x)$. Thus, squaring becomes

$$a^2(x) = \sum_{i=0}^{m-1} a_i x^{2i} \bmod p(x) \tag{2}$$

and it essentially requires only the reduction part of the multiplication (if implemented in hardware). The reduction is computed by taking the remainder after a division with $p(x)$ in $\mathbb{F}_2[x]$.

Squaring, $b(x) = a^2(x)$, can be seen as a linear transformation described by the matrix multiplication $\mathbf{b} = \mathbf{Q}\mathbf{a}$ where $\mathbf{a} = \begin{bmatrix} a_0 a_1 \cdots a_{m-1} \end{bmatrix}^T$ represents $a(x)$, $\mathbf{b} = \begin{bmatrix} b_0 b_1 \cdots b_{m-1} \end{bmatrix}^T$ represents the result $b(x)$, and the $m \times m$ matrix $\mathbf{Q}$ with elements in $\mathbb{F}_2$ is given by [4]:

$$\mathbf{Q} = \begin{bmatrix} 1 & q_{0,1} & q_{0,2} & \cdots & q_{0,m-1} \\ 0 & q_{1,1} & q_{1,2} & \cdots & q_{1,m-1} \\ \vdots & \vdots & \vdots & \ddots & \vdots \\ 0 & q_{m-1,1} & q_{m-1,2} & \cdots & q_{m-1,m-1} \end{bmatrix}. \tag{3}$$

See, e.g., [4] for description of the calculation of the coefficients $q_{i,j} \in \mathbb{F}_2$. A repeated squaring, $b(x) = a^{2^e}(x)$ with $e \geq 1$, is given by: $\mathbf{b} = \mathbf{Q}^e \mathbf{a}$ [4,11].

Using *normal basis* is another way of representing the elements of $\mathbb{F}_{2^m}$. A normal basis is constructed by taking a normal element in $\mathbb{F}_{2^m}$, i.e., an element $\alpha \in \mathbb{F}_{2^m}$ for which $\alpha, \alpha^2, \alpha^{2^2}, \ldots, \alpha^{2^{m-1}}$ are linearly independent. Then, an element $a \in \mathbb{F}_{2^m}$ is represented by

$$a = \sum_{i=0}^{m-1} a_i \alpha^{2^i} \tag{4}$$

Because $\alpha^{2^m} = \alpha$, squaring in normal basis is simply a rotation of the bitvector and, as a consequence, essentially free in hardware. Similarly, a repeated squaring is free if e is fixed (rotation by e bits). A repeated squaring in normal basis can be described with a squaring matrix defined by $q_{i,j} = 1$ if $i \equiv (j + e) \bmod m$, else $q_{i,j} = 0$. Despite the efficiency of (repeated) squarings in hardware, normal bases are less frequently used in contemporary cryptosystems than polynomial bases. The main reasons are their inefficiency on software and the complexity of multiplications. It should be also noted that if e varies, then squaring is not free in normal basis either, but still cheaper than in polynomial basis.

2.1   Inversion with Fermat's Little Theorem

The element $b \in \mathbb{F}_q$ is the multiplicative inverse of an element $a \in \mathbb{F}_q$ if they satisfy $a \times b = 1$. *Inversion* is a common problem in both cryptography and codes. There are essentially two ways to compute it: Extended Euclidean Algorithm and Fermat's Little Theorem. In this paper, we discuss the latter one.

Fermat's Little Theorem states that $a^p \equiv a \pmod{p}$ and it follows that:

$$a^{-1} = a^{q-2} \tag{5}$$

for all $a \in \mathbb{F}_q$. Hence, an inversion in $\mathbb{F}_{2^m}$ is an exponentiation to the power $2^m - 2$. Because $2^m - 2 = \langle 11 \ldots 1110 \rangle$, the *binary method* (see, e.g, [12]) requires $m - 2$ multiplications and $m - 1$ squarings, none of which are repeated squarings. A more efficient method, here referred to as Itoh-Tsujii inversion, was proposed in [3] requiring $\lfloor \log_2(m-1) \rfloor + w(m-1) - 1$ multiplications and $m - 1$ squarings where $w(\cdot)$ is the *Hamming weight*, i.e., the number of nonzeros in the expansion. Itoh-Tsujii inversion requires only $\lfloor \log_2(m-1) \rfloor + w(m-1)$ repeated squarings [4].

2.2   Elliptic Curve Cryptography

Elliptic curve cryptosystems [1,2] are build around an operation called *scalar multiplication*, kP, where P is a point on the curve and k is an integer. Scalar multiplication is computed with algorithms that are analogous to exponentiation algorithms (see, e.g., [12] for a review) with the exception that multiplications are replaced with operations on the curve. For instance, the binary method for exponentiation has the following analogue on an elliptic curve: k is represented using binary expansion and scanned one bit, k_i, at a time. Each bit requires a point doubling and $k_i = 1$ yields a point addition. An ℓ-bit k hence requires ℓ point doublings and $w(k)$ point additions.

Koblitz curves [5] are appealing because point doublings can be replaced with cheap *Frobenius endomorphisms* which map a point (x, y) to the point (x^2, y^2). Thus, on Koblitz curves a string of $e - 1$ zeros in the expansion of k results in e consecutive Frobenius maps which are performed by computing (x^{2^e}, y^{2^e}), i.e., with two repeated squarings. A common way to compute scalar multiplications on Koblitz curves is to represent k in a width-w τ-adic non-adjacent form (τNAF) with $w(k) \approx m/(w+1)$ so that there are no adjacent nonzeros [13]. For example, if k (or part of k) in width-2 τNAF is $\langle 100\bar{1}0001 \rangle$, it results in (from left to right) point addition, two repeated squarings (x^{2^3} and y^{2^3}), point subtraction, two repeated squarings (x^{2^4} and y^{2^4}), and point addition. Hence, the total cost is

$w(k)$ point additions(/subtractions) and $2w(k)$ or $2(w(k)-1)$ repeated squarings depending on whether the last digit is zero or nonzero, respectively. Points are often represented in projective coordinates, e.g. [14], and, in that case, Frobenius involves three squarings changing the costs accordingly. Repeated squarings can be useful also on general curves because scalar multiplication always requires at least one inversion.

The binary method is inherently vulnerable to *power analysis side-channel attacks* [15] because different operations are performed depending on whether a bit is zero or nonzero. Hasan [16] noted that using repeated squarings in normal basis is an efficient countermeasure because then the number of consecutive Frobenius maps, and thus the number of consecutive zeros in k, is not distinguishable with the power analysis. In polynomial basis, however, such solutions have not existed prior to this paper.

3 Analysis of Repeated Squarings

We begin with definitions and theorems which are used in analyzing implementation aspects of repeated squarings.

Definition 1. *Weight, $\mathcal{W}(\mathbf{Q}^e)$, is the number of ones in $\mathbf{Q}^e$.*

Fig. 1 plots $\mathcal{W}(\mathbf{Q}^e)$ with $1 \leq e \leq 25$ for all fields defined by NIST in [10]. Fig. 1 shows that repeated squarings with small(ish) exponents, e, have small weights, but when e grows, the weights quickly become approximately $m(m-1)/2$ [11]. The benefits gained from the sparseness of irreducible polynomials are clearly visible in the figure: the weights increase considerably slower for the fields defined by trinomials ($\mathbb{F}_{2^{233}}$ and $\mathbb{F}_{2^{409}}$) than pentanomials ($\mathbb{F}_{2^{163}}$, $\mathbb{F}_{2^{283}}$, and $\mathbb{F}_{2^{571}}$). In the following, we focus on the NIST field, arguably, having the most contemporary relevance, $\mathbb{F}_{2^{233}} : \mathbb{F}_2[x]/x^{233} + x^{74} + 1$, but the results and conclusions that follow are easy to generalize also for other NIST fields (or to any field defined by a sparse irreducible polynomial).

While the weight gives some insight into the actual cost, it is far too simplistic for our purposes. For example, it does not take the characteristics of an implementation platform into account, and neither does it say anything about the delay of the computation, both of which are highly relevant for an implementor.

Definition 2. *Row-weight, $\mathcal{W}_i(\mathbf{Q}^e)$, is the number of ones on the ith row of $\mathbf{Q}^e$.*

The row-weights reflect the costs of computing individual bits of the result. The row-weight is a valuable tool for computing the cost of a repeated squaring circuitry, as will be shown in the following.

Definition 3. *Area, $\mathcal{A}_n(\mathbf{Q}^e)$, is the number of n-LUTs required to implement $\mathbf{Q}^e$.*

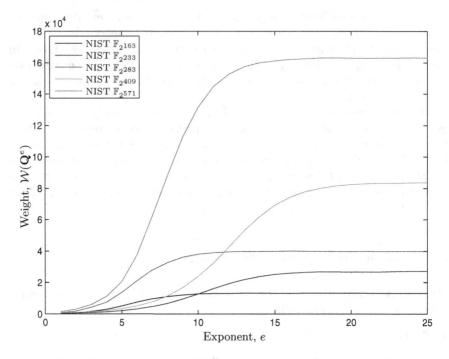

Fig. 1. The weights $\mathcal{W}(\mathbf{Q}^e)$ for the fields defined by NIST

Theorem 1. *It is possible to implement* $\mathbf{Q}^e$ *with a circuit whose area* $\mathcal{A}_n(\mathbf{Q}^e)$ *satisfies*

$$\mathcal{A}_n(\mathbf{Q}^e) \leq \sum_{i=1}^{m} \left\lceil \frac{\mathcal{W}_i(\mathbf{Q}^e) - 1}{n - 1} \right\rceil. \qquad (6)$$

Proof. (Sketch) Let L be the number of n-LUTs in a tree computing xor of its inputs. The number of inputs in the tree is $L(n-1)+1$ (If $L=1$, all n inputs of the LUT are available. After that, the output of each new LUT consumes one of the existing inputs, thus, increasing the number of inputs by only $n-1$). Because computing the ith row of $\mathbf{Q}^e$ requires a xor of $\mathcal{W}_i(\mathbf{Q}^e)$ bits, it follows directly from the previous that the number of LUTs is given by

$$L = \left\lceil \frac{\mathcal{W}_i(\mathbf{Q}^e) - 1}{n - 1} \right\rceil. \qquad (7)$$

Summing over all rows gives the upper bound of (6). It might be possible to share resources between rows and, as a result, produce an even smaller circuit; hence, the inequality. ☐

Definition 4. *Critical path,* $\mathcal{D}_n(\mathbf{Q}^e)$, *is the length of the longest path of consecutive* n-LUTs *in the circuit computing* $\mathbf{Q}^e$.

Theorem 2. *Critical path, $\mathcal{D}_n(\mathbf{Q}^e)$, is bounded by*

$$\mathcal{D}_n(\mathbf{Q}^e) \leq \max_i \lceil \log_n \mathcal{W}_i(\mathbf{Q}^e) \rceil. \tag{8}$$

Proof. (Sketch) As mentioned in the proof of Theorem 1, the tree corresponding to the ith row of $\mathbf{Q}^e$ must have $\mathcal{W}_i(\mathbf{Q}^e)$ inputs and it is easy to show that the tree has the depth $\lceil \log_n \mathcal{W}_i(\mathbf{Q}^e) \rceil$. The maximum is taken because, by Definition 4, the critical path is the longest path required by the repeated squaring. Again, resource sharing may enable even shorter critical paths. □

Example 1. Consider repeated squaring in $\mathbb{F}_2[x]/x^4 + x + 1$. Let us analyze the case $e = 2$ which gives us the following repeated squaring matrix:

$$\mathbf{Q}^2 = \begin{bmatrix} 1\,1\,1\,1 \\ 0\,1\,0\,1 \\ 0\,0\,1\,1 \\ 0\,0\,0\,1 \end{bmatrix}. \tag{9}$$

Clearly, we have $\mathcal{W}(\mathbf{Q}^2) = 9$ and $\mathcal{W}_1(\mathbf{Q}^2) = 4$, $\mathcal{W}_2(\mathbf{Q}^2) = 2$, $\mathcal{W}_3(\mathbf{Q}^2) = 2$, and $\mathcal{W}_4(\mathbf{Q}^2) = 1$. Using (6) we find out that the number of 2-LUTs (xor gates) required in implementation is bounded by $\mathcal{A}_2(\mathbf{Q}^2) \leq 5$. Indeed, we can reduce the cost to $\mathcal{A}_2(\mathbf{Q}^2) = 4$ by reusing either the xor of the 2nd row or the 3rd row on the 1st row. With 4-LUTs used in many FPGAs, we have the minimum cost: $\mathcal{A}_4(\mathbf{Q}^2) = 3$. (8) gives the critical paths $\mathcal{D}_2(\mathbf{Q}^2) = 2$ and $\mathcal{D}_4(\mathbf{Q}^2) = 1$.

Example 2. Table 1 shows the upper bounds of (6) and (8) for NIST $\mathbb{F}_{2^{233}}$ with $n \in [2,7]$ and $e \in [1,6]$. Squaring ($e = 1$) in NIST $\mathbb{F}_{2^{233}}$ requires 153 LUTs with all n. The benefits of larger n are clear: e.g., if $n = 2$, repeated squaring with $e = 4$ consumes 7.5 times more area and 4.0 times more delay than a squaring but, if $n = 7$, the differences are only 1.9 and 2.0 times[1].

Table 1. Area and delay estimates given by (6) and (8) for $\mathbb{F}_{2^{233}}$

	$\mathcal{A}_n(\mathbf{Q}^e)$						$\mathcal{D}_n(\mathbf{Q}^e)$					
	$n=2$	$n=3$	$n=4$	$n=5$	$n=6$	$n=7$	$n=2$	$n=3$	$n=4$	$n=5$	$n=6$	$n=7$
$e=1$	153	153	153	153	153	153	1	1	1	1	1	1
$e=2$	361	245	230	230	230	230	2	2	2	1	1	1
$e=3$	676	385	349	238	233	233	3	2	2	2	1	1
$e=4$	1141	616	466	358	349	291	4	3	2	2	2	2
$e=5$	1844	973	699	550	466	396	4	3	2	2	2	2
$e=6$	2892	1511	1035	812	663	580	5	3	3	2	2	2

Next, we show how to implement efficient repeated squarers by exploiting the fact that repeated squarings are cheap with small e (as demonstrated in Table 1).

---

[1] Notice that this does not imply that the time required in computation with $n = 7$ is shorter than with $n = 2$ because the delays of LUTs with different n are not the same; i.e., a LUT with small n is probably faster than one with large n.

4   Architectures for Repeated Squarers

4.1   Fixed Exponent

First, we consider the case where the exponent e is fixed; i.e, the problem is to produce a circuitry that computes only $a^{2^e}(x)$ optimally with respect to some metric, such as, area, delay, or area-delay product.

There are two simple approaches to compute a repeated squaring with e:

1. *Direct* where one produces a circuitry directly from the matrix $\mathbf{Q}^e$; and
2. *Square chain* where one produces a combinatorial circuitry of e successive squarings[2].

Let $\|$ denote a *concatenation* of two circuits, i.e., the outputs of the circuit on the left are fed to the inputs of the circuit on the right. Because $a^{2^e}(x) = (a^{2^{e_1}})^{2^{e_2}}(x)$ if $e = e_1 + e_2$, a repeated squaring with an exponent e can be implemented with a chain $\mathbf{Q}^{e_1}\|\mathbf{Q}^{e_2}\|\dots\|\mathbf{Q}^{e_N}$ where $\sum_{i=1}^{N} e_i = e$. The direct approach is the special case where $N = 1$ and $e_1 = e$ and the square chain is the special case where $N = e$ and $e_i = 1$ for all i.

Example 3. $\mathbf{Q}^3\|\mathbf{Q}^2$ denotes a circuit that given an input $a(x)$, first, computes $a^{2^3}(x)$ with a direct approach using $\mathbf{Q}^3$ and then feeds its result $b(x)$ to a circuit that computes $b^{2^2}(x)$ with $\mathbf{Q}^2$. Hence, the result from the entire circuit is $a^{2^5}(x)$.

Clearly, area and delay of concatenated circuits are bounded by

$$\mathcal{A}_n(\mathbf{Q}^{e_1}\|\dots\|\mathbf{Q}^{e_N}) \leq \sum_{i=1}^{N} \mathcal{A}_n(\mathbf{Q}^{e_i}) \quad \text{and} \tag{10}$$

$$\mathcal{D}_n(\mathbf{Q}^{e_1}\|\dots\|\mathbf{Q}^{e_N}) \leq \sum_{i=1}^{N} \mathcal{D}_n(\mathbf{Q}^{e_i}). \tag{11}$$

The upper bounds reflect the case where the circuits are concatenated as such without any optimizations between blocks. In practice, optimizations between concatenated circuits can reduce both area and delay. Nevertheless, in the following analysis, we assume the worst case, i.e., the equality in (10) and (11). We also assume the following ordering for the exponents: $e_1 \geq e_2 \geq \dots \geq e_N$.

Remark 1. Although synthesis usually manages to perform optimizations so that the resulting area and delay satisfy the bounds of (10) and (11), it is also possible that the synthesis fails and results in a larger area and/or delay. Even in that case, the above bounds are always achievable by preventing synthesis from performing optimizations between blocks.

Next, setups optimized for delay, area, and their product are discussed in more detail. In all cases, the task is to find a concatenation $\mathbf{Q}^{e_1}\|\mathbf{Q}^{e_2}\|\dots\|\mathbf{Q}^{e_N}$ that minimizes the metric under optimization.

---

[2] Notice that this is different from the repeated squarer of [9] which iterates a squarer for e clock cycles.

Minimum Area. The task is to find $\{e_1, e_2, \ldots, e_N\}$ with $e = \sum_{i=1}^{N} e_i$ that minimizes $\sum_{i=1}^{N} \mathcal{A}_n(\mathbf{Q}^{e_i})$.

Using only two inputs of an n-LUT costs as much as using all n inputs. As a consequence, even though $\mathcal{W}(\mathbf{Q}^e)$ grows rapidly when e increases (see Fig. 1), taking unused inputs in use attenuates the growth of the number of LUTs. The number of LUTs hence grows only moderately with small e as shown in Table 1. In the following, the goal is to utilize the region of moderate growth also for large e by using concatenations.

In order to design a repeated squarer for a large e with minimal area, it is critical to minimize the area used per squaring. We call the exponent, $\hat{e}$, that minimizes $\mathcal{A}_n(\mathbf{Q}^{\hat{e}})/\hat{e}$ with $\hat{e} \leq e$ the *optimal exponent*, e_{opt}.

Example 4. Fig. 2 plots $\mathcal{A}_n(\mathbf{Q}^{\hat{e}})/\hat{e}$ for NIST $\mathbb{F}_{2^{233}}$ with $2 \leq n \leq 7$ and $\hat{e} \leq 8$. It shows that the number of LUTs required per squaring first decreases if $n > 2$. Consider the case $n = 4$ (e.g., Spartan 3-A) and $e = 6$. Then, $e_{\text{opt}} = 2$ and it is more area efficient to use a concatenation $\mathbf{Q}^2 || \mathbf{Q}^2 || \mathbf{Q}^2$ than any other concatenation or direct $\mathbf{Q}^6$ (actually, direct $\mathbf{Q}^6$ requires the largest area because $\mathcal{A}_4(\mathbf{Q}^6)/6 > \mathcal{A}_4(\mathbf{Q}^{\hat{e}})/\hat{e}$ for all $1 \leq \hat{e} < 6$).

The phenomenon of Example 4 happens for all fields defined by NIST in [10], but the benefits are expectedly larger for the fields defined by trinomials. Notice that

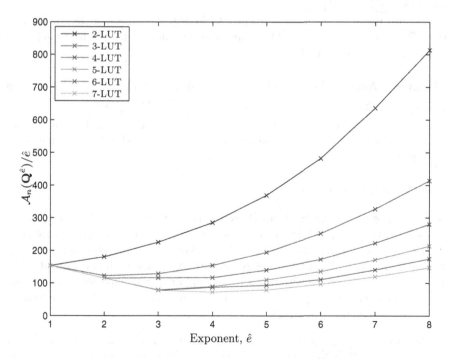

Fig. 2. Area per exponent, $\mathcal{A}_n(\mathbf{Q}^{\hat{e}})/\hat{e}$, with different LUT sizes for NIST $\mathbb{F}_{2^{233}}$

because $\mathcal{A}_2(\mathbf{Q}^{\hat{e}})/\hat{e}$ is strictly increasing, the phenomenon remains unobserved if one considers only 2-to-1 bit gates (2-LUTs).

A concatenation with minimal area can be found with an *exhaustive search* through all possible concatenations. Another approach to find a concatenation with a minimal or near-minimal area is a straightforward application of the *greedy algorithm* which results in a concatenation of $t = \lfloor e/e_{\text{opt}} \rfloor$ instances of $\mathbf{Q}^{e_{\text{opt}}}$ followed by $\mathbf{Q}^{e-te_{\text{opt}}}$ if $e_{\text{opt}} \nmid e$. The greedy algorithm guarantees an optimal concatenation if $e_{\text{opt}} \mid e$. However, if $e_{\text{opt}} \nmid e$, a concatenation with $\hat{e} > e_{\text{opt}}$ may result in a smaller area. Obviously, the greedy algorithm is computationally much simpler than the exhaustive search, especially, with large e; however, the computational complexity of the exhaustive search is insignificant if it is performed offline, as it typically is.

Example 5. Consider using 6-LUTs (e.g., Virtex-5) for implementing a repeated squarer for NIST $\mathbb{F}_{2^{233}}$ with a fixed exponent $e = 9$. Fig. 2 shows that $e_{\text{opt}} = 3$. Because $3 \mid 9$, both exhaustive search and the greedy algorithm return the same optimal concatenation: $\mathbf{Q}^3 \| \mathbf{Q}^3 \| \mathbf{Q}^3$, which has an area estimate of 699 LUTs. However, if $e = 10$, exhaustive search gives $\mathbf{Q}^4 \| \mathbf{Q}^3 \| \mathbf{Q}^3$ with an area estimate of 757 LUTs but, because $3 \nmid 10$, the greedy algorithm fails to find the optimal concatenation and returns $\mathbf{Q}^3 \| \mathbf{Q}^3 \| \mathbf{Q}^3 \| \mathbf{Q}$ with an estimated area of 852 LUTs.

Minimum Delay. The task is to find $\{e_1, e_2, \ldots, e_N\}$ with $e = \sum_{i=1}^{N} e_i$ that minimizes $\sum_{i=1}^{N} \mathcal{D}_n(\mathbf{Q}^{e_i})$.

Because delay grows logarithmically to $\mathcal{W}_i(\mathbf{Q}^e)$ as shown in (8), it is clear that $\mathcal{D}_n(\mathbf{Q}^e) \leq \mathcal{D}_n(\mathbf{Q}^{e_1} \| \ldots \| \mathbf{Q}^{e_N})$, and the minimum delay is always achieved with a direct circuit for $\mathbf{Q}^e$.

Minimum Area-Delay Product. The task is to find $\{e_1, e_2, \ldots, e_N\}$ with $e = \sum_{i=1}^{N} e_i$ that minimizes $\left(\sum_{i=1}^{N} \mathcal{A}_n(\mathbf{Q}^{e_i}) \right) \left(\sum_{i=1}^{N} \mathcal{D}_n(\mathbf{Q}^{e_i}) \right)$.

These optimizations are analogous to the minimum area optimizations; rather than minimizing $\mathcal{A}_n(\mathbf{Q}^e)$, one minimizes the product $\mathcal{A}_n(\mathbf{Q}^e)\mathcal{D}_n(\mathbf{Q}^e)$. Hence, we omit further analysis.

4.2   Varying Exponent

In Sec. 4.1, e was assumed fixed. However, many practical applications require support for varying e and, in the following, we discuss two solutions for providing such a support. The *first solution* is a simple generalization of the fixed exponent repeated squarer and suits for cases requiring support for distinct exponents. The *second solution* targets to situations where support for all exponents in a certain range is needed (for simplicity and without any loss of generality[3], we assume that the range is $0 \leq e \leq e_{\text{max}}$).

---

[3] The general case, $e_{\text{min}} \leq e \leq e_{\text{max}}$, can be realized by using a fixed exponent squarer with e_{min} to reach the lower bound and then the second solution for the range $0 \leq e \leq e_{\text{max}} - e_{\text{min}}$.

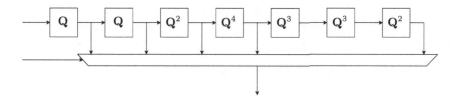

Fig. 3. The repeated squarer of Example 6

The First Solution. Let $E = \{e_1, \ldots, e_\ell\}$, where $e_i > e_{i-1}$ for all i, denote the set of exponents supported by a repeated squarer and let $\Delta_i = e_i - e_{i-1}$ (with $e_0 = 0$). The first solution is to produce fixed exponent circuits for each Δ_i by using the tools of Sec. 4.1 and to concatenate the resulting circuits in increasing order starting from Δ_1. A multiplexer with i as the selector is used to collect the wanted result from the chain.

Example 6. Consider using 6-LUTs (e.g., Virtex-5) for implementing an area optimized repeated squarer for NIST $\mathbb{F}_{2^{233}}$ that supports the exponents: $E = \{1, 2, 4, 8, 16\}$. We get $\Delta_1 = 1$, $\Delta_2 = 1$, $\Delta_3 = 2$, $\Delta_4 = 4$, and $\Delta_5 = 8$. Using exhaustive search for each Δ_i with area minimization as an optimization strategy and concatenating the resulting circuits and a multiplexer gives the circuit depicted in Fig. 3. Area and delay estimates (without the multiplexer) are 1600 LUTs and 8 LUTs, respectively.

Remark 2. The first solution is a generalization of the circuitry presented by Rebeiro and Mukhopadhyay [7]. Their circuitry, called quad-block[4], computes $a^{4^s}(x)$ with $s \in \{2, 3, 4, 5, 7, 9\}$. It is a special case of the first solution which is implemented for an exponent set $E = \{4, 6, 8, 10, 14, 18\}$ with $\mathbf{Q}^2$ blocks only. They showed that the circuit can be efficiently used for accelerating inversions. They used the addition chain $(1, 2, 3, 6, 7, 14, 28, 29, 58, 116, 232)$ for the inversion whereas $(1, 2, 4, 8, 16, 32, 64, 128, 192, 224, 232)$, where e is a power of two in all repeated squarings, would make the circuit simpler, i.e., the same latency could be achieved with $E = \{1, 2, 4, 8, 16\}$ (the circuit of Example 6).

The Second Solution. If a squarer must support all exponents from a given range, then $\Delta_i = 1$ for all i and the first solution results in a squaring chain from which the output is selected with a multiplexer from the outputs of each squarer in the chain. This is clearly very inefficient and we present the following second solution to overcome this problem.

The second solution splits computation for two chains, the first of which is a concatenation $\mathbf{Q}^{e_{opt}} || \mathbf{Q}^{e_{opt}} || \ldots || \mathbf{Q}^{e_{opt}}$ with a length of $\lfloor e_{max}/e_{opt} \rfloor$ and the

---

[4] Squaring and quading are not supported by the quad-block, but they are available elsewhere in the processor.

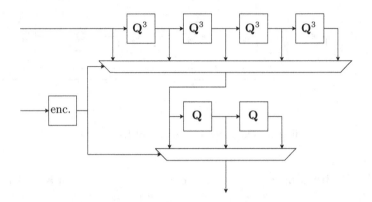

Fig. 4. The repeated squarer of Example 7

second is a square chain $(\mathbf{Q}||\mathbf{Q}||\ldots||\mathbf{Q})$ with a length of $e_{opt} - 1$. The input to the second chain is selected from the input, the intermediate values, and the output of the first chain with a $(\lfloor e_{max}/e_{opt} \rfloor + 1)$-to-1 multiplexer. The output of the entire repeated squarer is obtained with an e_{opt}-to-1 multiplexer from the square chain. The select signals of the multiplexers are derived from e with a simple encoder.

Example 7. Consider using 6-LUTs (e.g. Virtex-5) for implementing a repeated squarer for NIST $\mathbb{F}_{2^{233}}$ that supports exponents in the range $0 \le e \le e_{max} = 14$. Using the tools of Sec. 3 with area optimization, we get $e_{opt} = 3$. Thus, the first chain consists of four repeated squarers for $\mathbf{Q}^3$ and the second chain is a square chain with two squarers. The repeated squarer is depicted in Fig. 4. Area and delay estimates (without the multiplexers and encoder) are 1238 LUTs and 6 LUTs, respectively.

Remark 3. The second solution results in an exponent range $0 \le e \le e_{upper} = (\lfloor e_{max}/e_{opt} \rfloor + 1) e_{opt} - 1$ from which it follows that $e_{upper} = e_{max}$ if and only if $e_{opt} \mid e_{max} + 1$, else $e_{max} < e_{upper} < e_{max} + e_{opt}$. If a lower bound starting from one is wanted, it can be easily realized either by designing the encoder so that it does not allow exponent $e = 0$ or by attaching a squarer in front of the repeated squarer. However, the feature that also $a^{2^0}(x) = a(x)$ is supported can be very useful because it allows using the repeated squarer directly as a dummy operation.

Remark 4. Delay can be reduced by replacing the square chain with direct repeated squarers. For instance, in the case of Example 7 (Fig. 4), the second $\mathbf{Q}$ would be replaced by $\mathbf{Q}^2$ taking its input directly from the multiplexer. This would reduce the delay to 5 LUTs but increase area by 77 LUTs.

5   Implementation Results

All VHDL was generated automatically with Matlab scripts[5] (Matlab 7.7.0.471 (R2008b)). The VHDL is device independent. We compiled the code for two Xilinx FPGAs: Spartan-3A 3S1400A-5 ($n = 4$) and Virtex-5 5VLX50-3 ($n = 6$) which represent low-cost and high-end FPGAs, respectively. Synthesis and place&route were performed with Xilinx ISE 10.1.03 WebPACK using default options. The inputs and outputs were registered, otherwise the whole chip area was devoted for the repeated squarers. Several different designs were compiled and Table 2 collects the results. The areas in Table 2 represent the areas of combinatorial parts and the delays give the critical paths from the input registers to the output registers as reported by Xilinx ISE.

5.1   Discussion on the Results

In order to be feasible in practice, the repeated squarers should consume only moderate area and have a delay that is shorter than the critical paths of existing elliptic curve cryptography processors (this ensures that repeated squarings do not become the bottleneck for clock frequency). To the best of our knowledge, the two fastest FPGA-based elliptic curve processors have been presented by Chelton and Benaissa [17] for general binary curves and Järvinen and Skyttä [6] for Koblitz curves. The areas of the processors are 26364 4-LUTs (Virtex-4 4VLX200-11) and 34604 7-LUTs (Adaptive LUTs of Stratix II S180C3), respectively [17,6]. Compared to these, the areas listed in Table 2 are small. The critical paths of the processors are 6.50 ns and 5.33 ns, respectively [17,6]. The delays in Table 2 are of the same magnitude. However, both processors use NIST $\mathbb{F}_{2^{163}}$, whereas we provided results for a larger field, $\mathbb{F}_{2^{233}}$. It is likely that both area and critical paths of the processors would be significantly larger with the larger field. Hence, it is safe to say that the proposed repeated squarers are, indeed, feasible components for elliptic curve cryptography processors.

The delays of the circuits computing $\mathbf{Q}^e$ directly are only slightly faster (and in some cases even slower) than the delays of concatenated circuits. The reasons for this originate from the more difficult place&route of direct circuits that degrades their results. Hence, the proposed repeated squarers can be very competitive against direct circuits even with respect to delay.

Table 2 shows that (with only few exceptions) the setup declared as area optimal by the analysis, indeed, is the best concatenation also after the compilation. Of course, the number of compiled concatenations for each set of exponents is rather small; hence, it is not clear whether a concatenation resulting in an even smaller area exists or not. Most of the exceptions are for Spartan-3A. This is due to the fact that based on the analysis $e_{opt} = 2$ (see Table 1) whereas in reality based on the values from Table 2 it is $e_{opt} = 3$. This difference is caused by resource sharing between rows that was not considered by the analysis. The resource sharing between rows is not the only synthesis optimization which is not

---

[5] The scripts are available at http://www.tcs.hut.fi/~kjarvine/codes/

Table 2. Results for $\mathbb{F}_{2^{233}}$ on Spartan-3A 3S1400A-5 and Virtex-5 5VLX50-3

e	Concatenation[1]	Spartan-3A Area (LUTs)	Delay (ns)	Virtex-5 Area (LUTs)	Delay (ns)
1	1 $_{(4,6)}$	153	3.45	153	2.78
2	2 $_{(4,6)}$	230	4.61	230	3.01
3	3 $_{(4,6)}$	289	5.82	233	2.52
4	2,2 $_{(4)}$	436	5.99	299	2.86
4	4 $_{(6)}$	433	6.09	330	2.89
5	3,2 $_{(4,6)}$	546	7.01	372	3.58
5	5	656	6.52	476	3.63
6	2,2,2 $_{(4)}$	711	6.78	651	3.51
6	3,3 $_{(6)}$	637	6.91	434	4.06
6	6	996	6.59	659	3.55
8	2,2,2,2 $_{(4)}$	939	6.62	1415	4.49
8	3,3,2 $_{(6)}$	916	7.80	587	5.00
8	8	2090	7.27	1576	4.47
16	2,...,2 $_{(4)}$	1691	12.30	1525	8.25
16	4,3,...,3 $_{(6)}$	1733	12.35	1368	8.86
16	16	5727	12.89	4398	6.48
$\{1,2,4\}$	1,1,2 $_{(4,6)}$	810	5.77	580	4.07
$\{1,2,4,8\}$	1,1,2,2,2 $_{(4)}$	1350	8.85	1227	5.30
$\{1,2,4,8\}$	1,1,2,4 $_{(6)}$	1396	7.80	1189	5.83
$\{1,2,4,8,16\}$	1,1,2,...,2 $_{(4)}$	2694	14.82	2015	8.07
$\{1,2,4,8,16\}$	1,1,2,4,3,3,2 $_{(6)}$	2784	15.68	1823	8.23
$0 \le e \le 7$	2 $_{(4)}$	1365	8.40	1319	6.03
$0 \le e \le 11$	2 $_{(4)}$	2525	12.65	2113	7.72
$0 \le e \le 11$	3 $_{(6)}$	2005	11.00	1809	8.10
$0 \le e \le 13$	2 $_{(4)}$	2412	15.78	2694	8.66
$0 \le e \le 14$	3 $_{(6)}$	2773	13.58	1911	8.67

[1] The subscript shows n for which the concatenation is area optimal based on the analysis.

considered in the analysis. The synthesis also performs optimizations between different concatenated blocks.

In general, compensating these deficiencies in the analysis without any feedback from the synthesis is extremely difficult because optimizations depend heavily on the target device (on more than simply n) and the synthesis program as well as on the options given for the synthesis. However, the following procedure using feedback from the synthesis compensates the resource sharing between rows of $\mathbf{Q}^e$ nearly completely:

1. Synthesize repeated squaring matrices $\mathbf{Q}^{\hat{e}}$ starting from $\hat{e} = 1$ until e_{opt} is found.
2. Collect the areas from the results of the synthesis and use them as $\mathcal{A}_n(\mathbf{Q}^e)$ in the analysis instead of the estimates given by (6).

This procedure takes resource sharing between rows into account in the analysis and makes the predicted values considerably more accurate. Synthesis programs typically include options that prevent optimizations between design blocks and they can be used for improving the accuracy of the analysis. However, we did not use such options, because the optimizations between blocks typically reduce both area and delay considerably from the predicted values and, hence, preventing them reduces the quality of results.

6   Conclusions

Previously in the paper, we analyzed repeated squarings and presented several possibilities of how to realize repeated squarers. The results presented in Sec. 5 proved the feasibility of repeated squarers by showing that they can be implemented with reasonable area and that they are fast enough not to become the bottleneck.

We conclude by discussing the following benefits of using repeated squares compared to the existing solutions:

Faster Inversions in Binary Fields. Repeated squarers (the first solution) offer a relatively cheap way to implement fast inversions. For example, an Itoh-Tsujii inversion in NIST $\mathbb{F}_{2^{233}}$ requires 10 multiplications and 232 squarings. These squarings can be computed with only 19 repeated squarings using a repeated squarer (first solution) with $E = \{1, 2, 4, 8, 16\}$ (a component that was shown practical in Sec. 5). Speedups up to 88% can be achieved with this repeated squarer compared to an iterative repeated squarer. Naturally, the faster the multiplications are and the larger is the ratio of squarings compared to multiplications, the larger are the speedups. See Appendix A.1 for more information on the computation model, latencies, and speedups with different repeated squarers.

Faster General Exponentiations in Binary Fields. Inversion using Fermat's Little Theorem is simply an exponentiation to the power $2^m - 2$ in the field. Obviously, repeated squarers can be used for accelerating any exponentiation in a similar way and, again, the larger is the ratio of squarings compared to multiplications, the better improvements are achievable.

Faster Scalar Multiplications on Koblitz Curves. With an iterative squarer computing e successive Frobenius maps requires either $2e$ or $3e$ clock cycles depending on the coordinate system. Repeated squarers (the second solution) reduce this to $2\lceil e/e_{\text{max}} \rceil$ or $3\lceil e/e_{\text{max}} \rceil$ clock cycles. Defining speedups in the case of scalar multiplications is not as straightforward as it was for Itoh-Tsujii inversions, because they depend on k that varies. We ran 100000 experiments with random width-2 τNAFs using two computation models based on existing elliptic curve processors in order to determine the speedups. According to these experiments, repeated squarers can lead to average speedups of over 13% in the latency of scalar multiplication. Expectedly, the faster are the multipliers the larger are the speedups also in this case. Increasing the width ω of τNAF also increases the speedups. Furthermore, methods that reduce either the memory consumption of window methods [18] or the weight $w(k)$ [19] by using more Frobenius maps have been proposed recently and efficient computation of repeated squarings is essential for them. Appendix A.2 presents details on the experiments.

Improved Side-Channel Resistivity. Replacing an iterative repeated squarer with a repeated squarer (the second solution) makes attacking scalar multiplications on Koblitz curves with side-channel attacks considerably harder. Instead of counting clock cycles from the power trace, the adversary must be able to distinguish the exponent e from a single clock cycle in the power trace, i.e, after observing a single clock cycle in the trace, the adversary only knows that the number of Frobenius maps is $\leq e_{\max}$. Therefore, launching a successful side-channel attack is considerably more difficult. If $e_{\max}$ is small and/or ω large, situations where $e > e_{\max}$ occur commonly (see Appendix A.2) which may lead to certain side-channel weaknesses. It may also be possible to learn some information about e from the power consumption of the repeated squarer, e.g., with differential power analysis. Hence, it is clear that this approach does not remove the possibility of a successful power analysis attack entirely, but it is equally clear that it makes attacking significantly harder.

6.1   Future Research

Designing repeated squarers still requires some trial-and-error type optimizations, mainly because resource sharing is hard to incorporate into the analysis. However, the trials are easy and fast to do thanks to the automated VHDL generation provided by the Matlab scripts. Nonetheless, we are searching for heuristics compensating resource sharing.

Modern FPGAs commonly have a structure which cannot be modelled accurately with simple n-LUTs. As a consequence, optimizations for more advanced LUT structures, such as, Stratix ALUTs, 6-to-2 bit LUTs, etc., will be a topic for future research.

Square root can be implemented with fewer resources than squaring if the irreducible is a trinomial [20]. Hence, it might be possible to reduce the complexity of a repeated squaring (especially, varying exponent with the second solution) by, first, "shooting over" the required exponent e with an e_{opt} chain and, then, reversing back with (repeated) square roots, rather than using fewer e_{opt}'s and then reaching e with a few repeated squarings.

The effects of the LUT size, in general, have been surprisingly little studied considering how popular FPGAs are for implementing finite field arithmetic. As shown in this paper, the LUT structure may have significant consequences and could, therefore, open ways to further optimize existing designs on FPGAs;

hence, also other operations, such as finite field multiplications, should be studied from this point of view.

Acknowledgments

This work was supported by the European Commission's 7th Framework Programme (FP7) under contract number ICT-2007-216499 (CACE). The author would like to thank Billy Bob Brumley and the anonymous reviewers for valuable comments and improvement suggestions.

References

1. Miller, V.: Use of elliptic curves in cryptography. In: Williams, H.C. (ed.) CRYPTO 1985. LNCS, vol. 218, pp. 417–426. Springer, Heidelberg (1986)
2. Koblitz, N.: Elliptic curve cryptosystems. Math. Comput. 48, 203–209 (1987)
3. Itoh, T., Tsujii, S.: A fast algorithm for computing multiplicative inverses in $GF(2^m)$ using normal bases. Inf. Comput. 78, 171–177 (1988)
4. Guajardo, J., Paar, C.: Itoh-Tsujii inversion in standard basis and its application in cryptography and codes. Designs Codes Cryptogr. 25, 207–216 (2002)
5. Koblitz, N.: CM-curves with good cryptographic properties. In: Feigenbaum, J. (ed.) CRYPTO 1991. LNCS, vol. 576, pp. 279–287. Springer, Heidelberg (1992)
6. Järvinen, K., Skyttä, J.: Fast point multiplication on Koblitz curves: Parallelization method and implementations. Microprocess. Microsyst. 33, 106–116 (2009)
7. Rebeiro, C., Mukhopadhyay, D.: High speed compact elliptic curve cryptoprocessor for FPGA platforms. In: Chowdhury, D.R., Rijmen, V., Das, A. (eds.) INDOCRYPT 2008. LNCS, vol. 5365, pp. 376–388. Springer, Heidelberg (2008)
8. Wollinger, T., Guajardo, J., Paar, C.: Security on FPGAs: State-of-the-art implementations and attacks. ACM Trans. Embedd. Comput. Syst. 3, 534–574 (2004)
9. Lutz, J., Hasan, A.: High performance FPGA based elliptic curve cryptographic co-processor. In: International Conference on Information Technology: Coding and Computing, vol. 2, pp. 486–492. IEEE Computer Society, Los Alamitos (2004)
10. National Institute of Standards and Technology (NIST): Digital signature standard (DSS). Federal Information Processing Standard, FIPS PUB 186-2 (2000)
11. Ahmadi, O., Hankerson, D., Rodríguez-Henríquez, F.: Parallel formulations of scalar multiplication on Koblitz curves. J. Univers. Comput. Sci. 14, 481–504 (2008)
12. Gordon, D.M.: A survey of fast exponentiation methods. J. Algorithms 27, 129–146 (1998)
13. Solinas, J.A.: Efficient arithmetic on Koblitz curves. Designs Codes Cryptogr. 19, 195–249 (2000)
14. López, J., Dahab, R.: Improved algorithms for elliptic curve arithmetic in $GF(2^m)$. In: Tavares, S., Meijer, H. (eds.) SAC 1998. LNCS, vol. 1556, pp. 201–212. Springer, Heidelberg (1999)
15. Kocher, P., Jaffe, J., Jun, B.: Differential power analysis. In: Wiener, M. (ed.) CRYPTO 1999. LNCS, vol. 1666, pp. 388–397. Springer, Heidelberg (1999)
16. Hasan, M.A.: Power analysis attacks and algorithmic approaches to their countermeasures for Koblitz curve cryptosystems. IEEE Trans. Comput. 50, 1071–1083 (2001)
17. Chelton, W.N., Benaissa, M.: Fast elliptic curve cryptography on FPGA. IEEE Trans. Very Large Scale Integr (VLSI) Syst. 16, 198–205 (2008)

18. Vuillaume, C., Okeya, K., Takagi, T.: Short-memory scalar multiplication for Koblitz curves. IEEE Trans. Comput. 57, 481–489 (2008)
19. Dimitrov, V.S., Järvinen, K.U., Jacobson, M.J., Chan, W.F., Huang, Z.: Provably sublinear point multiplication on Koblitz curves and its hardware implementation. IEEE Trans. Comput. 57, 1469–1481 (2008)
20. Rodríguez-Henríquez, F., Morales-Luna, G., López, J.: Low-complexity bit-parallel square root computation over $GF(2^m)$ for all trinomials. IEEE Trans. Comput. 57, 471–480 (2008)
21. Al-Daoud, E., Mahmod, R., Rushdan, M., Kilicman, A.: A new addition formula for elliptic curves over $GF(2^n)$. IEEE Trans. Comput. 51, 972–975 (2002)

A   Speedup Evaluations

A.1   Itoh-Tsujii Inversion

We consider Itoh-Tsujii inversion in NIST $\mathbb{F}_{2^{233}}$ with the addition chain $\{1, 2, 4, 8, 16, 32, 64, 128, 192, 224, 232\}$ and the following computation model. Let multiplication require M clock cycles and repeated squaring one clock cycle. Assuming that the latency consists of only multiplications and repeated squarings (easily achievable, for example, with the architecture from [6]), an Itoh-Tsujii inversion in NIST $\mathbb{F}_{2^{233}}$ requires $10M + R$ clock cycles where R is the number of repeated squarings. Table 3 lists speedups compared to an iterative squarer.

Table 3. Latencies and speedups of Itoh-Tsujii inversion in $\mathbb{F}_{2^{233}}$ with different repeated squarers and multiplication latencies

E	R	$M = 233$	$M = 18$	$M = 6$	$M = 1$
$\{1\}$	232	2562	412	292	242
$\{1, 2\}$	117	2447(-4.5%)	297(-27.9%)	177(-39.4%)	127(-47.5%)
$\{1, 2, 4\}$	60	2390(-6.7%)	240(-41.7%)	120(-58.9%)	70(-71.1%)
$\{1, 2, 4, 8\}$	32	2362(-7.8%)	212(-48.5%)	92(-68.5%)	42(-82.6%)
$\{1, 2, 4, 8, 16\}$	19	2349(-8.3%)	199(-51.7%)	79(-72.9%)	29(-88.0%)
$\{1, 2, 4, 8, 16, 32\}$	13	2343(-8.5%)	193(-53.2%)	73(-75.0%)	23(-90.5%)
$\{1, 2, 4, 8, 16, 32, 64\}$	11	2341(-8.6%)	191(-53.6%)	71(-75.7%)	21(-91.3%)

A.2   Scalar Multiplication on Koblitz Curves

We consider scalar multiplication on a Koblitz curve NIST K-233 [10] with width-2 τNAF using two computation models. Model 1 represents a generic elliptic curve processors (similar, e.g., to [17]) having one multiplier with a latency M and all other operations having a latency of one. We assume that point additions are computed as proposed in [21] requiring 8 multiplications, 5 squarings, and 8 additions on K-233. Thus, we assume a latency of $(w(k) - 1)(8M + 13) + 3R$ for scalar multiplication (without the final inversion), where R is the number of repeated squarings per coordinate required in computation of Frobenius maps. Model 2 is taken directly from [6] representing the fastest elliptic curve processor available in the literature. With that processor the latency of scalar multiplication is $(w(k) - 1)(2M + 2) + R$. Table 4 presents results after evaluating both models with 100000 random width-2 τNAFs (obtained as proposed in [13]).

Table 4. Average latencies and speedups in scalar multiplications on NIST K-233 and width-2 τNAF with different repeated squarers and multiplication latencies using two computation models. Coverage gives the percentage of Frobenius maps where one coordinate can be mapped with a single repeated squaring; the value in parentheses gives the percentage of scalars where all Frobenius maps had this property.

e_{max}	Model	Coverage	$M = 17$	$M = 12$	$M = 8$	$M = 5$
1	1	0.33% (0.00%)	12133	9062	6604	4762
2	1	50.40% (0.00%)	11826(-2.5%)	8754(-3.4%)	6297(-4.7%)	4454 (-6.5%)
3	1	75.31% (0.00%)	11738(-3.3%)	8667(-4.4%)	6210(-6.0%)	4367 (-8.3%)
7	1	98.48%(29.57%)	11676(-3.8%)	8605(-5.0%)	6148(-6.9%)	4305 (-9.6%)
11	1	99.91%(93.07%)	11673(-3.8%)	8602(-5.1%)	6145(-7.0%)	4302 (-9.7%)
14	1	99.99%(99.15%)	11673(-3.8%)	8601(-5.1%)	6144(-7.0%)	4301 (-9.7%)
1	2	0.33% (0.00%)	2995	2227	1613	1152
2	2	50.40% (0.00%)	2893(-3.4%)	2125(-4.6%)	1510(-6.4%)	1050 (-8.9%)
3	2	75.31% (0.00%)	2863(-4.4%)	2095(-5.9%)	1481(-8.2%)	1020(-11.4%)
7	2	98.48%(29.57%)	2843(-5.1%)	2075(-6.8%)	1461(-9.4%)	1000(-13.2%)
11	2	99.91%(93.07%)	2842(-5.1%)	2074(-6.9%)	1459(-9.5%)	999(-13.3%)
14	2	99.99%(99.15%)	2842(-5.1%)	2074(-6.9%)	1459(-9.5%)	999(-13.3%)

Highly Regular m-Ary Powering Ladders

Marc Joye

Thomson R&D, Security Competence Center,
1 avenue de Belle Fontaine, 35576 Cesson-Sévigné Cedex, France
marc.joye@thomson.net
http://joye.site88.net/

Abstract. This paper describes new exponentiation algorithms with applications to cryptography. The proposed algorithms can be seen as m-ary generalizations of the so-called Montgomery ladder. Both left-to-right and right-to-left versions are presented.

Similarly to Montgomery ladder, the proposed algorithms always repeat the same instructions in the same order, without inserting dummy operations, and so offer a natural protection against certain implementation attacks. Moreover, as they are available in any radix m and in any scan direction, the proposed algorithms offer improved performance and greater flexibility.

Keywords: Exponentiation algorithms, Montgomery ladder, SPA-type attacks, safe-error attacks.

1 Introduction

We consider the general problem of evaluating $y = g^d$ in a (multiplicatively written) group $\mathbb{G}$ with identity element $1_{\mathbb{G}}$, on input $g \in \mathbb{G}$ and $d \in \mathbb{Z}_{>0}$. The m-ary expansion of d is given by $d = \sum_{i=0}^{\ell-1} d_i \, m^i$ with $0 \leqslant d_i < m$ and $d_{\ell-1} \neq 0$. Integer $\ell = \ell(m)$ represents the number of digits (in radix m) for the m-ary representation of d and is called the m-ary length of d.

1.1 Left-to-Right Algorithms

The most widely used exponentiation algorithm is the *binary method* (a.k.a. "square-and-multiply" algorithm) [15, Section 4.6.3]. It relies on the simple observation that $g^d = \left(g^{d/2}\right)^2$ if d is even, and $g^d = \left(g^{(d-1)/2}\right)^2 \cdot g$ if d is odd.

The binary method extends easily to any radix m. Let $H_i = \sum_{j=i}^{\ell-1} d_j \, m^{j-i}$. Since $H_i = \left(\sum_{j=i+1}^{\ell-1} d_j \, m^{j-i}\right) + d_i = m H_{i+1} + d_i$, we get

$$g^{H_i} = \begin{cases} \left(g^{H_{i+1}}\right)^m & \text{if } d_i = 0, \\ \left(g^{H_{i+1}}\right)^m \cdot g^{d_i} & \text{otherwise .} \end{cases} \tag{1}$$

Noting that $g^d = g^{H_0}$, the previous relation gives rise to an exponentiation algorithm. It can be readily programmed by scanning the m-ary representation

M.J. Jacobson Jr., V. Rijmen, and R. Safavi-Naini (Eds.): SAC 2009, LNCS 5867, pp. 350–363, 2009.
© Springer-Verlag Berlin Heidelberg 2009

of d from left to right. As, at iteration i, for $\ell - 2 \geqslant i \geqslant 0$, the method requires a multiplication by g^{d_i} when $d_i \neq 0$, the values of g^j with $1 \leqslant j \leqslant m - 1$ are precomputed and stored in $(m - 1)$ temporary variables; namely, $\mathsf{R}[j] \leftarrow g^j$ for $1 \leqslant j \leqslant m - 1$. If the successive values of g^{H_i} are kept track of in an accumulator A, Equation (1) then translates into

$$\mathsf{A} \leftarrow \begin{cases} \mathsf{A}^m & \text{if } d_i = 0 \\ \mathsf{A}^m \cdot \mathsf{R}[d_i] & \text{otherwise} \end{cases} \quad (\text{for } \ell - 2 \geqslant i \geqslant 0)$$

and where A is initialized to $\mathsf{R}[d_{\ell-1}]$. The corresponding algorithm is referred to as the *(left-to-right) m-ary algorithm*.

1.2 Right-to-Left Algorithms

It is also possible to devise a similar algorithm based on a right-to-left scan of exponent d. This may be convenient when the m-ary length of d is unknown in advance. In the binary case (i.e., when $m = 2$), letting $d = \sum_{i=0}^{\ell-1} d_i \, 2^i$ the binary expansion of d, the method makes use of the relation $g^d = \prod_{\substack{0 \leqslant i \leqslant \ell-1 \\ d_i \neq 0}} g^{2^i}$.
An accumulator A is initialized to g and squared at each iteration, so that it contains g^{2^i} at iteration i. Another accumulator, say $\mathsf{R}[1]$, initialized to $1_{\mathbb{G}}$, is multiplied with A if $d_i \neq 0$. Hence, we see that at iteration $\ell - 1$, accumulator $\mathsf{R}[1]$ contains the value of $\prod_{\substack{0 \leqslant i \leqslant \ell-1 \\ d_i \neq 0}} g^{2^i} = g^d$.

Although less known than its left-to-right counterpart, as shown by Yao [29], this method can be extended to higher radices. The basic idea remains the same. If $d = \sum_{i=0}^{\ell-1} d_i \, m^i$ denotes the m-ary expansion of d, we can write

$$g^d = \prod_{\substack{0 \leqslant i \leqslant \ell-1 \\ d_i=1}} g^{m^i} \cdot \prod_{\substack{0 \leqslant i \leqslant \ell-1 \\ d_i=2}} g^{2 \cdot m^i} \cdots \prod_{\substack{0 \leqslant i \leqslant \ell-1 \\ d_i=m-1}} g^{(m-1) \cdot m^i}$$

$$= \prod_{j=1}^{m-1} (L_j)^j \quad \text{where } L_j = \prod_{\substack{0 \leqslant i \leqslant \ell-1 \\ d_i=j}} g^{m^i} \, . \tag{2}$$

Hence, using $(m-1)$ accumulators, $\mathsf{R}[1], \ldots, \mathsf{R}[m-1]$, to keep track of the values of L_j, $1 \leqslant j \leqslant m - 1$, and an accumulator A that stores the successive values of g^{m^i}, at iteration i, the accumulators are updated as

$$\begin{cases} \mathsf{R}[d_i] \leftarrow \mathsf{R}[d_i] \cdot \mathsf{A} & \text{if } d_i \neq 0 \\ \mathsf{A} \leftarrow \mathsf{A}^m \end{cases} \quad (\text{for } 1 \leqslant i \leqslant \ell - 1)$$

where A is initialized to g and $\mathsf{R}[1], \ldots, \mathsf{R}[m-1]$ are initialized to $1_{\mathbb{G}}$. Equation (2) says that g^d is then given by $\mathsf{A} \leftarrow \prod_{j=1}^{m-1} \mathsf{R}[j]^j$. The so-obtained algorithm, also known as *Yao's algorithm*, is referred to as the *right-to-left m-ary algorithm*.

1.3 Implementation Attacks

If not properly implemented, exponentiation algorithms may be vulnerable to *side-channel attacks* [16,17] (see also [6,19]). Another threat against implementations of exponentiation algorithms resides in fault attacks [5] (see also [2,11]).

Two implementation attacks, namely *SPA-type attacks* and *safe-error attacks*, are particularly relevant in the context of exponentiation.

SPA-Type Attacks. By observing a suitable side channel, such as the power consumption [16] or electromagnetic emanations [10,24], an attacker may recover secret information. For exponentiation-based cryptosystems, the goal of the attacker is to recover the value of exponent d (or a part thereof) used in the computation of g^d in some group $\mathbb{G}$. *SPA-type attacks*[1] assume that the attacker infers secret information (typically one or several bits of d) from a single execution of g^d.

Consider for example the square-and-multiply algorithm (that is, the left-to-right m-ary algorithm with $m = 2$).[2]

Algorithm 1. Square-and-Multiply Algorithm

Input: $g \in \mathbb{G}$, $d = \sum_{i=0}^{\ell-1} d_i \, 2^i$
Output: g^d

1 $R[1] \leftarrow g$; $A \leftarrow 1_{\mathbb{G}}$
2 **for** $i = \ell - 1$ **down to** 0 **do**
3      $A \leftarrow A^2$
4   **if** $(d_i \neq 0)$ **then** $A \leftarrow A \cdot R[1]$
5 **end**
6 **return** A

Each iteration comprises a 'square' and, when the bit exponent is non-zero, a subsequent 'multiply'. Since the algorithm behaves differently depending on the bit values, this may be observed from a suitable side channel. The information thus gleaned may enable the attacker to deduce one or more bits of exponent d.

One way of preventing an attacker from recovering the bit values is to execute the same instructions regardless of the value of input bit d_i. Such an algorithm is said to be *regular*. There are several implementations of this idea.

– The test of whether a digit is nonzero may be removed if Line 1 in Algorithm 1 is replaced with $A \leftarrow A \cdot R[d_i]$ and where temporary variable $R[0]$ is initialized to $1_{\mathbb{G}}$. Alternatively, a fake multiply may be performed when $d_i = 0$, as suggested in [9]. Doing so, there will be no longer conditional branchings: at each iteration, there is a square always followed by a multiply.

---

[1] SPA stands for "Simple Power Analysis."

[2] We slightly differ from the presentation of Section 1.1 and initialize accumulator A with $1_{\mathbb{G}}$. This prevents the necessity of requiring $d_{\ell-1} \neq 0$ and therefore ℓ may denote any upper bound on the binary length of d. If $\ell' \leqslant \ell$ is the exact binary length of d, observe that accumulator A is correctly set in the for-loop to $g^{d_{\ell'-1}}$, as required.

This algorithm is known as the "square-and-multiply-always" algorithm. However, as will be explained in a moment, the resulting implementation now becomes vulnerable to safe-error attacks.

- Another possibility to get a regular exponentiation is to recode exponent d in such a way that none of the digits are zero [21,23,27,28]. As exemplified in [26], this however supposes that the recoding algorithm itself is resistant to SPA-type attacks.

The above analysis is not restricted to the square-and-multiply algorithm and generalizes to the m-ary exponentiation algorithms mentioned in Sections 1.1 and 1.2. While it may argued that, for larger m, m-ary exponentiation algorithms are more regular and therefore more resistant to SPA-type attacks, these algorithms are not entirely regular since two cases are to be distinguished: $d_i = 0$ and $d_i \neq 0$.

Safe-Error Attacks. By timely inducing a fault during the execution of an instruction, an attacker may deduce whether the targeted instruction is fake: if the final result is correct then the instruction is indeed fake (or dummy); if not, the instruction is effective. This knowledge may then be used to obtain one or more bits of exponent d. Such attacks are referred to as *safe-error attacks* [30,31].

Back to the "square-and-multiply-always" algorithm, an attacker can induce a fault during a multiply. If the final result is correct then the attacker may deduce that the corresponding exponent bit is a zero (i.e., fake multiply); otherwise, the attacker may deduce that the exponent bit is a one. Safe-error attacks apply likewise to higher-radix similar m-ary methods to distinguish zero digits.

1.4 Our Contributions

Using the terminology of [12], we deal in this paper with *highly* regular exponentiation algorithms, that is, exponentiation algorithms that

- are regular; i.e., always repeat the same instructions in the same order for any inputs;
- do not insert dummy operations.

Highly regular exponentiation algorithms protect against SPA-type attacks *and* safe-error attacks, at the same time [14]. Examples of such algorithms include the so-called Montgomery ladder [22] and a recent powering ladder presented at CHES 2007 [12, Algorithm 1″]. These two algorithms are depicted below.

Algorithm 2. Montgomery Ladder

Input: $g \in \mathbb{G}$, $d = \sum_{i=0}^{\ell-1} d_i \, 2^i$
Output: g^d

1 $R[0] \leftarrow 1_{\mathbb{G}}$; $R[1] \leftarrow g$
2 **for** $i = \ell - 1$ **down to** 0 **do**
3      $R[1 - d_i] \leftarrow R[1 - d_i] \cdot R[d_i]$
4      $R[d_i] \leftarrow R[d_i]^2$
5 **end**

6 **return** $R[0]$

Algorithm 3. Joye's Square-Multiply Ladder

Input: $g \in \mathbb{G}$, $d = \sum_{i=0}^{\ell-1} d_i \, 2^i$
Output: g^d

1 $R[0] \leftarrow 1_{\mathbb{G}}$; $R[1] \leftarrow g$
2 **for** $i = 0$ **to** $\ell - 1$ **do**
3      $R[1 - d_i] \leftarrow R[1 - d_i]^2 \cdot R[d_i]$
4 **end**

5 **return** $R[0]$

Montgomery ladder and Joye's square-multiply ladder both rely on specific properties of the binary representation. In particular, it is unclear how to generalize these two algorithms to higher radices.

In this paper, we present a new method to derive highly regular exponentiation algorithms by considering a representation of $d-1$ rather than that of plain exponent d. The proposed method is independent of the radix representation and of the scan direction (left-to-right or right-to-left). Interestingly, when particularized to $m = 2$, the method yields algorithms dual to Algorithms 3 and 2; i.e., similar algorithms but with the opposite scan direction.

Outline of the Paper. The rest of this paper is organized as follows. The next section is the core of our paper. We describe our new exponentiation algorithms. In Section 3, we present some applications thereof. Finally, we conclude in Section 4.

2   New Exponentiation Algorithms

As aforementioned, the goal is to evaluate $y = g^d$ given an element $g \in \mathbb{G}$ and an ℓ-digit exponent $d = \sum_{i=0}^{\ell-1} d_i \, m^i$. Our algorithms rely on the following proposition.

Proposition 1. Let $d = \sum_{i=0}^{\ell-1} d_i \, m^i$ denote the m-ary expansion of d. Then

$$d = (d_{\ell-1} - 1)m^{\ell-1} + \left(\sum_{i=0}^{\ell-2} (d_i + m - 1)m^i \right) + 1 \ .$$

Proof. Straightforward by noting that $\sum_{i=0}^{\ell-2}(d_i + m - 1)m^i = \sum_{i=0}^{\ell-2} d_i \, m^i + \sum_{i=0}^{\ell-2}(m-1)m^i = (d - d_{\ell-1} \, m^{\ell-1}) + (m^{\ell-1} - 1)$.   □

2.1   General Case

Proposition 1 can be rewritten as

$$d - 1 = \sum_{i=0}^{\ell-1} d_i^* \, m^i \quad \text{where } d_i^* = \begin{cases} d_i + m - 1 & \text{for } 0 \leqslant i \leqslant \ell - 2 \\ d_{\ell-1} - 1 & \text{for } i = \ell - 1 \end{cases} . \tag{3}$$

Left-to-Right Algorithm. If $d > 0$, it follows that $d_{\ell-1} \geqslant 1$ and so $d^*_{\ell-1} \geqslant 0$. Remember that the m-ary algorithm can accommodate a leading zero digit (i.e., when $d^*_{\ell-1} = 0$); see Footnote 2. It is also important to note that all the subsequent digits are nonzero (i.e., $d^*_i > 0$ for $i \leqslant \ell-2$). We can therefore devise a *regular* method to get the value of g^{d-1} for some $d > 0$. The value of $y = g^d$ is then obtained as $y = g^{d-1} \cdot g$.

The algorithm is an adaptation of the m-ary algorithm, as described in Section 1.1. It makes use of an accumulator A, initialized to $g^{d^*_{\ell-1}}$. At each iteration of the main loop, accumulator A is raised to the power of m and then always multiplied by $g^{d^*_i}$ (remember that $d^*_i \neq 0$). Since $d^*_i \in \{m-1, \ldots, 2m-2\}$, the values of $g^{m-1}, \ldots, g^{2m-2}$ are precomputed and stored in temporary variables R[1] $\ldots$, R[m]. At the end of the main loop, the accumulator is multiplied by g to get the correct result.

Precomputation & Initialization. Accumulator A has to be initialized to $g^{d^*_{\ell-1}}$ with $d^*_{\ell-1} = (d_{\ell-1} - 1)$ in $\{0, \ldots, m-2\}$ and this must be done in a regular manner. Moreover, since *(i)* the values of $g^{m-1}, \ldots, g^{2m-2}$ have to be precomputed and stored in registers R[1], $\ldots$, R[$m-1$] before entering the main loop and *(ii)* $d_{\ell-1} \in \{1, \ldots, m-1\}$, it is possible to

1. write g^{j-1} in R[j] for $1 \leqslant j \leqslant m$,
2. assign A to the corresponding register so that it contains $g^{d_{\ell-1}-1}$ (i.e., A $\leftarrow$ R[$d_{\ell-1}$]), and
3. multiply registers R[1], $\ldots$, R[m] by g^{m-1} so that they contain $g^{m-1}, \ldots, g^{2m-2}$, respectively;

or algorithmically, we replace Lines 1 and 2 in Algorithm 4 with

    ▷ Precomputation & Initialization
  1   R[1] $\leftarrow$ $1_{\mathbb{G}}$; R[2] $\leftarrow$ g; for $i = 3$ to m do R[i] $\leftarrow$ R[$i-1$] $\cdot$ R[2]
  2   A $\leftarrow$ R[$d_{\ell-1}$]; for $i = 1$ to m do R[i] $\leftarrow$ R[i] $\cdot$ R[m]

Doing so, the evaluation of $g^{d_{\ell-1}-1}$ is regular.

Algorithm 4. Regular Left-to-Right Exponentiation (General description)

Input: $g \in \mathbb{G}$, $d = \sum_{i=0}^{\ell-1} d_i m^i$ $(d > 0)$
Output: g^d
Uses: A and R[1], $\ldots$, R[m]

    ▷ Precomputation & Initialization
1   for $i = 1$ to m do R[i] $\leftarrow$ g^{m+i-2}
2   A $\leftarrow$ $g^{d_{\ell-1}-1}$
    ▷ Main loop
3   for $i = \ell - 2$ down to 0 do
4   A $\leftarrow$ A^m $\cdot$ R[$1 + d_i$]
5   end
    ▷ Final correction
6   A $\leftarrow$ A $\cdot$ g
7   return A

Yet another way of obtaining a regular evaluation is to force the leading digit to a predetermined value by adding to d a suitable multiple of the order of g prior to the exponentiation. When applicable, this method should be preferred. Furthermore, it nicely combines with the classical DPA countermeasure consisting in adding to d a random multiple of the order of g [9].

Final correction. The final correction can be avoided by replacing d with $d + 1$ prior to the exponentiation, $d \leftarrow d + 1$. This may be useful when the memory is scarce and that the value of g is not available in memory. Note also that this step may be combined with the addition of a multiple of the order of g.

Right-to-Left Algorithm. We can likewise devise a right-to-left m-ary exponentiation algorithm. We follow the presentation of Section 1.2. From Equation (3), we have

$$g^{d-1} = \left(g^{m^{\ell-1}}\right)^{d^*_{\ell-1}} \cdot \prod_{j=1}^{m-1} (L^*_j)^{m+j-2} \quad \text{where } L^*_j = \prod_{\substack{0 \leqslant i \leqslant \ell-2 \\ d^*_i = j}} g^{m^i} . \qquad (4)$$

The algorithm makes use of m accumulators, $R[1], \ldots, R[m]$, to keep track of the values of L^*_j, $1 \leqslant j \leqslant m$, and an accumulator that keeps track of the successive values of g^{m^i}. Accumulators $R[1], \ldots, R[m]$ are initialized to $1_{\mathbb{G}}$ and accumulator A is initialized to g. Again, it is to be noted that all digits d^*_i are nonzero (i.e., $d^*_i \in \{m-1, \ldots, 2m-2\}$ for $0 \leqslant i \leqslant \ell-2$). As a consequence, at each iteration i, an accumulator $R[j]$ is updated (namely, $R[d^*_i] \leftarrow R[d^*_i] \cdot A$) and accumulator A is updated as $A \leftarrow A^m$. Hence, we see that the evaluation of L^*_j is regular. It then remains to evaluate the above relation in a regular manner to obtain a *regular* right-to-left m-ary exponentiation algorithm to get g^{d-1} and thus $y = g^d$ as $g^{d-1} \cdot g$.

Initialization. In certain groups, the neutral element $1_{\mathbb{G}}$ requires special treatment (e.g., elliptic curves given by the Weierstraß form).[3] In such groups, the multiplication of two elements A and B is typically implemented by checking whether A or B is $1_{\mathbb{G}}$: if this is the case, then the other element is returned; if not, the 'regular' multiplication, $A \cdot B$, is evaluated and returned. As this may be observed through SPA, this can leak the first occurrence of a digit in $\{0, \ldots, m-1\}$ in the m-ary representation of d. One way to prevent this leakage is to initialize $R[1], \ldots, R[m]$ to values different from $1_{\mathbb{G}}$.

---

[3] By special treatment, we mean that the group operation is not unified. The usual addition formulas obtained by the chord-and-tangent rule on Weierstraß elliptic curves are not valid for $1_{\mathbb{G}}$ (i.e., the point at infinity). In contrast, in $\mathbb{G} = \mathbb{Z}^*_N$, neutral element $1_{\mathbb{G}} = 1$ does not require a special treatment. Further, in this latter case, it is easy to get SPA-resistance even if multiplication by 1 modulo N may be observed through some side channel. For example, this can be achieved by working in $\mathbb{Z}^*_{2^w N}$ and replacing 1 with an equivalent representation $1 + \alpha N$; the correct result is then obtained by reducing the final output modulo N.

Algorithm 5. Regular Right-to-Left Exponentiation (General description)

Input: $g \in \mathbb{G}$, $d = \sum_{i=0}^{\ell-1} d_i\, m^i$ $(d > 0)$
Output: g^d
Uses: A and R[1], ..., R[m]

▷ Initialization
1 **for** $i = 1$ **to** m **do** R[i] $\leftarrow 1_{\mathbb{G}}$
▷ Main loop
2 A $\leftarrow g$
3 **for** $i = 0$ **to** $\ell - 2$ **do**
4   R[$1 + d_i$] $\leftarrow$ R[$1 + d_i$] $\cdot$ A
5   A $\leftarrow$ A^m
6 **end**
▷ Aggregation
7 A $\leftarrow$ A$^{d_{\ell-1}-1} \cdot \prod_{i=1}^{m}$ R[i]$^{m+i-2}$
▷ Final correction
8 A $\leftarrow$ A $\cdot g$
9 **return** A

As an example R[1], ..., R[m] are initialized to g. Since each R[i] will be raised to the power of $(m + i - 2)$ during the aggregation step, we subtract $\sum_{i=1}^{m}(m + i - 2) = \frac{3m(m-1)}{2}$ from d prior to the exponentiation. In more detail, we replace Line 1 in Algorithm 5 with

▷ Initialization
1a **for** $i = 1$ **to** m **do** R[i] $\leftarrow g$
1b $d \leftarrow d - 3m(m-1)/2$

In groups where inverses can be easily obtained (e.g., on elliptic curves), another option is to keep the value of d unchanged but to correct the result at the end of the computation. This can be for example achieved by replacing Line 8 in Algorithm 5 with

▷ Final correction
8 A $\leftarrow$ A $\cdot g^{3m(m-1)/2+1}$

Alternatively, R[1], ..., R[m] can be initialized to elements of small order in $\mathbb{G}$. Suppose that R[1], ..., R[m] are all initialized to h with $\mathrm{ord}_{\mathbb{G}}(h) = t$. Define $b = 3m(m - 1)/2 \bmod t$. At the end of the computation, accumulator A then contains a multiplicative surplus factor of h^b. Hence, the correct result is obtained by multiplying A by h^{t-b}. For example, in RSA groups $\mathbb{G} = \mathbb{Z}_N^*$, we can take $h = N - 1$, which is of order $t = 2$.

Aggregation. If done naively, the aggregation step at Line 5 (i.e., the evaluation of $\prod_{i=1}^{m}$ R[i]$^{m+i-2}$) can be somewhat expensive. We extend a technique described in [15, p. 634] to suit our present needs. It requires an accumulator A initialized to R[m]. If we set R[i] $\leftarrow$ R[i] $\cdot$ R[$i + 1$] and A $\leftarrow$ A $\cdot$ R[i] for $i = m - 1, ..., 1$, we end up with $R[1] \leftarrow \prod_{1 \leqslant i \leqslant m} R[i]$ and $A \leftarrow \prod_{i=1}^m R[i]^i$. Therefore, writing $\prod_{i=1}^m R[i]^{m+i-2}$ as $\prod_{i=1}^m R[i]^i \cdot (\prod_{i=1}^m R[i])^{m-2}$, we can use the above technique to get it as $A \cdot R[1]^{m-2}$. In our case, accumulator A is initialized to $A^{d_\ell-1-1} \cdot R[m]$ to get the value of g^{d-1} as per Eq. (4).

> ▷ Aggregation
> 7a $A \leftarrow A^{d_\ell-1-1}$; $A \leftarrow A \cdot R[m]$
> 7b for $i = m - 1$ down to 1 do
> 7c $\quad R[i] \leftarrow R[i] \cdot R[i+1]$; $A \leftarrow A \cdot R[i]$
> 7d end
> 7e $A \leftarrow A \cdot R[1]^{m-2}$

The initialization of accumulator A (i.e., $A \leftarrow A^{d_\ell-1-1}$) must be performed in a regular manner. An easy way to do so is to add to d a suitable multiple of the order of g so as to force the leading digit of the resulting d to a predetermined value. An alternative method is described in Appendix A.

Final correction. As for the left-to-right version, the final correction can be avoided by replacing d with $d + 1$. Again, this step can be combined with other steps, including the initialization step when neutral element needs a special treatment or the initialization of accumulator A in the aggregation step to force the leading digit.

2.2   Binary Case

The m-ary algorithms we developed are subject to numerous variants. We present now algorithms tailored to the binary case.

In the binary case, we have $m = 2$ and thus, provided that $d > 0$, $d_{\ell-1} = 1$. Equation (3) then simplifies to $d - 1 = \sum_{i=0}^{\ell-2} d_i^* 2^i$ with $d_i^* = d_i + 1$.

Left-to-Right Algorithm. We can use Algorithm 4 as is, where m is set to 2 and accumulator A is initialized to $g^{d_\ell-1-1} = 1_{\mathbb{G}}$. Alternatively, assuming $d > 1$ (and thus $\ell \geqslant 2$), we can initialize the accumulator to $g^{d_\ell^*-2}$ and start the loop at index $\ell - 3$; this avoids dealing with neutral element $1_{\mathbb{G}}$.

Algorithm 6. Regular Left-to-Right Binary Exponentiation

Input: $g \in \mathbb{G}$, $d = \sum_{i=0}^{\ell-1} d_i 2^i$ $(d > 1)$
Output: g^d

1 $R[1] \leftarrow g$; $R[2] \leftarrow R[1]^2$
2 $A \leftarrow R[1 + d_{\ell-2}]$
3 for $i = \ell - 3$ down to 0 do
4 $\quad A \leftarrow A^2 \cdot R[1 + d_i]$
5 end
6 $A \leftarrow A \cdot R[1]$
7 return A

Right-to-Left Algorithm. A direct application of Algorithm 5 with $m = 2$ yields a regular right-to-left algorithm. To prevent the final correction,[4] assuming $d > 1$, we can initialize accumulators R[1] to g^{d_0} and R[2] to g. We then swap the order of squaring and multiplication and start the loop at index 1.

Algorithm 7. Regular Right-to-Left Binary Exponentiation

Input: $g \in \mathbb{G}$, $d = \sum_{i=0}^{\ell-1} d_i\, 2^i$ $(d > 1)$
Output: g^d

1 R[1] $\leftarrow g^{d_0}$; R[2] $\leftarrow g$
2 A $\leftarrow$ R[2]
3 **for** $i = 1$ **to** $\ell - 2$ **do**
4   A $\leftarrow$ A^2
5   R[1 + d_i] $\leftarrow$ R[1 + d_i] $\cdot$ A
6 **end**
7 A $\leftarrow$ R[1] $\cdot$ R[2]^2

8 **return** A

Implementation notes. In some cases, exponent d is known to be odd (this is for example the case in RSA [25]). If so, R[1] can be initialized to g. When the least significant bit of d is arbitrary, R[1] and R[2] can be initialized as R[1] $\leftarrow$ 1$_{\mathbb{G}}$; R[2] $\leftarrow g$; R[1] $\leftarrow$ R[1] $\cdot$ R[1 + d_0]. Yet another strategy, provided that the order of g is odd, is to add a suitable multiple thereof to force the parity of d.

Comparison. It is striking to see the resemblance between the so-obtained algorithms (i.e., Algorithms 6 and 7) and Algorithms 3 and 2, respectively. For Algorithm 7 and Montgomery ladder, this is even more apparent from the general description (i.e., when the multiply is performed prior the squaring). Actually, our algorithms when $m = 2$ may be considered as dual of Algorithms 3 and 2 in the sense that they execute similar instructions but scan the exponent in the opposite direction.

3 Further Results

The proposed exponentiation algorithms apply to any group $\mathbb{G}$. In this section, we exploit some of their features to get faster yet secure implementations in certain groups. Our focus will be on the group of points of an elliptic curve over a large prime field. We note however that similar speed-ups may be available in other groups.

Composite Group Operations. Elliptic curves over prime field $\mathbb{F}_p$ are usually implemented using Jacobian coordinates. A point P on elliptic curve E given by

$$E_{/\mathbb{F}_p} : Y^2 = X^3 + a_4 X Z^4 + a_6 Z^6$$

---

[4] Note that, contrarily to the left-to-right version, the value of g is not readily available from R[1].

is then represented as a triple $(X_1 : Y_1 : Z_1)$. Such a representation is not unique: $(X_2 : Y_2 : Z_2) \sim (X_1 : Y_1 : Z_1)$ if $X_2 = \lambda^2 X_1$, $Y_2 = \lambda^3 Y_1$ and $Z_2 = \lambda Z_1$ for some nonzero $\lambda \in \mathbb{F}_p$. We refer the reader to [3,4] for state-of-the-art formulas for point addition and point doubling in Jacobian coordinates.

In [20], Meloni developed new point addition formulas for points with the same Z-coordinate. This technique was successfully applied in [18] to derive efficient composite point addition formulas of the form $kP + Q$ for some $k \geqslant 2$. The key observation is that the intermediate calculations in the computation of $P + Q = (X_3 : Y_3 : Z_3)$ with $Z_3 = \alpha Z_1$ involve quantities $\alpha^2 X_1$ and $\alpha^3 Y_1$. Initial point P can then be viewed as $(\alpha^2 X_1 : \alpha^3 Y_1 : Z_3)$ and the evaluation of $2P + Q$ can be done as $(P + Q) + P$ where P and $P + Q$ have the same Z-coordinate. This technique can be used recursively to obtain the value of $kP + Q$.

As the main loop of our regular left-to-right exponentiation algorithm (Algorithm 4) consists of evaluating such a composite operation (i.e., $mA + R[1 + d_i]$ in additive notation), it can benefit from these improved formulas for faster computation of a point multiple on E.

Repeated Powerings. Building on [7], Cohen et al. [8] suggested considering mixed coordinate systems for representing points. An interesting case for point doubling is when curve parameter a_4 is equal to -3 as it saves some multiplications (in $\mathbb{F}_p$). Similar performance for an *arbitrary* parameter a_4 can be achieved by representing points in modified Jacobian coordinates, namely tuples of the form $(X_1 : Y_1 : Z_1 : W_1)$ where $W_1 = a_4 Z_1^4$.

For efficiency purposes, m is usually chosen as a power of 2, say $m = 2^k$, in m-ary exponentiation algorithms. Raising to the power of m (resp. multiplying by scalar m, in additive notation) then amounts to computing k squarings (resp. k doublings). As in [13], our right-to-left m-ary algorithm (Algorithm 5) repeatedly updates accumulator A as $A \leftarrow A^m$ (resp. $A \leftarrow mA$). The key observation here is that accumulator A is only modified in this step during the main loop (i.e., Line 5 in Algorithm 5).

As a consequence, back to elliptic curves, dP can be evaluated using mixed coordinate systems: $R[1], \ldots, R[m]$ are tuples $(X : Y : Z)$ representing points in Jacobian coordinates and A is a tuple $(X : Y : Z : W)$ representing a point in modified Jacobian coordinates. Line 5 (i.e., $R[1+d_i] \leftarrow R[1+d_i]+A$ using additive notation) only use the three first coordinates of A to evaluate a regular Jacobian point addition whereas Line 5 (i.e., $A \leftarrow 2^k A$ in additive notation) updates accumulator A as a series of k doublings in modified Jacobian coordinates. This allows one to have a fast point doubling without increasing the cost of a point addition, regardless of the value of a_4. More precisely, the evaluation of dP can be implemented using the fastest formulas [3] for both point doubling (i.e., the same speed as when $a_4 = -3$ even if $a_4 \neq -3$) and point addition.

Other improvements using different mixed coordinate systems for right-to-left algorithms can be found in [1].

4 Conclusion

In this paper, we developed new m-ary exponentiation algorithms. Remarkably, the proposed algorithms are highly regular: they always repeat the same (effective) instructions in the same order. This feature is useful in the implementation of exponentiation-based cryptosystems protected against SPA-type attacks and safe-error attacks. Contrary to previous regular exponentiation algorithms, our algorithms are not restricted to radix 2 but are available in any radix m. They can also accommodate a left-to-right or a right-to-left exponent scanning. Both scan directions have their own advantages. Furthermore, being generic, we note that the proposed algorithms can easily be combined with other known countermeasures to protect against other classes of attacks, including DPA-type attacks and fault attacks.

Acknowledgments. I am grateful to the anonymous referees for useful comments.

References

1. Avanzi, R.M.: Delaying and merging operations in scalar multiplication: Applications to curve-based cryptosystems. In: Biham, E., Youssef, A.M. (eds.) SAC 2006. LNCS, vol. 4356, pp. 203–219. Springer, Heidelberg (2007)
2. Bar-El, H., Choukri, H., Naccache, D., Tunstall, M., Whelan, C.: The sorcerer's apprentice guide to fault attacks. Proceedings the IEEE 94(2), 370–382 (2004); Earlier version in Proc. of FDTC 2004
3. Bernstein, D.J., Lange, T.: Explicit-formulas database, http://www.hyperelliptic.org/EFD/jacobian.html
4. Bernstein, D.J., Lange, T.: Faster addition and doubling on elliptic curves. In: Kurosawa, K. (ed.) ASIACRYPT 2007. LNCS, vol. 4833, pp. 29–50. Springer, Heidelberg (2007)
5. Boneh, D., DeMillo, R.A., Lipton, R.J.: On the importance of eliminating errors in cryptographic computations. Journal of Cryptology 14(2), 110–119 (2001); Extended abstract in Proc. of EUROCRYPT 1997
6. Koç, Ç.K. (ed.): Cryptographic Engineering. Springer, Heidelberg (2009)
7. Chudnovsky, D.V., Chudnovsky, G.V.: Sequences of numbers generated by addition in formal groups and new primality and factorization tests. Advances in Applied Mathematics 7(4), 385–434 (1986)
8. Cohen, H., Miyaji, A., Ono, T.: Efficient elliptic curve exponentiation using mixed coordinates. In: Ohta, K., Pei, D. (eds.) ASIACRYPT 1998. LNCS, vol. 1514, pp. 51–65. Springer, Heidelberg (1998)
9. Coron, J.-S.: Resistance against differential power analysis for elliptic curve cryptosystems. In: Koç, Ç.K., Paar, C. (eds.) CHES 1999. LNCS, vol. 1717, pp. 292–302. Springer, Heidelberg (1999)
10. Gandolfi, K., Mourtel, C., Olivier, F.: Electromagnetic analysis: Concrete results. In: Koç, Ç.K., Naccache, D., Paar, C. (eds.) CHES 2001. LNCS, vol. 2162, pp. 251–261. Springer, Heidelberg (2001)
11. Giraud, C., Thiebeauld, H.: A survey on fault attacks. In: Quisquater, J.-J., et al. (eds.) Smart Card Research and Advanced Applications VI (CARDIS 2004), pp. 159–176. Kluwer, Dordrecht (2004)

12. Joye, M.: Highly regular right-to-left algorithms for scalar multiplication. In: Paillier, P., Verbauwhede, I. (eds.) CHES 2007. LNCS, vol. 4727, pp. 135–147. Springer, Heidelberg (2007)
13. Joye, M.: Fast point multiplication on elliptic curves without precomputation. In: von zur Gathen, J., Imaña, J.L., Koç, Ç.K. (eds.) WAIFI 2008. LNCS, vol. 5130, pp. 36–46. Springer, Heidelberg (2008)
14. Joye, M., Yen, S.-M.: The Montgomery powering ladder. In: Kaliski Jr., B.S., Koç, Ç.K., Paar, C. (eds.) CHES 2002. LNCS, vol. 2523, pp. 291–302. Springer, Heidelberg (2003)
15. Knuth, D.E.: The Art of Computer Programming, 2nd edn. Seminumerical Algorithms, vol. 2. Addison-Wesley, Reading (1981)
16. Kocher, P., Jaffe, J., Jun, B.: Differential power analysis. In: Wiener, M. (ed.) CRYPTO 1999. LNCS, vol. 1666, pp. 388–397. Springer, Heidelberg (1999)
17. Kocher, P.C.: Timing attacks on implementations of Diffie-Hellman, RSA, DSS, and other systems. In: Koblitz, N. (ed.) CRYPTO 1996. LNCS, vol. 1109, pp. 104–113. Springer, Heidelberg (1996)
18. Longa, P., Miri, A.: New composite operations and precomputation scheme for elliptic curve cryptosystems over prime fields. In: Cramer, R. (ed.) PKC 2008. LNCS, vol. 4939, pp. 229–247. Springer, Heidelberg (2008)
19. Mangard, S., Oswald, E., Popp, T.: Power Analysis Attacks: Revealing the Secrets of Smart Cards. Springer, Heidelberg (2007)
20. Meloni, N.: New point addition formulæ for ECC applications. In: Carlet, C., Sunar, B. (eds.) WAIFI 2007. LNCS, vol. 4547, pp. 189–201. Springer, Heidelberg (2007)
21. Möller, B.: Securing elliptic curve point multiplication against side-channel attacks. In: Davida, G.I., Frankel, Y. (eds.) ISC 2001. LNCS, vol. 2200, pp. 324–334. Springer, Heidelberg (2001)
22. Montgomery, P.L.: Speeding the Pollard and elliptic curve methods of factorization. Mathematics of Computation 48(177), 243–264 (1987)
23. Okeya, K., Takagi, T.: The width-w NAF method provides small memory and fast elliptic scalar multiplications secure against side channel attacks. In: Joye, M. (ed.) CT-RSA 2003. LNCS, vol. 2612, pp. 328–342. Springer, Heidelberg (2003)
24. Quisquater, J.-J., Samyde, D.: Electromagnetic analysis (EMA): Measures and counter-measures for smart cards. In: Attali, S., Jensen, T. (eds.) E-smart 2001. LNCS, vol. 2140, pp. 200–210. Springer, Heidelberg (2001)
25. Rivest, R.L., Shamir, A., Adleman, L.M.: A method for obtaining digital signatures and public-key cryptosystems. Communications of the ACM 21(2), 120–126 (1978)
26. Sakai, Y., Sakurai, K.: A new attack with side channel leakage during exponent recoding computations. In: Joye, M., Quisquater, J.-J. (eds.) CHES 2004. LNCS, vol. 3156, pp. 298–311. Springer, Heidelberg (2004)
27. Thériault, N.: SPA resistant left-to-right integer recodings. In: Preneel, B., Tavares, S. (eds.) SAC 2005. LNCS, vol. 3897, pp. 345–358. Springer, Heidelberg (2006)
28. Vuillaume, C., Okeya, K.: Flexible exponentiation with resistance to side channel attacks. In: Zhou, J., Yung, M., Bao, F. (eds.) ACNS 2006. LNCS, vol. 3989, pp. 268–283. Springer, Heidelberg (2006)
29. Yao, A.C.: On the evaluation of powers. SIAM Journal on Computing 5(1), 100–103 (1976)
30. Yen, S.-M., Joye, M.: Checking before output may not be enough against fault-based cryptanalysis. IEEE Transactions on Computers 49(9), 967–970 (2000)
31. Yen, S.-M., Kim, S.-J., Lim, S.-G., Moon, S.-J.: A countermeasure against one physical cryptanalysis may benefit another attack. In: Kim, K.-c. (ed.) ICISC 2001. LNCS, vol. 2288, pp. 414–427. Springer, Heidelberg (2002)

A   Regular Aggregation

In the general description of the regular right-to-left exponentiation algorithm (i.e., Algorithm 5), the aggregation step consists in evaluating the product $A^{d_{\ell-1}-1} \cdot \prod_{i=1}^{m} R[i]^{m+i-2}$. When multiplication by 1_G can be distinguished through SPA, the initialization of $A \leftarrow A^{d_{\ell-1}-1}$ is not sufficient to prevent the leakage of $d_{\ell-1}$. We present here an alternative method for evaluating $A^{d_{\ell-1}-1} \cdot \prod_{i=1}^{m} R[i]^{m+i-2}$ in such a case. For concreteness, we detail it for the ternary case (i.e., $m = 3$) but it can easily be extented to other radices. The case $m = 2$ is treated in § 2.2.

For $m = 3$, the aggregation step becomes $A^{d_{\ell-1}-1} \cdot \prod_{i=1}^{m} R[i]^{m+i-2}$ with $d_{\ell-1} \in \{1, 2\}$. To ease the presentation, we let $R[0]$ denote the accumulator. So, we need to evaluate

$$\begin{cases} R[0] \leftarrow R[1]^2 \cdot R[2]^3 \cdot R[3]^4 & \text{if } d_{\ell-1} = 1 \\ R[0] \leftarrow R[0] \cdot R[1]^2 \cdot R[2]^3 \cdot R[3]^4 & \text{if } d_{\ell-1} = 2 \end{cases}.$$

The idea is to rewrite the product so that the different cases appear as a same series of squarings and multiplications. For example, we can write

$$\begin{cases} B \leftarrow R[1]^2 \text{ and } R[0] \leftarrow (B \cdot R[2]) \cdot (R[3] \cdot R[2] \cdot R[3])^2 \\ B \leftarrow R[3]^2 \text{ and } R[0] \leftarrow (R[0] \cdot R[2]) \cdot (R[1] \cdot R[2] \cdot B)^2 \end{cases}$$

respectively. Moreover, in order not to introduce an additional temporary variable (B in the above description), we make use of $R[1]$ and $R[3]$, respectively. We have:

$$d \leftarrow d_{\ell-1} - 1$$
$$R[1 + 2d] \leftarrow R[1 + 2d]^2$$
$$R[0] \leftarrow R[2] \cdot R[1 - d]$$
$$R[2] \leftarrow R[2] \cdot R[3 - 2d]; \ R[2] \leftarrow R[2] \cdot R[3]; \ R[2] \leftarrow R[2]^2$$
$$R[0] \leftarrow R[0] \cdot R[2]$$

There are many possible variants of this methodology; the proposed implementation can be modified to better suit a given architecture.

An Efficient Residue Group Multiplication for the η_T Pairing over $\mathbb{F}_{3^m}$

Yuta Sasaki, Satsuki Nishina, Masaaki Shirase, and Tsuyoshi Takagi

Future University Hakodate

Abstract. When we implement the η_T pairing, which is one of the fastest pairings, we need multiplications in a base field $\mathbb{F}_{3^m}$ and in a group G. We have previously regarded elements in G as those in $\mathbb{F}_{3^{6m}}$ to implement the η_T pairing. Gorla et al. proposed a multiplication algorithm in $\mathbb{F}_{3^{6m}}$ that takes 5 multiplications in $\mathbb{F}_{3^{2m}}$, namely 15 multiplications in $\mathbb{F}_{3^m}$. This algorithm then reaches the theoretical lower bound of the number of multiplications. On the other hand, we may also regard elements in G as those in the residue group $\mathbb{F}_{3^{6m}}^* / \mathbb{F}_{3^m}^*$ in which βa is equivalent to a for $a \in \mathbb{F}_{3^{6m}}^*$ and $\beta \in \mathbb{F}_{3^m}^*$. This paper proposes an algorithm for computing a multiplication in the residue group. Its cost is asymptotically 12 multiplications in $\mathbb{F}_{3^m}$ as $m \to \infty$, which reaches beyond the lower bound the algorithm of Gorla et al. reaches. The proposed algorithm is especially effective when multiplication in the finite field is implemented using a basic method such as shift-and-add.

Keywords: Finite field multiplication, pairing, residue group, Vandermonde matrix.

1 Introduction

Most public key cryptosystems (PKCs) are mainly computed using multiplications in finite fields, thus polynomial multiplications are important to efficiently implement PKCs because elements in the finite fields are represented as polynomials. The algorithms that most efficiently compute polynomial multiplications are those derived by Karatsuba [12], Toom-Cook [4, 10, 17], Cantor [9], and Schönhage [15]. Karatsuba's algorithm is suitable for polynomial multiplications of small and medium degrees, Toom-Cook's algorithm is suitable for those of medium degrees, and Cantor's and Schönhage's algorithms are suitable for those of large degrees. Brent et al. [8] inclusively improved these algorithms for $\mathbb{F}_2[x]$.

Recently, pairing based cryptosystems (PBCs) such as an identity-based encryption [6], an efficient broadcast encryption [7], and a keyword searchable encryption [5] have been attracting attention. For PBCs, we need multiplications in a base field $\mathbb{F}_q$ and in a group G. We have regarded elements in G as those in $\mathbb{F}_{q^k}$ to implement pairings, where k is an integer called the embedding degree. PBCs are practical when k is small. Thus multiplications in $\mathbb{F}_{q^k}$ are generally implemented using Karatsuba's algorithm.

The η_T pairing proposed by Barreto et al. [1] is one of the fastest pairings. It is defined over $\mathbb{F}_{3^m}$ or $\mathbb{F}_{2^m}$, and the embedding degrees become 6 or 4, respectively,

M.J. Jacobson Jr., V. Rijmen, and R. Safavi-Naini (Eds.): SAC 2009, LNCS 5867, pp. 364–375, 2009.

where m has to be a prime number for PBC security. This paper focuses on multiplications on $\mathbb{F}_{3^{6m}}$ to efficiently implement the η_T pairing. Arithmetic in $\mathbb{F}_{3^{6m}}$ is generally implemented using a tower of extensions $\mathbb{F}_{3^m} \subset \mathbb{F}_{3^{2m}} \subset \mathbb{F}_{3^{6m}}$ that Kerins et al. [13], Gorla et al. [11] and Beuchat et al. [3] used.

Using Karatsuba's algorithm, a multiplication in $\mathbb{F}_{3^{2m}}$ is computed by 3 multiplications and a multiplication in $\mathbb{F}_{3^{6m}}$ is computed by 6 multiplications. Then 18 multiplications in $\mathbb{F}_{3^m}$ are needed. Additionally, a polynomial multiplication of degree t needs at least $2t + 1$ multiplications according to the theory of multiplicative complexity (see Lempel et al. [14] and Winograd [18]). Then a multiplication in $\mathbb{F}_{3^{2m}}$, which needs 3 multiplications in $\mathbb{F}_{3^m}$, reaches the lower bound because elements in $\mathbb{F}_{3^{2m}}$ are represented as the polynomials degree of 1. On the other hand, a multiplication in $\mathbb{F}_{3^{6m}}$, which needs 6 multiplications in $\mathbb{F}_{3^{2m}}$, does not yet reach the lower bound, which is 5. Gorla et al. proposed a multiplication algorithm for $\mathbb{F}_{3^{6m}}$ that takes 5 multiplications in $\mathbb{F}_{3^{2m}}$, namely 15 multiplications in $\mathbb{F}_{3^m}$, using the 4×4 Vandermende matrix, all the coefficients of which are the fourth roots of unity. Thus this algorithm reaches the lower bound.

In this paper, we regard elements in G as those in the residue group $\mathcal{G} = \mathbb{F}_{3^{6m}}^* / \mathbb{F}_{3^m}^*$. In $\mathcal{G}$, βa is equivalent to a for $a \in \mathbb{F}_{3^{6m}}^*$ and $\beta \in \mathbb{F}_{3^m}^*$. The aim of this paper is to propose a *residue group multiplication* (RGM) algorithm in $\mathcal{G}$, which is a modification of an algorithm of Shirase et al. [16] to the case of characteristic 3. The cost of the proposed RGM algorithm is asymptotically 12 multiplications in $\mathbb{F}_{3^m}$ as $m \to \infty$, which reaches beyond the lower bound the algorithm of Gorla et al. reaches. In the proposed RGM algorithm, $\mathbb{F}_{3^{6m}}$ is directly represented as the sixth extension of $\mathbb{F}_{3^m}$ unlike current implementation as done by Kerins et al. [13], Gorla et al. [11], and Beuchat et al [3]. Consequently we can use a Vandermonde matrix (8×8) bigger than that used in the algorithm of Gorla et al. (4×4). This bigger Vandermond matrix reduces the cost of the proposed RGM algorithm.

Moreover, we implemented the η_T pairing over $\mathbb{F}_{3^{97}}$, which for security had 1,024-bit RSA on a Core 2 Duo E6320 1.86GHz with 1GB RAM using gcc 3.4.4. Using the algorithm of Gorla et al. and the proposed RGM algorithm, we then compared timings of the η_T pairings. Consequently, the timing of the η_T pairing using the proposed RGM algorithm was almost 5 percent faster than that using the algorithm of Gorla et al.

This paper is organized as follows: In Section 2 we explain the η_T pairing over $\mathbb{F}_{3^m}$. We explain multiplication algorithm in $\mathbb{F}_{3^{6m}}$ in Section 3. In Section 4 we present our proposed RGM algorithm. Lastly, we conclude this paper in Section 5.

2   Implementation of the η_T Pairing over $\mathbb{F}_{3^m}$

In this section, we explain implementations of finite fields $\mathbb{F}_{3^m}$ and $\mathbb{F}_{3^{6m}}$, and the η_T pairing over $\mathbb{F}_{3^m}$.

2.1   Finite Field $\mathbb{F}_{3^m}$ and Extension Field $\mathbb{F}_{3^{6m}}$

Let $\mathbb{F}_3 = \{0, 1, 2\}$ be the prime field with characteristic 3. First, we explain how $\mathbb{F}_3$ is represented on computers by following the method of Kerins et al. [13]. An element $a \in \mathbb{F}_3$ is represented by two bits such as $a = (a_{hi}, a_{lo})$ for $a_{hi}, a_{lo} \in \{0, 1\}$, specifically, $(0, 0), (0, 1), (1, 0)$, mean 0, 1, 2, respectively. Note that the negative $-a$ for $a \in \mathbb{F}_3$ is replaced by $2a$, and it is represented by $-a = (a_{lo}, a_{hi})$ for $a = (a_{hi}, a_{lo})$.

Let $\mathbb{F}_3[x]$ be a set of polynomials with coefficients in $\mathbb{F}_3$. Then a finite field $\mathbb{F}_{3^m}$ is represented as

$$\mathbb{F}_{3^m} = \mathbb{F}_3[x] \,/\, f(x),$$

where $f(x)$ is an irreducible polynomial of degree m. Let A be an element in $\mathbb{F}_{3^m}$. A can be represented as the polynomial of degree at most $m - 1$ as

$$A = a_{m-1}x^{m-1} + a_{m-2}x^{m-2} + ... + a_1 x + a_0.$$

$\mathbb{F}_{3^{6m}}$ is the sixth extension field of $\mathbb{F}_{3^m}$. Let $g(\sigma)$ and $h(\rho)$ be irreducible polynomials with $g(\sigma) = \sigma^2 + 1$ over $\mathbb{F}_{3^m}$ and $h(\rho) = \rho^3 - \rho - 1$ over $\mathbb{F}_{3^{2m}}$. We then follow the tower field representation of Kerins et al. [13],

$$\mathbb{F}_{3^{2m}} = \mathbb{F}_{3^m}[\sigma] \,/\, g(\sigma),$$
$$\mathbb{F}_{3^{6m}} = \mathbb{F}_{3^{2m}}[\rho] \,/\, h(\rho).$$

Let A_0, A_1, A_2 be elements in $\mathbb{F}_{3^{2m}}$ with $A_0 = a_1\sigma + a_0$, $A_1 = a_3\sigma + a_2$ and $A_2 = a_5\sigma + a_4$. Then $A' \in \mathbb{F}_{3^{6m}}$ is represented as

$$A' = A_2\rho^2 + A_1\rho + A_0 = a_5\sigma\rho^2 + a_4\rho^2 + a_3\sigma\rho + a_2\rho + a_1\sigma + a_0.$$

Then a set $\{\sigma\rho^2, \rho^2, \sigma\rho, \rho, \sigma, 1\}$ forms a base of $\mathbb{F}_{3^{6m}}$ over $\mathbb{F}_{3^m}$. We call it $\sigma\rho$ base in this paper. Note that the roots of $\sigma^2 + 1$ are the primitive fourth roots of unity since $\sigma^2 = -1$ and $\sigma \neq \pm 1$.

Let $\mathbb{F}_{3^m}^*$ be the multiplicative group of $\mathbb{F}_{3^m}$, and let $\mathbb{F}_{3^{6m}}^*$ be the multiplicative group of $\mathbb{F}_{3^{6m}}$. That is, $\mathbb{F}_{3^m}^* = \mathbb{F}_{3^m} - \{0\}$ and $\mathbb{F}_{3^{6m}}^* = \mathbb{F}_{3^{6m}} - \{0\}$.

2.2 η_T Pairing over $\mathbb{F}_{3^m}$

Let E be a supersingular curve $E : y^2 = x^3 - x + b, b = \pm 1$ over $\mathbb{F}_{3^m}$. Then the η_T pairing is a bilinear map

$$\eta_T : E(\mathbb{F}_{3^m})[r] \times E(\mathbb{F}_{3^m})[r] \rightarrow \mathbb{F}_{3^{6m}}^* \,/\, (\mathbb{F}_{3^{6m}}^*)^r,$$

where r is the largest prime number such that $r \mid \#E(\mathbb{F}_{3^m})$, and 6 is the embedding degree. The η_T pairing satisfies the equation $\eta_T(aP, Q) = \eta_T(P, aQ) = \eta_T(P, Q)^a$ for any integer $a \neq 0$.

We used an algorithm for computing the η_T pairing used in [3], which is efficient due to it not having a cube root operation (Algorithm 1).

Algorithm 1. The η_T pairing algorithm without a cube root operation [3]

INPUT: $P(x_p, y_p), Q(x_q, y_q) \in E(\mathbb{F}_{3^m})[r]$
OUTPUT: $\eta_T(P, Q) \in \mathbb{F}_{3^{6m}}$
1: $y_p \leftarrow -y_p, d \leftarrow 1$
2: $R_0 \leftarrow -y_p(x_p + x_q + 1) + y_q\sigma + y_p\rho$
3: **for** $i \leftarrow 0$ to $(n-1)/2$ **do**
4:     $v \leftarrow x_p + x_q + d$
5:     $R_1 \leftarrow -v^2 + y_p y_q\sigma - v\rho - \rho^2$
6:     $R_0 \leftarrow R_0 R_1$
7:     $y_p \leftarrow -y_p$
8:     $x_q \leftarrow x_q^9, y_q \leftarrow y_q^9$
9:     $d \leftarrow ((d-1) \bmod 3)$
10:    $R_0 \leftarrow R_0^3$
11: **end for**
12: **return** R_0

Remark 1. Elements in $\mathbb{F}_{3^{6m}}^* / (\mathbb{F}_{3^{6m}}^*)^r$ have previously been regarded as those in $\mathbb{F}_{3^{6m}}$ to implement the η_T pairing. However, note that elements in $\mathbb{F}_{3^{6m}}^* / (\mathbb{F}_{3^{6m}}^*)^r$ may be regarded as those in the residue group $\mathcal{G} = \mathbb{F}_{3^{6m}}^* / \mathbb{F}_{3^m}^*$ because $(\mathbb{F}_{3^{6m}}^*)^r$ is a subgroup of $\mathbb{F}_{3^m}^*$. In $\mathcal{G}$, βa is equivalent to a for $a \in \mathbb{F}_{3^{6m}}^*$ and $\beta \in \mathbb{F}_{3^m}^*$.

3   Multiplication Algorithm in $\mathbb{F}_{3^{6m}}$

In this section, we explain Karatsuba's algorithm [12] and the multiplication algorithm by Gorla et al. [11].

3.1   Karatsuba's Algorithm [12]

Karatsuba's algorithm is generally used in a multiplication algorithm in $\mathbb{F}_{3^{6m}}$.

We consider a case in which $\mathbb{F}_{3^{6m}}$ is implemented using a tower of extensions $\mathbb{F}_{3^m} \subset \mathbb{F}_{3^{2m}} \subset \mathbb{F}_{3^{6m}}$ that Kerins et al. [13], Gorla et al. [11], and Beuchat et al. [3] used.

Let $A(\rho), B(\rho)$ be elements in $\mathbb{F}_{3^{6m}}$ with $A(\rho) = a_2\rho^2 + a_1\rho + a_0$ and $B(\rho) = b_2\rho^2 + b_1\rho + b_0$. Multiplication in $\mathbb{F}_{3^{6m}}$ is defined by $A(\rho) \cdot B(\rho) \bmod h(\rho)$. Let $P(\rho) = A(\rho) \cdot B(\rho) \bmod h(\rho)$. Let

$$t_1 = a_2(b_0 + b_2), \quad t_2 = a_1(b_1 + b_2), \quad t_3 = a_0(b_0 - b_1),$$
$$t_4 = b_2(a_0 - a_1), \quad t_5 = b_1(a_0 - a_1 + a_2), \quad t_6 = b_0(a_1 - a_2).$$

$P(\rho)$ is computed by Karatsuba's algorithm as follows:

$$P(\rho) = (t_1 + t_2 + t_4)\rho^2 + (t_1 + t_2 + t_5 + t_6)\rho + (t_2 + t_3 + t_5).$$

Thus, multiplication in $\mathbb{F}_{3^{6m}}$ is computed by 6 multiplications in $\mathbb{F}_{3^{2m}}$.

Next, let $A'(\sigma), B'(\sigma)$ be elements in $\mathbb{F}_{3^{2m}}$ with $A'(\sigma) = a_1'\sigma + a_0'$ and $B'(\sigma) = b_1'\sigma + b_0'$. Let $Q(\sigma) = A'(\sigma) \cdot B'(\sigma) \bmod g(\sigma)$. $Q(\sigma)$ is computed by Karatsuba's algorithm by

$$Q(\sigma) = (u_1 + u_3)\sigma + (u_2 + u_3),$$

where $u_1 = a_1'(b_0' + b_1')$, $u_2 = a_0'(b_0' - b_1')$, $u_3 = b_1'(a_0' - a_1')$. Thus, multiplication in $\mathbb{F}_{3^{2m}}$ is computed by 3 multiplications in $\mathbb{F}_{3^m}$. Therefore, multiplication in $\mathbb{F}_{3^{6m}}$ can be obtained by 18 multiplications in $\mathbb{F}_{3^m}$.

3.2 Multiplication Algorithm of Gorla et al. [11]

The algorithm of Gorla et al. [11] can compute a multiplication in $\mathbb{F}_{3^{6m}}$ most efficiently. Indeed it computes a multiplication in $\mathbb{F}_{3^{6m}}$ with 5 multiplications in $\mathbb{F}_{3^{2m}}$, which theoretically reaches the lower bound [14, 18] because a polynomial multiplication of degree m needs at least $2m + 1$ multiplications according to the theory of multiplicative complexity.

The algorithm of Gorla et al. uses the primitive fourth root of unity and the Vandermonde matrix. Let $A(\rho), B(\rho)$ be elements in $\mathbb{F}_{3^{6m}}$ with $A(\rho) = a_2\rho^2 + a_1\rho + a_0$ and $B(\rho) = b_2\rho^2 + b_1\rho + b_0$. Let $C(\rho)$ be a product $A(\rho)$ and $B(\rho)$,

$$C(\rho) = A(\rho) \cdot B(\rho) = c_4\rho^4 + c_3\rho^3 + c_2\rho^2 + c_1\rho + c_0.$$

Note that we refer to σ as the primitive fourth root of unity in Section 2.1 and σ as the generator of the base of $\mathbb{F}_{3^{2m}}$ over $\mathbb{F}_{3^m}$. Let $Y = (1, \sigma^1, \sigma^2, \sigma^3) = (1, \sigma, -1, -\sigma)$, and let V_σ be the Vandermonde matrix for Y as follows:

$$V_\sigma = \begin{pmatrix} 1 & 1 & 1 & 1 \\ 1 & \sigma & -1 & -\sigma \\ 1 & -1 & 1 & -1 \\ 1 & -\sigma & -1 & \sigma \end{pmatrix}. \tag{1}$$

Coefficients of $C(\rho)$ satisfy the following matrix.

$$\begin{pmatrix} 1 & 1 & 1 & 1 \\ 1 & \sigma & -1 & -\sigma \\ 1 & -1 & 1 & -1 \\ 1 & -\sigma & -1 & \sigma \end{pmatrix} \begin{pmatrix} c_0 \\ c_1 \\ c_2 \\ c_3 \end{pmatrix} = \begin{pmatrix} c_0 + c_1 + c_2 + c_3 \\ c_0 + c_1\sigma - c_2 - c_3\sigma \\ c_0 - c_1 + c_2 - c_3 \\ c_0 - c_1\sigma - c_2 + c_3\sigma \end{pmatrix}$$

$$= \begin{pmatrix} C(1) - c_4 \\ C(\sigma) - c_4 \\ C(-1) - c_4 \\ C(-\sigma) - c_4 \end{pmatrix} = \begin{pmatrix} A(1)B(1) - c_4 \\ A(\sigma)B(\sigma) - c_4 \\ A(-1)B(-1) - c_4 \\ A(-\sigma)B(-\sigma) - c_4 \end{pmatrix}, \tag{2}$$

where $c_4 = a_2 b_2$. Let

$$P_0 = A(1)B(1), P_1 = A(\sigma)B(\sigma), P_2 = A(-1)B(-1), P_3 = A(-\sigma)B(-\sigma), \atop P_4 = c_4. \tag{3}$$

Then we arrive at

$$
\begin{pmatrix} c_0 \\ c_1 \\ c_2 \\ c_3 \end{pmatrix} = V_\sigma^{-1} \begin{pmatrix} P_0 - P_4 \\ P_1 - P_4 \\ P_2 - P_4 \\ P_3 - P_4 \end{pmatrix} = \begin{pmatrix} 1 & 1 & 1 & 1 \\ 1 & -\sigma & -1 & \sigma \\ 1 & -1 & 1 & -1 \\ 1 & \sigma & -1 & -\sigma \end{pmatrix} \begin{pmatrix} P_0 - P_4 \\ P_1 - P_4 \\ P_2 - P_4 \\ P_3 - P_4 \end{pmatrix}
$$

$$
= \begin{pmatrix} P_0 + P_1 + P_2 + P_3 - P_4 \\ P_0 - P_1\sigma - P_2 + P_3\sigma \\ P_0 - P_1 + P_2 - P_3 \\ P_0 + P_1\sigma - P_2 - P_3\sigma \end{pmatrix}
$$

using (2) and (3). In the above algorithm, the cost of the multiplication of V_σ and $(c_0, c_1, c_2, c_3)^T$ is virtually free. Thus, the above algorithm can compute a multiplication in $\mathbb{F}_{3^{6m}}$ by 5 multiplications in $\mathbb{F}_{3^{2m}}$, namely 15 multiplications in $\mathbb{F}_{3^m}$. Therefore, the algorithm of Gorla et al. [11] reduces the number of multiplications in $\mathbb{F}_{3^m}$. Note that the algorithm of Gorla et al. theoretically reaches the lower bound.

4   Proposed Residue Group Multiplication and Timing of the η_T Pairing

In this section, we present a residue group multiplication (RGM) algorithm in $\mathcal{G} = \mathbb{F}_{3^{6m}}^* / \mathbb{F}_{3^m}^*$. Its cost becomes 12 multiplications in $\mathbb{F}_{3^m}$ as $m \to \infty$, which reaches beyond the lower bound of the algorithm of Gorla et al. The proposed algorithm is effective when multiplication in the finite field is implemented using a basic method such as shift-and-add. Note that we can use RGMs at step 6 of Algorithm 1 due to what we described in Remark 1. Note that m is a prime for security of the pairing-based cryptography.

Moreover, we compared the timing of the η_T pairings using the algorithm of Gorla et al. and the proposed algorithm to verify whether the proposed algorithm is effective.

4.1 z Base

In the proposed RGM algorithm, $\mathbb{F}_{3^{6m}}$ directly represents the sixth extension of $\mathbb{F}_{3^m}$ unlike previous representations in Kerins et al. [13], Gorla et al. [11], and Beuchat et al. [3]. In other words, elements in $\mathbb{F}_{3^{6m}}$ are represented as polynomials with coefficients in $\mathbb{F}_{3^m}$ of one variable z. Although the m has to be co-prime to 6, it is satisfied because the m is a prime number. Consequently we can use a Vandermonde matrix (8×8) bigger than that of the algorithm of Gorla et al. (4×4) (see (1)) to compute multiplications. The bigger Vandermond matrix reduces the cost of RGMs. Therefore, $\mathbb{F}_{3^{6m}}$ is represented as $\mathbb{F}_{3^m}[z]/k(z)$, where $k(z)$ is an irreducible polynomial with $k(z) = z^6 + z - 1$. Let V be an element in $\mathbb{F}_{3^{6m}}$. Then V can be represented by z base as follows:

$$V = v_5 z^5 + v_4 z^4 + v_3 z^3 + v_2 z^2 + v_1 z + v_0.$$

Then a set $\{z^5, z^4, z^3, z^2, z, 1\}$ forms a base of $\mathbb{F}_{3^{6m}}$ over $\mathbb{F}_{3^m}$. We call it z base in this paper. Let W be an element in $\mathbb{F}_{3^{6m}}$ represented by $\sigma\rho$ base with $W = w_5 \sigma\rho^2 + w_4 \rho^2 + w_3 \sigma\rho + w_2 \rho + w_1 \sigma + w_0$. If $V = W$ then we can convert between V and W as follows:

$$
\begin{pmatrix} v_0 \\ v_1 \\ v_2 \\ v_3 \\ v_4 \\ v_5 \end{pmatrix}
=
\begin{pmatrix}
1 & 0 & 0 & 0 & 0 & 0 \\
0 & 1 & 2 & 1 & 0 & 1 \\
0 & 1 & 1 & 2 & 0 & 1 \\
2 & 0 & 2 & 0 & 0 & 2 \\
1 & 2 & 0 & 1 & 0 & 1 \\
2 & 0 & 2 & 1 & 1 & 0
\end{pmatrix}
\begin{pmatrix} w_0 \\ w_1 \\ w_2 \\ w_3 \\ w_4 \\ w_5 \end{pmatrix},
\qquad
\begin{pmatrix} w_0 \\ w_1 \\ w_2 \\ w_3 \\ w_4 \\ w_5 \end{pmatrix}
=
\begin{pmatrix}
1 & 0 & 0 & 0 & 0 & 0 \\
0 & 2 & 1 & 2 & 2 & 0 \\
2 & 0 & 2 & 1 & 2 & 0 \\
2 & 2 & 0 & 1 & 2 & 0 \\
1 & 1 & 2 & 0 & 0 & 1 \\
0 & 0 & 1 & 1 & 1 & 0
\end{pmatrix}
\begin{pmatrix} v_0 \\ v_1 \\ v_2 \\ v_3 \\ v_4 \\ v_5 \end{pmatrix}
$$

$\sigma\rho$ base $\rightarrow z$ base z base $\rightarrow \sigma\rho$ base

Note that these conversions need no multiplications.

4.2   Proposed RGM Algorithm

The proposed RGM algorithm in $\mathcal{G} = \mathbb{F}_{3^{6m}}^* / \mathbb{F}_{3^m}^*$ uses the Vandermonde matrix in the same way as the algorithm of Gorla et al. [11]. Let $A(z), B(z)$ be elements in $\mathbb{F}_{3^{6m}}$ with $A(z) = a_5 z^5 + a_4 z^4 + a_3 z^3 + a_2 z^2 + a_1 z + a_0$ and $B(z) = b_5 z^5 + b_4 z^4 + b_3 z^3 + b_2 z^2 + b_1 z + b_0$. Let $D(z)$ be an element in $\mathbb{F}_{3^m}[z]$ with

$$D(z) = A(z) \cdot B(z) = d_{10} z^{10} + d_9 z^9 + \cdots + d_1 z + d_0.$$

Note that there are relationships $d_0 = a_0 b_0$, $d_9 = a_4 b_5 + a_5 b_4$, and $d_{10} = a_5 b_5$. We then have to compute $D' = (d_1, d_2, d_3, d_4, d_5, d_6, d_7, d_8)$. Let $Z = (z_1, z_2, z_3, z_4, z_5, z_6, z_7, z_8) = (1, 2, x, x+1, x+2, -x, -(x+1), -(x+2))$, where x is the generator of the polynomial base of $\mathbb{F}_{3^m}$ over $\mathbb{F}_3$. Let V_z be the Vandermonde matrix for Z. Then we have the following matrix equation.

$$
V_z D'^T =
\begin{pmatrix}
z_1 & z_1^2 & \cdots & z_1^7 & z_1^8 \\
z_2 & z_2^2 & \cdots & z_2^7 & z_2^8 \\
\vdots & \vdots & \ddots & \vdots & \vdots \\
z_7 & z_7^2 & \cdots & z_7^7 & z_7^8 \\
z_8 & z_8^2 & \cdots & z_8^7 & z_8^8
\end{pmatrix}
\begin{pmatrix} d_1 \\ d_2 \\ \vdots \\ d_7 \\ d_8 \end{pmatrix}
=
\begin{pmatrix}
D(z_1) - d_0 - d_9 z_1^9 - d_{10} z_1^{10} \\
D(z_2) - d_0 - d_9 z_2^9 - d_{10} z_2^{10} \\
\vdots \\
D(z_7) - d_0 - d_9 z_7^9 - d_{10} z_7^{10} \\
D(z_8) - d_0 - d_9 z_8^9 - d_{10} z_8^{10}
\end{pmatrix}
$$

$$
=
\begin{pmatrix}
A(z_1)B(z_1) - d_0 - d_9 z_1^9 - d_{10} z_1^{10} \\
A(z_2)B(z_2) - d_0 - d_9 z_2^9 - d_{10} z_2^{10} \\
\vdots \\
A(z_7)B(z_7) - d_0 - d_9 z_7^9 - d_{10} z_7^{10} \\
A(z_8)B(z_8) - d_0 - d_9 z_8^9 - d_{10} z_8^{10}
\end{pmatrix}
$$

$$
=
\begin{pmatrix}
A(1)B(1) - d_0 - d_9 - d_{10} \\
A(2)B(2) - d_0 + d_9 - d_{10} \\
\vdots \\
A(-(x+1))B(-(x+1)) - d_0 + d_9(x+1)^9 - d_{10}(x+1)^{10} \\
A(-(x+2))B(-(x+2)) - d_0 + d_9(x+2)^9 - d_{10}(x+2)^{10}
\end{pmatrix}. \qquad (4)
$$

Let

$$
\begin{aligned}
P_0 &= d_0 = a_0 b_0, & P_1 &= A(1)B(1), \\
P_2 &= A(2)B(2), & P_3 &= A(x)B(x), \\
P_4 &= A(x+1)B(x+1), & P_5 &= A(x+2)B(x+2), \\
P_6 &= A(-x)B(-x), & P_7 &= A(-(x+1))B(-(x+1)), \\
P_8 &= A(-(x+2))B(-(x+2)), & P_9 &= d_9 = a_4 b_5 + a_5 b_4, \\
P_{10} &= d_{10} = a_5 b_5. &
\end{aligned}
\tag{5}
$$

Using (4) and (5) we get

$$
D'^T = V_z^{-1}
\begin{pmatrix}
P_1 - P_0 - P_9 - P_{10} \\
P_2 - P_0 + P_9 - P_{10} \\
P_3 - P_0 - P_9 x^9 - P_{10} x^{10} \\
P_4 - P_0 - P_9(x+1)^9 - P_{10}(x+1)^{10} \\
P_5 - P_0 - P_9(x+2)^9 - P_{10}(x+2)^{10} \\
P_6 - P_0 + P_9 x^9 - P_{10} x^{10} \\
P_7 - P_0 + P_9(x+1)^9 - P_{10}(x+1)^{10} \\
P_8 - P_0 + P_9(x+2)^9 - P_{10}(x+2)^{10}
\end{pmatrix}
= V_z^{-1} P'.
\tag{6}
$$

The matrix V_z^{-1} is explicitly represented as $\frac{1}{\beta}(\gamma_{ij})$, where $\beta = x^6 + x^4 + x^2 \in \mathbb{F}_{3^m}$ and each γ_{ij} is presented in Appendix A.

Recall that we may compute $(\beta d_0, \beta d_1, \cdots, \beta d_{10})$ instead of $(d_0, d_1, \cdots, d_{10})$ to compute $D(z) = A(z) \cdot B(z)$ in $\mathcal{G}$ due to Remark 1. Let $E(z) = \beta D(z) = e_{10} z^{10} + e_9 z^9 + \cdots + e_1 z + e_0$. We can compute coefficients of $E(z)$ as follows:

$$
e_0 = \beta P_0, \quad (e_1, e_2, e_3, e_4, e_5, e_6, e_7, e_8)^T = (\gamma_{ij}) P', \quad e_9 = \beta P_9, \quad e_{10} = \beta P_{10}.
$$

Next, we consider the cost of the proposed RGM algorithm in a number of multiplications. Let M be the cost of a multiplication in $\mathbb{F}_{3^m}$. Then we assume that the cost of multiplication in a polynomial of degree $(m-1)$ and a polynomial of degree k is $(\frac{k+1}{m})M$ with $m > k$ using the shift-and-add method. The cost of (5) is then $12M$. We will explain the cost of (6). First, let α_1 be the cost of computation of the vector P'. Multiplications that needs to compute P' are $P_9 x^9, P_9(x+1)^9, P_9(x+2)^9, P_{10} x^{10}, P_{10}(x+1)^{10}, P_{10}(x+2)^{10}$. The degree of P_9 and P_{10} is $(m-1)$. Then α_1 is $(\frac{10+10+10+11+11+11}{m})M = (\frac{63}{m})M$. Next, let α_2 be the cost of computation of $(\gamma_{ij}) P'$. α_2 depends on the highest order of the column in (γ_{ij}). Then, α_2 is $(\frac{7+7+6+6+6+6}{m})M = (\frac{48}{m})M$. Last, let α_3 be costs of e_0, e_9, e_{10}. β is the polynomial of degree 6. Then α_3 is $(\frac{7+7+7}{m})M = (\frac{21}{m})M$. Therefore, the cost of a multiplication in the proposed algorithm is $(12 + \alpha_1 + \alpha_2 + \alpha_3)M = (12 + \frac{132}{m})M$. If $m \to \infty$, then the multiplication cost is $12M$.

The proposed algorithm is especially effective when multiplication in the finite field is implemented using a basic method, such as shift-and-add, so we used the

Table 1. Multiplication cost estimation

Multiplication method	The number of multiplications in $\mathbb{F}_{3^m}$
Gorla et al. algorithm [11]	15
Proposed RGM algorithm	$12 + \frac{132}{m}$

Table 2. Timing on a Core 2 Duo E6320 1.86GHz ($m = 97$)

Multiplication method	Multiplication in $\mathbb{F}_{3^{6m}}$	η_T pairing
Gorla et al. algorithm [11]	$73.2\mu s$	5.01ms
Proposed RGM algorithm	$\mathbf{68.3\mu s}$	$\mathbf{4.76ms}$

shift-and-add method to estimate the cost of (6). We estimate multiplication costs in Table 1.

4.3   Timing of the η_T Pairing

Algorithm 2 is an algorithm for computing the η_T pairing modyfing Algorithm 1 with the propsed RGM alorithm and the z-base. We implemented the η_T pairing over $\mathbb{F}_{3^{97}}$ on a Core 2 Duo E6320 1.86GHz with 1GB RAM using gcc 3.4.4. We show the timing of the multiplication in $\mathbb{F}_{3^{6m}}$ and the η_T pairing in Table 2.

The number of multiplications for the proposed algorithm is $(12 + \frac{132}{m})$. If $m = 97$, then the number of multiplications is $(12 + \frac{132}{97}) = (13 + \frac{35}{97})$. However, in the proposed RGM the number of additions increases by 212 for one multiplication compared to the algorithm of Gorla et al. in our implementation. Then the

---

Algorithm 2. The η_T pairing algorithm using RGMs and the z base

---

INPUT: $P(x_p, y_p), Q(x_q, y_q) \in E(\mathbb{F}_{3^m})[r]$
OUTPUT: $\eta_T(P, Q) \in \mathbb{F}_{3^{6m}}$
1: $y_p \leftarrow -y_p, d \leftarrow 1$
2: $v \leftarrow x_p + x_q + 1$
3: $R_0 \leftarrow -y_p v + (y_q - y_p)z + (y_q + y_p)z^2 + y_p(x_p + x_q)z^3 - (y_p v + y_q)z^4 + y_p(x_p + x_q)z^5$
4: **for** $i \leftarrow 0$ to $(n-1)/2$ **do**
5:     $v \leftarrow x_p + x_q + d$
6:     $R_1 \leftarrow -v^2 + (y_p y_q + v)z + (y_p y_q - v)z^2 + v(v+1)z^3 - (v^2 + y_p y_q)z^4 + (v^2 + v + 1)z^5$
7:     $R_0 \leftarrow R_0 R_1$ **(RGM)**
8:     $y_p \leftarrow -y_p$
9:     $x_q \leftarrow x_q^9, y_q \leftarrow y_q^9$
10:    $d \leftarrow ((d-1) \bmod 3)$
11:    $R_0 \leftarrow R_0^3$
12: **end for**
13: **return** R_0

---

timing of the multiplication in $\mathbb{F}_{3^{6m}}$ is almost 7 percent faster than that of the multiplication algorithm by Gorla et al., which is $68.3\mu s$. Moreover, the timing of the η_T pairing was almost 5 percent faster than that of the multiplication algorithm of Gorla et al., which is 4.76ms.

Remark 2. The loop unrolling technique has been adopted to implement parings as [3], which can reduce the number of multiplications needed to compute a pairing. We can use the proposed algorithm together with the loop unrolling technique.

5 Conclusion

In this study, we developed a residue group multiplication (RGM) algorithm in $\mathbb{F}_{3^{6m}}^{*}/\mathbb{F}_{3^m}^{*}$ to compute the η_T pairing. The proposed RGM algorithm takes $12 + \frac{132}{m}$ multiplications in $\mathbb{F}_{3^m}$, which reaches beyond the lower bound of the algorithm for multiplication in $\mathbb{F}_{3^{6m}}^{*}$ by Gorla et al. We can use the proposed RGM algorithm to compute the η_T pairing. Moreover, we implemented the η_T pairing on a Core 2 Duo E6320 1.86GHz with 1GB RAM using gcc 3.4.4 using the proposed RGM algorithm. The timing of the η_T pairing was almost 5 percent faster than that of the multiplication algorithm by Gorla et al., which is 4.76ms.

We expect that RGMs are applicable to other pairings, for example the Ate pairing using the Barreto-Naehrig curve [2], which has the embedding degree $k = 12$ defined over a large prime field.

References

1. Barreto, P., Galbraith, S., O'hEigeartaigh, C., Scott, S.: Efficient pairing computation on supersingular Abelian varieties. Designs, Codes and Cryptography 42(3), 239–271 (2007)
2. Barreto, P., Naehrig, M.: Pairing-friendly elliptic curves of prime order. In: Preneel, B., Tavares, S. (eds.) SAC 2005. LNCS, vol. 3897, pp. 319–331. Springer, Heidelberg (2006)
3. Beuchat, J.-L., Brisebarre, N., Detrey, J., Okamoto, E., Shirase, M., Takagi, T.: Algorithms and arithmetic operators for computing the η_T pairing in characteristic three. IEEE Transactions on Computers 57(11), 1454–1468 (2008)
4. Bodrato, M.: Towards optimal Toom-Cook multiplication for univariate and multivariate polynomials in characteristic 2 and 0. In: Carlet, C., Sunar, B. (eds.) WAIFI 2007. LNCS, vol. 4547, pp. 116–133. Springer, Heidelberg (2007)
5. Boneh, D., Di Crescenzo, G., Ostrovsky, R., Persiano, G.: Public key encryption with keyword search. In: Cachin, C., Camenisch, J.L. (eds.) EUROCRYPT 2004. LNCS, vol. 3027, pp. 506–522. Springer, Heidelberg (2004)
6. Boneh, D., Franklin, M.: Identity based encryption from the Weil pairing. SIAM Journal of Computing 32(3), 586–615 (2003)

7. Boneh, D., Gentry, C., Waters, B.: Collusion resistant broadcast encryption with short ciphertexts and private keys. In: Shoup, V. (ed.) CRYPTO 2005. LNCS, vol. 3621, pp. 258–275. Springer, Heidelberg (2005)

8. Brent, R., Gaudry, P., Thomé, E., Zimmermann, P.: Faster multiplication in GF(2)[x]. In: van der Poorten, A.J., Stein, A. (eds.) ANTS-VIII 2008. LNCS, vol. 5011, pp. 153–166. Springer, Heidelberg (2008)

9. Cantor, D.: On arithmetical algorithms over finite fields. J. Combinatorial Theory, Series A-50, 285–300 (1989)

10. Cook, S.: On the minimum computation time of functions. PhD thesis, Harvard University (1966)

11. Gorla, E., Puttmann, C., Shokrollahi, J.: Explicit formulas for efficient multiplication in $\mathbb{F}_{3^{6m}}$. In: Adams, C., Miri, A., Wiener, M. (eds.) SAC 2007. LNCS, vol. 4876, pp. 173–183. Springer, Heidelberg (2007)

12. Karatsuba, A., Ofman, Y.: Multiplication of multidigit numbers on automata. Soviet Physics-Doklady 7, 595–596 (1963)

13. Kerins, T., Marnane, W., Popovici, E., Barreto, P.: Efficient hardware for the Tate pairing calculation in characteristic three. In: Rao, J.R., Sunar, B. (eds.) CHES 2005. LNCS, vol. 3659, pp. 412–426. Springer, Heidelberg (2005)

14. Lempel, A., Winograd, S.: A new approach to error-correcting codes. IEEE Transactions on Information Theory IT-23, 503–508 (1977)

15. Schönhage, A.: Schnelle multiplikation von polynomen über körpern der Charakteristik 2. Acta Inf. 7, 395–398 (1977)

16. Shirase, M., Takagi, T., Choi, D., Han, D.-H., Kim, H.: Efficient computation of Eta pairing over binary field with Vandermonde matrix. ETRI Journal 31(2), 129–139 (2009)

17. Toom, A.: The complexity of a scheme of functional elements realizing the multiplication of integers. Soviet Mathematics 3, 714–716 (1963)

18. Winograd, S.: Arithmetic complexity of computations. SIAM, Philadelphia (1980)

A   Elements of Matrix (γ_{ij})

When a matrix V_z^{-1} in (6) is represented as $V_z^{-1} = \frac{1}{\beta}(\gamma_{ij})$, each element γ_{ij} is provided as follows, where x is the generator of the base of $\mathbb{F}_{3^m}$ over $\mathbb{F}_3$ and $\beta = x^6 + x^4 + x^2$.

$$V_z^{-1} = \frac{1}{\beta} \begin{pmatrix} \gamma_{11} & \gamma_{12} & \gamma_{13} & \gamma_{14} & \gamma_{15} & \gamma_{16} & \gamma_{17} & \gamma_{18} \\ \gamma_{21} & \gamma_{22} & \gamma_{23} & \gamma_{24} & \gamma_{25} & \gamma_{26} & \gamma_{27} & \gamma_{28} \\ \gamma_{31} & \gamma_{32} & \gamma_{33} & \gamma_{34} & \gamma_{35} & \gamma_{36} & \gamma_{37} & \gamma_{38} \\ \gamma_{41} & \gamma_{42} & \gamma_{43} & \gamma_{44} & \gamma_{45} & \gamma_{46} & \gamma_{47} & \gamma_{48} \\ \gamma_{51} & \gamma_{52} & \gamma_{53} & \gamma_{54} & \gamma_{55} & \gamma_{56} & \gamma_{57} & \gamma_{58} \\ \gamma_{61} & \gamma_{62} & \gamma_{63} & \gamma_{64} & \gamma_{65} & \gamma_{66} & \gamma_{67} & \gamma_{68} \\ \gamma_{71} & \gamma_{72} & \gamma_{73} & \gamma_{74} & \gamma_{75} & \gamma_{76} & \gamma_{77} & \gamma_{78} \\ \gamma_{81} & \gamma_{82} & \gamma_{83} & \gamma_{84} & \gamma_{85} & \gamma_{86} & \gamma_{87} & \gamma_{88} \end{pmatrix}$$

$$\gamma_{11} = -(x^6 + x^4 + x^2)$$
$$\gamma_{21} = -(x^6 + x^4 + x^2)$$
$$\gamma_{31} = 1$$
$$\gamma_{41} = 1$$
$$\gamma_{51} = 1$$
$$\gamma_{61} = 1$$
$$\gamma_{71} = 1$$
$$\gamma_{81} = 1$$

$$\gamma_{12} = x^6 + x^4 + x^2$$
$$\gamma_{22} = -(x^6 + x^4 + x^2)$$
$$\gamma_{32} = -1$$
$$\gamma_{42} = 1$$
$$\gamma_{52} = -1$$
$$\gamma_{62} = 1$$
$$\gamma_{72} = -1$$
$$\gamma_{82} = 1$$

$$\gamma_{13} = -(x^5 + x^3 + x)$$
$$\gamma_{23} = -(x^4 + x^2 + 1)$$
$$\gamma_{33} = x^5$$
$$\gamma_{43} = x^4$$
$$\gamma_{53} = x^3$$
$$\gamma_{63} = x^2$$
$$\gamma_{73} = x$$
$$\gamma_{83} = 1$$

$$\gamma_{14} = -(x^5 + 2x^4 + 2x^3 + x^2)$$
$$\gamma_{24} = -(x^4 + x^3 + x^2)$$
$$\gamma_{34} = (x + 1)^5$$
$$\gamma_{44} = (x + 1)^4$$
$$\gamma_{54} = (x + 1)^3$$
$$\gamma_{64} = (x + 1)^2$$
$$\gamma_{74} = (x + 1)$$
$$\gamma_{84} = 1$$

$$\gamma_{15} = -(x^5 + x^4 + 2x^3 + 2x^2)$$
$$\gamma_{25} = -(x^4 + 2x^3 + x^2)$$
$$\gamma_{35} = (x + 2)^5$$
$$\gamma_{45} = (x + 2)^4$$
$$\gamma_{55} = (x + 2)^3$$
$$\gamma_{65} = (x + 2)^2$$
$$\gamma_{75} = (x + 2)$$
$$\gamma_{85} =$$

$$\gamma_{16} = x^5 + x^3 + x$$
$$\gamma_{26} = -(x^4 + x^2 + 1)$$
$$\gamma_{36} = -x^5$$
$$\gamma_{46} = x^4$$
$$\gamma_{56} = -x^3$$
$$\gamma_{66} = x^2$$
$$\gamma_{76} = -x$$
$$\gamma_{86} = 1$$

$$\gamma_{17} = x^5 + 2x^4 + 2x^3 + x^2$$
$$\gamma_{27} = -(x^4 + x^3 + x^2)$$
$$\gamma_{37} = -(x + 1)^5$$
$$\gamma_{47} = (x + 1)^4$$
$$\gamma_{57} = -(x + 1)^3$$
$$\gamma_{67} = (x + 1)^2$$
$$\gamma_{77} = -(x + 1)$$
$$\gamma_{87} = 1$$

$$\gamma_{18} = x^5 + x^4 + 2x^3 + 2x^2$$
$$\gamma_{28} = -(x^4 + 2x^3 + x^2)$$
$$\gamma_{38} = -(x + 2)^5$$
$$\gamma_{48} = (x + 2)^4$$
$$\gamma_{58} = -(x + 2)^3$$
$$\gamma_{68} = (x + 2)^2$$
$$\gamma_{78} = -(x + 2)$$
$$\gamma_{88} = 1$$

Compact McEliece Keys from Goppa Codes

Rafael Misoczki and Paulo S.L.M. Barreto*

Departamento de Engenharia de Computação e Sistemas Digitais (PCS),
Escola Politécnica, Universidade de São Paulo, Brazil
{rmisoczki,pbarreto}@larc.usp.br

Abstract. The classical McEliece cryptosystem is built upon the class of Goppa codes, which remains secure to this date in contrast to many other families of codes but leads to very large public keys. Previous proposals to obtain short McEliece keys have primarily centered around replacing that class by other families of codes, most of which were shown to contain weaknesses, and at the cost of reducing in half the capability of error correction. In this paper we describe a simple way to reduce significantly the key size in McEliece and related cryptosystems using a subclass of Goppa codes, while also improving the efficiency of cryptographic operations to $\tilde{O}(n)$ time, and keeping the capability of correcting the full designed number of errors in the binary case.

1 Introduction

Quantum computers can potentially break most if not all conventional cryptosystems actually deployed in practice, namely, all systems based on the integer factorization problem (like RSA) or the discrete logarithm problem (like traditional or elliptic curve Diffie-Hellman and DSA, and also all of pairing-based cryptography).

Certain classical cryptosystems, inspired on computational problems of a nature entirely different from the above and potentially much harder to solve, remain largely unaffected by the threat of quantum computing, and have thus been called quantum-resistant or, more suggestively, 'post-quantum' cryptosystems. These include lattice-based cryptosystems and syndrome-based cryptosystems like McEliece [16] and Niederreiter [19]. Such systems usually have even a speed advantage over conventional schemes; for instance, both McEliece and Niederreiter encryption over a code of length n has time complexity $O(n^2)$, while Diffie-Hellman/DSA and (private exponent) RSA with n-bit keys have time complexity $O(n^3)$. On the other hand, they are plagued by very large keys compared to their conventional counterparts.

It is therefore of utmost importance to seek ways to reduce the key sizes for post-quantum cryptosystems while keeping their security level. The first steps

---

* Supported by the Brazilian National Council for Scientific and Technological Development (CNPq) under research productivity grant 312005/2006-7 and universal grant 485317/2007-9, and by the Science Foundation Ireland (SFI) as E. T. S. Walton Award fellow under grant 07/W.1/I1824.

M.J. Jacobson Jr., V. Rijmen, and R. Safavi-Naini (Eds.): SAC 2009, LNCS 5867, pp. 376–392, 2009.
© Springer-Verlag Berlin Heidelberg 2009

toward this goal were taken by Monico et al. using low density parity-check codes [18], by Gaborit using quasi-cyclic codes [8], and by Baldi and Chiaraluce using a combination of both [1].

However, these proposals were all shown to contain weaknesses [22]. In those proposals the trapdoor is protected essentially by no other means than a private permutation of the underlying code. The attack strategy consists of obtaining a solvable system of linear equations that the components of the permutation matrix must satisfy, and was successfully mounted due to the very constrained nature of the secret permutation (since it has to preserve the quasi-cyclic structure of the result) and the fact that the secret code is a subcode of a public code.

A dedicated fix to the problems in [1] is proposed in [2]. More recently, Berger et al. [3] showed how to circumvent the drawbacks of Gaborit's original scheme and remove the weaknesses pointed out in [22] by means of two techniques:

1. Extracting block-shortened public codes from very large private codes, exploiting Wieschebrink's theorem on the NP-completeness of distinguishing punctured codes [29];
2. Working with subfield subcodes over an intermediate subfield between the base field and the extension field of the original code.

These two techniques were successfully applied to quasi-cyclic codes, yet we will see that their applicability is not restricted to that class.

Our contribution: In this paper we propose the class of *quasi-dyadic* Goppa codes, which admit a very compact parity-check or a generator matrix representation, for efficiently instantiating syndrome-based cryptosystems. We stress that we are not proposing any new cryptosystem, but rather a technique to obtain efficient parameters and algorithms for such systems, current or future. In contrast to many other proposed families of codes [10,11,22,27], Goppa codes have withstood cryptanalysis quite well, and despite considerable progress in the area [14,26] (see also [6] for a survey) they remain essentially unscathed since they were suggested with the very first syndrome-based cryptosystem known, namely, the original McEliece scheme. Our method produces McEliece-type keys that are up to a factor $t = \tilde{O}(n)$ smaller than keys produced from generic t-error correcting Goppa codes of length n in characteristic 2. In the binary case it also retains the ability of correcting the full designed number of errors rather than just half as many, a feature that is missing in all previous attempts at constructing compact codes for cryptographic purposes, including [3]. Moreover, the complexity of all typical cryptographic operations become $\tilde{O}(n)$; specifically, under the common cryptographic setting $t = O(n/\lg n)$, code generation, encryption and decryption all have asymptotic complexity $O(n \lg n)$.

The remainder of this paper is organized as follows. Section 2 introduces some basic concepts of coding theory. In section 3 we describe our proposal of using binary Goppa codes in quasi-dyadic form, and how to build them. We consider hardness issues in Section 4, and efficiency issues, including guidelines on how to choose parameters, in Section 5. We conclude in Section 6.

2   Preliminaries

In what follows all vector and matrix indices are numbered from zero onwards.

Definition 1. *Given a ring $\mathcal{R}$ and a vector $h = (h_0, \ldots, h_{n-1}) \in \mathcal{R}^n$, the* dyadic *matrix $\Delta(h) \in \mathcal{R}^{n \times n}$ is the symmetric matrix with components $\Delta_{ij} = h_{i \oplus j}$ where $\oplus$ stands for bitwise exclusive-or on the binary representations of the indices. The sequence h is called its* signature. *The set of dyadic $n \times n$ matrices over $\mathcal{R}$ is denoted $\Delta(\mathcal{R}^n)$. Given $t > 0$, $\Delta(t, h)$ denotes $\Delta(h)$ truncated to its first t rows.*

One can recursively characterize a dyadic matrix when n is a power of 2: any 1×1 matrix is dyadic, and for $k > 0$ any $2^k \times 2^k$ dyadic matrix M has the form

$$M = \begin{bmatrix} A & B \\ B & A \end{bmatrix}$$

where A and B are $2^{k-1} \times 2^{k-1}$ dyadic matrices. It is not hard to see that the signature of a dyadic matrix coincides with its first row. Dyadic matrices form a commutative subring of $\mathcal{R}^{n \times n}$ as long as $\mathcal{R}$ is commutative [12].

Definition 2. *A* dyadic permutation *is a dyadic matrix $\Pi^i \in \Delta(\{0, 1\}^n)$ whose signature is the i-th row of the identity matrix.*

A dyadic permutation is clearly an involution, i.e. $(\Pi^i)^2 = I$. The i-th row (or equivalently the i-th column) of the dyadic matrix defined by a signature h can be written $\Delta(h)_i = h\Pi^i$.

Definition 3. *A* quasi-dyadic *matrix is a (possibly non-dyadic) block matrix whose component blocks are dyadic submatrices.*

Quasi-dyadic matrices are at the core of our proposal. We will be mainly concerned with the case $\mathcal{R} = \mathbb{F}_q$, the finite field with q (a prime power) elements.

Definition 4. *Given two disjoint sequences $z = (z_0, \ldots, z_{t-1}) \in \mathbb{F}_q^t$ and $L = (L_0, \ldots, L_{n-1}) \in \mathbb{F}_q^n$ of distinct elements, the* Cauchy matrix $C(z, L)$ *is the $t \times n$ matrix with elements $C_{ij} = 1/(z_i - L_j)$, i.e.*

$$C(z, L) = \begin{bmatrix} \dfrac{1}{z_0 - L_0} & \cdots & \dfrac{1}{z_0 - L_{n-1}} \\ \vdots & \ddots & \vdots \\ \dfrac{1}{z_{t-1} - L_0} & \cdots & \dfrac{1}{z_{t-1} - L_{n-1}} \end{bmatrix}.$$

Cauchy matrices have the property that all of their submatrices are nonsingular [25]. Notice that, in general, Cauchy matrices are not dyadic and vice-versa, although the intersection of these two classes is non-empty in characteristic 2.

Definition 5. *Given $t > 0$ and a sequence $L = (L_0, \ldots, L_{n-1}) \in \mathbb{F}_q^n$, the* Vandermonde *matrix $\mathrm{vdm}(t, L)$ is the $t \times n$ matrix with elements $V_{ij} = L_j^i$.*

Definition 6. *Given a sequence $L = (L_0, \ldots, L_{n-1}) \in \mathbb{F}_q^n$ of distinct elements and a sequence $D = (D_0, \ldots, D_{n-1}) \in \mathbb{F}_q^n$ of nonzero elements, the* General-ized Reed-Solomon code $GRS_r(L, D)$ *is the* $[n, k, r]$ *linear error-correcting code defined by the parity-check matrix*

$$H = \text{vdm}(r - 1, L) \cdot \text{diag}(D).$$

An alternant code is a subfield subcode of a Generalized Reed-Solomon code.

Let p be a prime power, let $q = p^d$ for some d, and let $\mathbb{F}_q = \mathbb{F}_p[x]/b(x)$ for some irreducible polynomial $b(x) \in \mathbb{F}_p[x]$ of degree d. Given a code specified by a parity-check matrix $H \in \mathbb{F}_q^{t \times n}$, the *trace construction* derives from it an $\mathbb{F}_p$-subfield subcode by writing the $\mathbb{F}_p$ coefficients of each $\mathbb{F}_q$ component of H onto d successive rows of a parity-check matrix $T_d(H) \in \mathbb{F}_p^{dt \times n}$ for the subcode. The related *co-trace* parity-check matrix $T'_d(H) \in \mathbb{F}_p^{dt \times n}$, equivalent to $T_d(H)$ by a left permutation, is obtained from H by writing the $\mathbb{F}_p$ coefficients of terms of equal degree from all components on a column of H onto successive rows of $T'_d(H)$.

Thus, given elements $u_i(x) = u_{i,0} + \cdots + u_{i,d-1}x^{d-1} \in \mathbb{F}_q = \mathbb{F}_p[x]/b(x)$, the trace construction maps a column $(u_0, \ldots, u_{t-1})^\mathsf{T}$ from H to the column $(u_{0,0}, \ldots, u_{0,d-1}; \ldots; u_{t-1,0}, \ldots, u_{t-1,d-1})^\mathsf{T}$ on the trace matrix $T_d(H)$, and to the column $(u_{0,0}, \ldots, u_{t-1,0}; \ldots; u_{0,d-1}, \ldots, u_{t-1,d-1})^\mathsf{T}$ on the co-trace matrix $T'_d(H)$.

Finally, one of the most important families of linear error-correcting codes for cryptographic purposes is that of Goppa codes:

Definition 7. *Given a prime power p, $q = p^d$ for some d, a sequence $L = (L_0, \ldots, L_{n-1}) \in \mathbb{F}_q^n$ of distinct elements and a polynomial $g(x) \in \mathbb{F}_q[x]$ of degree t such that $g(L_i) \neq 0$ for $0 \leqslant i < n$, the* Goppa code $\Gamma(L, g)$ *over $\mathbb{F}_p$ is the alternant code over $\mathbb{F}_p$ corresponding to $GRS_t(L, D)$ where $D = (g(L_0)^{-1}, \ldots, g(L_{n-1})^{-1})$, and its minimum distance is at least $2t + 1$.*

An irreducible Goppa code in characteristic 2 can correct up to t errors us-ing Patterson's algorithm [23], or slightly more using Bernstein's list decoding method [5], and t errors can still be corrected by suitable decoding algorithms if the generator $g(x)$ is not irreducible[1]. In all other cases no algorithm is known that can correct more than $t/2$ errors (or just a few more).

3   Goppa Codes in Cauchy and Dyadic Form

A property of Goppa codes that is central to our proposal is that they admit a parity-check matrix in Cauchy form:

---

[1] For instance, one can equivalently view the binary Goppa code as the alternant code defined by the generator polynomial $g^2(x)$, in which case any alternant decoder will decode t errors. We are grateful to Nicolas Sendrier for pointing this out.

Theorem 1 ([28]). *The Goppa code generated by a monic polynomial $g(x) = (x - z_0) \ldots (x - z_{t-1})$ without multiple zeros admits a parity-check matrix of the form $H = C(z, L)$, i.e. $H_{ij} = 1/(z_i - L_j)$, $0 \leqslant i < t$, $0 \leqslant j < n$.*

This theorem (also appearing in [15, Ch. 12, §3, Pr. 5]) is entirely general when one considers the factorization of the Goppa polynomial over its splitting field, in which case a single root of g is enough to completely characterize the code. For simplicity, we will restrict our attention to the case where all roots of that polynomial are in the field $\mathbb{F}_q$ itself.

3.1   Building a Binary Goppa Code in Dyadic Form

We now show how to build a binary Goppa code that admits a parity-check matrix in dyadic form. To this end we seek a way to construct dyadic Cauchy matrices. The following theorem characterizes all matrices of this kind.

Theorem 2. *Let $H \in \mathbb{F}_q^{n \times n}$ with $n > 1$ be simultaneously a dyadic matrix $H = \Delta(h)$ for some $h \in \mathbb{F}_q^n$ and a Cauchy matrix $H = C(z, L)$ for two disjoint sequences $z \in \mathbb{F}_q^n$ and $L \in \mathbb{F}_q^n$ of distinct elements. Then $\mathbb{F}_q$ is a field of characteristic 2, h satisfies*

$$\frac{1}{h_{i \oplus j}} = \frac{1}{h_i} + \frac{1}{h_j} + \frac{1}{h_0}, \tag{1}$$

and $z_i = 1/h_i + \omega$, $L_j = 1/h_j + 1/h_0 + \omega$ for some $\omega \in \mathbb{F}_q$.

Proof. Since a dyadic matrix is symmetric, the sequences that define it must satisfy $1/(z_i - L_j) = 1/(z_j - L_i)$, hence $L_j = z_i + L_i - z_j$ for all i and j. Then $z_i + L_i$ must be a constant α, and taking $i = 0$ in particular this simplifies to $L_j = \alpha - z_j$. Substituting back into the definition $M_{ij} = 1/(z_i - L_j)$ one sees that $H_{ij} = 1/(z_i + z_j + \alpha)$. But dyadic matrices also have constant diagonal, namely, $H_{ii} = 1/(2z_i + \alpha) = h_0$. This is only possible if all z_i are equal (contradicting the definition of a Cauchy matrix), or else if the characteristic of the field is 2, as claimed.

In this case we see that $\alpha = 1/h_0$, and hence $H_{ij} = 1/(z_i + z_j + 1/h_0)$. Plugging in the definition $H_{ij} = h_{i \oplus j}$ we get $1/H_{ij} = 1/h_{i \oplus j} = z_i + z_j + 1/h_0$, and taking $j = 0$ in particular this yields $1/h_i = z_i + z_0 + 1/h_0$, or simply $z_i = 1/h_i + 1/h_0 + z_0$. Substituting back one obtains $1/h_{i \oplus j} = z_i + z_j + 1/h_0 = 1/h_i + 1/h_0 + z_0 + 1/h_j + 1/h_0 + z_0 + 1/h_0 = 1/h_i + 1/h_j + 1/h_0$, as expected.

Finally, define $\omega = 1/h_0 + z_0$ and substitute into the derived relations $z_i = 1/h_i + 1/h_0 + z_0$ and $L_j = \alpha - z_j$ to get $z_i = 1/h_i + \omega$ and $L_j = 1/h_j + 1/h_0 + \omega$, as desired.   □

Therefore all we need is a method to solve Equation 1. The technique we propose consists of simply choosing distinct nonzero h_0 and h_i at random where i scans all powers of two smaller than n, and setting all other values as

$$h_{i+j} \leftarrow \frac{1}{\dfrac{1}{h_i} + \dfrac{1}{h_j} + \dfrac{1}{h_0}}$$

for $0 < j < i$ (so that $i + j = i \oplus j$), as long as this value is well-defined. Algorithm 1 captures this idea. Since each element of the signature h is assigned a value exactly once, its running time is $O(n)$ steps. The notation $u \xleftarrow{\$} U$ means that variable u is uniformly sampled at random from set U. For convenience we also define the *essence* of h to be the sequence $\eta_s = 1/h_{2^s} + 1/h_0$ for $s = 0, \ldots, \lceil \lg n \rceil - 1$ together with $\eta_{\lceil \lg n \rceil} = 1/h_0$, so that, for $i = \sum_{k=0}^{\lceil \lg n \rceil - 1} i_k 2^k$, $1/h_i = \eta_{\lceil \lg n \rceil} + \sum_{k=0}^{\lceil \lg n \rceil - 1} i_k \eta_k$.

Algorithm 1. Constructing a binary Goppa code in dyadic form

INPUT: q (a power of 2), $n \leqslant q/2$, t.

OUTPUT: Support L, generator polynomial g, dyadic parity-check matrix H for a binary Goppa code $\Gamma(L, g)$ of length n and design distance $2t + 1$ over $\mathbb{F}_q$, and the essence η of the signature of H.

1: $U \leftarrow \mathbb{F}_q \setminus \{0\}$
   ▷ Choose the dyadic signature $(h_0, \ldots, h_{n-1})$. N.B. Whenever h_j with $j > 0$ is taken from U, so is $1/(1/h_j + 1/h_0)$ to prevent a potential spurious intersection between z and L.
2: $h_0 \xleftarrow{\$} U$
3: $\eta_{\lceil \lg n \rceil} \leftarrow 1/h_0$
4: $U \leftarrow U \setminus \{h_0\}$
5: **for** $s \leftarrow 0$ **to** $\lceil \lg n \rceil - 1$ **do**
6:     $i \leftarrow 2^s$
7:     $h_i \xleftarrow{\$} U$
8:     $\eta_s \leftarrow 1/h_i + 1/h_0$
9:     $U \leftarrow U \setminus \{h_i, 1/(1/h_i + 1/h_0)\}$
10:     **for** $j \leftarrow 1$ **to** $i - 1$ **do**
11:        $h_{i+j} \leftarrow 1/(1/h_i + 1/h_j + 1/h_0)$
12:        $U \leftarrow U \setminus \{h_{i+j}, 1/(1/h_{i+j} + 1/h_0)\}$
13:     **end for**
14: **end for**
15: $\omega \xleftarrow{\$} \mathbb{F}_q$
   ▷ Assemble the Goppa generator polynomial:
16: **for** $i \leftarrow 0$ **to** $t - 1$ **do**
17:     $z_i \leftarrow 1/h_i + \omega$
18: **end for**
19: $g(x) \leftarrow \prod_{i=0}^{t-1} (x - z_i)$
   ▷ Compute the support:
20: **for** $j \leftarrow 0$ **to** $n - 1$ **do**
21:     $L_j \leftarrow 1/h_j + 1/h_0 + \omega$
22: **end for**
23: $h \leftarrow (h_0, \ldots, h_{n-1})$
24: $H \leftarrow \Delta(t, h)$
25: **return** L, g, H, η

Theorem 3. *Algorithm 1 produces up to $\prod_{i=0}^{\lceil \lg n \rceil} (q - 2^i)$ Goppa codes in dyadic form.*

Proof. Each dyadic signature produced by Algorithm 1 is entirely determined by the values h_0 and h_{2^s} for $s = 0, \ldots, \lceil \lg n \rceil - 1$ chosen at steps 2 and 7 (ω only produces equivalent codes). Along the loop at line 5, exactly $2i = 2^{s+1}$ elements are erased from U, corresponding to the choices of $h_{2^s} \ldots h_{2^{s+1}-1}$. At the end of that loop, $2 + 2\sum_{\ell=0}^{s} 2^\ell = 2^{s+2}$ elements have been erased in total. Hence at the beginning of each step of the loop only 2^{s+1} elements had been erased from U, i.e. there are $q - 2^{s+1}$ elements in U to choose h_{2^s} from, and $q - 1$ possibilities for h_0. Therefore this construction potentially yields up to $(q - 1) \prod_{s=0}^{\lceil \lg n \rceil - 1} (q - 2^{s+1}) = \prod_{i=0}^{\lceil \lg n \rceil} (q - 2^i)$ possible codes.   □

Theorem 3 actually establishes the number of distinct essences of dyadic signatures corresponding to Cauchy matrices. The roots of the Goppa polynomial are completely specified by the first $\lceil \lg t \rceil$ elements of the essence η together with $\eta_{\lceil \lg n \rceil}$, namely, $z_i = \eta_{\lceil \lg n \rceil} + \sum_{k=0}^{\lceil \lg t \rceil - 1} i_k \eta_k$, disregarding the ω term which is implicit in the choice of $\eta_{\lceil \lg n \rceil}$. We see that any permutation of the essence elements $\eta_0, \ldots, \eta_{\lceil \lg t \rceil - 1}$ only changes the order of those roots. Since the Goppa polynomial itself is defined by its roots regardless of their order, the total number of possible Goppa polynomials is therefore $\left(\prod_{i=0}^{\lceil \lg t \rceil} (q - 2^i) \right) / \lceil \lg t \rceil! \approx (q-t) \binom{q}{\lceil \lg t \rceil}$.

For $n \approx q/2$ the number of dyadic codes can be approximated by $q^m Q = 2^{m^2} Q$ where $Q = \prod_{i=1}^{\infty} (1 - 1/2^i) \approx 0.2887881$. We will also see that the number of quasi-dyadic codes, which we describe next and propose for cryptographic applications, is larger than this. Before we proceed, however, it is interesting to notice that one of the reasons the attack proposed in [22] succeeds against certain quasi-cyclic codes, besides the constrained structure of the applied permutation, is that those schemes start from a known BCH or Reed-Solomon code which is unique up to the choice of a primitive element from the underlying finite field. Thus, in those proposals an initial code over $\mathbb{F}_{2^m}$ is at best chosen from a set of $O(2^m)$ codes. In comparison, we start from a secret code sampled from a much larger family of $O(2^{m^2})$ codes. For instance, while those proposals have only 2^{15} starting points over $\mathbb{F}_{2^{16}}$, our scheme can sample a family with more than 2^{254} codes over the same field. The main protection of the hidden trapdoor is, of course, the block puncturing process and the more complex blockwise permutation of the initial secret code, as detailed next.

3.2   Constructing Quasi-Dyadic, Permuted Subfield Subcodes

To complete the construction it is necessary to choose a compact generator matrix for the subfield subcode. Although the parity check matrix H built by Algorithm 1 is dyadic over $\mathbb{F}_q$, the usual trace construction leads to a generator of the dual code that most probably violates the dyadic symmetry. However, by representing each field element to a basis of $\mathbb{F}_q$ over the subfield $\mathbb{F}_p$, one can view H as a superposition of $d = [\mathbb{F}_q : \mathbb{F}_p]$ distinct dyadic matrices over $\mathbb{F}_p$, and each of them can be stored in a separate dyadic signature.

A cryptosystem cannot be securely defined on a Goppa code specified directly by a parity-check matrix in Cauchy form, since this would immediately reveal the Goppa polynomial $g(x)$: it suffices to solve the overdefined linear system $z_i - L_j = 1/H_{ij}$ consisting of tn equations in $t + n$ unknowns.

Algorithm 1 generates fully dyadic codes. We now show how to integrate the techniques of Berger et al. with Algorithm 1 so as to build quasi-dyadic subfield subcodes whose parity-check matrix is a non-dyadic matrix composed of blocks of dyadic submatrices. The principle to follow here is to *select, permute, and scale* the columns of the original parity-check matrix so as to preserve quasi-dyadicity in the target subfield subcode and the distribution of introduced errors in cryptosystems. A similar process yields a generator matrix in convenient quasi-dyadic, systematic form.

For the desired security level (see the discussion in Section 5.1), choose $p = 2^s$ for some s, $q = p^d = 2^m$ for some d with $m = ds$, a code length n and a design number of correctable errors t such that $n = \ell t$ for some $\ell > d$. For simplicity we assume that t is a power of 2, but the following construction method can be modified to work with other values.

Run Algorithm 1 to produce a code over $\mathbb{F}_q$ whose length $N \gg n$ is a large multiple of t not exceeding the largest possible length $q/2$, so that the constructed $t \times N$ parity-check matrix $\hat{H}$ can be viewed as a sequence of N/t dyadic blocks $[B_0 \mid \cdots \mid B_{N/t-1}]$ of size $t \times t$ each. Select uniformly at random ℓ distinct blocks $B_{i_0}, \ldots, B_{i_{\ell-1}}$ in any order from $\hat{H}$, together with ℓ dyadic permutations $\Pi^{j_0}, \ldots, \Pi^{j_{\ell-1}}$ of size $t \times t$ and ℓ nonzero scale factors $\sigma_0, \ldots, \sigma_{\ell-1} \in \mathbb{F}_p$. Let $\hat{H}' = [B_{i_0}\Pi^{j_0} \mid \cdots \mid B_{i_{\ell-1}}\Pi^{j_{\ell-1}}] \in (\mathbb{F}_q^{t \times t})^\ell$ and $\Sigma = \mathrm{diag}(\sigma_0 I_t, \ldots, \sigma_{\ell-1} I_t) \in (\mathbb{F}_p^{t \times t})^{\ell \times \ell}$. Compute the co-trace matrix $H' = T'_d(\hat{H}'\Sigma) = T'_d(\hat{H}')\Sigma \in (\mathbb{F}_p^{t \times t})^{d \times \ell}$ and finally the systematic form H of H'. Notice that, if the systematic form of $T'_d(\hat{H}')$ is H_0, then $H = U^{-1}H_0 V$ where $U = \mathrm{diag}(\sigma_0 I_t, \ldots, \sigma_{\ell-d-1} I_t)$ and $V = \mathrm{diag}(\sigma_{\ell-d} I_t, \ldots, \sigma_{\ell-1} I_t)$.

The resulting parity-check matrix defines a code of length n and dimension $k = n - dt$ over $\mathbb{F}_p$, and since all block operations performed during the Gaussian elimination are carried out in the ring $\Delta(\mathbb{F}_p^t)$, the result still consists of dyadic submatrices which can be represented by a signature of length t. Hence the whole matrix can be stored in an area a factor t smaller than a general matrix. However, the dyadic submatrices that appear in this process are not necessarily nonsingular, as they are not associated to a Cauchy matrix anymore; should all the submatrices on a column be found to be singular (above or below the diagonal, according to the direction of this process) so that no pivot is possible, the whole block containing that column may be replaced by another block $B_{j'}$ chosen at random from $\hat{H}$ as above.

The trapdoor information consisting of the essence η of h, the sequence $(i_0, \ldots, i_{\ell-1})$ of blocks, the sequence $(j_0, \ldots, j_{\ell-1})$ of dyadic permutation identifiers, and the sequence of scale factors $(\sigma_0, \ldots, \sigma_{\ell-1})$, relates the public code defined by H with the private code defined by $\hat{H}$. The space occupied by the trapdoor information is thus $m^2 + \ell \lg N + \ell s$ bits. If one starts with the largest possible $N = 2^{m-1}$, this simplifies to the maximal size of $m^2 + \ell(m - 1 + s)$ bits.

The total space occupied by the essential part of the resulting generator (or parity-check) matrix over $\mathbb{F}_p$ is $dt \times (n - dt)/t = dk$ $\mathbb{F}_p$ elements, or mk bits – a factor t better than plain Goppa codes, which occupy $k(n - k) = mkt$ bits. Had t not been chosen to be a power of 2, say, $t = 2^u v$ where $v > 1$ is odd, the cost of multiplying $t \times t$ matrices would be in general $O(2^u uv^3)$ rather than simply $O(2^u u)$, and the final parity-check matrix would be compressed by only a factor 2^u.

For each code produced by Algorithm 1, the number of codes generated by this construction is $\binom{N/t}{\ell} \times \ell! \times t^\ell \times (r - 1)^\ell$, hence $\binom{N/t}{\ell} \times \ell! \times t^\ell \times (r - 1)^\ell \times \prod_{i=0}^{\lceil \lg N \rceil} (q - 2^i)$ codes are possible in principle.

3.3   A Toy Example

Let $\mathbb{F}_{2^5} = \mathbb{F}_2[u]/(u^5 + u^2 + 1)$. The dyadic signature

$$h = (u^{20}, u^3, u^6, u^{28}, u^9, u^{29}, u^4, u^{22}, u^{12}, u^5, u^{10}, u^2, u^{24}, u^{26}, u^{25}, u^{15})$$

and the offset $\omega = u^{21}$ define a 2-error correcting binary Goppa code of length $N = 16$ with $g(x) = (x - u^{12})(x - u^{15})$ and support $L = (u^{21}, u^{29}, u^{19}, u^{26}, u^6, u^{16}, u^7, u^5, u^{25}, u^3, u^{11}, u^{28}, u^{27}, u^9, u^{22}, u^2)$. The associated parity-check matrix built according to Theorem 1 is

$$\hat{H} = \begin{bmatrix} u^{20} & u^3 & u^6 & u^{28} & u^9 & u^{29} & u^4 & u^{22} & u^{12} & u^5 & u^{10} & u^2 & u^{24} & u^{26} & u^{25} & u^{15} \\ u^3 & u^{20} & u^{28} & u^6 & u^{29} & u^9 & u^{22} & u^4 & u^5 & u^{12} & u^2 & u^{10} & u^{26} & u^{24} & u^{15} & u^{25} \end{bmatrix},$$

with eight 2×2 blocks $B_0, \ldots, B_7$ as indicated. From this we extract the shortened, rearranged and permuted sequence $\hat{H}' = [B_7 \Pi^0 \mid B_5 \Pi^1 \mid B_1 \Pi^0 \mid B_2 \Pi^1 \mid B_3 \Pi^0 \mid B_6 \Pi^1 \mid B_4 \Pi^0]$ (because in this example the subfield is the base field itself, all scale factors have to be 1), i.e.:

$$\hat{H} = \begin{bmatrix} u^{25} & u^{15} & u^2 & u^{10} & u^6 & u^{28} & u^{29} & u^9 & u^4 & u^{22} & u^{26} & u^{24} & u^{12} & u^5 \\ u^{15} & u^{25} & u^{10} & u^2 & u^{28} & u^6 & u^9 & u^{29} & u^{22} & u^4 & u^{24} & u^{26} & u^5 & u^{12} \end{bmatrix},$$

whose co-trace matrix over $\mathbb{F}_2$ has the systematic form:

$$H = \begin{bmatrix} 0 & 1 & 0 & 1 & 1 & 0 & 0 & 0 & 0 & 0 & 0 & 0 & 0 & 0 \\ 1 & 0 & 1 & 0 & 0 & 1 & 0 & 0 & 0 & 0 & 0 & 0 & 0 & 0 \\ 0 & 1 & 0 & 0 & 0 & 0 & 1 & 0 & 0 & 0 & 0 & 0 & 0 & 0 \\ 1 & 0 & 0 & 0 & 0 & 0 & 0 & 1 & 0 & 0 & 0 & 0 & 0 & 0 \\ 0 & 0 & 1 & 1 & 0 & 0 & 0 & 0 & 1 & 0 & 0 & 0 & 0 & 0 \\ 0 & 0 & 1 & 1 & 0 & 0 & 0 & 0 & 0 & 1 & 0 & 0 & 0 & 0 \\ 0 & 1 & 1 & 0 & 0 & 0 & 0 & 0 & 0 & 0 & 1 & 0 & 0 & 0 \\ 1 & 0 & 0 & 1 & 0 & 0 & 0 & 0 & 0 & 0 & 0 & 1 & 0 & 0 \\ 1 & 1 & 0 & 0 & 0 & 0 & 0 & 0 & 0 & 0 & 0 & 0 & 1 & 0 \\ 1 & 1 & 0 & 0 & 0 & 0 & 0 & 0 & 0 & 0 & 0 & 0 & 0 & 1 \end{bmatrix} = [M^\mathsf{T} \mid I_{n-k}],$$

from which one readily obtains the $k \times n = 4 \times 14$ generator matrix in systematic form:

$$G = \begin{bmatrix} 1 & 0 & 0 & 0 & 0 & 1 & 0 & 1 & 0 & 0 & 0 & 1 & 1 & 1 \\ 0 & 1 & 0 & 0 & 1 & 0 & 1 & 0 & 0 & 0 & 1 & 0 & 1 & 1 \\ 0 & 0 & 1 & 0 & 0 & 1 & 0 & 0 & 1 & 1 & 1 & 0 & 0 & 0 \\ 0 & 0 & 0 & 1 & 1 & 0 & 0 & 0 & 1 & 1 & 0 & 1 & 0 & 0 \end{bmatrix} = [I_k \mid M],$$

where both G and H share the essential part M:

$$M = \begin{bmatrix} \mathbf{0}\,\mathbf{1}\,\mathbf{0}\,\mathbf{1}\,0\,0\,0\,1\,1\,1 \\ 1\,0\,1\,0\,0\,0\,1\,0\,1\,1 \\ \mathbf{0}\,\mathbf{1}\,\mathbf{0}\,\mathbf{0}\,1\,1\,1\,0\,0\,0 \\ 1\,0\,0\,0\,1\,1\,0\,1\,0\,0 \end{bmatrix},$$

which is entirely specified by the elements in boldface and can thus be stored in 20 bits instead of, respectively, $4 \cdot 14 = 56$ and $10 \cdot 14 = 140$ bits.

4 Assessing the Hardness of Decoding Quasi-Dyadic Codes

The original McEliece (or, for that matter, the original Niederreiter) schemes are perhaps better described as a candidate *trapdoor one-way functions* rather than full-fledged public-key encryption schemes. Such functions are used in cryptography in many different settings, each with different security requirements, and we do not consider such applications in this paper. Instead we focus purely on the question of inverting the trapdoor function, in other words, decoding.

As we pointed out in Section 1, the well-studied class of Goppa codes remains one of the best choices to instantiate McEliece-like schemes. Although our proposal is ultimately based on Goppa codes, one may wonder whether or not the highly composite nature of the Goppa generator polynomial $g(x)$, or the peculiar structure of the quasi-dyadic parity-check and generator matrices, leak any information that might facilitate decoding without knowledge of the trapdoor.

Yet, any alternant code can be written in Goppa-like fashion by using the diagonal component of its default parity-check matrix (see Definition 6) to interpolate a generating polynomial (not necessarily of degree t) that is composite with high probability. We are not aware of any way this fact could be used to facilitate decoding without full knowledge of the code structure, and clearly any result in this direction would affect most of the alternant codes proposed for cryptographic purposes to date.

Otmani et al.'s attack against quasi-cyclic codes [22] could be modified to work against Goppa codes in dyadic form. For this reason we adopt the same countermeasures proposed by Berger et al. to thwart it for cyclic codes, namely, working with a block-shortened subcode of a very large code as described in Section 3.2. This idea also build upon the work of Wieschebrink [29] who proved that deciding whether a code is equivalent to a shortened code is NP-complete. In our case, the result is to hide the Cauchy structure of the private code in a general dyadic structure, rather than disguising a quasi-cyclic code as another one with the same symmetry.

We now give a reduction of the problem of decoding the particular class of quasi-dyadic codes to the well-studied syndrome decoding problem, classical in coding theory and known to be NP-complete [4].

Definition 8 (Syndrome decoding). *Let $\mathbb{F}_q$ be a finite field, and let (H, w, s) be a triple consisting of a matrix $H \in \mathbb{F}_q^{r \times n}$, an integer $w < n$, and a vector*

$s \in \mathbb{F}_q^r$. Does there exist a vector $e \in \mathbb{F}_q^n$ of Hamming weight $\mathsf{wt}(e) \leqslant w$ such that $He^\mathsf{T} = s^\mathsf{T}$?

The corresponding problem for quasi-dyadic matrices reads:

Definition 9 (Quasi-Dyadic Syndrome Decoding). *Let $\mathbb{F}_q$ be a finite field, and let (H, w, s) be a triple consisting of a quasi-dyadic matrix $H \in \Delta(\mathbb{F}_q^\ell)^{r \times n}$, an integer $w < \ell n$, and a vector $s \in \mathbb{F}_q^{\ell r}$. Does there exist a vector $e \in \mathbb{F}_q^{\ell n}$ of Hamming weight $\mathsf{wt}(e) \leqslant w$ such that $He^\mathsf{T} = s^\mathsf{T}$?*

Theorem 4. *The quasi-dyadic syndrome decoding problem (QD-SDP) is polynomially equivalent to the syndrome decoding problem (SDP). In other words, decoding quasi-dyadic codes is as hard in the worst case as decoding general codes.*

Proof. The QD-SDP, being an instance of the SDP restricted to a particular class of codes, is clearly a decision problem in NP.

Consider now a generic instance $(H', w', s') \in \mathbb{F}_q^{r \times n} \times \mathbb{Z} \times \mathbb{F}_q^r$ of the SDP. Assume one is given an oracle that solves the QD-SDP over $\Delta(\mathbb{F}_q^\ell)$ for some given $\ell > 0$. Let $v_\ell \in \mathbb{F}_q^\ell$ be the all-one vector, i.e. $(v_\ell)_j = 1$ for all j. Define the quasi-dyadic matrix $H = H' \otimes I_\ell \in \Delta(\mathbb{F}_q^\ell)^{r \times n}$ with blocks $H_{ij} = H'_{ij} I_\ell$, the vector $s = s' \otimes v_\ell \in (\mathbb{F}_q^\ell)^r$ with blocks $s_i = s'_i v_\ell$, and $w = \ell w'$. It is evident that the instance $(H, w, s) \in \Delta(\mathbb{F}_q^\ell)^{r \times n} \times \mathbb{Z} \times (\mathbb{F}_q^\ell)^r$ of the QD-SDP can be constructed in polynomial time.

Assume now that there exists $e \in \mathbb{F}_q^{\ell n}$ of Hamming weight $\mathsf{wt}(e) \leqslant w$ such that $He^\mathsf{T} = s^\mathsf{T}$. For all $0 \leqslant i < \ell$, let $e'_i \in \mathbb{F}_q^n$ be the vector with elements $(e'_i)_j = e_{i+j\ell}$, $0 \leqslant j < n$, so that the e'_i are interleaved to compose e. Obviously at least one of the e'_i has Hamming weight not exceeding $w/\ell = w'$, and by the construction of H any of them satisfies $He'^\mathsf{T}_i = s'^\mathsf{T}$, constituting a solution to the given instance of the SDP. This effectively reduces the SDP to the QD-SDP for any given ℓ in polynomial time. Thus, the QD-SDP itself is NP-complete. $\square$

Although this theorem does not say anything about hardness in the average case, it nevertheless strengthens our claim that the family of codes we propose is in principle no less suitable for cryptographic applications than a generic code, in the sense that, should the QD-SDP problem turn out to be feasible in the worst case, then *all* coding-based cryptosystems would definitely be ruled out, regardless of which code is used to instantiate them. Incidentally, the expected running time of all known algorithms for the SDP (and the QD-SDP) is exponential, so there is empirical evidence that the average case is also very hard. We stress, however, that particular cryptosystems based on quasi-dyadic codes will usually depend on more specific security assumptions, whose assessment transcends the scope of this paper.

5   Efficiency Considerations

Due to their simple structure the matrices in our proposal can be held on a simple vector not only for long-term storage or transmission, but for processing as well.

The operation of multiplying a vector by a (quasi-)dyadic matrix is at the core of McEliece encryption. The fast Walsh-Hadamard transform (FWHT) [12] approach for dyadic convolution via lifting[2] to characteristic 0 leads to the asymptotic complexity $O(n \lg n)$ for this operation and hence also for encoding. Sarwate's decoding method [24] sets the asymptotic cost of that operation at roughly $O(n \lg n)$ as well for the typical cryptographic setting $t = O(n/\lg n)$.

Inversion, on the other hand, can be carried out in $O(n)$ steps: one can show by induction that a binary dyadic matrix $\Delta(h)$ of dimension n satisfies $\Delta^2 = (\sum_i h_i)^2 I$, and hence its inverse, when it exists, is $\Delta^{-1} = (\sum_i h_i)^{-2} \Delta$, which can be computed in $O(n)$ steps since it is entirely determined by its first row.

Converting a quasi-dyadic matrix to systematic (echelon) form involves a Gaussian elimination incurring about $d^2 \ell$ products of dyadic $t \times t$ submatrices, implying a complexity $O(d^2 \ell t \lg t) = O(d^2 n \lg n)$, and hence the overall cost of formatting is $O(n \lg n)$ as long as d is a small constant, which is indeed the case in practice since maximum size reduction is achieved when $\mathbb{F}_p$ is a large proper subfield of $\mathbb{F}_q$ (see Section 5.1). Notice that, contrary to systems based on quasi-circulant matrices [8, Proposition 3.4], our proposal does not require a lengthy process, involving expensive $O(n^3)$ matrix rank computations to construct a generator matrix in suitable form, often larger than one would expect for a code of the given dimension.

Table 1 summarizes the asymptotic complexities of code generation (mainly due to systematic formatting), encoding and decoding, which coincide with the complexities of key generation, encryption and decryption of typical cryptosystems based on codes.

Table 1. Operation complexity relative to the code length n

operation	generic	ours
Code generation	$O(n^3)$	$O(n \lg n)$
Encode/Decode	$O(n^2)$	$O(n \lg n)$

5.1   Suggested Parameters

Several trade-offs are possible when choosing parameters for a particular application. One may wish to minimize the key size, or increase speed, or simplify the underlying arithmetic, or attaining a balance between them. We present here some non-exhaustive combinations. The number of errors is always a power of 2 to enable maximum size reduction.

Table 2 shows the influence of varying the subfield degree while keeping fixed the approximate security level and the number of design errors. In general, codes over larger subfields allow for smaller keys as already indicated in [3]. For these parameters the number of possible codes ranges from 2^{392} to 2^{731}.

---

[2] We are grateful to Dan Bernstein for suggesting the lifting technique to emulate the FWHT in characteristic 2.

Table 2. Sample parameters for a fixed number of errors ($t = 128$) and approximately 128-bit security level, using a subcode over the subfield $\mathbb{F}_{2^s}$ of $\mathbb{F}_{2^{16}}$

s	n	k	size (bits)
1	4096	2048	32768
2	2560	1536	24576
4	1408	896	14336
8	768	512	8192

Table 3 displays a different trade-off whereby the key size and the subfield are kept constant at the cost of varying the number of errors and the code length. The estimated security level on column 'level' refers to the approximate logarithmic cost of the best known attack according to the guidelines in [7].

Table 3. Sample parameters for a fixed key size (8192 bits, corresponding to $k = 512$), using a subcode over the subfield $\mathbb{F}_{2^8}$ of $\mathbb{F}_{2^{16}}$

n	t	level
640	64	102
768	128	136
1024	256	168

One more trade-off is obtained by defining the subfield subcode over the base field itself, following the common practice for generic codes. The corresponding settings[3] are summarised on Table 4.

Table 4. Sample parameters for a subcode over the base subfield $\mathbb{F}_2$ of $\mathbb{F}_{2^{16}}$

level	n	k	t	size (bits)
80	2304	1280	64	20480
112	3584	1536	128	24576
128	4096	2048	128	32768
192	7168	3072	256	49152
256	8192	4096	256	65536

Table 5 contains a variety of balanced parameters for practical security levels. Although we do not recommend these for actual deployment before further analysis is carried out, these parameters were chosen to stress the possibilities of our proposal while giving a realistic impression of what one might indeed

---

[3] The actual security levels computed according to the attack strategy in [7] for the parameters suggested in Table 4 are, respectively, 84.3, 112.3, 136.5, 216.0, and 265.1. We are grateful to Christiane Peters for kindly providing these estimates.

adopt in practice. The target security level, roughly corresponding to the estimated logarithmic cost of the best known attack according to the guidelines in [7], is shown on the 'level' column. The 'size' column contains the amount of bits effectively needed to store a quasi-dyadic generator or parity-check matrix in systematic form. The size of a corresponding systematic matrix for a generic Goppa code at roughly the same security level as suggested in [7] is given on column 'generic'. The 'shrink' column contains the size ratio between such a generic matrix and a matching quasi-dyadic matrix. The 'RSA' column lists the typical size of a (quantum-susceptible) RSA modulus at the specified security level (more accurate RSA estimates can be found in [20,21]). To assess our results against what can be achieved by other post-quantum settings, column 'QC' lists key sizes for quasi-cyclic codes of approximately the specified security level (although not necessarily for the same code length, dimension, and distance) as suggested in [3], column 'LDPC' does the same for (quasi-cyclic) low-density parity-check codes as discussed in [2], and finally the 'NTRU' column contains the range (from size-optimal to speed-optimal) of NTRU key sizes as suggested in the draft IEEE 1363.1 standard [13]. For these very compact parameters the number of possible codes ranges between 2^{346} and 2^{392}, less than those of Table 2 but still very large.

Table 5. Sample parameters for a subcode over the subfield $\mathbb{F}_{2^8}$ of $\mathbb{F}_{2^{16}}$

level	n	k	t	size	generic	shrink	RSA	QC	LDPC	NTRU
80	512	256	128	4096	460647	112	1024	6750	49152	–
112	640	384	128	6144	1047600	170	2048	14880	–	4411–7249
128	768	512	128	8192	1537536	188	3072	20400	–	4939–8371
192	1280	768	256	12288	4185415	340	7680	–	–	7447–11957
256	1536	1024	256	16384	7667855	468	15360	–	–	11957–16489

For the parameters on Table 5, we observed the timings on Table 6 (measured in ms) for generic Goppa codes and quasi-dyadic (QD) codes, and also for RSA to assess the efficiency relative to a very common pre-quantum cryptosystem. We made no serious attempt at optimizing the implementation, which was done in C++ and tested on an AMD Turion 64X2 2.4 GHz. Benchmarks for RSA-15360 were omitted due to the enormous time needed to generate suitable parameters.

Table 6. Benchmarks for typical parameters

level	generation			encoding			decoding		
	RSA	generic	QD	RSA	generic	QD	RSA	generic	QD
80	563	375	17.2	0.431	0.736	0.817	15.61	1.016	3.685
112	1971	1320	18.7	1.548	1.696	1.233	110.34	2.123	4.463
128	4998	2196	20.5	3.467	2.433	1.575	349.91	3.312	5.261
192	628183	13482	47.6	22.320	6.872	4.695	5094.10	8.822	17.783
256	–	27161	54.8	–	12.176	6.353	–	15.156	21.182

6   Conclusion and Further Research

We have described how to generate Goppa codes in quasi-dyadic form suitable for cryptographic applications. Key sizes for a typical McEliece-like cryptosystem are roughly a factor $t = \tilde{O}(n)$ smaller than generic Goppa codes, and keys can be kept in this compact size not only for storing and transmission but for processing as well. In the binary case these codes can correct the full design number of errors. This brings the size of cryptographic keys to within a factor 4 or less of equivalent RSA keys, comparable to NTRU keys. Our work provides an alternative to conventional cyclic and quasi-cyclic codes, and benefits from the same trapdoor-hiding techniques proposed by Wieschebrink in general [29], and by Berger et al. for that family of codes [3].

The complexity of all operations in McEliece and related cryptosystems is reduced to $O(n \lg n)$. Other cryptosystems can also benefit from dyadic codes, e.g. entity identification and certain digital signatures for which double circulant codes have been proposed [9] could use dyadic codes instead, even random ones without a Goppa trapdoor. One further line of research is whether one can securely combine the techniques in [2] with ours to define quasi-dyadic, low-density parity-check (QD-LDPC) codes that are suitable for cryptographic purposes and potentially even shorter than plain quasi-dyadic codes.

Interestingly, it is equally possible to define *lattice*-based cryptosystems with short keys using dyadic lattices entirely analogous to ideal (cyclic) lattices as proposed by Micciancio [17], and achieving comparable size reduction. We leave this line of inquiry for future research since it falls outside the scope of this paper.

Acknowledgments

We are most grateful and deeply indebted to Marco Baldi, Dan Bernstein, Pierre-Louis Cayrel, Philippe Gaborit, Steven Galbraith, Robert Niebuhr, Christiane Peters, Nicolas Sendrier, and the anonymous reviewers for their valuable comments and feedback during the preparation of this work.

References

1. Baldi, M., Chiaraluce, F.: Cryptanalysis of a new instance of McEliece cryptosystem based on QC-LDPC code. In: IEEE International Symposium on Information Theory – ISIT 2007, Nice, France, pp. 2591–2595. IEEE, Los Alamitos (2007)
2. Baldi, M., Chiaraluce, F., Bodrato, M.: A new analysis of the mcEliece cryptosystem based on QC-LDPC codes. In: Ostrovsky, R., De Prisco, R., Visconti, I. (eds.) SCN 2008. LNCS, vol. 5229, pp. 246–262. Springer, Heidelberg (2008)
3. Berger, T.P., Cayrel, P.-L., Gaborit, P., Otmani, A.: Reducing key length of the McEliece cryptosystem. In: Preneel, B. (ed.) AFRICACRYPT 2009. LNCS, vol. 5580, pp. 77–97. Springer, Heidelberg (2009),
   http://www.unilim.fr/pages_perso/philippe.gaborit/reducing.pdf

4. Berlekamp, E., McEliece, R., van Tilborg, H.: On the inherent intractability of certain coding problems. IEEE Transactions on Information Theory 24(3), 384–386 (1978)
5. Bernstein, D.J.: List decoding for binary Goppa codes (2008) (preprint), http://cr.yp.to/papers.html#goppalist
6. Bernstein, D.J., Buchmann, J., Dahmen, E.: Post-Quantum Cryptography. Springer, Heidelberg (2008)
7. Bernstein, D.J., Lange, T., Peters, C.: Attacking and defending the mcEliece cryptosystem. In: Buchmann, J., Ding, J. (eds.) PQCrypto 2008. LNCS, vol. 5299, pp. 31–46. Springer, Heidelberg (2008), http://www.springerlink.com/content/68v69185x478p53g
8. Gaborit, P.: Shorter keys for code based cryptography. In: International Workshop on Coding and Cryptography – WCC 2005, Bergen, Norway, pp. 81–91. ACM Press, New York (2005)
9. Gaborit, P., Girault, M.: Lightweight code-based authentication and signature. In: IEEE International Symposium on Information Theory – ISIT 2007, Nice, France, pp. 191–195. IEEE, Los Alamitos (2007)
10. Gibson, J.K.: Severely denting the Gabidulin version of the McEliece public key cryptosystem. Designs, Codes and Cryptography 6(1), 37–45 (1995)
11. Gibson, J.K.: The security of the Gabidulin public key cryptosystem. In: Maurer, U.M. (ed.) EUROCRYPT 1996. LNCS, vol. 1070, pp. 212–223. Springer, Heidelberg (1996)
12. Gulamhusein, M.N.: Simple matrix-theory proof of the discrete dyadic convolution theorem. Electronics Letters 9(10), 238–239 (1973)
13. IEEE P1363 Working Group. IEEE 1363-1: Standard Specifications for Public-Key Cryptographic Techniques Based on Hard Problems over Lattices, Draft (2009), http://grouper.ieee.org/groups/1363/lattPK/index.html
14. Loidreau, P., Sendrier, N.: Some weak keys in McEliece public-key cryptosystem. In: IEEE International Symposium on Information Theory – ISIT 1998, Boston, USA, p. 382. IEEE, Los Alamitos (1998)
15. MacWilliams, F.J., Sloane, N.J.A.: The theory of error-correcting codes. North-Holland Mathematical Library, vol. 16 (1977)
16. McEliece, R.: A public-key cryptosystem based on algebraic coding theory. The Deep Space Network Progress Report, DSN PR 42–44 (1978), http://ipnpr.jpl.nasa.gov/progressreport2/42-44/44N.PDF
17. Micciancio, D.: Generalized compact knapsacks, cyclic lattices, and efficient one-way functions. Computational Complexity 16(4), 365–411 (2007)
18. Monico, C., Rosenthal, J., Shokrollahi, A.: Using low density parity check codes in the McEliece cryptosystem. In: IEEE International Symposium on Information Theory – ISIT 2000, Sorrento, Italy, p. 215. IEEE, Los Alamitos (2000)
19. Niederreiter, H.: Knapsack-type cryptosystems and algebraic coding theory. Problems of Control and Information Theory 15(2), 159–166 (1986)
20. European Network of Excellence in Cryptology (ECRYPT). ECRYPT yearly report on algorithms and keysizes (2007-2008). D.SPA.28 Rev. 1.1, IST-2002-507932 ECRYPT, 07/2008 (2008), http://www.ecrypt.eu.org/ecrypt1/documents/D.SPA.28-1.1.pdf
21. National Institute of Standards and Technology (NIST). Recommendation for key management – part 1: General (2007), http://csrc.nist.gov/publications/nistpubs/800-57/sp800-57-Part1-revised2_Mar08-2007.pdf

22. Otmani, A., Tillich, J.-P., Dallot, L.: Cryptanalysis of two McEliece cryptosystems based on quasi-cyclic codes (2008) (preprint),
    http://arxiv.org/abs/0804.0409v2
23. Patterson, N.J.: The algebraic decoding of Goppa codes. IEEE Transactions on Information Theory 21(2), 203–207 (1975)
24. Sarwate, D.V.: On the complexity of decoding Goppa codes. IEEE Transactions on Information Theory 23(4), 515–516 (1977)
25. Schechter, S.: On the inversion of certain matrices. Mathematical Tables and Other Aids to Computation 13(66), 73–77 (1959),
    http://www.jstor.org/stable/2001955
26. Sendrier, N.: Finding the permutation between equivalent linear codes: the support splitting algorithm. IEEE Transactions on Information Theory 46(4), 1193–1203 (2000)
27. Sidelnikov, V., Shestakov, S.: On cryptosystems based on generalized Reed-Solomon codes. Discrete Mathematics 4(3), 57–63 (1992)
28. Tzeng, K.K., Zimmermann, K.: On extending Goppa codes to cyclic codes. IEEE Transactions on Information Theory 21, 721–726 (1975)
29. Wieschebrink, C.: Two NP-complete problems in coding theory with an application in code based cryptography. In: IEEE International Symposium on Information Theory – ISIT 2006, Seattle, USA, pp. 1733–1737. IEEE, Los Alamitos (2006)

Herding, Second Preimage and Trojan Message Attacks beyond Merkle-Damgård

Elena Andreeva[1], Charles Bouillaguet[2], Orr Dunkelman[2], and John Kelsey[3]

[1] ESAT/SCD — COSIC, Dept. of Electrical Engineering,
Katholieke Universiteit Leuven and IBBT
elena.andreeva@esat.kuleuven.be
[2] Ecole Normale Supérieure
{charles.bouillaguet,orr.dunkelman}@ens.fr
[3] National Institute of Standards and Technology
john.kelsey@nist.gov

Abstract. In this paper we present new attack techniques to analyze the structure of hash functions that are not based on the classical Merkle-Damgård construction. We extend the herding attack to concatenated hashes, and to certain hash functions that process each message block several times. Using this technique, we show a second preimage attack on the folklore "hash-twice" construction which process two concatenated copies of the message. We follow with showing how to apply the herding attack to tree hashes. Finally, we present a new type of attack — the trojan message attack, which allows for producing second preimages of unknown messages (from a small known space) when they are appended with a fixed suffix.

Keywords: Herding attack, Second preimage attack, Trojan message attack, Zipper hash, Concatenated hash, Tree hash.

1   Introduction

The works of Dean [6] showed that fixed points of the compression function can be transformed into a long message second preimage attack on the Merkle-Damgård functions in $\mathcal{O}\left(2^{n/2}\right)$ time (n is the size in bits of the chaining and digest values). Later, the seminal work by Joux [8] suggested a new method to efficiently construct *multicollisions*, by turning ℓ pairs of colliding message blocks into 2^ℓ colliding messages.[1] In [10], Kelsey and Schneier applied the multicollision ideas of Joux to Dean's attack, and eliminated the need for finding fixed points in the compression function by building *expandable messages*, which are a set of 2^ℓ colliding messages each of a distinct length.

Another result in the same line of research is the *herding attack* by Kelsey and Kohno in [9]. The attack is a chosen-target prefix attack, *i.e.*, the adversary commits to a digest value h and is then presented with a challenge prefix P. Now, the adversary efficiently computes a suffix S, such that $H(P\|S) = h$.

---

[1] We note that in [5] the same basic idea was used for a dedicated preimage attack.

M.J. Jacobson Jr., V. Rijmen, and R. Safavi-Naini (Eds.): SAC 2009, LNCS 5867, pp. 393–414, 2009.

The underlying technique uses a *diamond structure*, a precomputed data structure, which allows 2^ℓ sequences of message blocks to iteratively converge to the same final digest value. The latter result, together with the long message second preimage attack, was used in [1] to exhibit a new type of second preimage attack that allows to construct second preimages differing from the original messages only by a small number of message blocks.

We note that the work of Joux in [8] also explores concatenated hashing, *i.e.*, the hash function $H(M) = H_1(M)||H_2(M)$. It appears that if one of the underlying hash functions is iterative, then the task of finding a collision (resp., a preimage) in $H(M)$ is only polynomially harder than finding a collision (resp., a preimage) for any of the $H_i(M)$.

1.1   Our Contributions

In this section we summarize our results on herding, long-message second preimage and trojan message attacks and present the complexities of these attacks in Tables 1, 2, and 3.

Herding Attacks. We introduce herding attacks on four non-Merkle-Damgård hash constructions: the concatenated, zipper [11], "hash-twice" and tree hash [12] functions.

Long-Message Second Preimage Attacks. By reusing the newly presented herding techniques we apply long-message second preimage attacks on the "hash twice". We also show that in tree hashes it is possible to find long-message second preimages using time-memory-data tradeoff attack.

Trojan Message Attacks. We introduce a new kind of attack, called the trojan message attack. This involves an attacker producing a poisoned suffix S (the "Trojan message") such that, when the victim prepends one of a constrained set of possible prefixes to it, producing the message $P||S$, the attacker can produce a second preimage for that message. The attack comes in a less and more powerful form, as is described in Section 9. Note that for both forms of the attack, the number of message blocks in the Trojan message is at least as large as the number of possible prefixes.

1.2   Organization of the Paper

Section 2 outlines the definitions used in the paper. Prior work is surveyed in Section 3. In Section 4 we introduce the diamond structures on κ pipes and use them to apply the herding attack to concatenated hashes. We use the same ideas to present herding attacks on the hash-twice and zipper hash in Section 5. These herding attacks are used in Section 6 to present a second preimage attack on hash-twice. The new results on herding tree hashes are presented in Section 7, and in Section 8 we present second preimage attacks on tree hashes. We follow to introduce the new trojan message attack, in Section 9 and conclude with Section 10.

Table 1. Complexities of the Suggested Herding Attacks

Construction	Precomputation	Online	128-bit hash
Concatenated hash	$\mathcal{O}\left(\frac{n\ell\kappa}{2}\cdot(n/2)^{\kappa-2}\cdot 2^{n/2}+\kappa\cdot 2^{\frac{n+\ell}{2}+2}\right)$	$\mathcal{O}\left(\kappa\cdot 2^{n-\ell}\right)$	$3\cdot 2^{88\ddagger}$
Hash twice	$\mathcal{O}\left(2^{(n+\ell)/2+2}\right)$	$\mathcal{O}\left(2^{n-\ell}\right)$	$3\cdot 2^{88}$
Zipper hash	$\mathcal{O}\left(2^{(n+\ell)/2+2}+(n-\ell+\frac{n\ell}{2})\cdot 2^{n/2}\right)$	$\mathcal{O}\left(2^{n-\ell}\right)$	$2\cdot 2^{88}$
Tree hash[†]	$15.08\cdot 2^{\frac{n+\ell}{2}}$	$2^{n-(\ell-1)}$	2^{99*}
Merkle-Damgård [9]	$2^{\frac{n+\ell}{2}+2}$	$2^{n-\ell}$	2^{88}

κ — number of hash functions (chains)
[†] — for tree hashes, the size of the diamond is restricted by the message length
[‡] — for $\kappa=2$
[*] — for message of length $\ell=2^{30}$ blocks, and a 128-bit digest size

Table 2. Complexities of the Suggested Second Preimage Attacks

Construction	Precomputation	Online	128-bit hash
Hash twice	$\mathcal{O}\left(2^{(n+\ell)/2+2}\right)$	$\mathcal{O}\left(2^{n-\ell}+2^{n-\kappa}\right)$	$2^{88\dagger}$
Tree hash[†]	$\mathcal{O}\left(2^{n-\kappa+1}\right)$	$2^{2(n-\kappa+1)}$	2^{99*}
Merkle-Damgård [10]	$2^{n+\ell)/2+2}$	$2^{n-\kappa}$	$2^{78\ddagger}$

2^{κ} — message length
[†] — for messages of length greater than 2^{42} blocks
[‡] — for messages of length 2^{50} blocks
[*] — for message of length $\ell=2^{30}$ blocks, and a 128-bit digest size

Table 3. Complexities of the Suggested Trojan Message Attacks

Trojan message attack	Precomputation	Online	128-bit hash
Collision	$N\cdot 2^{n/2}$	negl.	$2^{74\dagger}$
Herding	$2^{\frac{n+\ell}{2}+2}+N\cdot 2^{n/2}$	$2^{n-\ell}$	$2^{88\dagger}$

N — number of possible prefixes
[†] — for $N=1024$
[*] — for message of length $\ell=2^{30}$ blocks, and a 128-bit digest size

In all the tables, n denotes the digest size, 2^{ℓ} is the size of the diamond structures, and the figures for the 128-bit hash are derived for the case where the time of the preprocessing and the online time complexity is the same.

2   Background

GENERAL NOTATION. Let n be a positive integer, then $\{0,1\}^n$ denotes the set of all bitstrings of length n, and $\{0,1\}^*$ be the set of all bitstrings. If x,y are strings, then $x\|y$ is the concatenation of x and y. We denote by $|x|$ the length of the bitstring x, but in some places we use $|M|$ to denote the length of the message M in blocks. For $M = m_1\|m_2\|\dots\|m_L$ we define $\widetilde{M} = m_L\|m_{L-1}\|\dots\|m_1$.

Let $f : \{0,1\}^n \times \{0,1\}^b \to \{0,1\}^n$ be a compression function taking as inputs bitstrings of length n and b, respectively. Unless stated explicitly we denote the n-bit input values by h (chaining values) and the b-bit values by m (message blocks). A function $H : \{0,1\}^* \to \{0,1\}^n$ built on top of some fixed compression function f is denoted by H^f. To indicate the use of fixed initialization vectors for some hash functions, with $H^f(IV, M)$ we explicitly denote the result of evaluating H^f on inputs the message M and the initialization vector IV.

MERKLE-DAMGÅRD HASH FUNCTION. $\mathcal{MDH}^f : \{0,1\}^* \to \{0,1\}^n$ takes a message $M \in \{0,1\}^*$ as an input to return a digest value of length n bits. Given a fixed initialization vector IV and a padding function pad_{MD} that appends to M fixed bits (a single 1 bit and sufficiently many 0 bits) together with the message length encoding of M to obtain a message multiple of the blocksize b, the $\mathcal{MDH}^f$ function is defined as:

1. $m_1, \dots, m_L \leftarrow \text{pad}_{MD}(M)$.
2. $h_0 = IV$.
3. For $i = 1$ to L compute $h_i = f(h_{i-1}, m_i)$.
4. $\mathcal{MDH}^f(M) \triangleq h_L$.

Often in the sequel we consider chaining values obtained by hashing a given prefix of the message. When P is a message whose size is a multiple of b bits, we denote by $f^*(P)$ the chaining value resulting from the Merkle-Damgård iteration of f without the padding scheme being applied.

CONCATENATED HASH. $\mathcal{CH}$ of $\kappa \geq 2$ pipes is defined as:

$$\mathcal{CH}(M) = H^{f_1}(IV_1, M) \| H^{f_2}(IV_2, M) \| \dots \| H^{f_\kappa}(IV_\kappa, M).$$

"HASH-TWICE". This is a folklore hashing mode of operation that hashes two consecutive copies of the (padded) message. Formally, it is defined by:

$$\mathcal{HT}(M) \triangleq H^f(H^f(IV, M), M).$$

ZIPPER HASH. It is proposed in [11] and is proven indifferentiable from a random oracle up to the birthday bound even if the compression functions in use, f_1 and f_2, are weak (i.e., they can be inverted and collisions can be found efficiently). The zipper hash $\mathcal{ZH}$ is defined as:

$$\mathcal{ZH}(M) \triangleq H^{f_2}\left(H^{f_1}(IV, M), \widetilde{M}\right).$$

Throughout the paper, we assume that all H^{f_i} are applications of the Merkle-Damgård mode of operation and thus the respective padding pad_{MD} is present in the syntax of M, and also $\widetilde{M}$ in the case of Zipper hash.

TREE HASH. The first suggested tree hash construction dates back to [12]. Let $f : \{0,1\}^n \times \{0,1\}^n \to \{0,1\}^n$ be a compression function and pad_{TH} be a padding function that appends a single 1 bit and as many 0 bits as needed to the message M to obtain $\text{pad}_{\text{TH}}(M) = m_1\|m_2\|\ldots\|m_L$, where $|m_i| = n$, $L = 2^d$ for $d = \lceil\log_2(|M| + 1)\rceil$. Then, the tree hash function $\textit{Tree}$ is defined as:

1. $m_1\|m_2\|\ldots\|m_L \leftarrow \text{pad}_{\text{TH}}(M)$
2. For $j = 1$ to 2^{d-1} compute $h_{1,j} = f(m_{2j-1}, m_{2j})$
3. For $i = 2$ to d:
   - For $j = 1$ to 2^{d-i} compute $h_{i,j} = f(h_{i-1,2j-1}, h_{i-1,2j})$
4. $\textit{Tree}(M) \triangleq f(h_{d,1}, |M|)$.

3 Existing Attack Techniques

3.1 Herding Attack

The herding attack is a chosen-target preimage attack on Merkle-Damgård constructions [9]. In the attack, an adversary commits to a public digest value h_T. After the commitment phase, the adversary is challenged with a prefix P which she has no control over, and she is to produce a suffix S for which $h_T = H^f(P\|S)$. Of course, h_T is specifically chosen after a precomputation phase by the adversary. The main idea behind this attack is to store 2^ℓ possible chaining values $D = \{h_i\}$ from which the adversary knows how to reach h_T. To construct this data structure, which is in fact a tree, the adversary picks about $2^{n/2-\ell/2+1/2}$ single-block messages m_j, and evaluates $f(h_i, m_j)$ for all i and j. Due to the large number of values, it is expected that collisions occur, and it is expected that the adversary knows for all 2^ℓ values of h_i a corresponding message block $m_{\alpha(i)}$ such that the set $\{f(h_i, m_{\alpha(i)})\}$ contains only $2^{\ell-1}$ distinct values. The process is then repeated $\ell-1$ more times until a final digest value is found. These values (and the corresponding message blocks) are called a diamond structure, as presented in Figure 1.

In the online phase of the attack, the adversary tries at random message blocks m^* until $f^*(P\|m^*) \in D$. Once such a value is found, it is possible to follow the

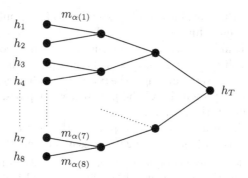

Fig. 1. A Diamond Structure

path connecting this value to the committed hash (which is at the "root" of the diamond) and produce the required suffix S.

The total time complexity of the attack is about $2^{n/2+\ell/2+2}$ offline compression function evaluations, and $2^{n-\ell}$ online compression function evaluations.

3.2   Collisions on Concatenated Hashes

We describe the collision attack of [8] against the concatenated hash $\mathcal{CH}$ with two pipes. Starting from two fixed chaining values IV_1 and IV_2 in the two pipes, the adversary first finds a $2^{n/2}$-multicollision for the first function f_1. The adversary then evaluates f_2 on the $2^{n/2}$ messages of the multicollision, all yielding the same chaining value for f_1, while yielding a set of $2^{n/2}$ chaining values for f_2, as shown in figure 2. The adversary then looks for the expected collision in this set. To construct a 2^ℓ-*multicollision* on the two pipes, just replay Joux's attack using the two-pipe collision finding algorithm described above ℓ times.

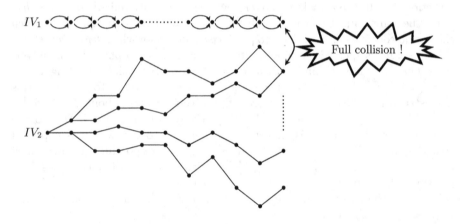

Fig. 2. Joux's attack against concatenated hashes

Joux also shows that this idea can be extended to find (multi)collisions in the concatenation of κ hash functions. To build a collision on κ parallel pipes, the adversary proceeds inductively: first construct a $2^{n/2}$-multicollision on the first $\kappa - 1$ pipes and hash the $2^{n/2}$ messages in the last pipe. Then, by the birthday bound, a collision is expected amongst the set of $2^{n/2}$ values generated in the last pipe. This collision is present in all the previous $\kappa - 1$ pipes, and hence results in a full collision on all the κ pipes.

The cost of building a collision on κ pipes is the cost of building the multicollision, plus the cost of hashing the $2^{n/2}$ messages of length $(n/2)^{\kappa-1}$. Solving the recurrence yields a time complexity of $\kappa \cdot \left(\frac{n}{2}\right)^{\kappa-1} \cdot 2^{\frac{n}{2}}$ compression function calls. More generally, the complexity of building a 2^ℓ-multicollision on κ pipes is exactly ℓ times the preceding expression, or $\ell \cdot \kappa \cdot \left(\frac{n}{2}\right)^{\kappa-1} \cdot 2^{\frac{n}{2}}$.

4 Herding Attack on Concatenated Hashes

We start by showing how to adapt the herding attack to concatenated hashes. The main idea behind the new attack is to construct *multi-pipe diamonds*, which can be done on top of multicollisions. We recall that a multicollision on $(\kappa - 1)$ pipes can be used to construct a collision on κ pipes. In the same vein, we succeed in herding κ pipes by building a κ-pipe diamond using a $(\kappa - 1)$-pipe diamond and $(\kappa - 1)$-pipe multicollision.

Assume that the adversary succeeded in herding $\kappa - 1$ pipes. Then, she faces the problem of herding the last pipe. Now, if the adversary tries to connect in the κ-th pipe with a random block, she is very likely to lose the control over the previous pipes. However, if she uses a "block" which is part of a multicollision on the first $\kappa - 1$ pipes, she still maintains the control over the previous pipes, while offering enough freedom for herding the last pipe.

4.1 Precomputation Phase

In the precomputation phase, the adversary starts with the $(\kappa - 1)$-diamond which is already known. The first step is the randomization step: given the concatenated chaining value the adversary constructs a $2^{n-\ell}$-multicollision on the first $\kappa - 1$ pipes. Let the resulting chaining value be $(h^1, h^2, \ldots, h^{\kappa-1})$.

The second step is the actual diamond construction. The adversary picks at random 2^ℓ values for $D_\kappa = \{h_i^\kappa\}$. Then, she generates a set of further (in addition to the preceding $2^{n-\ell}$ multicollisions) $2^{n/2}$-multicollisions[2] on the first $\kappa-1$ pipes, starting from $(h^1, h^2, \ldots, h^{\kappa-1})$. For each possible message in the multicollision, and any starting point $(h^1, h^2, \ldots, h^{\kappa-1}, h_i^\kappa)$, the adversary computes the new chaining values, expecting to reach enough collisions, such that for any h_i^κ, there exists a "message" m_i (*i.e.*, a sequence of message blocks in the multicollision) where $\#\{f_\kappa^*(h_i^\kappa, m_i)\} = 2^{\ell-1}$. After this step, the same process is repeated. Figure 3 depicts the process for $\kappa = 2$.

The running time is dominated by the generation of the last diamond structure. First, we need to generate $2^{n-\ell+\frac{n\ell}{2}}$-multicollisions on $\kappa - 1$ pipes, which takes $(n - \ell + \frac{n\ell}{2}) \cdot (\kappa - 1) \cdot (n/2)^{\kappa-2} \cdot 2^{n/2}$ compression function calls. Then, we need to "hash" 2^ℓ values under $2^{\frac{n-\ell+1}{2}}$ message sequences (for the last layer of the diamond structure). While at a first glance it may seem that we need a very long time for each message sequence, it can be done efficiently if we take into consideration the fact that there is no need to recompute all the chaining values only if the last block was changed. Hence, the actual time required to construct the diamond structure is $2 \cdot 2^{\frac{n+\ell}{2}+2}$ (twice the time needed for a classic diamond structure). In total, the preprocessing takes

---

[2] We note that fewer multicollisions are needed (herding the first layers takes less messages). However, for the ease of description we shall assume all layers of the diamond structure require the same number of multicollisions. Hence, the total of $2^{n\ell/2}$-multicollisions, can be reduced to $2^{\ell(n+1-\ell)/2}$-multicollisions.

$$\left(n - \ell + \frac{n\ell}{2}\right) \cdot (\kappa - 1) \cdot (n/2)^{\kappa-2} \cdot 2^{n/2} + (2 \cdot \kappa - 1) \cdot 2^{\frac{n+\ell}{2}+2}.$$

One may ask what is the reason for the randomization step. As demonstrated in the online phase of the attack, the need arises from the fact that herding the values in the first $\kappa - 1$ pipes fixes the value in the κ-th pipe. Hence, we need enough "freedom" to randomize this chaining value, without affecting the already solved pipes.

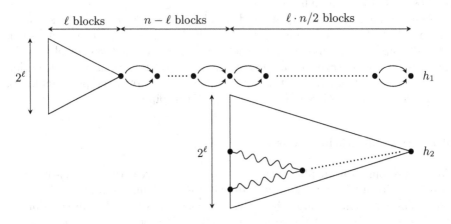

Fig. 3. Diamond Structure on two pipes

4.2   Online Phase

Given a precomputed κ-diamond structure, it is possible to apply the herding attack to κ concatenated hash functions. The adversary is given a prefix P, and tries various message blocks m^* until $f_1^*(P\|m^*)$ gives one of the 2^ℓ values in D_1 of the diamond structure on the first pipe. Then, the adversary traverses the first diamond structure to its root, finding the first part of the suffix S_1 (so far all computations are done in the first pipe). At this point, the adversary starts computing $f_2^*(P\|m^*\|S_1)$, and for all $2^{n-\ell}$ paths of the multicollision in the randomization path, until one of them hits one of the 2^ℓ values in D_2. At this point, the adversary can use the paths inside this second diamond (built upon a multicollision). This process can start again (with a randomization part, and traversing the diamond structure) until all κ pipes were herded correctly. We outline the process for $\kappa = 2$ in Figure 4.

We note that once a pipe is herded, there is no longer a need to compute it (as the multicollision predicts its value), and then it is possible to start analyzing the next pipe. In each new pipe, we need to evaluate $2^{n-\ell}$ "messages" (for all pipes but the first one, these messages are multicollisions on the previous pipes), each takes on average (in an efficient implementation) two compression function calls (besides the first layer). Hence, the online time complexity of the attack is

$$2^{n-\ell} \cdot [1 + 2 \cdot (\kappa - 1)] = (2\kappa - 1) \cdot 2^{n-\ell}$$

compression function calls.

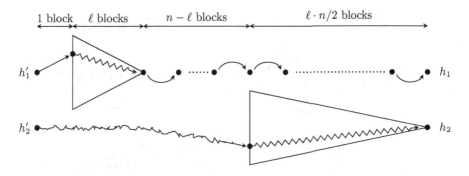

1 block ℓ blocks $n - \ell$ blocks $\ell \cdot n/2$ blocks

Fig. 4. The Online Phase of the Herding Attack for $\kappa = 2$

5   Herding beyond Merkle-Damgård

In this section we show that the previous technique can be applied to two other hash constructions which previously appeared to be immune to herding attacks — the Hash Twice construction and Zipper Hash. Both attacks make use of the two-pipe diamond structure described above.

5.1   Herding the Hash-Twice Function

It follows from the very general result of [7,13] that it is possible to build Joux-style multicollisions efficiently on the hash-twice construction. In this section, we extend their results by describing a herding attack against the hash-twice construction. This attack can then be adapted into a full second-preimage attack, following the ideas of [1,10] (as described in Section 6).

Because each message block enters the hashing process twice, choosing a message block in the second pass may change not only the chaining value going *out* but also the chaining value going *into* the second pass. Choices of the message intended to affect the second pass must thus be done in a way that does not randomize the result of the first pass.

Apart from this technicality, the attack is essentially the same as the one against the concatenation of two hash functions, as shown in figure 5. The adversary commits to h_3, and is then being challenged with an unknown prefix P. Hashing the prefix yields a chaining value h_c. Starting from h_c, she chooses a message block m^* connecting to a chaining value h_e which is one of the starting points of the first diamond, then a path S_1 inside it yields the chaining value h_1 on the first pass, from which we traverse a precomputed $2^{n-\ell+n\ell/2}$-multicollision, producing h_2 as the input chaining value to the second pass. Starting from h_2, the challenge prefix P leads to a random chaining value $h_{c'}$ in the second pass. Then, the second pass can be herded without losing control of the chaining value in the first pipe thanks to the diamond built on top of a multicollision. Amongst the $2^{n-\ell}$ messages in the multicollision following the first diamond, we expect one to connect to the chaining value $h_{e'}$ in the starting points of the second

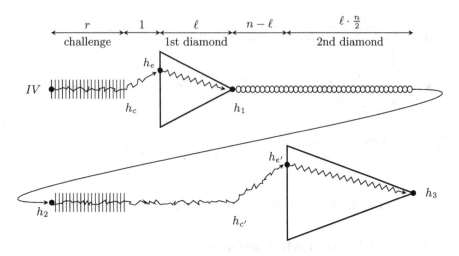

Fig. 5. Herding the Hash-Twice construction

diamond. We can then follow a path inside the second diamond, which is also a path in the multicollision of the first pipe, that yields the chaining value at the root of the second diamond, namely h_3.

The offline complexity of the attack is the time required for generating a diamond structure of 2^ℓ starting points (which takes $2^{(n+\ell)/2+2}$), finding $(n - \ell) + n \cdot \ell/2$ collisions (which takes $[(n - \ell) + n \cdot \ell/2] \cdot 2^{n/2}$), and constructing a two-pipe diamond (which takes $2 \cdot 2^{(n+\ell)/2+2}$). The total offline complexity is therefore $3 \cdot 2^{(n+\ell)/2+2}$.

The online complexity is composed of finding two connecting "messages". The first search takes $2^{n-\ell}$, while the second takes $2 \cdot 2^{n-\ell}$, or a total of $3 \cdot 2^{n-\ell}$.

ATTACKS ON HASH-THRICE. It is relatively clear that the attack can be generalized to the case where the message is hashed three or more times (by using multicollisions on 3 pipes, or the respective number of passes). The complexity of the attack becomes polynomially higher, though.

5.2   Herding the Zipper Hash Function

It is also possible to mount a modified herding attack against the zipper-hash. The regular herding attack is not feasible, because the last message block going into the compression function is the first message block of the challenge. Therefore, an adversary who is capable of doing the herding attack can be used to invert the compression function. We therefore consider a variant of the herding attack where the challenge is placed at the end: the adversary commits to a hash value h_T, then she is challenged with a suffix S, and has to produce a prefix P such that $\mathcal{ZH}(P \,\|\, S) = h_T$.

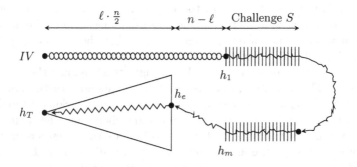

Fig. 6. Herding the Zipper Hash

The attack is relatively similar to the hash-twice case. The offline part is as follows:

1. Starting from the IV, build a $2^{n\ell/2+n-\ell}$-multicollision that yields a chaining value h_1.
2. Build a diamond structure on top of the *reversed* multicollision (*i.e.*, where the order of colliding messages in the multicollision is reversed). The chaining value at the root of the second diamond is h_T.
3. Commit to h_T.

And the online part:

1. Given a challenge suffix S, compute the chaining value after the two copies of the challenge: $h_m = f_2^* \left(f_1^* (h_1, S), \widetilde{S} \right)$.
2. From h_m, find a connecting path in the part of (reversed) multicollision that is just before the diamond (in the second run) yielding a chaining value $h_e \in D_1$ of the diamond structure. Then find a path inside the (reversed) diamond structure towards the committed hash h_T.

We note that the fact that two different hash functions f_1 and f_2 are used in the two passes has no impact on our results, as the attack technique is independent of the actual functions used. The precomputation phase takes $2 \cdot 2^{(n+\ell)/2+2} + (n-\ell+\frac{n\ell}{2}) \cdot 2^{n/2}$, and the online computation takes $2 \cdot 2^{n-\ell}$ compression function calls.

6 From Herding to Second Preimages: Hash-Twice

If a construction is susceptible to the herding attack, then it is natural to ask whether the second preimage attack of [1] is applicable. The general idea of this attack is to connect the root of the diamond structure to some chaining value encountered during the hashing of the target message, and then connect into the diamond structure (either from the corresponding location in the original message or from a random prefix). This ensures that the new message has the same length (foiling the Merkle-Damgård strengthening).

In this section, we present a second preimage attack against the Hash-Twice construction. The general strategy is to build a diamond structure, and try to connect it to the challenge message (in the second pass). Some complications appear, because the connection may happen anywhere, and the diamond only works on top of a multicollision that has to be located somewhere in the first pass. However, we can use an expandable message [10] to move the multicollision (and therefore the diamond) around. Here is a complete description of the attack. Let us assume that the adversary is challenged with a message M of 2^κ blocks.

The offline processing is as follows:

1. Generate a Kelsey-Schneier expandable message which can take any length between κ and $2^\kappa + \kappa - 1$, starting from the IV yielding a chaining value h_a.
2. Starting from h_a, generate a multicollision of length $(n - \ell) + \ell \cdot n/2$ blocks, that yields a chaining value h_b.
3. Build a diamond structure on top of the multicollision. It yields a chaining value h_x. It is used to herd the second pass.

The online phase, given a message M, is as follows (depicted in Figures 7 and 8):

1. Given h_x, select at random message blocks m^* until $f(h_x, m^*)$ equals to a chaining value h_{i_0} appearing in the second pass of the hashing of M. Let us denote by m the right message block.
2. To position the end of the diamond at the $i_0 - 1$-th block of M, instantiate the expandable message in length of $i_0 - 1 - n \cdot \ell/2 - (n - \ell)$ blocks.
3. Let $h_c = f^*(h_b, m_{i_0} || m_{i_0+1} || \ldots || m_{2^\kappa})$. Compute the second pass on the expandable message, until h_d is reached. Now, using the freedom in the first $n - \ell$ blocks of the multicollision, find a message that sends h_d to a chaining value h_e occurring in the starting points of the diamond in the second pass.
4. Find a path inside the diamond in the second pass (this is also a path inside the multicollision of the first pass). It yields the chaining value h_x at the root of the diamond in the second pipe.
5. Append the connection block m and the suffix of M to obtain the second preimage.

Note that the message forged by assembling the right parts has the same length as M, therefore the padding scheme act the same way on both.

The offline complexity of the attack is the mostly dominated by the need to construct a diamond structure on two pipes, $i.e.$, $2 \cdot 2^{(n+\ell)/2+2}$. The online time complexity is $2^{n-\kappa}$ for finding m, and $2 \cdot 2^{n-\ell}$ connecting to the diamond structure. Hence, the total online time is $2^{n-\kappa} + 2^{n+1-\ell}$.

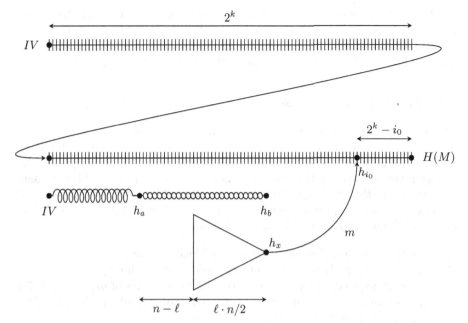

Fig. 7. Second preimage attack on Hash-Twice: first online step

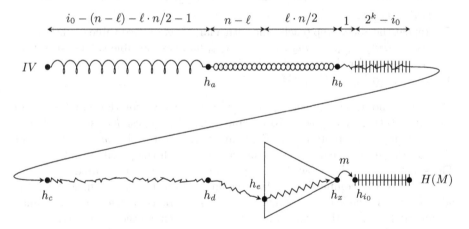

Fig. 8. Second preimage attack on Hash-Twice: online steps 2 to 5

7   Herding Tree Hashes

In this section we introduce a new method for mounting herding attacks on tree hashes. As in the previous attacks, our method is composed of two steps: offline computation (presented in Section 7.1) and online computation (presented in Section 7.2).

The main differences with the regular herding attacks is the fact that in the case of tree hashes the adversary may suggest embedding the challenge in any block she desires (following the precomputation step). Moreover, the adversary, may publish in advance a great chunk of the answer to the challenge.

7.1   Precomputation Phase

In the offline precomputation phase of the herding attack, the adversary determines the position for inserting the challenge block and commits to the digest value h_T. The diamond-like structure built in this attack allows for freedom in choosing the location of the challenge. Let the length of the padded message (the answer to the challenge, after the embedding of it) be 2^ℓ n-bit blocks, and assume that the compression function is $f : \{0,1\}^n \times \{0,1\}^n \to \{0,1\}^n$. The details of the offline computation are as follows:

1. Determine the location for the challenge block, *i.e.*, m_3.
2. Choose some set A_1 of $2^{\ell-1}$ arbitrary chaining values for $h_{1,2}$.
3. Fix the message block m_2 (alternatively fix m_1, or parts of m_1 and m_2). For arbitrary m_1^j, and compute $h_{1,1}^j = f(m_1^j, m_2)$. For each $h_{1,2}^i$ find a value $h_{1,1}^i$, such that $\#A_2 \triangleq \{h_{2,1} = f(h_{1,1}^i, h_{1,2}^j)\} = 2^{\ell-2}$.
4. Fix m_6, m_7 and m_8. For arbitrary m_5^j compute $h_{2,2}^j = f(f(m_5^j, m_6), f(m_7, m_8))$. For each $h_{2,1}^i$, find a value $h_{2,2}^i$, such that $\#A_3 \triangleq \{h_{3,1} = f(h_{2,1}^i, h_{2,2}^j)\} = 2^{\ell-3}$.
5. Repeat the above step (each time with a larger set of fixed values), until fixing $m_{2^{\ell-1}+2}, m_{2^{\ell-1}+3}, \ldots, m_{2^\ell}$. For $m_{2^{\ell-1}+1}$, find two possible values, such that $h_{\ell,1}$ collides for the two values in $A_{\ell-1}$.
6. Commit to $h_T = f(h_{\ell,1}, |M|)$.

The chosen points in the set A_1 serve as target values for the online stage of the computation. The goal is to compute the hash digest h_T, such that it is reachable from all points in A_1. For that we reduce the size of A_1 by a factor of 2 to form A_2 by means of collision search through the possible values of m_1. The same principle is followed until the root hash value is computed.

To reduce the complexity of the precomputation it is more efficient for the adversary to fix the known message blocks from the precomputation to constants, rather than to store the exact values needed for each collision in the tree. For a tree of depth ℓ, the adversary can fix all but $\ell + 1$ message blocks, leaving one message block for the challenge, and controlling the paths in the tree through the remaining ℓ blocks. The adversary also can publish the fixed message blocks in advance. However, this is not a strict requirement since these message blocks are already under the control of the adversary.

The time complexity of the precomputation with $2^{\ell-1}$ starting points is about $2 \cdot 2^{\frac{n+(\ell-1)+1}{2}}$ for finding the first layer of collisions. This follows from the fact that we need to try about $2^{\frac{n-(\ell-1)+1}{2}}$ possible message blocks for m_1 (or its equivalent) to find collisions between any pair in the target set A_1, and we need to perform

2 compression function calls to evaluate $h_{2,1}$. For the collisions on the second level $3 \cdot 2^{\frac{n+(\ell-1)-1+1}{2}} + 1$ compression function calls are needed (the last term is due to the computation of $f(m_7, m_8)$ which can be done once). The third level requires $4 \cdot 2^{\frac{n+(\ell-1)-2+1}{2}} + 4$ compression function calls. Hence, in total we have

$$\sum_{j=1}^{\ell-1} \left[(j+1) \cdot 2^{\frac{n+\ell-j+1}{2}} + 2^{j+1} - (j+2) \right] \leq \left(\sum_{j=1}^{\ell-1} (j+1) \cdot 2^{-j/2} \right) \cdot 2^{\frac{n+\ell+1}{2}} + 2^{\ell+1} - \frac{\ell^2}{2}$$

The sum in the right-hand side admits $\frac{1-2\sqrt{2}}{2\sqrt{2}-3} \leq 10.66$ as a limit when ℓ goes to infinity, which yields an approximate offline complexity of $15.08 \cdot 2^{\frac{n+\ell}{2}}$ compression function calls. The space complexity here is $2^\ell - 1$ and is determined by the amount of memory blocks that are required for the storage of the target points in A_1 and the precomputed values for the non-fixed message blocks (in this example chosen to be $m_1^*, m_5^*, m_9^*, \ldots$).

7.2   Online Phase

Here the adversary obtains the challenge P, and has to:

1. Find m_4^*, such that $f(P, m_4^*) = h_{1,2}$ where $h_{1,2} \in A_1$. Note that $h_{1,2}$ fixes the rest of the message blocks $m_1^*, m_5^*, m_9^*, \ldots, m_{2^{\ell-1}+1}^*$.
2. Retrieve the stored value for m_1^* for which $f(f(m_1^*, m_2), h_{2,1}) \in A_2$. Tracing the correct intermediate chaining values, arrive to the correct value for $m_{2^{\ell-1}+1}^*$ which leads to $h_{\ell,1}$ and h_T.
3. Output $m_1^*, m_2, P, m_4^*, m_5^*, m_6, m_7, m_8, m_9^*, m_{10}, \ldots, m_\ell$ as the answer.

The workload in the online phase of the computation reflects the cost of linking to a point contained in the set A_1. Approximately $2^{n-(\ell-1)}$ compression function calls are required to link correctly to one of the $2^{(\ell-1)}$ points in A_1.

7.3   Variants and Applications of the Herding Attack on Tree Hash Functions

PRECOMPUTED CHALLENGE MESSAGES. If there exists a limited set of possible challenges, it is possible to precompute the points in A_1. This allows for a very efficient connection in the online stage, however, at the cost of losing flexibility– only the precomputed message blocks can be "herded" to h_T.

HERDING SEQUENCES OF ADJACENT MESSAGE BLOCKS. The herding attack also allows for inserting sequences (instead of a single block) of adjacent challenge blocks. In this case the set of target chaining values A_1 has to be embedded deeper in the tree structure. This results in larger online complexity due to the evaluation of additional nodes on the path to the target linking set A_1.

HERDING BOTH SIDES OF THE HASH TREE. The diamond-like structure used for herding trees can accommodate the insertion of a challenge message blocks on both sides of the hash tree due to symmetry of the structure. It is thus no more expensive to construct a diamond structure that allows $2^{\ell-1}$ choices on both sides of the root. This means that an adversary can either herd one message block on the left half of the message, and another on the right, or satisfying two challenges simultaneously with the same diamond structure.

APPLICABILITY. We note that the proposed herding attack is applicable to other variants of the tree hash function. Even if the employed compression functions in the tree are distinct (e.g., as considered in MD6 [15]), it is still possible to apply the attack, because an adversary knows (and controls) the location of the challenges.[3] The attack also works irrespective of the known random XOR masks (e.g., tree constructions of the XOR tree type [2,16]) applied on the chaining values at each level.

8   Long-Message Second Preimages in Tree Hashes

Tree hashes that apply the same compression function to each message block (i.e., the only difference between $f(m_{2i-1}, m_{2i})$ and $f(m_{2j-1}, m_{2j})$ is the position of the resulting node in the tree) are vulnerable to a long-message second preimage attack which changes at most two blocks of the message.

We know that $h_{1,j} = f(m_{2j-1}, m_{2j})$ for $j = 1$ to $L/2$ for a message M of length $L = 2^\kappa$ blocks. Then given the target message M, there are $2^{\kappa-1}$ chaining values $h_{1,j}$ that can be targeted. If the adversary is able to invert even one of these chaining values, i.e., to produce (m', m'') such that $f(m', m'') = h_{1,j}$ for some $1 \leq j \leq 2^{\kappa-1}$, then he has successfully produced a second preimage M'. Note, however that (m', m'') shall differ than the corresponding pair of message blocks in the original target message M. Thus, a long-message second preimage attack on message of length 2^κ requires about $2^{n-\kappa+1}$ trial inversions for $f(\cdot)$.

More precisely, the adversary just tries message pairs (m', m''), until $f(m', m'') = h_{1,j}$ for some $1 \leq j \leq 2^{\kappa-1}$. Then, the adversary replaces $(m_{2j-1}\|m_{2j})$ with $m'\|m'$ without affecting the computed hash value for M. Note that the number of modified message blocks is only two. This result also applies to other parallel modes where the exact position has no effect on the way the blocks are compressed.

Furthermore, it is also possible to model the inversion of f as a task for a time-memory-data attack [4]. The $h_{1,j}$ values are the multiple targets, which compose $D = 2^{\kappa-1}$ data points. Using the time-memory-data curve of the attack from [4], it is possible to have an inversion attack which satisfy the relation $N^2 = TM^2D^2$, where N is the size of the output space of f, T is the online computation, and M is the number of memory blocks used to store the tables of the attack. As $N = 2^n$, we obtain that the curve for this attack is $2^{2(n-\kappa+1)} = TM^2$ (with preprocessing of $2^{n-\kappa+1}$). We note that the trade-off curve can be used

---

[3] Still, note that the herding attack on MD6 has increased offline complexity (compared to our estimates) because of its large internal state and subsequent truncation in the final output transformation.

as long as $M < N, T < N$, and $T \geq D^2$ (see [3] for more details about the last constraint). Thus, for $\kappa < n/3$, it is possible to choose $T = M$, and obtain the curve $T = M = 2^{2(n-\kappa+1)/3}$. For $n = 128$ with $\kappa = 30$, one can apply the time-memory-data tradeoff attack using 2^{99} pre-processing time and 2^{66} memory blocks, and find a second preimage in 2^{66} online computation.

The described long message second preimage attack on trees applies to not only strengthened Merkle trees, but also to XOR-Trees [2] and optimized variants of these hash functions [16].

9 New Trojan Message Attacks on Merkle-Damgård Hash Functions

"Do not trust the horse, Trojans. Whatever it is, I fear the Greeks even when they bring gifts" (Virgil's Aeneid, Book 2, 19 BC)

In this section, we introduce a new generic attack on many hash function constructions, called the Trojan Message attack. A Trojan message is a string S which is produced offline by an attacker, and is then provided to a victim. The victim then selects some prefix P from a constrained set of choices, and produces the message $P \| S$. However, due to the way S was chosen, the attacker is now able to find a second preimage for $P \| S$.

Given a Merkle-Damgård hash for which collisions may be found, Trojan messages may be produced. In general, the Trojan message requires at least one message input block, and one collision search, per possible value of P. If there are 1024 possible values of P, an attacker may produce a 1024-block Trojan message, requiring 1024 collision searches.

One can imagine a Trojan message attack being practical against applications which use MD5, and which permit an attacker to provide some victim with "boilerplate" text for the end of his document, while imposing a relatively constrained set of choice for his part of the document.

Against Merkle-Damgård hashes, Trojan message attacks take two forms:

1. If only straightforward collisions of the compression function are possible, second preimages for the full message keep the victim's choice of P, but introduce a limited change in S. That is, the attacker finds $S' \neq S$ such that $H(P \| S) = H(P \| S')$.
2. If collisions of the compression function starting from different chaining values are possible, second preimages for the full message give the attacker a choice of P, and leave S mostly unchanged. That is, the attacker finds P' and S' such that $H(P \| S) = H(P' \| S')$.

Let $\mathcal{P} = \{P_1, \ldots, P_N\}$ be a set of N known prefix messages and h_0^i be the intermediate chaining value resulting from the computation of $f^*(P_i)$. Note, that without loss of generality, we can assume that all the prefixes have the same length (otherwise, we just consider padded versions). Therefore, we safely disregard strengthening and padding issues.

9.1   The Collision Trojan Attack

The collision variant of the trojan message attack makes use of a collision finding algorithm **IndenticalPrefixCollision** which takes a chaining value as parameter and produces a pair of messages colliding from this chaining value. The attack proceeds as follows:

1. $\mathcal{A}$ computes N colliding message pairs (S_i, T_i) using the algorithm of figure 9.
2. $\mathcal{A}$ sends $\mathcal{B}$ a suffix message $S = S_1 \,||\, \ldots \,||\, S_N$.
3. $\mathcal{B}$ commits to $h = H^f(P_i \,||\, S)$ where P_i is in $\mathcal{P}$.
4. $\mathcal{A}$ finds out P_i through exhaustive search amongst the N possible choices and outputs:
$$M' = P_i \,||\, S_1 \,||\, \ldots \,||\, T_i \,||\, \ldots \,||\, S_N$$

We have that $H^f(M') = h$. The hashing of $P_i \,||\, S$ and $P_i \,||\, S'$ differs only when T_i replaces S_i, but because these two blocks collide, both hash processes do not diverge.

The only non-trivial part of the attack for $\mathcal{A}$ is the first step where $\mathcal{A}$ precomputes N collisions for each prefix from the set $\mathcal{P}$ (in time $N \cdot 2^{n/2}$), and evaluates the compression function N^2 times. If finding a collision for the hash function is easy, e.g., like the legacy hash function MD5 [14] the attack can be even practical. It has recently been shown that finding a collision in MD5 takes about 2^{16} evaluations of the compression function [17]. For instance, one can forge in a matter of seconds a suffix S of 46720 bytes permitting to find second preimages for MD5 if the prefix set $\mathcal{P}$ is the set of the days of the year.

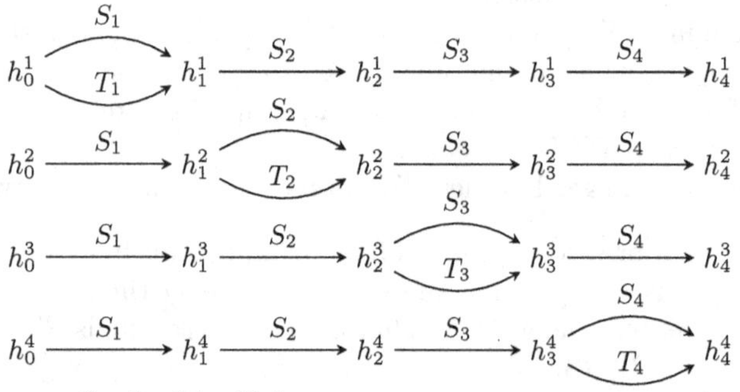

for $i = 1$ to N do
$\quad (S_i, T_i) \leftarrow$ **IndenticalPrefixCollision** (h^i_{i-1})
$\quad$ for $j = 1$ to N do
$\quad\quad h^j_i \leftarrow f\left(h^j_{i-1}, S_i\right)$
$\quad$ end for
end for

Fig. 9. Trojan Message Attack, Collision Variant

9.2   The Herding Trojan Attack

The herding variant of the trojan message attack is stronger, and allows for more freedom for the attacker. In exchange, the preprocessing and the online running times are larger.

Let K denote the length of all possible prefixes in $\mathcal{P}$. We can extend K to be as large as we wish. The herding variant of the trojan message attack makes use of a different, more sophisticated "chosen-prefix" collision finding algorithm **ChosenPrefixCollision**(h_1, h_2) that returns the messages m_1 and m_2, such that $f(h_1, m_1) = f(h_2, m_2)$. In some specific cases this collision is harder to find (for instance in MD5, such collision takes 2^{41} compression function evaluations [17]).

Another difference between this variant and the previous one, is that in this variant, the adversary is challenged by a second prefix P', not controlled by him, which he has to herd to the same value as $H^f(P_i \,||\, S)$. The attack proceeds as follows:

1. $\mathcal{A}$ computes a diamond structure with 2^ℓ entry points, denoted by $D_1 = \{h_i\}$, converging to the hash value h_0^D, with the constraint that $\ell < K - 2$.
2. $\mathcal{A}$ generates N colliding message pairs using the algorithm of figure 10.
3. $\mathcal{A}$ sends $\mathcal{B}$ a suffix message $S = S_1 \,||\, \ldots \,||\, S_N$.
4. $\mathcal{B}$ commits to $h = H^f(P_i \,||\, S)$ where $P_i \in \mathcal{P}$.
5. $\mathcal{A}$ is challenged with an arbitrary prefix P' of size at most $K - \ell - 1$ blocks, not necessarily in the known prefix set.
6. $\mathcal{A}$ finds (by random trials) a connecting message C of size $K - \ell - |P'|$ blocks such that $h_{i_0} = f^*(P' \,||\, C) \in D_1$.
7. $\mathcal{A}$ forges a new prefix $Q = P' \,||\, C \,||\, m_{i_0}^D$, which is such that $f^*(Q) = h_0^D$.
8. As in the collision version, $\mathcal{A}$ outputs $Q \,||\, S'$, where $S' = S_1 \,||\, \ldots \,||\, T_i \,||\, \ldots \,||\, S_N$.

As in the collision variant, we have that $H^f(Q \,||\, S') = h$. The reasoning to establish this fact is essentially the same.

The workload of the attack is step one where $\mathcal{A}$ constructs a diamond structure with 2^ℓ starting points and N collisions for each prefix from the set P. Thus, the precomputation complexity is of order $2^{n/2+\ell/2+2} + N \cdot 2^{n/2}$. The online cost is the connection step for computing the prefix P' and is of order $2^{n-\ell}$.

9.3   Applications of the Trojan Attacks

The trojan attack can is highly useful in instances with a set of predictable prefixes, and where the attacker is able to suggest a suffix to introduce to the message. Such a case is the X.509 certificate, where the adversary may generate a second certificate (with the same identification) but with different public keys. Another possible application is a time stamping service, which signs $\mathcal{MDH}(ts, M)$ where ts is a time stamp and M is the document.

TROJAN ATTACKS ON TREE HASHES. The processing of the prefix in tree hashes is independent of the suffix processing. Thus, $\mathcal{A}$ computes independent collisions for each message input node. In fact, it is only enough that $\mathcal{A}$ produces a single

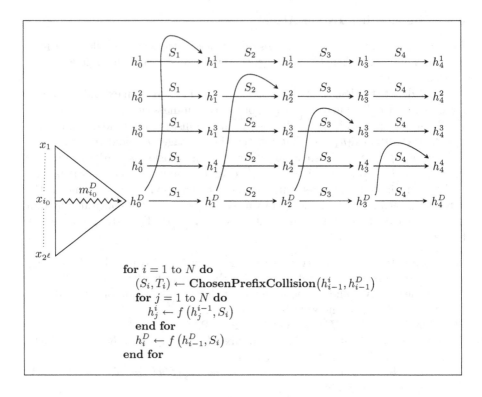

$$\begin{aligned}
&\textbf{for } i = 1 \textbf{ to } N \textbf{ do}\\
&\quad (S_i, T_i) \leftarrow \textbf{ChosenPrefixCollision}\big(h^i_{i-1}, h^D_{i-1}\big)\\
&\quad \textbf{for } j = 1 \textbf{ to } N \textbf{ do}\\
&\qquad h^i_j \leftarrow f\big(h^{i-1}_j, S_i\big)\\
&\quad \textbf{end for}\\
&\quad h^D_i \leftarrow f\big(h^D_{i-1}, S_i\big)\\
&\textbf{end for}
\end{aligned}$$

Fig. 10. Trojan Message Attack, Herding Variant

colliding suffix block $f(S_i) = f(T_i)$ in the first level of the tree evaluation. Then, for all $P_i \in \mathcal{P}$, $\mathcal{A}$ can compute $\mathit{Tree}(P_i\|S) = \mathit{Tree}(P_i\|S')$ where $S = S_1\|\ldots\|S_i\|\ldots\|S_L$, $S' = S_1\|\ldots\|T_i\|\ldots\|S_L$ and L may be different than $|\mathcal{P}|$. The latter is true, because as opposed to Merkle-Damgård, here the length of the suffix is independent of the size of the prefix set $\mathcal{P}$. This completes the collision variant of the trojan message attack on tree hashes.

The herding trojan message attack on tree hashes could be applied as follows. $\mathcal{A}$ executes first the herding attack on tree hashes. Instead of selecting arbitrary chosen target set, here $\mathcal{A}$ fixes the target set T to consist of all intermediate hash values of the known prefixes $P_i \in \mathcal{P}$. Then, as in the collision tree variant of the attack, $\mathcal{A}$ computes S' by creating collision(s) on the top tree node(s) (distinct from the target set). Now, challenged on P', $\mathcal{A}$ finds P'_s, such that $f(P'\|P'_s) \in T$ where $|P'\|P'_s| = |P_i|$ and $|P'_s|$ is at least $\log_2(2^n/|\mathcal{P}|)$. Then $\mathit{Tree}(P'\|P'_s\|S') = \mathit{Tree}(P_i\|S)$, which concludes the herding variant of the attack.

10   Summary and Conclusions

Our results enhance the understanding of the multi-pipe and multi-pass modes of iteration, such as concatenated hashes, zipper hash, hash-twice, and tree hash functions. The presented attacks reconfirm the knowledge that there is only a limited gain by concatenating the output of hash functions when it comes to security, and that the hash twice construction is not secure.

Moreover, we show that all of the investigated constructions suffer the herding attack. An interesting result is that domain separation (equivalent to distinct internal compression function evaluation, e.g., by means of a counter separation) does not protect any of the existing hash functions against herding attacks. And while domain separation often does offer protection against second preimage attacks, it appears to be unable to also mitigate herding attacks. An open question remains to exhibit either a generic herding protective mechanism or a mode of operation optimally secure against standard and herding attacks.

Acknowledgments

We would like to thank Lily Chen, Barbara Guttman and the anonymous referees for their useful feedback. This work was supported in part by the IAP Programme P6/26 BCRYPT of the Belgian State (Belgian Science Policy), and in part by the European Commission through the ICT programme under contract ICT-2007-216676 ECRYPT II. The first author is supported by a Ph.D. Fellowship from the Flemish Research Foundation (FWO–Vlaanderen). The third author was supported by the France Telecom Chaire.

References

1. Andreeva, E., Bouillaguet, C., Fouque, P.A., Hoch, J.J., Kelsey, J., Shamir, A., Zimmer, S.: Second Preimage Attacks on Dithered Hash Functions. In: Smart, N.P. (ed.) EUROCRYPT 2008. LNCS, vol. 4965, pp. 270–288. Springer, Heidelberg (2008)
2. Bellare, M., Rogaway, P.: Collision-Resistant Hashing: Towards Making UOWHFs Practical. In: Kaliski Jr., B.S. (ed.) CRYPTO 1997. LNCS, vol. 1294, pp. 470–484. Springer, Heidelberg (1997)
3. Biryukov, A., Mukhopadhyay, S., Sarkar, P.: Improved time-memory trade-offs with multiple data. In: Preneel, B., Tavares, S. (eds.) SAC 2005. LNCS, vol. 3897, pp. 110–127. Springer, Heidelberg (2006)
4. Biryukov, A., Shamir, A.: Cryptanalytic time/memory/data tradeoffs for stream ciphers. In: Okamoto, T. (ed.) ASIACRYPT 2000. LNCS, vol. 1976, pp. 1–13. Springer, Heidelberg (2000)
5. Coppersmith, D.: Another birthday attack. In: Williams, H.C. (ed.) CRYPTO 1985. LNCS, vol. 218, pp. 14–17. Springer, Heidelberg (1986)
6. Dean, R.D.: Formal Aspects of Mobile Code Security. PhD thesis, Princeton University (January 1999)
7. Hoch, J.J., Shamir, A.: Breaking the ice - finding multicollisions in iterated concatenated and expanded (ice) hash functions. In: Robshaw, M.J.B. (ed.) FSE 2006. LNCS, vol. 4047, pp. 179–194. Springer, Heidelberg (2006)
8. Joux, A.: Multicollisions in Iterated Hash Functions. Application to Cascaded Constructions. In: Franklin, M. (ed.) CRYPTO 2004. LNCS, vol. 3152, pp. 306–316. Springer, Heidelberg (2004)

9. Kelsey, J., Kohno, T.: Herding Hash Functions and the Nostradamus Attack. In: Vaudenay, S. (ed.) EUROCRYPT 2006. LNCS, vol. 4004, pp. 183–200. Springer, Heidelberg (2006)

10. Kelsey, J., Schneier, B.: Second Preimages on n-Bit Hash Functions for Much Less than 2^n Work. In: Cramer, R. (ed.) EUROCRYPT 2005. LNCS, vol. 3494, pp. 474–490. Springer, Heidelberg (2005)

11. Liskov, M.: Constructing an Ideal Hash Function from Weak Ideal Compression Functions. In: Biham, E., Youssef, A.M. (eds.) SAC 2006. LNCS, vol. 4356, pp. 358–375. Springer, Heidelberg (2007)

12. Merkle, R.C.: One Way Hash Functions and DES. In: Brassard, G. (ed.) CRYPTO 1989. LNCS, vol. 435, pp. 428–446. Springer, Heidelberg (1990)

13. Nandi, M., Stinson, D.R.: Multicollision attacks on some generalized sequential hash functions. IEEE Transactions on Information Theory 53(2), 759–767 (2007)

14. Rivest, R.L.: The MD5 message-digest algorithm. RFC1321 (April 1992)

15. Rivest, R.L., Agre, B., Bailey, D.V., Crutchfield, C., Dodis, Y., Fleming, K.E., Khan, A., Krishnamurthy, J., Lin, Y., Reyzin, L., Shen, E., Sukha, J., Sutherland, D., Tromer, E., Yin, Y.L.: The md6 hash function, a proposal to nist for sha-3 (2008)

16. Sarkar, P.: Construction of universal one-way hash functions: Tree hashing revisited. Discrete Applied Mathematics 155(16), 2174–2180 (2007)

17. Stevens, M., Sotirov, A., Appelbaum, J., Lenstra, A., Molnar, D., Osvik, D.A., de Weger, B.: Short chosen-prefix collisions for md5 and the creation of a rogue ca certificate. Cryptology ePrint Archive, Report 2009/111 (2009), http://eprint.iacr.org/

Cryptanalysis of Dynamic SHA(2)

Jean-Philippe Aumasson[1,*], Orr Dunkelman[2,**], Sebastiaan Indesteege[3,4,***],
and Bart Preneel[3,4]

[1] FHNW, Windisch, Switzerland
[2] École Normale Supérieure, INRIA, CNRS, Paris, France
[3] Department of Electrical Engineering ESAT/COSIC,
Katholieke Universiteit Leuven, Belgium
[4] Interdisciplinary Institute for BroadBand Technology (IBBT), Belgium

Abstract. In this paper, we analyze the hash functions Dynamic SHA
and Dynamic SHA2, which have been selected as first round candidates
in the NIST hash function competition. These hash functions rely heavily on data-dependent rotations, similar to certain block ciphers, e.g.,
RC5. Our analysis suggests that in the case of hash functions, where the
attacker has more control over the rotations, this approach is less favorable than in block ciphers. We present practical, or close to practical,
collision attacks on both Dynamic SHA and Dynamic SHA2. Moreover,
we present a preimage attack on Dynamic SHA that is faster than exhaustive search.

Keywords: Dynamic SHA, Dynamic SHA2, SHA-3 candidate, hash
function, collision attack.

1 Introduction

New generic cryptanalytic techniques for hash functions [1, 2] and the recent
results on MD5 and SHA-1 [3,4,5], along with the fact that the SHA-2 family of
hash functions was designed with a similar structure, have led to the initiation
of the NIST hash function competition [6], a public competition to develop a
new hash standard, which will be called SHA-3.

The competition has sparked a great deal of submissions: 64 new hash function proposals were submitted to the competition, of which 51 were accepted as
meeting the submission criteria for the first round. Among the 51 candidates,
Dynamic SHA and Dynamic SHA2 stand out as a combination of the SHA family
design with data-dependent rotations.

The concept of data-dependent rotations has been explored for block ciphers in
several constructions, most notably in the RC5 and RC6 block ciphers [7,8]. The
security of such block ciphers has been challenged many times, and a majority
of attacks is based on guessing the distances of the rotations. In cryptanalysis of

---

* Supported by the Swiss National Science Foundation, project no. 113329.
** This author was supported by the France Telecom chaire.
*** F.W.O. Research Assistant, Fund for Scientific Research — Flanders (Belgium).

M.J. Jacobson Jr., V. Rijmen, and R. Safavi-Naini (Eds.): SAC 2009, LNCS 5867, pp. 415–432, 2009.

hash functions, however, the internal state is known. The attacker even has control over (parts of) the internal state, including rotations, though sometimes this control is only indirect. For example, Mendel et al. [9] exploited data-dependent rotations to find collisions for the hash function of Shin et al. [10]. Our attacks on Dynamic SHA and Dynamic SHA2 also exploit data-dependent rotations, to find (second) preimages and collisions.

2   Brief Description of Dynamic SHA and Dynamic SHA2

Dynamic SHA and Dynamic SHA2 use similar building blocks, but have different compression functions. This section gives a brief description of these algorithms.

Dynamic SHA and Dynamic SHA2 follow a classical Merkle-Damgård construction, based on a compression function that maps an 8-word chaining value and a 16-word message to a new 8-word chaining value. The 256-bit versions use 32-bit words, and the 512-bit versions use 64-bit words. We focus on the 256-bit versions, also called Dynamic SHA-256 and Dynamic SHA2-256. See [11,12] for details on the 512-bit versions, Dynamic SHA-512 and Dynamic SHA2-512. The following presents a bottom-up description of the compression function, thus starting with its building blocks.

The symbol $\oplus$ stands for exclusive OR (XOR), $\wedge$ for logical AND, $\vee$ for logical OR, and $+$ for integer addition. Numbers in hexadecimal basis are written in typewriter font (e.g., $\mathtt{FF} = 255$). We count bit indices starting from zero at the least significant bit (LSB). Thus, the first bit of a word w is written as w^0, and more generally we use the notation w^i for the bit i of the word w. The most significant bit (MSB) of w is thus w^{31} for Dynamic SHA-256, and w^{63} for Dynamic SHA-512. Note that the i-th bit of a word corresponds to the bit number $i - 1$, since we start counting from zero.

2.1   Building Blocks

The function G takes as input three words x_1, x_2, x_3 and an integer $t \in \{0, 1, 2, 3\}$, and returns one word, computed as follows:

$$
G_t(x_1, x_2, x_3) = \begin{cases} x_1 \oplus x_2 \oplus x_3 & \text{if } t = 0 \\ (x_1 \wedge x_2) \oplus x_3 & \text{if } t = 1 \\ (x_1 \wedge x_2) \oplus x_3 \oplus \neg x_1 & \text{if } t = 2 \\ (x_1 \wedge x_2) \oplus x_3 \oplus \neg x_2 & \text{if } t = 3 \end{cases} .
$$

Note that this definition is simplified, but equivalent to the original in [11,12].

The function R takes as input eight words $x_1, \ldots, x_8$ and an integer t, and returns one word computed as follows:

$$
R(x_1, \ldots, x_8, t) = (((((((x_1 \oplus x_2) + x_3) \oplus x_4) + x_5) \oplus x_6) + x_7) \oplus x_8) \ggg t .
$$

The function $R1$ takes as input eight words $x_1, \ldots, x_8$ and returns one word computed as follows (in the 256-bit versions):

$$t_0 \leftarrow (((((x_1 + x_2) \oplus x_3) + x_4) \oplus x_5) + x_6) \oplus x_7$$
$$t_1 \leftarrow ((t_0 \gg 17) \oplus t_0) \wedge \text{0001FFFF}$$
$$t_2 \leftarrow ((t_1 \gg 10) \oplus t_1) \wedge \text{000003FF}$$
$$t_3 \leftarrow ((t_2 \gg 5) \oplus t_2) \wedge \text{0000001F}$$
$$\textbf{return } x_8 \ggg t_3$$

Finally, the COMP function takes as input eight words $a, \ldots, h$ representing the internal state, eight message words $w_0, \ldots, w_7$, or $w_8, \ldots, w_{15}$, and an integer t. COMP updates the internal state as follows (in the 256-bit versions):

$T \leftarrow R(a, \ldots, h, w_t \bmod 32)$	$T \leftarrow R(a, \ldots, h, (w_t \gg 15) \bmod 32)$
$h \leftarrow g$	$h \leftarrow g + w_{t+7}$
$g \leftarrow f \ggg ((w_t \gg 5) \bmod 32)$	$g \leftarrow f \ggg ((w_t \gg 20) \bmod 32)$
$f \leftarrow e + w_{t+3}$	$f \leftarrow e + w_{t+6}$
$e \leftarrow d \ggg ((w_t \gg 10) \bmod 32)$	$e \leftarrow d \ggg ((w_t \gg 25) \bmod 32)$
$d \leftarrow G_{w_t \gg 30}(a, b, c) + w_{t+2}$	$d \leftarrow G_{t \bmod 4}(a, b, c) + w_{t+5}$
$c \leftarrow b$	$c \leftarrow b + w_t$
$b \leftarrow a$	$b \leftarrow a$
$a \leftarrow T + w_{t+1}$	$a \leftarrow T + w_{t+4}$

2.2 Compression Functions

Given a chaining value $h_0, \ldots, h_7$ and a message block $w_0, \ldots, w_{15}$, the compression function of Dynamic SHA (Dynamic SHA2, respectively) produces a new chaining value, as described in Algorithm 1 (Algorithm 2, resp.).

The compression function of Dynamic SHA is composed of an initialization, an iterative part of 48 rounds, and a feedforward of the initial chaining value. It uses three constants TT_0, TT_1, TT_2.

The compression function of Dynamic SHA2 is composed of an initialization followed by three iterative parts, and finally by a feedforward. Note that, when calling COMP with the message words $w_8, \ldots, w_{15}$ and an integer t, w_t stands for w_8, w_{t+1} stands for w_9, etc. Dynamic SHA2 surprisingly enough, uses no constants.

3   Collision Attack on Dynamic SHA

This section describes a practical collision attack on Dynamic SHA. It builds on a 9-step local collision that exploits an important differential property of the function $R1$, which we introduce first. The same local collision pattern is repeated three times to find collisions for the entire compression function. Furthermore, these three instances of the local collision pattern can be decoupled, which drastically reduces the attack complexity. We present the attack on Dynamic SHA-256 here. We could adapt it to Dynamic SHA-512 with only minimal changes, as detailed in Appendix C.

Algorithm 1. Compression function of Dynamic SHA

Initialization

$$a = h_0 \quad b = h_1 \quad c = h_2 \quad d = h_3 \quad e = h_4 \quad f = h_5 \quad g = h_6 \quad h = h_7$$

Iterative part

  for $t = 0, 1 \ldots, 47$

$$T \leftarrow R1(a, b, c, d, e, f, g, h)$$
$$U \leftarrow G(a, b, c, t \bmod 4) + w_{t \bmod 16} + TT_{t \ggg 4}$$
$$(a, b, c, d, e, f, g, h) \leftarrow (T, a, b, U, d, e, f, g)$$

Feedforward

$$h_0 \leftarrow h_0 + a \quad h_1 \leftarrow h_1 + b \quad h_2 \leftarrow h_2 + c \quad h_3 \leftarrow h_3 + d$$
$$h_4 \leftarrow h_4 + e \quad h_5 \leftarrow h_5 + f \quad h_6 \leftarrow h_6 + g \quad h_7 \leftarrow h_7 + h$$

Algorithm 2. Compression function of Dynamic SHA2

Initialization

$$a = h_0 \quad b = h_1 \quad c = h_2 \quad d = h_3 \quad e = h_4 \quad f = h_5 \quad g = h_6 \quad h = h_7$$

First iterative part

$$\mathsf{COMP}\,(a, b, c, d, e, f, g, h, w_0, w_1, \ldots, w_7, 0)$$
$$\mathsf{COMP}\,(a, b, c, d, e, f, g, h, w_8, w_9, \ldots, w_{15}, 0)$$

Second iterative part

  for $t = 0, 1 \ldots, 8$

$$T \leftarrow R1(a, b, c, d, e, f, g, h)$$
$$(a, b, c, d, e, f, g, h) \leftarrow (T, a, b, c, d, e, f, g)$$

Third iterative part

  for $t = 1, 2 \ldots, 7$

$$\mathsf{COMP}\,(a, b, c, d, e, f, g, h, w_0, w_1, \ldots, w_7, t)$$
$$\mathsf{COMP}\,(a, b, c, d, e, f, g, h, w_8, w_9, \ldots, w_{15}, t)$$

Feedforward

$$h_0 \leftarrow h_0 + a \quad h_1 \leftarrow h_1 + b \quad h_2 \leftarrow h_2 + c \quad h_3 \leftarrow h_3 + d$$
$$h_4 \leftarrow h_4 + e \quad h_5 \leftarrow h_5 + f \quad h_6 \leftarrow h_6 + g \quad h_7 \leftarrow h_7 + h$$

3.1 A Differential Property of the Function $R1$

To overcome the obstacle of data-dependent rotation, our attack ensures that no difference occurs in any of the data-dependent rotation amounts. This section clarifies how to achieve this.

The data-dependent rotations are located in the 8-input function $R1$. For Dynamic SHA-256, consider the difference $\Delta = $ 80004000, i.e., only bits 31 and 14 are set. Let one of the first seven inputs to the function $R1$ have this difference, i.e., one of $x_1, \ldots, x_7$. In the first step of $R1$, an intermediary word t_0 is computed as follows:

$$t_0 \leftarrow (((((x_1 + x_2) \oplus x_3) + x_4) \oplus x_5) + x_6) \oplus x_7 .$$

The difference in the MSB always propagates to t_0. Assuming that no carry occurs for bit 14, the intermediary t_0 also has the difference Δ. If t_0 has a difference Δ, this difference is then absorbed by the rest of the function $R1$. Indeed, the next step computes the intermediary word t_1 as

$$t_1 \leftarrow ((t_0 \ggg 17) \oplus t_0) \wedge \text{0001FFFF} .$$

Note that $(\Delta \ggg 17) \oplus \Delta = $ 80000000, which is absorbed by the logical AND operation. We note that there are other differences of Hamming weight 2 that exhibit the same property and may be used without any change in the attack, e.g., $\Delta = $ 80000010.

We now estimate the probability that a single Δ-difference in one of the first seven inputs of the function $R1$ is absorbed. As a Δ-difference in t_0 is absorbed with certainty, it suffices that a Δ-difference in one of the seven first inputs propagates to t_0. This happens when no carry difference occur for bit 14 in any of the modular additions. The probability that a one-bit difference in one of the summands in an addition does not cause a carry difference is $1/2$. Thus, the probability that a Δ-difference is absorbed by the function $R1$ can be estimated to 2^{-k}, where k is the number of modular additions the difference propagates through. For instance, a difference in x_3 activates two modular additions, so $k = 2$.

However, the actual probability is higher, as the undesirable effects of a carry difference in one modular addition can be reverted by another carry difference in a subsequent addition. The combination of modular additions and XOR can be represented compactly in a trellis, and a variant of the Viterbi algorithm can be used to efficiently count the probability that a Δ-difference is passed to t_0 unchanged. Our computer aided research revealed that this is indeed an important effect: For a difference in x_3 or x_4, the actual probability is $2^{-1.58}$ rather than 2^{-2}, and for a difference in x_1 or x_2, the actual probability is $2^{-2.07}$ rather than 2^{-3}. For differences in the other words, only one modular addition is affected, so no carry differences can be canceled. Hence, in those cases, the simple estimation is correct.

Table 1. A 9-step local collision for Dynamic SHA. The difference at step t is the difference in the state *before* computing step t.

t	a	b	c	d	e	f	g	h	w	Pr
0	0	0	0	0	0	0	0	0	Δ	2^{-1}
1	0	0	0	Δ	0	0	0	0	0	$2^{-1.58}$
2	0	0	0	0	Δ	0	0	0	0	2^{-1}
3	0	0	0	0	0	Δ	0	0	0	2^{-1}
4	0	0	0	0	0	0	Δ	0	0	1
5	0	0	0	0	0	0	0	Δ	0	2^{-5}
6	Δ	0	0	0	0	0	0	0	0	$2^{-2.07} \cdot 2^{-2}$
7	0	Δ	0	0	0	0	0	0	0	$2^{-2.07} \cdot 2^{-2}$
8	0	0	Δ	0	0	0	0	0	Δ	$2^{-1.58} \cdot 2^{-1}$
	0	0	0	0	0	0	0	0		

3.2   A 9-Step Local Collision

We present a simple 9-step local collision for Dynamic SHA in Table 1. A difference of $\Delta = 80004000$ is introduced, then, all further diffusion of this difference is avoided. After seven more steps, the difference has rotated through the internal state of Dynamic SHA once, and can be canceled via an appropriate difference in the message word. The characteristic has probability $2^{-20.3}$.

In step 0, a Δ-difference is introduced via the message word. Note that the message word itself can contain any additive difference that can cause a Δ-difference in the state. In steps 1 to 4, the Δ-difference in one of the state variables is absorbed by the function $R1$, as described in Section 3.1. Then, at the beginning of step 5, there is a Δ-difference in the internal state word h. This word is rotated by a data-dependent amount, and thus we can require that it is rotated by zero bits, i.e., not rotated at all. In steps 6 and 7, the Δ-difference should be absorbed by the G-functions. Any G-function except XOR absorbs differences in its first two inputs with probability $1/2$ per bit. Also, $R1$ should absorb the differences in these steps. Finally, in step 8, the difference in the state variable c is canceled by another Δ-difference coming from the message word.

The probability that the local collision pattern is followed is estimated by simply multiplying the probabilities of all the events discussed above. The probabilities of each step are indicated in Table 1. This yields an overall probability of $2^{-20.3}$ for the entire 9-step local collision.

3.3   The Attack

Our attack repeats the 9-step collision three times. This made possible by the simple message schedule, which consists of a simple repetition of the 16 words in a message block. Thus, the only message words that have a difference are w_0, which introduces the differences, and w_8, which cancels them.

A straightforward attack would consist of choosing an arbitrary message block, and applying a difference of $\Delta = 80004000$ to w_0 and w_8. As the local collision is repeated three times, the complexity of this attack would be approximately $(2^{20.3})^3 = 2^{61}$. This can be improved tremendously by making the three local collisions independent. Then, the three local collision complexities can be added rather than multiplied.

The first two local collisions can be decoupled in a straightforward manner as only the message words w_0 to w_8 influence the first local collision. Therefore, once suitable values for these message words have been found, there is still enough freedom remaining in the other message words. The words w_0 to w_8 can thus be kept constant, while values for w_9 to w_{15} are searched such that the second local collision is also achieved.

Controlling Internal State Values. In each step of Dynamic SHA, the new value of the internal state word d is found as the modular addition of a message word and an intermediate depending on the internal state words a, b and c. Full control over message words allows an adversary to give the internal state word d any desired value. Indeed, it holds that

$$w_{t \bmod 16} = d_{\text{new}} - G(a, b, c, t \bmod 4) - TT_{t \ggg 4} \ .$$

Applying this to eight consecutive steps allows one to almost fully control the final internal state. In every step, the new value of d is fixed to some desired value. These values then shift through the internal state words a number of times, to end up as one of the internal state words after the eighth step. However, a complication arises with the first three steps, which ends up in the state words a, b and c. Before a controlled value from d ends up in one of these three state words, it is be rotated by a data-dependent amount. An obvious way to sidestep this issue is to choose a rotation-invariant value for these three words, i.e., 00000000 or FFFFFFFF. Then, the data-dependent rotations have no influence.

Decoupling All Three Local Collisions. Our attack consists of three phases, each dealing with one local collision. The first phase satisfies the first local collision, using the message words w_0 to w_8. It would be possible to use message modification techniques here to find a conforming message pair quicker, but as the later phases of the attack dominate the overall complexity anyway, no significant gains can be made in this way.

To satisfy the second local collision, we use the freedom in the remaining message words. However, we do not choose the remaining message words directly, but rather choose the internal state after step 15. We then use the words w_8 to w_{15} to connect to this state, using the technique outlined earlier. We fix the values of a, b and c to zero, to make them rotation-invariant, and choose the remaining five words arbitrarily. Note that w_8 was already determined in phase 1, so it should not be modified again, but w_8 is used here to force a zero value, which ends up in the internal state word d after step 15. This issue is solved by shifting this condition on w_8 to phase 1. Instead of arbitrarily choosing w_8 there, it is computed such that the required zero is generated. This does not change the complexity of the first phase.

Finally, to satisfy the third local collision, we modify w_7. Then, only d changes after step seven. As the value in w_8, which should force d to zero after step eight, depends only on the internal state words a, b and c before step eight, modifying w_7 does not require a correction in w_8. Thus, such modifications do not change the fact that the first local collision pattern is followed. The values of w_9 to w_{15} are then updated such that the internal state after step 15 is unchanged, and so the start of the second local collision will be unaltered. For the same reasons as before, the change in w_7 also does not affect the end of the second local collision pattern.

Hence, we dispose of a modification algorithm that leaves the first two local collisions unaffected, but changes the internal state values before the third local collision randomly. This provides the required freedom to also satisfy this third and final local collision. Hence, the overall attack complexity can be estimated at about 2^{21} Dynamic SHA compression function computations. Appendix A reports on our implementation of the attack, with an example of collision.

4   Preimage Attack on Dynamic SHA

This section describes (first and second) preimage attacks on Dynamic SHA. We first describe how to find preimages for the compression function of Dynamic SHA, and then explain how to extend this to first and second preimage attacks. on the Dynamic SHA hash function. We describe how to attack Dynamic SHA-256 here, and refer to Appendix C for details on how to adapt the attack to Dynamic SHA-512.

Conceptually, our preimage attack bears some similarity to the work on SHA-0 and SHA-1 by De Cannière and Rechberger [13], for it finds a preimage bit slice per bit slice. If all data-dependent rotation amounts in Dynamic SHA are assumed to be zero, then a bit of any intermediate word cannot be influenced by any other bit of higher position. This is because, besides rotations, all operations are either bit-wise or modular additions.

4.1   Preimage Attack on the Compression Function

Assume that the rotations in a block of Dynamic SHA are all zero. Then, all words in Dynamic SHA can be divided into bit slices, as all computations are now T-functions [14]. As noted above, bit i of each word can only be influenced by bits 0 to i of other words. When bits 0 to $(i - 1)$ of each word are known, bit i of all words can be determined.

In a preimage attack on the Dynamic SHA compression function, the internal state is given before step 0 and after step 47. Our attack starts by determining the LSB of each word. To determine this bit of all of the internal state words in every step, only the LSBs of the 16 message words need to be known. There are 2^{16} choices for these 16 bits. Then, it can be verified whether the LSBs of the eight internal state words after step 47 are correct. This occurs with probability 2^{-8}, so 2^8 choices are expected to survive.

We then proceed to the next bit slice. Keeping the choice for the LSB slice fixed, the same procedure can be repeated. For each choice of the LSB slice again 2^8 choices for the second LSB are expected to survive. For Dynamic SHA-256, this procedure is repeated until the 28 LSBs (bits 0–27) have been determined. At that point, one of the bits of each of the 48 rotation constants can be determined, as it does not depend on the higher bits of any word. Now, it can be verified if the initial assumption that all rotation constants are zero indeed holds. This corresponds to a 48-bit condition, i.e., for all rotation constants to be zero, surely this single bit of each rotation constant has to be zero. Any choices that do not satisfy this condition are eliminated. Then, the next bit is determined as before, after which another bit of each rotation constant can be verified. This is repeated until all bits have been determined.

4.2 Complexity Evaluation

The attack can be described as a simple tree search, where a tree level corresponds to a bit slice, and a node represents an assignment for all bits in the slice under consideration, and all LSB slices. To expand a node in the tree, one guesses the 16 message bits of the next slice, and checks that the conditions on the state words after step 47 are satisfied. As explained above, on average about 2^8 choices are expected to survive, i.e., the tree has a branching factor of 2^8. When the 28 LSB slices are known, however, the average number of child nodes drops by 2^{-48} due to the additional filtering. The cost of expanding one node is about 2^{16} Dynamic SHA compression function evaluations, as 2^{16} choices have to be investigated. The expected number of solutions is equal to the expected number of nodes at the deepest level of the tree, which is $2^{8\cdot32}\cdot2^{-48\cdot5} = 2^{16}$. This agrees with the observation that for a given input/output chaining values of the compression function, there are expected to be 2^{256} message blocks that conform to this combination. For each of these, the probability that all the rotations are by 0 positions is 2^{-240}, so about 2^{16} remain.

As we aim to find just one solution, i.e., any node on the deepest level of the tree, a depth-first search is well suited to our application. It requires only negligible memory and can easily be parallelised. Since, for Dynamic SHA-256, 2^{16} solutions are expected, the depth-first search needs to search only about a fraction 2^{-16} of the entire tree before encountering the first solution. Due to the large branching factor, the total number of nodes in the tree is well approximated by the number of nodes on the widest level of the tree, which has $2^{8\cdot27} = 2^{216}$ nodes for Dynamic SHA-256. The search is thus expected to expand about 2^{200} nodes, each of which costs 2^{16} Dynamic SHA-256 compression function evaluations, resulting in a total attack complexity of 2^{216} Dynamic SHA-256 compression function evaluations.

4.3 Application to the Hash Function

Our preimage attack on the compression function directly gives a second preimage attack on the Dynamic SHA hash function with the same complexity, provided that there is at least one message block that does not contain any padding in the challenge message.

For a first preimage, the padding bits limit the control an attacker has over the message bits. It is not possible to simply copy the padding as in a second preimage attack. Thus, we use the following approach instead. First, choose a message length such that the last padded message block only contains 65 bits of padding, which is the minimum. Then, choose an arbitrary message for all but the last message block. Finally, a modified version of the attack in Section 4.1 is used to determine the last message block.

The main difference is that the last 65 bits of the message block can not be chosen by the adversary, as they are padding bits. Their contents are fixed by the choice of the message length. However, the same approach as in Section 4.1 can still be applied, except that fewer bits can be chosen in each bit slice. For Dynamic SHA-256, the expected number of solutions in the search tree now becomes $2^{6 \cdot 27} \cdot 2^{-42 \cdot 4} \cdot 2^{-43 \cdot 1} = 2^{-49}$. A solution is thus only expected to exist with probability 2^{-49}, thus the attack is repeated sufficiently many times with a different message length. The number of nodes at the widest level of the tree is $2^{6 \cdot 27}$, and the cost for expanding a single node at this level is 2^{14} Dynamic SHA compression function calls. Thus, the total attack complexity becomes approximately $2^{49} \cdot 2^{6 \cdot 27} \cdot 2^{14} = 2^{225}$ Dynamic SHA compression function evaluations.

5   Collision Attack on Dynamic SHA2

To attack Dynamic SHA2, we use similar ideas as for Dynamic SHA. Specifically, we use the control of the message to ensure that as many rotations as possible are by the amounts that we need. Moreover, as many of the rotations amounts are directly determined by the message, our task becomes easier. Our attack is based on introducing a difference in the most significant bit of two message words, w_8 and w_{14}. As a 32-bit condition is imposed on the chaining value, a two-block collision finding technique is used, where the first block is searched until a suitable chaining value is encountered. We describe our attack on Dynamic SHA2-256 here. It can be adapted to Dynamic SHA2-512, as Appendix C shows.

5.1   First Iterative Part

Given an initial value $a, \ldots, h$, the first iterative part of the compression function of Dynamic SHA2 updates the chaining value words $a, \ldots, h$ by computing

$$\mathsf{COMP}(a, b, \ldots, h, w_0, w_1, \ldots, w_7) \;,$$

Since there is no difference in the message words $w_0, \ldots, w_7$ nor in the initial value, we have no difference at this stage.

Then, Dynamic SHA2 computes

$$\mathsf{COMP}(a, b, \ldots, h, w_8, w_9, \ldots, w_{15}) \;.$$

To follow our characteristic, the difference in w_8 and in w_{14} should lead to a difference $\Delta = $ 80000000 in c and in f. Below, we show that, to obtain these differences, it suffices to set $w_8^{30} = 1$ and to ensure that b equals FFFFFFFF after the first COMP. These conditions are easily satisfied, and do not increase the complexity of our attack.

We note that w_{14} is used only once in the first iterative part. Thus the difference Δ in w_{14} only propagates to f, when COMP sets $f \leftarrow e + w_{14}$. The word w_8, however, is used eight times, but as only the MSB has a difference, only two of these require our attention: first, when setting $c \leftarrow b + w_8$ (which gives the difference Δ in c with probability one), and second when setting

$$d \leftarrow G_{w_8 \ggg 30}(a, b, c) + w_{10} .$$

Here, the two MSBs of w_8 encode the index of the function used in G. Since we have a difference in the MSB of w_8, different functions are applied to (a, b, c). To obtain the same output, we require that the functions G_1 and G_3 are used, that is, we set the bit $w_8^{30} = 1$. The reason for this is that, when b equals FFFFFFFF, it is ensured that the outputs of both functions are equal, as can readily be seen from the definition of the G-functions in Section 2.1.

To summarize, a difference Δ in w_8 and w_{14} yields a difference Δ in c and f after the first iterative part. To have $b = $ FFFFFFFF, it is sufficient to start from a chaining values that gives at the very first COMP a T such that $T + w_1 = $ FFFFFFFF. Such a chaining value can be reached in about 2^{32} trials, and needs to be precomputed only once. That is, one first needs to find a message block leading to a chaining value that satisfies $T + w_1 = $ FFFFFFFF, before starting the actual differential attack with a second block. Actually, by using the freedom in w_0 and w_1 rather than fixing them a priori, this step can be accelerated further. However, as the other parts of the attack dominate the overall complexity, no significant gains can be made in this way.

5.2 Second Iterative Part

Table 2 describes our differential characteristic for the second iterative part of Dynamic SHA2. Note that no message word enters this part. A set of conditions that ensure that this characteristic is followed, is relatively simple. Indeed, except when $t = 2$ and $t = 5$, the two differences Δ vanish in the first step of the computation of $R1$, namely when computing

$$(((((a + b) \oplus c) + d) \oplus e) + f) \oplus g .$$

Therefore, particular conditions are only required for $t = 2$ and $t = 5$.

When $t = 2$, the difference in e gives a difference of 16 in the rotation amounts, and so the function $R1$ returns $h \ggg r$ and $(h \oplus \Delta) \ggg (r + 16 \bmod 32)$, respectively. In order to obtain, as required by our differential characteristic, the relation

$$(h \ggg r) \oplus \Delta = (h \oplus \Delta) \ggg (r + 16 \bmod 32) ,$$

Table 2. Differential characteristic for the second iterative part of Dynamic SHA2. The difference at step t is the difference in the state *before* computing step t.

t	a	b	c	d	e	f	g	h
0	0	0	Δ	0	0	Δ	0	0
1	0	0	0	Δ	0	0	Δ	0
2	0	0	0	0	Δ	0	0	Δ
3	Δ	0	0	0	0	Δ	0	0
4	0	Δ	0	0	0	0	Δ	0
5	0	0	Δ	0	0	0	0	Δ
6	Δ	0	0	Δ	0	0	0	0
7	0	Δ	0	0	Δ	0	0	0
8	0	0	Δ	0	0	Δ	0	0

a sufficient condition is to have $r = 16$, and h invariant under 16-bit rotation, i.e., $(h \ggg 16) = h$. This means that h should be of the form XYZTXYZT, which we call *symmetric*. When $t = 5$, we require similar conditions.

Now, observe that the words that should be symmetric are c and f obtained after the first iterative part. The values of c and f then directly depend on w_8 and w_{14} (see description of COMP in Section 2). We now have to find values of w_8 and of w_{14} that give symmetric c and f.

Such w_8 and w_{14} can be found as follows: first fix w_{14} to some arbitrary value, and search for a w_8 that gives a symmetric c, in 2^{16} trials. Then, fix w_8 to the value found, and search for a pair (w_5, w_{14}) that gives a symmetric f after the first iterative part. Here we need w_5 to have enough freedom, since for certain choices of w_5, there does not exist a suitable w_{14}. Again, 2^{16} trials are expected. Then we are enough degrees of freedom in the message words that do not affect c and f to find rotation $r = 16$.

Assuming symmetric c and f after the first iterative part, the characteristic is followed with probability 2^{-10}, since the condition $r = 16$ is satisfied for both $t = 2$ and $t = 5$ with probability $2^{-5} \times 2^{-5}$. By trying several values of, for example, w_9, and leaving the other message words fixed, one can thus find a conforming message pair for the first two iterative parts in about 2^{10} trials.

5.3 Third Iterative Part

Given the final difference of the second iterative part, we found a characteristic for the second round that yields no difference in the final state, thus given a collision. Table 6 in Appendix B describes our differential characteristic. Appendix B also explains in detail why the characteristic can be followed with probability 2^{-42}, given some conditions on the input.

Combining our differential characteristics with their respective conditions on the message, we obtain a method for finding a 2-block collision in about $2^{42+10} = 2^{52}$ trials. The attack succeeds with probability close to one.

Table 3. Summary of our results

Hash Function	Attack	Complexity	Section
Dynamic SHA-256	Collision	2^{21}	3
Dynamic SHA-512	Collision	2^{22}	3,C
Dynamic SHA-256	Second preimage	2^{216}	4
Dynamic SHA-512	Second preimage	2^{256}	4,C
Dynamic SHA-256	First preimage	2^{225}	4
Dynamic SHA-512	First preimage	2^{262}	4,C
Dynamic SHA2-256	Collision	2^{52}	5
Dynamic SHA2-512	Collision	2^{85}	5,C

6 Conclusion

In this paper we have discussed the security of the two SHA-3 candidates Dynamic SHA and Dynamic SHA2. We have analyzed their security, and found out that, despite their reliance on data-dependent rotations and in the case of Dynamic SHA2 even data-dependent functions, their security is subverted by the vast control and knowledge the adversary has while attacking a hash function. We also showed that neither Dynamic SHA nor Dynamic SHA2 are suitable to be selected as SHA-3, following their lack of security. Table 3 summarizes our results.

Acknowledgements

The research presented in this paper was performed in part while the authors were visiting Schloss Dagstuhl (http://www.dagstuhl.de/) in January 2009.

This work was supported in part by the IAP Programme P6/26 BCRYPT of the Belgian State (Belgian Science Policy), and in part by the European Commission through the ICT programme under contract ICT-2007-216676 ECRYPT II.

References

1. Kelsey, J., Schneier, B.: Second preimages on n-bit hash functions for much less than 2^n work. In: [15], pp. 474–490
2. Kelsey, J., Kohno, T.: Herding hash functions and the Nostradamus attack. In: Vaudenay, S. (ed.) EUROCRYPT 2006. LNCS, vol. 4004, pp. 183–200. Springer, Heidelberg (2006)
3. Wang, X., Yu, H.: How to break MD5 and other hash functions. In: [15], pp. 19–35
4. De Cannière, C., Rechberger, C.: Finding SHA-1 characteristics: General results and applications. In: Lai, X., Chen, K. (eds.) ASIACRYPT 2006. LNCS, vol. 4284, pp. 1–20. Springer, Heidelberg (2006)
5. Stevens, M., Lenstra, A.K., de Weger, B.: Chosen-prefix collisions for MD5 and colliding X.509 certificates for different identities. In: Naor, M. (ed.) EUROCRYPT 2007. LNCS, vol. 4515, pp. 1–22. Springer, Heidelberg (2007)

6. National Institute of Standards and Technology. Cryptographic hash algorithm competition, http://www.nist.gov/hash-competition

7. Rivest, R.L.: The RC5 encryption algorithm. In: Preneel, B. (ed.) FSE 1994. LNCS, vol. 1008, pp. 86–96. Springer, Heidelberg (1995)

8. Rivest, R.L., Robshaw, M.J.B., Yin, Y.L.: RC6 as the AES. In: AES Candidate Conference, pp. 337–342 (2000)

9. Mendel, F., Pramstaller, N., Rechberger, C.: Improved collision attack on the hash function proposed at PKC 1998. In: Rhee, M.S., Lee, B. (eds.) ICISC 2006. LNCS, vol. 4296, pp. 8–21. Springer, Heidelberg (2006)

10. Shin, S.U., Rhee, K.H., Ryu, D., Lee, S.: A new hash function based on MDx-family and its application to MAC. In: Imai, H., Zheng, Y. (eds.) PKC 1998. LNCS, vol. 1431, pp. 234–246. Springer, Heidelberg (1998)

11. Xu, Z.: Dynamic SHA. Submission to NIST (2008)

12. Xu, Z.: Dynamic SHA2. Submission to NIST (2008)

13. De Cannière, C., Rechberger, C.: Preimages for reduced SHA-0 and SHA-1. In: Wagner, D. (ed.) CRYPTO 2008. LNCS, vol. 5157, pp. 179–202. Springer, Heidelberg (2008)

14. Klimov, A., Shamir, A.: Cryptographic applications of t-functions. In: Matsui, M., Zuccherato, R.J. (eds.) SAC 2003. LNCS, vol. 3006, pp. 248–261. Springer, Heidelberg (2004)

15. Cramer, R. (ed.): EUROCRYPT 2005. LNCS, vol. 3494. Springer, Heidelberg (2005)

A   Practical Results

We have implemented our collision attack on Dynamic SHA. Collisions for Dynamic SHA-256 and Dynamic SHA-512 are found in a matter of seconds on an average desktop PC. A collision example for Dynamic SHA-256 is given in Table 4. An all-zero block was appended to both messages to circumvent an error in the padding routine of the Dynamic SHA reference implementation, which causes part of the last message block to be reused in the padding block.

Table 4. Collision example for Dynamic SHA-256: two messages and their common digest

34BC5378	1150D86C	3085EB92	7538ECEE	199FB91A	5A9614EC	4D21FB88	728FF21E
22FBFA2E	08CE50DF	95CDE61F	71E5F222	3D30C361	EB7676B8	F1AE9728	758B70AF
00000000	00000000	00000000	00000000	00000000	00000000	00000000	00000000
00000000	00000000	00000000	00000000	00000000	00000000	00000000	00000000
B4BC9378	1150D86C	3085EB92	7538ECEE	199FB91A	5A9614EC	4D21FB88	728FF21E
A2FBBA2E	08CE50DF	95CDE61F	71E5F222	3D30C361	EB7676B8	F1AE9728	758B70AF
00000000	00000000	00000000	00000000	00000000	00000000	00000000	00000000
00000000	00000000	00000000	00000000	00000000	00000000	00000000	00000000
703C40F7	9DDFE2C6	8298F6D0	8D2B45B6	664CBB71	8BAB1BE3	DD563F77	0D0901E6

B Differential Characteristic for Dynamic SHA2

This appendix describes the differential characteristic for the third iterative part of Dynamic SHA2, used in our collision attack presented in Section 5.

A transition in Table 6 has probability $1/2$ when there is a difference in a or b and G_1, G_2 or G_3 is used. In this case, the difference does (not) propagate with probability $1/2$. When there is a difference only in c, it always propagates to the output of the G function, independent of the function used. We also note that a difference Δ in one operand of R is always transferred to T, and thus to a, except when w_{t+1} or w_{t+4} are w_8 or w_{14}, in which case the differences vanish. When two operands of T have a difference Δ, they cancel out and yield no difference in T.

The probabilities for each step assume some conditions on the message. We will take as example the first COMP when $t = 2$: we start with a difference

$$0 \quad \Delta \quad 0 \quad \Delta \quad 0 \quad 0 \quad 0 \quad 0$$

in the chaining value $a, b, \ldots, h$. In the computation of COMP (first half), there is no difference in T, because the Δ difference in b cancels that of d. The assignment of the new values of f, g, h requires no condition on the message, for it only involves words with no difference. To obtain a difference Δ in e, we need that d is rotated by zero bit positions, that is, we need the bits 10 to 14 of w_2 to be zero. This is easy as we have direct control over w_2. Then, to obtain no difference in d, we require that the difference in b does not propagate in G. This is only possible if the Boolean function in G is not $x_1 \oplus x_2 \oplus x_3$ (see Section 2.1). Since the Boolean function is determined by the last two bits of w_2, we require $w_2^{30} \vee w_2^{31} = 1$, i.e., these bits should not be both zero. Now, the difference will not propagate in G with probability $1/2$. Finally, we get a difference Δ in c with probability 1.

By applying a similar reasoning to all the steps of our differential characteristic, we obtain conditions on the message $w_0, \ldots, w_{15}$ that are sufficient to conform to the characteristic with probability 2^{-42}. Table 5 summarizes these conditions, along with the conditions for the other iterative parts.

Conditions on $w_0, \ldots, w_7$ ensure that in the first COMP of each step the rotations are by bit zero positions, and thus the difference remains in the MSB. The probabilities smaller than one are the probabilities that the function G absorbs or passes a difference in a or b. In the second COMP, we need some rotations to be zero in order the difference to stay in the MSB. This is achieved by setting conditions on the message, for example at $t = 1$, the first ten bits of w_9 should be zero. Table 5 summarizes these conditions. After satisfying all these conditions, about 200 bits of freedom remain; indeed, besides w_8 and w_{14}, the message words w_1 to w_4 have to be fixed to let the symmetric c and f unchanged after the first iterative part.

At step $t = 6$, the difference in the MSB of w_{14} implies that G will apply different functions to (a, b, c). Similarly to Section 5.1, we will require $w_{14}^{30} = 1$ and $b = \text{EFFFFFFF}$, which will occur with probability 2^{-32}. The MSB of b should

Table 5. Conditions on the message words $w_0, \ldots, w_{15}$ sufficient to follow our differential characteristic

Word	Condition
w_0	–
w_1	$w_1 = 0$
w_2	$w_2^{10} = \cdots = w_2^{14} = 0$, $w_2^{25} = \cdots = w_2^{29} = 0$, $w_2^{30} \vee w_2^{31} = 1$
w_3	$w_3^{30} \vee w_3^{31} = 1$
w_4	$w_4^{20} = \cdots = w_4^{29} = 0$, $w_4^{30} \vee w_4^{31} = 1$
w_5	$w_5^{5} = \cdots = w_5^{9} = 0$
w_6	$w_6^{0} = \cdots = w_6^{4} = 0$, $w_6^{15} = \cdots = w_6^{19} = 0$, $w_6^{20} = \cdots = w_6^{29} = 0$
w_7	$w_7^{5} = \cdots = w_7^{14} = 0$, $w_7^{20} = \cdots = w_7^{24} = 0$
w_8	difference in w_8^{31}, $w_8^{30} = 1$
w_9	$w_9^{0} = \cdots = w_9^{9} = 0$
w_{10}	$w_{10}^{5} = \cdots = w_{10}^{14} = 0$
w_{11}	$w_{11}^{15} = \cdots = w_{11}^{29} = 0$, $w_{11}^{30} \vee w_{11}^{31} = 1$
w_{12}	$w_{12}^{10} = \cdots = w_{12}^{24} = 0$
w_{13}	$w_{13}^{0} = \cdots = w_{13}^{4} = 0$, $w_{13}^{15} = \cdots = w_{13}^{24} = 0$
w_{14}	difference in w_{14}^{31}, $w_{14}^{10} = \cdots = w_{14}^{14} = 0$, $w_{14}^{20} = \cdots = w_{14}^{29} = 0$, $w_{14}^{30} = 1$
w_{15}	$w_{15}^{0} = \cdots = w_{15}^{9} = 0$

be zero in order the difference to propagate, which will happen with probability $1/2$, thus the total probability for this step $1/2 \times 2^{-32} = 2^{-33}$.

C   Extensions to the 512-Bit Versions

The attacks presented in this paper can be extended to the 512-bit versions of Dynamic SHA and Dynamic SHA2 in a straightforward way. This appendix details how the attacks can be adapted.

Collision Attack on Dynamic SHA. The attack on Dynamic SHA-256 can be adapted to Dynamic SHA-512 with almost no change. Due to the different $R1$ function, the difference word is $\Delta = 8000000080000000$. Also, the probability of the local collision is lowered by about 2^{-1} compared to Dynamic SHA-256, as in the fifth step six rotation bits have to be fixed to zero instead of only five.

Preimage Attack on Dynamic SHA. The preimage attack on Dynamic SHA-512 is similar to that on Dynamic SHA-256, except that the 59 LSBs are determined, instead of the 28 LSBs. Then, when building the tree, 2^{224} solutions are expected,

Table 6. Differential characteristic for the third iterative part of Dynamic SHA2. The difference at step t is the difference in the state *before* computing step t. The column T indicates the difference in the temporary variable T. The probability on a line is the probability to reach the *next* difference, when conditions on the message are satisfied.

t	(message input)	a	b	c	d	e	f	g	h	T	prob.
1	$(w_1, \ldots, w_0)$	0	0	0	Δ	0	0	Δ	0	0	1
		0	0	0	0	Δ	0	0	Δ	0	1
1	$(w_9, \ldots, w_8)$	0	0	0	0	0	Δ	0	0	Δ	1
		Δ	0	0	0	0	0	Δ	0	0	2^{-1}
2	$(w_2, \ldots, w_1)$	0	Δ	0	Δ	0	0	0	0	0	2^{-1}
		0	0	Δ	0	Δ	0	0	0	0	1
2	$(w_{10}, \ldots, w_9)$	0	0	0	Δ	0	Δ	0	0	0	1
		0	0	0	0	Δ	0	Δ	0	0	1
3	$(w_3, \ldots, w_2)$	Δ	0	0	0	0	0	0	Δ	0	2^{-1}
		0	Δ	0	0	0	0	0	0	Δ	2^{-1}
3	$(w_{11}, \ldots, w_{10})$	Δ	0	Δ	0	0	0	0	0	0	2^{-1}
		0	Δ	0	Δ	0	Δ	0	0	Δ	2^{-1}
4	$(w_4, \ldots, w_3)$	Δ	0	Δ	0	Δ	0	Δ	0	0	2^{-1}
		0	Δ	0	Δ	0	Δ	0	Δ	0	1
4	$(w_{12}, \ldots, w_{11})$	0	0	Δ	Δ	Δ	0	Δ	0	0	1
		0	0	0	0	Δ	Δ	0	Δ	Δ	1
5	$(w_5, \ldots, w_4)$	0	0	0	0	0	Δ	Δ	0	0	1
		0	0	0	0	0	0	Δ	Δ	0	1
5	$(w_{13}, \ldots, w_{12})$	0	0	0	0	0	0	0	Δ	Δ	1
		0	0	0	0	0	Δ	0	0	Δ	1
6	$(w_6, \ldots, w_5)$	Δ	0	0	0	0	0	Δ	0	0	1
		0	Δ	0	Δ	0	0	0	Δ	Δ	2^{-1}
6	$(w_{14}, \ldots, w_{13})$	Δ	0	Δ	Δ	Δ	0	0	0	Δ	2^{-33}
		0	Δ	0	Δ	Δ	Δ	0	0	0	2^{-1}
7	$(w_7, \ldots, w_6)$	0	0	0	Δ	Δ	Δ	Δ	0	0	1
		0	0	0	0	Δ	Δ	Δ	Δ	0	1
7	$(w_{15}, \ldots, w_{14})$	0	0	0	0	0	Δ	Δ	Δ	Δ	1
		0	0	0	0	0	0	Δ	Δ	0	1
		0	0	0	0	0	0	0	0		

leading to an attack complexity of 2^{256} on the compression function. Calculations for preimages on the full hash function (with correct padding bits) give a cost of of 2^{262} compression function evaluations.

Collision Attack on Dynamic SHA2. To attack Dynamic SHA2-512 we use a similar differential path. The changes are that the condition on the first block is on 64 bits (starting from a chaining value with b = FFFFFFFFFFFFFFFF), the fact that in the second iterative part the probability is 2^{-6} for each of the

Table 7. Conditions on the message words $w_0, \ldots, w_{15}$ sufficient to follow our differential characteristic in Dynamic SHA2-512

Word	Condition
w_0	–
w_1	$w_1 = 0$, $w_1^{18} = \cdots = w_1^{23} = 0$, $w_1^{42} = \cdots = w_1^{47} = 0$, $w_1^{60} = 1, w_1^{61} = 0$
w_2	$w_2^{18} = \cdots = w_2^{29} = 0$, $w_2^{42} = \cdots = w_2^{47} = 0$, $w_2^{60} = 0, w_2^{61} = 1$, $w_2^{62} \vee w_2^{63} = 1$
w_3	$w_3^{54} = \cdots w_3^{59} = 0$, $w_3^{60} = w_3^{61} = 1$, $w_3^{62} \vee w_3^{63} = 1$
w_4	$w_4^{6} = \cdots = w_4^{11} = 0$, $w_4^{18} = \cdots = w_4^{23} = 0$, $w_4^{42} = \cdots = w_4^{47} = 0$, $w_4^{60} = w_4^{61} = 0$, $w_4^{62} \vee w_4^{63} = 1$
w_5	$w_5^{6} = \cdots = w_5^{11} = 0$, $w_5^{60} = 1, w_5^{61} = 1$
w_6	$w_6^{48} = \cdots = w_6^{53} = 0$, $w_6^{60} = 0, w_6^{61} = 1$
w_7	$w_7^{6} = \cdots = w_7^{23} = 0$, $w_7^{36} = \cdots = w_7^{53} = 0$, $w_7^{60} = w_7^{61} = 1$
w_8	difference in w_8^{63}, $w_8^{62} = 1$
w_9	$w_9^{12} = \cdots = w_9^{17} = 0$, $w_9^{36} = \cdot = w_9^{41} = 0$, $w_9^{60} = 1, w_9^{61} = 0$
w_{10}	$w_{10}^{6} = \cdots = w_{10}^{11} = 0$, $w_{10}^{18} = \cdots = w_{10}^{23} = 0$, $w_{10}^{42} = \cdots = w_{10}^{47} = 0$, $w_{10}^{60} = 0$, $w_{10}^{61} = 1$
w_{11}	$w_{11}^{36} = \cdots = w_{11}^{41} = 0$, $w_{11}^{48} = \cdots = w_{11}^{59} = 0$, $w_{11}^{60} = w_{11}^{61} = 1$, $w_{11}^{62} \vee w_{11}^{63} = 1$
w_{12}	$w_{12}^{12} = \cdots = w_{12}^{23} = 0$, $w_{12}^{36} = \cdots = w_{12}^{47} = 0$, $w_{12}^{60} = w_{12}^{61} = 0$
w_{13}	$w_{13}^{36} = \cdots = w_{13}^{41} = 0$, $w_{13}^{60} = 1, w_{13}^{61} = 0$
w_{14}	difference in w_{14}^{63}, $w_{14}^{12} = \cdots = w_{14}^{23} = 0$, $w_{14}^{36} = \cdots = w_{14}^{53} = 0$, $w_{14}^{60} = 0, w_{14}^{61} = 1$
w_{15}	$w_{15}^{6} = \cdots = w_{15}^{11} = 0$, $w_{15}^{36} = \cdots = w_{15}^{41} = 0$, $w_{15}^{60} = w_{15}^{61} = 1$

two transitions, the decrease in the probability only of the sixth COMP from 2^{-33} to 2^{-65}, and the different set of conditions on the message described in Table 7. Hence, the total time complexity of this attack is 2^{85}. We note that in this approach the attack fixes w_i^{60} and w_i^{61} to $i \bmod 4$ (which causes the same function to be used in this case as in the attack on Dynamic SHA2-256).

A New Approach for FCSRs[*]

François Arnault[1], Thierry Berger[1], Cédric Lauradoux[2],
Marine Minier[3], and Benjamin Pousse[1]

[1] XLIM (UMR CNRS 6172), Université de Limoges
123 avenue Albert Thomas, F-87060 Limoges Cedex - France
first name.name@xlim.fr
[2] Information Security Group
UCL / INGI / GSI
2, Place Saint Barbe
B-1348 Louvain-la-Neuve - Belgium
cedric.lauradoux@uclouvain.be
[3] Lyon University - CITI Laboratory - INSA de Lyon
6, avenue des arts, 69621 Villeurbanne Cedex - France
marine.minier@insa-lyon.fr

Abstract. The Feedback with Carry Shift Registers (FCSRs) have been
proposed as an alternative to Linear Feedback Shift Registers (LFSRs)
for the design of stream ciphers. FCSRs have good statistical properties and they provide a built-in non-linearity. However, two attacks have
shown that the current representations of FCSRs can introduce weaknesses in the cipher. We propose a new "ring" representation of FCSRs
based upon matrix definition which generalizes the Galois and Fibonacci
representations. Our approach preserves the statistical properties and
circumvents the weaknesses of the Fibonacci and Galois representations.
Moreover, the ring representation leads to automata with a quicker diffusion characteristic and better implementation results. As an application,
we describe a new version of F-FCSR stream ciphers.

Keywords: Stream cipher, FCSRs, ℓ-sequence, ring FCSRs.

1 Introduction

The FCSRs have been proposed by Klapper and Goresky [1,2,3] as an alternative to LFSRs for the design of stream ciphers. FCSRs share many of the good
properties of LFSRs: sequences with known period and good statistical properties. But unlike LFSRs, they provide an intrinsic resistance to algebraic and
correlation attacks because of their quadratic feedback function. However, two
recent results [4,5] have shown weaknesses in stream ciphers using either the
Fibonacci or Galois FCSR. Hell and Johansson [5] have exploited the bias in
the carries behaviour of a Galois FCSR to mount a very powerful attack against

---

[*] This work was partially supported by the french National Agency of Research:
ANR-06-SETI-013.

M.J. Jacobson Jr., V. Rijmen, and R. Safavi-Naini (Eds.): SAC 2009, LNCS 5867, pp. 433–448, 2009.
© Springer-Verlag Berlin Heidelberg 2009

the F-FCSR stream cipher [6,7]. Fisher *et al.* [4] have considered an equivalent of the F-FCSR stream cipher based upon a Fibonacci FCSR to study the linear behavior of the induced system.

We present a new approach for FCSRs, which we call the ring representation or ring FCSR. This representation is based on the adjacency matrix of the automaton graph. A ring FCSR can be viewed as a generalization of the Fibonacci and Galois representations. Similar structure has been widely studied for the LFSR case as in [8,9,10], and is a building block of the stream cipher Pomaranch where LFSRs are used [11]. However, we present here for the first time this structure in the FCSR case.

A Fibonacci FCSR, has a single feedback function which depends on multiple inputs. A Galois FCSR has multiple feedbacks which all share one common input. A ring FCSR can be viewed as a trade-off between the two extreme cases. It has several feedback functions with different inputs. An example of ring FCSR is shown in Fig. 1.

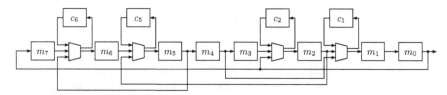

Fig. 1. An example of a ring FCSR ($q = -347$)

Ring FCSRs have many advantages, while preserving all the good and traditional properties of Galois/Fibonnacci FCSRs (known period, large entropy,...). The main one is that the attack of Hell and Johansson [5] does not work with Ring FCSR. Also, they have better diffusion properties. Moreover, ring representation allows fine tune in the implementation.

Section 2 gives an overview on FCSRs theory and classical representations. Section 3 presents ring FCSRs. We discuss implementation in Section 4 and a new version of F-FCSR is proposed in Section 5.

2   Theoretical Background

First, we will recall some basic properties of 2-adic integers. For a more theoretical approach the reader can refer to [1,2,12,13,14].

2.1   2-adic Numbers and Period

A 2-adic integer is formally a power series $s = \sum_{i=0}^{\infty} s_i 2^i$, $s_i \in \{0, 1\}$. This series always converges if we consider the 2-adic topology. The set of 2-adic integers is denoted by $\mathbb{Z}_2$. Addition and multiplication in $\mathbb{Z}_2$ can be performed by reporting the carries to the higher order terms, i.e. $2^n + 2^n = 2^{n+1}$ for all $n \in \mathbb{N}$. If there exists an integer N such that $\dot{s}_n = 0$ for all $n \geq N$, then s is a positive integer. Every odd integer q has an inverse in $\mathbb{Z}_2$.

The following property gives a complete characterization of eventually periodic binary sequences in terms of 2-adic integers (see [13] for the proof).

Property 1. *Let $S = (s_n)_{n \in \mathbb{N}}$ be a binary sequence and let $s = \sum_{i=0}^{\infty} s_i 2^i$ be the corresponding 2-adic integer. The sequence S is eventually periodic if and only if there exist two numbers p and q in $\mathbb{Z}$, q odd, such that $s = p/q$.*

Moreover, S is strictly periodic if and only if $pq \leq 0$ and $|p| \leq |q|$. In this case, we have the relation $s_n = (p \cdot 2^{-n} \bmod q) \bmod 2$.

The period of S is the order of 2 modulo q, i.e., the smallest integer T such that $2^T \equiv 1 \pmod{q}$. The period satisfies $T \leq |q| - 1$. If q is prime, then T divides $|q| - 1$. If $T = |q| - 1$, the sequence S is called an ℓ-sequence. As detailed in [1,2,13,15], ℓ-sequences have many proved properties that could be compared to the ones of m-sequences: known period, good statistical properties, fast generation, etc. In summary, FCSRs have almost the same properties as LFSRs but they have a nonlinear structure.

2.2   Galois FCSRs

A Galois FCSR (as shown in Fig. 2) consists of an n-bit main register $M = (m_0, \ldots, m_{n-1})$ with some fixed feedback positions $d_0, \ldots, d_{n-1}$. All the feedbacks are controlled by the cell m_0, and $n - 1$ binary carry cells $C = (c_0, \ldots, c_{n-2})$. At time t, an automaton in state (M, C) is updated in the following way:

1. Compute the sums $x_i = m_{i+1} + c_i d_i + m_0 d_i$ for all i such that $0 \leq i < n$ with $m_n = 0$ and $c_{n-1} = 0$ and where m_0 represents the feedback bit;
2. Update the state as follows: $m_i \leftarrow x_i \bmod 2$ for all $i \in [0..n-1]$ and $c_i \leftarrow x_i$ div 2 for $0 \leq i < n$ for all $i \in [0..n-2]$.

The reader can refer to [13] for a complete description of Galois FCSRs and some properties. We recall however a very important one, found in [16].

Property 2. *Let $q = 1 - 2\sum_{i=0}^{n-1} d_i 2^i$ and $r_i = \sum_{t=0}^{\infty} m_i(t) 2^t$ (for $0 \leq i < n$); r_i is the 2-adic integer corresponding to the sequence observed in the i-th cell of the main register M. Then, for all $0 \leq i < n$, there exists $p_i \in \mathbb{Z}$ such that $r_i = p_i / q$.*

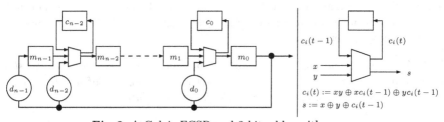

Fig. 2. A Galois FCSR and 2-bit adder with carry

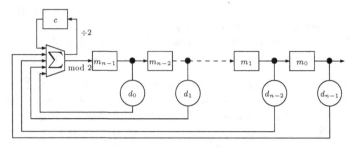

Fig. 3. A Fibonacci FCSR

In a Galois FCSR, a single cell controls all the feedbacks. As a consequence, there exist some correlations between the carries values and the feedback value. This fact is the basis of the attack presented in [5].

2.3   Fibonacci FCSRs

A Fibonacci FCSR (represented in Fig. 3) is composed of a main register $M = (m_0, \ldots, m_{n-1})$ with n binary cells. The binary feedback taps $(d_0, \ldots, d_{n-1})$ are associated to an additional carry register c of $w_H(d)$ binary cells, where $w_H(d)$ is the Hamming weight of $d = (1 + |q|)/2$.

An automaton in state (M, c) is updated in this way:

1. compute the sum $x = c + \sum_{i=0}^{n-1} m_i d_{n-1-i}$;
2. then, update the state: $M \leftarrow (m_1, \ldots, m_{n-1}, x \mod 2)$, $c \leftarrow x$ div 2.

As shown in [13], Property 2 also holds for Fibonacci FCSRs : the sequence observed in a cell m_i is a 2-adic integer.

The cell m_{n-1} is the only one with a non-linear behaviour in a Fibonacci FCSR. As shown in [4], an attack can be carried out if a linear filter is used with a Fibonacci FCSR.

3   A New Approach for FCSRs

Galois and Fibonacci FCSRs are two different automata with similar properties, as seen in the previous section. In a Galois FCSR, the first cell m_0 modifies $w_H(d)$ cells of the main register. In a Fibonacci FCSR, the cell m_{n-1} is modified by $w_H(d)$ cells of the main register. Ring representation of FCSRs is a trade-off between these extreme cases.

Definition 1. *A ring FCSR is an automaton composed of a main shift register of n binary cells $m = (m_0, \ldots, m_{n-1})$, and a carry register of n integer cells $c = (c_0, \ldots, c_{n-1})$. It is updated using the following relations:*

$$\begin{cases} m(t+1) = Tm(t) + c(t) \mod 2 \\ c(t+1) = Tm(t) + c(t) \quad \text{div } 2 \end{cases} \tag{1}$$

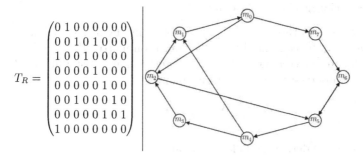

$$T_R = \begin{pmatrix} 0\,1\,0\,0\,0\,0\,0\,0 \\ 0\,0\,1\,0\,1\,0\,0\,0 \\ 1\,0\,0\,1\,0\,0\,0\,0 \\ 0\,0\,0\,0\,1\,0\,0\,0 \\ 0\,0\,0\,0\,0\,1\,0\,0 \\ 0\,0\,1\,0\,0\,0\,1\,0 \\ 0\,0\,0\,0\,0\,1\,0\,1 \\ 1\,0\,0\,0\,0\,0\,0\,0 \end{pmatrix}$$

Fig. 4. Matrix and graph representation of FCSR presented in Fig.1

where T is a $n \times n$ matrix with coefficients 0 or 1 in $\mathbb{Z}$, called transition matrix, of this form:

$$\begin{pmatrix} * & 1 & & & & \\ & * & 1 & & (*) & \\ & & * & 1 & & \\ & & & \ddots & \ddots & \\ & (*) & & & * & 1 \\ 1 & & & & & * \end{pmatrix}$$

Note that $\div 2$ is the traditional expression: $X \operatorname{div} 2 = \frac{X - (X \bmod 2)}{2}$.

Ring FCSRs differ from Fibonacci and Galois FCSRs in the fact that any cell can be used as a feedback for any other cell. A more convenient way to draw ring FCSRs is presented in Figure 4, which represents the same FCSR as the one in Figure 1.

3.1   Remarks on the Transition Matrix

According to Definition 1, we have the following property, where $t_{i,j}$ is the element at the i-th row and j-th column:

$$T = (t_{i,j})_{0 \le i,j < n} \text{ with } t_{i,j} = \begin{cases} 1 & \text{if cell } m_j \text{ is used to update cell } m_i, \\ 0 & \text{otherwise.} \end{cases}$$

As the main register of a ring FCSR is by definition a shift register, the over-diagonal of the transition matrix T is full of ones, i.e. for all $0 \le i < n$ we have $t_{i,i+1 \bmod n} = 1$. For example, the FCSR presented in Fig.1 has the following transition matrix T_R (and $q = -347$):

This notation agrees with the one proposed in [13]. In particular, Galois and Fibonacci FCSRs have respectively transition matrices T_G and T_F of the form:

$$T_G = \begin{pmatrix} d_0 & 1 & & & \\ d_1 & 0 & 1 & & (0) \\ d_2 & & 0 & 1 & \\ \vdots & & & \ddots & \ddots \\ d_{n-2} & (0) & & 0 & 1 \\ 1 & & & & 0 \end{pmatrix} \qquad T_F = \begin{pmatrix} 0 & 1 & & & \\ & 0 & 1 & & (0) \\ & & 0 & 1 & \\ (0) & & & \ddots & \ddots \\ & & & 0 & 1 \\ 1 & d_{n-2} & \dots & d_2 & d_1 & d_0 \end{pmatrix}$$

3.2   Characterizing the Cells Content

Definition 1 introduces the transition matrix of a ring FCSR. We explain now how the value q can be computed from the transition matrix T.

Let $m_i(t)$ denote the content of the i-th cell of the main register at time t and $M_i(t)$ the series observed in this cell, from time t:

$$M_i(t) = \sum_{k \in \mathbb{N}} m_i(t+k)2^k.$$

From Equation 1, we derive the following relation

$$M(t+1) = TM(t) + c(t) \tag{2}$$

where $M(t) = (M_0(t), \cdots, M_{n-1}(t))$, and $c(t) = (c_0(t), \cdots, c_{n-1}(t))$ is the content of the carry register at time t.

The series $M_i(t)$ and the vector $M(t)$ play a fundamental role in our approach. We have the following important generalisation of Property 2.

Theorem 1. *The series $M_i(t)$ observed in the cells of the main register are 2-adic expansion of p_i/q with $p_i \in \mathbb{Z}$ and with $q = \det(I - 2T)$.*

Proof. According to the definition of $M_i(t)$ and to Definition 1, we have $M(t) = 2M(t+1) + m(t)$ where $m(t)$ is a binary vector of size n. Using Equation 2, we get:

$$(I - 2T) \cdot M(t) - 2 \cdot c(t) - m(t) = 0.$$

Considering the adjugate of $I - 2T$, we obtain:

$$\det(I - 2T) \cdot M(t) = Adj(I - 2T)(m(t) + 2 \cdot c(t)).$$

In this relation, the right member is a vector of integers $(p_0(t), \ldots, p_{n-1}(t))$. Dividing by $\det(I - 2T)$, we obtain $M_i(t) = p_i(t)/\det(I - 2T)$.

Lemma 1. *With the notation of Theorem 1, if $q = \det(I - 2T)$ is prime, and if the order of 2 in $\mathbb{Z}/q\mathbb{Z}$ is maximal, then each M_i is an ℓ-sequence.*

4   Implementation Properties

We detail in this section the new implementation characteristics of ring FCSRs. All this section applies also to LFSRs by replacing addition with carry with addition modulo 2.

Path/fan-out – The Galois FCSR is considered in many works [13,17,18] as the best representation for hardware implementation. It has a better critical path, i.e, a shorter longest path, than a Fibonacci FCSR. A drawback of the Galois representation is that the fan-out of the feedback cell m_0 is $w_H(d)$ with $d = (1 + |q|)/2$. At the opposite, the Fibonacci representation has a fan-out of 2. A ring FCSR allows the designer to tune both the critical path and the fan-out through the choice of the transition matrix:

Table 1. Comparison of the different representations

	Fibonacci	Galois	Ring
Path	$\lceil \log_2(w_H(d)) \rceil$	1	$\max(\lceil \log_2(w_H(a_i)) \rceil)$
Fan-out	2	$w_H(d)$	$\max(w_H(b_i))$
Cost ($\#_{adders}$)	$w_H(d) - 1$	$w_H(d) - 1$	$w_H(T) - n$

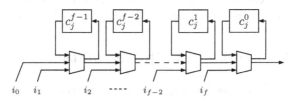

Fig. 5. A naive adder

- the critical path is given by the row a_i with the largest number of 1s;
- the fan-out is given by the column b_i with the largest number of 1s.

We compare in Table 1 the critical path, the fan-out and the cost of the different representations of an FCSR. We have expressed the critical path as the number of adders crossed. The choice of the adder has also an impact on the path of a ring FCSR. A naive adder (Fig. 5) composed of a serialisation of generic adder leads to a path of $max(w_H(a_i)) - 1$ adders. However, it is possible to exploit the commutativity to perform additions in parallel. This reduces the critical path to $max(\lceil \log_2(w_H(a_i)) \rceil)$ adders.

For each given q, it should be possible to find a transition matrix corresponding to a critical path with only one adder and a fan-out equal to 2. This is the case of the ring FCSR given in Fig. 1.

Cost – Ring FCSR have implementations which require fewer gates than Fibonacci/Galois equivalent ones. This possibility was first observed in [10] for LFSRs. However, the solution proposed in [10] is specific to LFSRs and cannot be applied systematically to FCSRs. The number $\#_{adders}$ of 2-bit adders required in the different representations of an n-bit FCSR is shown in Table 1. Ring representation is the only one that allows to find an implementation with less than $(w_H(d) - 1)$ 2-bit adders. For $q = -347$, a Galois or Fibonacci representation leads to $\#_{adders} = 4$. A ring representation with the following transition matrix T_R:

$$T_R = \begin{pmatrix} 0 & 1 & 0 & 0 & 0 & 0 & 0 & 0 \\ 0 & 0 & 1 & 0 & 0 & 0 & 0 & 0 \\ 0 & 0 & 0 & 1 & 0 & 0 & 0 & 1 \\ 0 & 1 & 0 & 0 & 1 & 0 & 0 & 0 \\ 0 & 0 & 0 & 0 & 0 & 1 & 0 & 0 \\ 0 & 0 & 0 & 0 & 0 & 0 & 1 & 0 \\ 0 & 0 & 0 & 0 & 0 & 0 & 0 & 1 \\ 1 & 0 & 1 & 0 & 0 & 0 & 0 & 0 \end{pmatrix}$$

leads to an implementation with $\#_{adders} = 3$, a fan-out of 2 and a critical path of 1 adder.

Side-channel attacks – It seems possible to work out an equivalent of the side-channel attack of Joux and Delaunay [18] on Galois FCSR using the results of Hell and Johansson [5]. Such an attack would exploit the power consumption to recover the feedback m_0 (because of the excessive fan-out of the feedback cell) and therefore how the carry cells are modified. As the ring FCSR has a reduced fan-out and uncorrelated carries, it is a better alternative to prevent side-channel attacks.

5   F-FCSR Based on Ring Representation

In this section, we propose a generic algorithm to construct F-FCSR stream ciphers based upon a ring FCSR with a linear filter. We give two particular examples which are F-FCSR-H v3 and F-FCSR-16 v3. Any designer using the proposed algorithm could generate its own stream cipher according to the following parameters:

- key length k and IV length v that will provide the corresponding size $n :=$ $k + v$ of the T matrix (usually $k = v$);
- the number u of bits output at each clock taken between 1 and $n/16$ to ensure a hard inversibility of the filter. Moreover for later design we require u to be a divisor of n;
- the number of willing feedbacks ℓ usually taken between $n/2 - 5$ and $n/2 + 5$ to ensure a sufficient non linear structure and a sufficiently weighted filter.

The algorithm is composed of 3 particular steps: the choice of the matrix T, the choice of the linear filter and the key/IV setup.

5.1   The Choice of the Matrix T

According to the remarks in Section 3, we pick a $n \times n$ random matrix T with the following requirements:

- the matrix must be composed of 0 and 1 and with a general weight equal to $n + \ell$;
- the over-diagonal must be full of 1 and $t_{n-1,0} = 1$ (to preserve the ring structure of the automaton);
- the number of ones for a given row or a given column must be at most two. This last condition allows a better diffusion by maximizing the number of cells reached by the feedbacks. It also provides uncorrelated carries and a fan-out bounded by 2.

For each picked matrix with the previous requirements, test if:

1. $\log_2(q) \geq n$; $\det(T) \neq 0$;
2. $q = \det(I - 2T)$ is prime; the order of 2 modulo q is $|q| - 1$.

The first condition ensures a non-degenerated matrix. The second ensures good statistical properties and a long period.

This matrix completely defines the ring FCSR. The diffusion speed (which is faster than in Galois/Fibonacci FCSRs) is related to the diameter d of the transition graph. This diameter is the maximal distance between two cells of the main register. In other words, d is the distance after which any cell of the main register have been influenced by any other cell through the feedbacks. It corresponds to the minimal number of clocks required to have all the cells of the main register influenced by any other cell. d should be small for better diffusion.

5.2   The Filter

As in the previous versions of F-FCSR [7], we use a linear filter to extract the keystream in order to break the 2-adic structure of the automaton. This also prevents linearization attacks over the set of 2-adic numbers. The filter includes the cells of the main register which receive a feedback to prevent correlation attacks. The periodic structure of the filter in the previous versions of F-FCSR has been exploited in [5] to speed up the linear part of the attack. We prefer now a non periodic structure:

- let $F = \{m_{f_0}, \cdots, m_{f_{\ell-1}}\}$ be the set of all the cells m_i that receive a feedback and indexed in this way: the row f_i of the matrix T has more than one 1 for $0 \leq i < \ell$, and $f_i < f_{i+1}$.
- The u bits of output are: $\forall\, 0 \leq i < u$, $z_i = \bigoplus_{j \equiv i \mod u} m_{f_j}$.

5.3   Key and IV Setup

As shown in [5], if at a given time, the FCSR is in a synchronized state (i.e. a state from which after a finite number of steps the automaton will return, i.e. a state belonging to the main cycle), adjacent states of the main cycle could be directly deduced using only multiplications over $\mathbb{Z}/q\mathbb{Z}$. Moreover, as shown in [16], a Galois FCSR is synchronized in at most $n + 4$ clocks, but in reality, few clocks are sufficient. So, to completely avoid the weakness of the key and IV setup used in [5], we prefer to maintain a non synchronized state during the key and IV setup. The new key and IV setup creates a transformation that is really hard to invert, in order to prevent a direct key recovery attack.

However, using a ring FCSR leads to a new problem: we can not ensure the entropy of the automaton. In the case of F-FCSR with Galois or Fibonacci structure, zeroing the content of the carry register prevents collisions (i.e. one point of the states graph with two preimages) and warrants a constant entropy. This particular property comes from the particular structure of the adjoint matrix $(I - 2T)^*$, which has successive powers of two in its first row in case of a Galois FSCR (in a Fibonacci FCSR, a similar property holds for the last row). In the ring case, no obvious structure exists in $(I - 2T)^*$. Note that in this case the collisions search becomes an instance of the subset sum problem, with a complexity equals to $2^{n/2}$ (if the carries are zeroes) or $2^{3n/2}$ (in the general case).

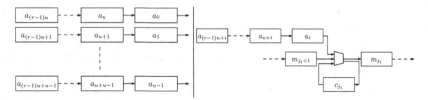

Fig. 6. Disposition of the cells $a_0, \ldots, a_{n-1}$ in u shift registers and connection of a shift register in position j_i

Thus, the new key and IV setup aims to stay on non-synchronized states as long as possible and to limit the entropy loss. We connect at u different places shift registers of length $r = n/u$ (this corresponds to adding n binary cells $a_0, \ldots, a_{n-1}$ at different places as shown in Fig. 6).

The u positions denoted by $J := \{j_0 < \cdots < j_{u-1}\}$ where the u shift registers are connected have been chosen such that, for all $0 \le i < u$, no adder exists between cells m_{j_i+1} and m_{j_i} (i.e. $w_H(R_{j_i}) = 1$ where R_{j_i} is the j_i^{th} row of the matrix T). Each shift register is connected using a 2-bit adder with carry (as shown in Fig. 6). The content of cell m_{j_i} after transition depends on the values of m_{j_i+1}, a_i and of the carry cell c_{j_i}.

With these u shift registers inserted in the ring FCSR, the key and IV setup works as follows:

- Initialize $(a_0, \ldots, a_{n-1})$ with $(K\|0^{n-k-v}\|IV)$, $M \leftarrow 0$, $C \leftarrow 0$.
- The FCSR is clocked r times. At each clock, the FCSR is filtered using F to produce a u bits vector $z_0, \ldots, z_{u-1}$ used to fill back $a_{(r-1)u}, \ldots, a_{(r-1)u+u-1}$: $a_{(r-1)u+i} \leftarrow z_i$ for $0 \le i < u$.
- The FCSR is clocked $\max(r, d+4)$ times discarding the output.

The first step of the key and IV setup allows an initial diffusion of the key through the simple shift registers. The next r clocks helps a complete diffusion of the IV and of the key in the FCSR. The diffusion is complete at the end of the key and IV setup. If an attacker is able to recover the state just at the end of the key and IV setup, he won't be able to use this information to recover the key because of the occurence of non-synchronized states that are hard to inverse: for a given m_{k+1} bit value of the main register, the values c_k and m_k producing m_{k+1} are not unique and this leads to a combinatorial explosion when an attacker wants to recover a previous state.

As previously mentioned, this construction does not provide a bijection and behaves more like a random function. From this point, two attacks are essentially possible: direct collisions search and time memory data trade-off attack for collisions search built upon entropy loss. As mentioned before, direct collisions search has a cost of $2^{(n/2)}$ if the attacker is able to clear the carry bits. With the use of a ring FCSR that does not allow a direct control of the carry bits through the feedback bit, the probability to force to 0 the carry bits is about $2^{-\ell}$. Thus such an attack is more expensive than a key exhaustive search. In the other cases, the corresponding complexity is $2^{(3n/2)}$ preventing collisions search.

TMDTO attacks are possible if a sufficient quantity of entropy is lost. As studied in [19], considering that the key and IV setup are random function, the induced entropy loss is about 1 bit, so considering an initial entropy equal to n bits, the entropy after the key and IV setup is close to $n - 1$ bits. Is it possible to exploit this entropy loss for a collisions search in a TMDTO attack? A well-known study case is the attack proposed in [20] by J. Hong and W.H. Kim against the stream cipher MICKEY. Even if this attack seems to work, A. Rock has shown in [19] that the query complexity in the initial states space could not be significantly reduced and that the attacks based on the problem of entropy loss are less efficient than expected especially regarding the query complexity. So, we conjecture, that our key and IV setup behaves as a random function, and that the induced entropy loss is not sufficient to mount a complete TMDTO attack for collisions search taking into account the query complexity.

5.4   F-FCSR-H v3 and F-FCSR-16 v3

The details of the two constructions, especially the corresponding T matrices, are given respectively in Appendix A and B. The respective parameters are the following ones:

- For F-FCSR-H v3: $k = 80$, $v = 80$, $\ell = 82$, $n = 160$, $u = 8$, $d = 24$;
- For F-FCSR-16 v3: $k = 128$, $v = 128$, $\ell = 130$, $n = 256$, $u = 16$, $d = 28$.

These two automata have been chosen with an additional property: $(|q| - 1)/2$ prime. This condition ensures maximal period for the output stream. However this condition is hard to fill. So we don't require this condition in the general case.

5.5   Resistance against Known Attacks

We do not discuss here resistance against traditional attacks such as correlation / fast correlation attacks, guess and determine attacks, algebraic attacks, etc. Some details about this can be found in [7]. Resistance against TMDTO attacks was considered in Section 5.3. We focus now on the two recent attacks [5] and [4] against FCSR and F-FCSR.

The attack presented in [5] against F-FCSR, which is based on a Galois FCSR, relies on the existence of correlations between the carries and the feedback values. More precisely, the control of the m_0 bit leads to the control of the feedback values. If the feedback can be forced to 0 during t consecutive clocks, the behavior of the stream cipher becomes linear, and its synthesis is possible by solving a really simple system. This linear behavior happens with a probability about 2^{-t} for a Galois FCSR. If instead a ring FCSR is used, this probability decreases to $2^{-t \cdot k}$ where k is the number of cells of the main register controlling a feedback. Thus, for k values corresponding to most ring FCSR, the linear behavior probability becomes so small that the cost of the corresponding attack becomes higher than an exhaustive search. Also the attack from [5] relies on situations where the carries remain constant during t consecutive clocks. We made an experiment with F-FCSR-H v3 to search for states for which carries does not change during transition. Looking over 2^{38} states, we found only 41 different states for which carries remains constant after one transition. We found none for which carries remains constant after two transitions.

In [4], the authors propose a linearization attack against a linearly filtered Fibonacci FCSR. This attack does not affect any version of F-FCSR. In a Fibonacci FCSR, the carries only influence one bit of the main register at each clock. Thus, if one could imagine to build a F-FCSR using a Fibonacci FCSR, such a generator would be subject to an attack where the control of the carries leads to the control of a part of the main register. Thus, we recommend to NOT use a Fibonacci FCSR in a linearly filtered stream cipher.

6   Conclusion and Future Work

In this paper, we have presented a new approach for FCSRs that unifies the two classical representations. We can obtain, with the ring representation, better diffusion characteristics and faster implementations. The recent attacks designed against F-FCSR are prevented, when using a ring FCSR, as shown in Section 5.

References

1. Klapper, A., Goresky, M.: 2-adic shift registers. In: Anderson, R. (ed.) FSE 1993. LNCS, vol. 809, pp. 174–178. Springer, Heidelberg (1994)
2. Klapper, A., Goresky, M.: Feedback shift registers, 2-adic span and combiners with memory. Journal of Cryptology 10(2), 111–147 (1997)
3. Klapper, A.: A survey of feedback with carry shift registers. In: Helleseth, T., Sarwate, D., Song, H.-Y., Yang, K. (eds.) SETA 2004. LNCS, vol. 3486, pp. 56–71. Springer, Heidelberg (2005)
4. Fischer, S., Meier, W., Stegemann, D.: Equivalent Representations of the F-FCSR Keystream Generator. In: ECRYPT Network of Excellence - SASC Workshop, pp. 87–94 (2008), http://www.ecrypt.eu.org/stvl/sasc2008/
5. Hell, M., Johansson, T.: Breaking the F-FCSR-H stream cipher in real time. In: Pieprzyk, J. (ed.) ASIACRYPT 2008. LNCS, vol. 5350, pp. 557–569. Springer, Heidelberg (2008)
6. Arnault, F., Berger, T.P.: F-FCSR: Design of a new class of stream ciphers. In: Gilbert, H., Handschuh, H. (eds.) FSE 2005. LNCS, vol. 3557, pp. 83–97. Springer, Heidelberg (2005)
7. Arnault, F., Berger, T.P., Lauradoux, C.: Update on F-FCSR Stream Cipher. ECRYPT - Network of Excellence in Cryptology (Call for stream Cipher Primitives - Phase 2 2006) (2006), http://www.ecrypt.eu.org/stream/
8. Roggeman, Y.: Varying feedback shift registers. In: Quisquater, J.-J., Vandewalle, J. (eds.) EUROCRYPT 1989. LNCS, vol. 434, pp. 670–679. Springer, Heidelberg (1990)
9. Jansen, C.J., Helleseth, T., Kholosha, A.: Cascade jump controlled sequence generator and pomaranch stream cipher (version 2). eSTREAM, ECRYPT Stream Cipher Project, Report 2006/006 (2006), http://www.ecrypt.eu.org/stream
10. Mrugalski, G., Rajski, J., Tyszer, J.: Ring generators - new devices for embedded test applications. IEEE Trans. on CAD of Integrated Circuits and Systems 23(9), 1306–1320 (2004)

11. Jansen, C.J., Helleseth, T., Kholosha, A.: Pomaranch version 3. eSTREAM, ECRYPT Stream Cipher Project (2006), http://www.ecrypt.eu.org/stream
12. Koblitz, N.: p-adic numbers, p-adic analysis and Zeta-Functions. Springer, Heidelberg (1997)
13. Goresky, M., Klapper, A.: Fibonacci and Galois representations of feedback-with-carry shift registers. IEEE Transactions on Information Theory 48(11), 2826–2836 (2002)
14. Arnault, F., Berger, T.P.: Design and Properties of a New Pseudorandom Generator Based on a Filtered FCSR Automaton. IEEE Transaction on Computers 54(11), 1374–1383 (2005)
15. Lauradoux, C., Röck, A.: Parallel generation of ℓ-sequences. In: Golomb, S.W., Parker, M.G., Pott, A., Winterhof, A. (eds.) SETA 2008. LNCS, vol. 5203, pp. 299–312. Springer, Heidelberg (2008)
16. Arnault, F., Berger, T.P., Minier, M.: Some Results on FCSR Automata With Applications to the Security of FCSR-Based Pseudorandom Generators. IEEE Transactions on Information Theory 54(2), 836–840 (2008)
17. Goldberg, I., Wagner, D.: Architectural considerations for cryptanalytic hardware. Technical report, Secrets of Encryption Research, Wiretap Politics & Chip Design (1996)
18. Joux, A., Delaunay, P.: Galois LFSR, embedded devices and side channel weaknesses. In: Barua, R., Lange, T. (eds.) INDOCRYPT 2006. LNCS, vol. 4329, pp. 436–451. Springer, Heidelberg (2006)
19. Röck, A.: Stream ciphers using a random update function: Study of the entropy of the inner state. In: Vaudenay, S. (ed.) AFRICACRYPT 2008. LNCS, vol. 5023, pp. 258–275. Springer, Heidelberg (2008)
20. Hong, J., Kim, W.H.: TMD-Tradeoff and State Entropy Loss Considerations of Streamcipher MICKEY. In: Maitra, S., Veni Madhavan, C.E., Venkatesan, R. (eds.) INDOCRYPT 2005. LNCS, vol. 3797, pp. 169–182. Springer, Heidelberg (2005)

A Description of the Transition Matrix for F-FCSR-H v3

Input parameters: $k = 80$ (key length), $v = 80$ (IV length), $\ell = 82$ (number of feedbacks), $n = 160$ (size of T), $u = 8$ (number of output bits), $d = 24$ (diameter of the graph).

We give here the description of the transition matrix $T = (t_{i,j})_{0 \le i,j < 160}$ (see Fig. 7 for graphic representations):

- For all $0 \le i < 160$, $t_{i,i+1 \mod 160} = 1$;
- For all $(i,j) \in S$, $t_{i,j} = 1$, where $S = \{$ (1, 121); (2, 133); (4, 44); (5, 82); (9, 38); (11, 40); (12, 54); (14, 105); (15, 42); (16, 63); (18, 80); (19, 136); (20, 2); (21, 35); (23, 28); (25, 137); (28, 131); (31, 102); (36, 41); (39, 138); (40, 31); (42, 126); (44, 127); (45, 77); (46, 110); (47, 86); (48, 93); (49, 45); (51, 17); (54, 8); (56, 7); (57, 150); (59, 25); (62, 51); (63, 129); (65, 130); (67, 122); (73, 148); (75, 18); (77, 46); (79, 26); (80, 117); (81, 1); (84, 72); (86, 60); (89, 15); (90, 89); (91, 73); (93, 12); (94, 84); (102, 141); (104, 142); (107, 71); (108, 152); (112, 92); (113, 83); (115, 23); (116, 32); (118, 50); (119, 43); (121, 34); (124, 13); (125, 74); (127, 149); (128, 90); (129, 57); (130, 103); (131, 134); (132, 155); (134, 98); (139, 24);

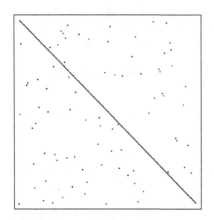 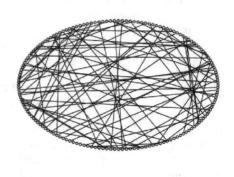

Fig. 7. Matrix representation and graph representation of the matrix T chosen for F-FCSR-H v3

(140, 61); (141, 104); (144, 48); (145, 14); (148, 112); (150, 59); (153, 39); (156, 22); (157, 107); (158, 30); (159, 78) };
- Otherwise, $t_{i,j} = 0$.

- The corresponding q value is (in decimal notation):

$$q = 174161873672323786281235399625569968955252\,6450883$$

- The set J (for the first part of the Key/IV setup) is:

$$J = \{3, 22, 43, 64, 83, 103, 123, 143\}$$

- The 8 subfilters $F_0, \cdots, F_7$ are given by:

$$F_0 = \{1, 15, 28, 46, 59, 79, 93, 115, 128, 141, 158\}$$
$$F_1 = \{2, 16, 31, 47, 62, 80, 94, 116, 129, 144, 159\}$$
$$F_2 = \{4, 18, 36, 48, 63, 81, 102, 118, 130, 145\}$$
$$F_3 = \{5, 19, 39, 49, 65, 84, 104, 119, 131, 148\}$$
$$F_4 = \{9, 20, 40, 51, 67, 86, 107, 121, 132, 150\}$$
$$F_5 = \{11, 21, 42, 54, 73, 89, 108, 124, 134, 153\}$$
$$F_6 = \{12, 23, 44, 56, 75, 90, 112, 125, 139, 156\}$$
$$F_7 = \{14, 25, 45, 57, 77, 91, 113, 127, 140, 157\}$$

B   Description of the Transition Matrix for F-FCSR-16 v3

Input parameters: $k = 128$, $v = 128$, $\ell = 130$, $n = 256$, $u = 16$, $d = 28$.

We give here a description of the transition matrix $T = (t_{i,j})_{0 \le i,j < 256}$ (see Fig. 8 for graphic representations):

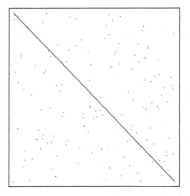 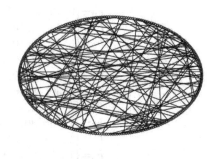

Fig. 8. Matrix representation and graph representation of the matrix T chosen for F-FCSR-16 v3

- For all $0 \leq i < 256$, $t_{i,i+1 \bmod 256} = 1$;
- For all $(i,j) \in S$, $t_{i,j} = 1$, where $S = \{$ (0, 52); (2, 150); (3, 2); (5, 169); (6, 89); (8, 100); (9, 1); (11, 156); (12, 9); (13, 46); (19, 146); (20, 206); (26, 204); (31, 254); (32, 151); (38, 144); (40, 108); (46, 167); (47, 198); (48, 70); (49, 98); (50, 213); (53, 214); (56, 87); (57, 55); (58, 162); (62, 160); (63, 13); (64, 192); (65, 59); (66, 12); (67, 207); (68, 209); (71, 229); (73, 84); (74, 199); (77, 168); (78, 122); (79, 35); (80, 154); (82, 153); (85, 188); (87, 51); (89, 4); (90, 49); (93, 231); (95, 224); (97, 249); (101, 208); (102, 120); (104, 218); (105, 8); (108, 77); (109, 68); (110, 250); (113, 237); (115, 252); (116, 17); (118, 73); (119, 182); (123, 29); (124, 234); (127, 138); (132, 190); (134, 244); (136, 219); (141, 228); (142, 205); (143, 58); (144, 230); (145, 210); (146, 44); (147, 137); (148, 130); (150, 79); (152, 111); (153, 172); (154, 141); (156, 78); (157, 131); (158, 110); (159, 127); (170, 189); (171, 112); (174, 217); (175, 7); (176, 187); (177, 40); (179, 118); (181, 195); (184, 48); (186, 64); (189, 246); (190, 47); (191, 37); (192, 211); (193, 85); (194, 181); (195, 61); (196, 54); (198, 222); (199, 83); (203, 105); (204, 201); (205, 43); (206, 139); (208, 20); (210, 242); (211, 124); (213, 253); (215, 243); (216, 69); (218, 176); (220, 30); (222, 19); (223, 232); (224, 239); (225, 220); (227, 102); (231, 185); (232, 15); (234, 152); (236, 62); (238, 245); (242, 197); (245, 235); (246, 171); (247, 67); (253, 26); (254, 202) $\}$;
- Otherwise, $t_{i,j} = 0$.

- The corresponding q value is (in hexadecimal notation):

$$q = (B085834B6BFAE1541C54F7D84F42084C$$
$$B0568496DDD0FEA5E99AA79C022023241)$$

- The set J (for the first part of the Key/IV setup) is:

$$J = \{10, 27, 43, 59, 75, 91, 107, 122, 139, 155, 172, 187, 202, 219, 235, 251\}$$

– The 16 subfilters $F_0, \cdots, F_{15}$ are given by:

$$F_0 = \{0, 40, 68, 101, 134, 158, 193, 218, 253\}$$
$$F_1 = \{2, 46, 71, 102, 136, 159, 194, 220, 254\}$$
$$F_2 = \{3, 47, 73, 104, 141, 170, 195, 222\}$$
$$F_3 = \{5, 48, 74, 105, 142, 171, 196, 223\}$$
$$F_4 = \{6, 49, 77, 108, 143, 174, 198, 224\}$$
$$F_5 = \{8, 50, 78, 109, 144, 175, 199, 225\}$$
$$F_6 = \{9, 53, 79, 110, 145, 176, 203, 227\}$$
$$F_7 = \{11, 56, 80, 113, 146, 177, 204, 231\}$$
$$F_8 = \{12, 57, 82, 115, 147, 179, 205, 232\}$$
$$F_9 = \{13, 58, 85, 116, 148, 181, 206, 234\}$$
$$F_{10} = \{19, 62, 87, 118, 150, 184, 208, 236\}$$
$$F_{11} = \{20, 63, 89, 119, 152, 186, 210, 238\}$$
$$F_{12} = \{26, 64, 90, 123, 153, 189, 211, 242\}$$
$$F_{13} = \{31, 65, 93, 124, 154, 190, 213, 245\}$$
$$F_{14} = \{32, 66, 95, 127, 156, 191, 215, 246\}$$
$$F_{15} = \{38, 67, 97, 132, 157, 192, 216, 247\}$$

New Cryptanalysis of Irregularly Decimated Stream Ciphers

Bin Zhang

Laboratory of Algorithmics, Cryptology and Security,
University of Luxembourg,
6, rue Coudenhove-Kalergi, L-1359, Luxembourg
bin.zhang@uni.lu

Abstract. In this paper we investigate the security of irregularly decimated stream ciphers. We present an improved correlation analysis of various irregular decimation mechanisms, which allows us to get much larger correlation probabilities than previously known methods. Then new correlation attacks are launched against the shrinking generator with Krawczyk's parameters, LILI-II, DECIM^{v2} and DECIM-128 to access the security margin of these ciphers. We show that the shrinking generator with Krawczyk's parameters is practically insecure; the initial internal state of LILI-II can be recovered reliably in $2^{72.5}$ operations, if $2^{24.1}$-bit keystream and $2^{74.1}$-bit memory are available. This disproves the designers' conjecture that the complexity of any divide-and-conquer attack on LILI-II is in excess of 2^{128} operations and requires a large amount of keystream. We also examine the main design idea behind DECIM, i.e., to filter and then decimate the output using the ABSG algorithm, by showing a class of correlations in the ABSG mechanism and mounting attacks faster than exhaustive search on a 160-bit (out of 192-bit) reduced version of DECIM^{v2} and on a 256-bit (out of 288-bit) reduced version of DECIM-128. Our result on DECIM is the first nontrivial cryptanalytic result besides the time/memory/data tradeoffs. While our result confirms the underlying design idea, it shows an interesting fact that the security of DECIM rely more on the length of the involved LFSR than on the ABSG algorithm.

1 Introduction

Irregular decimation (or irregular clocking) is a well-known strategy in hardware-oriented stream cipher design. It is commonly believed that such a mechanism can strengthen the security of the underlying pseudo-random bit generators with respect to correlation [4, 5, 20, 23, 27] and algebraic attacks [9, 10].

In this paper we consider four well-known stream ciphers using irregular decimation as the main protective mechanism, namely, the shrinking generator (SG) with Krawczyk's parameters [8, 19], LILI-II [7], DECIM^{v2} and DECIM-128 [1, 2]. So far, the best known attack on the SG with Krawczyk's parameters is a near-practical attack in [27]; the best known attack on LILI-II is a distinguishing attack which requires 2^{103} bits keystream and 2^{103} operations [11]. Previous work

M.J. Jacobson Jr., V. Rijmen, and R. Safavi-Naini (Eds.): SAC 2009, LNCS 5867, pp. 449–465, 2009.
© Springer-Verlag Berlin Heidelberg 2009

[3, 22, 25] on DECIM focused on the initialization phase which does not contain the ABSG algorithm and there are no known cryptanalytic result on DECIM exploiting the properties of the ABSG decimation. In order to get better attacks on these stream ciphers, we first present an improved correlation analysis of various irregular decimation mechanisms in a unified way, which allows us to get much larger correlation probabilities in these mechanisms than previously known results [16, 26]. Then new correlation attacks are launched against the SG with Krawczyk's parameters, LILI-II, DECIM^{v2} and DECIM-128 to precisely access the security margin of these ciphers. We show that the SG with Krawczyk's parameters is practically insecure. We also show that the initial internal state of LILI-II can be recovered in $2^{72.5}$ operations with a success rate of 93.4%, if $2^{24.1}$-bit keystream and $2^{74.1}$-bit memory are available. This disproves the designers' conjecture that the complexity of any divide-and-conquer attack on LILI-II is in excess of 2^{128} operations and requires a large amount of keystream. We also examine the main design idea behind DECIM, i.e., to filter and then decimate the output using the ABSG algorithm, by showing a class of rather large correlations in the ABSG mechanism and mounting attacks faster than exhaustive search on a 160-bit (out of 192-bit) reduced version of DECIM^{v2} and on a 256-bit (out of 288-bit) reduced version of DECIM-128. We then extend the attack to the full length versions and show that our attack, though slower than exhaustive key search, is 2^{35} times faster than the generic time/memory/data tradeoffs. This is the first nontrivial cryptanalytic result on DECIM besides the tricky time/memory/data tradeoffs. While our result confirms the underlying design idea, it shows an interesting fact that the security of DECIM rely more on the length of the involved LFSR than on the ABSG algorithm.

This paper is organized as follows. We first introduce the four target stream ciphers in Section 2. Then an improved correlation analysis of various irregular decimations in a unified way is presented in Section 3. The application of our method to the SG with Krawczyk's parameters, LILI-II, DECIM^{v2} and DECIM-128 are given in Section 4 respectively. Finally, some conclusions are provided in Section 5.

2   Four Irregularly Decimated Stream Ciphers

The shrinking generator (SG) was proposed in [8] at Crypto'93, which is considered as one of the simplest and strongest stream ciphers currently available. It consists of two LFSR's, say the data LFSR B and the control LFSR S. LFSR B is irregularly decimated by the regularly clocked LFSR S according to the following rule: the output bit of the data LFSR B is taken iff the current output bit of the control LFSR S is 1. In [19], Krawczyk suggested to use a SG with the following parameters, i.e., LFSR B of length 61 and LFSR S of similar length.

The diagrams of LILI and DECIM are presented in Figure 1. In LILI [6, 7], there are two components, i.e., the data generation subsystem and the clock control subsystem. The function f_c takes two stages from the regularly clocked LFSR c as inputs and produces an integer $c_j \in \{1, 2, 3, 4\}$ which defines the

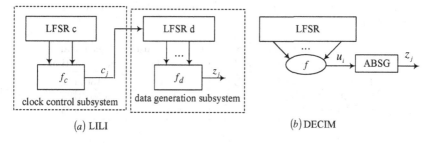

(a) LILI   (b) DECIM

Fig. 1. LILI and DECIM

clocking number of LFSR d at time j. After the clocking of LFSR d, some stages are taken as inputs to a non-linear boolean function f_d to generate the keystream bit. There are two algorithms in this family, i.e., LILI-128 and LILI-II. LILI-II is shown to be much stronger than its predecessor LILI-128 in security. We *stress* here that our target cipher is not the much weaker version LILI-128, but LILI-II.

DECIM contains a unique component in eSTREAM project [1] and is recognized as interesting additions to the field of stream ciphers.

The ABSG algorithm
Input: $(u_0, u_1, \cdots)$
Set: $i \leftarrow 0$ and $j \leftarrow 0$.
Repeat:
1. $e \leftarrow u_i$, $z_j \leftarrow u_{i+1}$
2. $i \leftarrow i + 1$
3. while $(u_i = \bar{e}) i \leftarrow i + 1$
4. $i \leftarrow i + 1$
5. output z_j
6. $j \leftarrow j + 1$

In DECIM, the ABSG algorithm (shown in the block diagram) is used to generate keystream $\{z_j\}$ by irregularly decimating an input stream $\{u_i\}$ from a filter generator. Previous work [3, 22, 25] on DECIM focused on the initialization phase which does not contain the irregular decimation.

DECIM has entered into the third and last phase of the eSTREAM project. The main reason that DECIM is not selected in the final portfolio is its performance compared to other phase 3 hardware candidates. There are no known cryptanalytic result on DECIM exploiting the properties of the ABSG algorithm. Since our results on the latter 3 ciphers are structural cryptanalysis, we only need the relevant parameters of the specific designs, see Table 1.

3   Correlation Analysis of Arbitrary Irregular Decimation Mechanism

In this section, we present some theoretical results on *arbitrary* irregular decimation mechanisms.

Table 1. Parameters of LILI-II, DECIM^{v2} and DECIM-128 that are relevant for our structural cryptanalysis

cipher	key size	clock control LFSR	data generation LFSR	best linear approximation
LILI-II	128-bit	128-bit	127-bit	0.513671875
DECIM^{v2}	80-bit	-	192-bit	0.5078125
DECIM-128	128-bit	-	288-bit	0.5078125

In irregular decimation, there is an input bit stream $U = \{u_i\}_{i\geq 0}$ and a selection function or rule $\mathcal{D} : i \rightarrow d_i$, unknown to the attacker, which defines a nonnegative integer sequence. By applying $\mathcal{D}$ to $\{u_i\}$, another stream $Z = \{z_i\}_{i\geq 0}$ with $z_i = u_{\sum_{t=0}^{i} d_t}$ is obtained. In practice, the nonnegative integer sequence $\{d_i\}_{i\geq 0}$ may be dependent on $\{u_i\}$, as in the ABSG case.

The starting point of our analysis is the following observation, i.e., any irregular decimation mechanism can be converted into a shrinking-like mechanism. More precisely, given an irregular decimation mechanism, we can construct a bit stream $\{s_i\}_{i\geq 0}$ as follows: for each i, let $s_{\sum_{t=0}^{i} d_t} = 1$, otherwise let $s_j = 0$. If we regard $\{u_i\}_{i\geq 0}$ as the data source bits, then we have a unified shrinking-like representation of different irregular decimation mechanisms. Please see the following example.

Example 1. Consider $U = \{0,1,0,1,1,1,0,0,0,0,0,1,0,0,1,0,0,0,0,1,1,0,1,0,$ $1,1,1,0\}$. If the ABSG algorithm is applied, we have

$$\underbrace{0,1,0,}_{1}\,\underbrace{1,1,}_{1}\,\underbrace{1,0,0,0,0,0,1,}_{0}\,\underbrace{0,0,}_{0}\,\underbrace{1,0,0,0,0,1,}_{0}\,\underbrace{1,0,}_{0}\,\underbrace{1,0,1,1,1,0.}_{1} \quad (1)$$

The output is $Z = \{1,1,0,0,0,0,1\}$ and the corresponding $\{d_i\}_{i\geq 0}$ is $\{1,3,2,7,2,$ $6,3\}$. If we construct a stream $\{s_i\}_{i\geq 0}$ as $\{0,1,0,0,1,0,1,0,0,0,0,0,0,1,0,1,0,0,$ $0,0,0,1,0,0,1,0,0,0\}$, then the shrinking-like representation is

$$\begin{cases} s_i : 0,1,0,0,1,0,1,0,0,0,0,0,0,1,0,1,0,0,0,0,0,1,0,0,1,0,0,0 \\ u_i : 0,1,0,1,1,1,0,0,0,0,0,1,0,0,1,0,0,0,0,1,1,0,1,0,1,1,1,0 \end{cases}$$

It can be easily checked that the output of the shrinking process is $\{1,1,0,0,0,$ $0,1\}$. □

The advantage of the above viewpoint is that we need *not* know the actual values of $\{s_i\}$ when determining the data source bits. What we really need is the probabilistic distribution of $\{s_i\}$, which can be determined according to the irregular decimation mechanism by experiments or by theoretical analysis. In the following, we always consider the shrinking-like representation and assume that the distribution of $\{s_i\}$ is known.

Denote the input to the irregular decimation by $\{u_i\}$ and the output keystream by $\{z_j\}$. It follows that $\{u_i\}$ is the data stream in the shrinking-like representation, but we do not know the corresponding $\{s_i\}$. Let $p_1 = P(s_i = 1)$ and

$p_0 = P(s_i = 0)$, then the index interval in $\{z_j\}$ that u_i probably falls into can be determined, i.e., if u_i $(i \geq 1)$ (without loss of generality, we assume that $s_0 = 1$ corresponds to u_0) was selected into the keystream, then we have $u_i = z_{\sum_{t=0}^{i-1} s_t}$.

By the central limit theorem, we have $\frac{\sum_{t=0}^{i-1} s_t - i \cdot p_1}{\sqrt{i \cdot p_0 p_1}} \rightarrow \mathcal{N}(0,1)$, where $\mathcal{N}(0,1)$ is the standard normal distribution. Thus, the probability that the index of u_i in the keystream belongs to the interval $I_\alpha = [ip_1 - \alpha \cdot \sqrt{i \cdot p_1 p_0}, ip_1 + \alpha \cdot \sqrt{i \cdot p_1 p_0}]$ is

$$P_{I_\alpha} = P(\sum_{t=0}^{i-1} s_t \in I_\alpha) = \frac{1}{\sqrt{2\pi}} \int_{-\alpha}^{\alpha} e^{-\frac{x^2}{2}} dx.$$

Let n_i be the closest integer to $\frac{2\alpha\sqrt{ip_1 p_0} - 1}{2}$, then there are approximately $2n_i + 1$ possible indices in the interval I_α. Thus, we can count the number of times that 0 and 1 appear in I_α and make a majority poll to construct a prediction stream $\{\tilde{u}_i\}$ satisfying $P(\tilde{u}_i = u_i) > 0.5$. More precisely, let $N_j = \#\{z_i | z_i = j, i \in I_\alpha\}$ for $j \in \{0,1\}$ and denote by $\bar{b}$ the complement of a bit b, then if $N_j > N_{\bar{j}}$, let $\tilde{u}_i = j$, otherwise let $\tilde{u}_i = \bar{j}$. Thus, we have a predicted stream $\{\tilde{u}_i\}$ of $\{u_i\}$ satisfying the following theorem, which is a generalization of the results in [12, 26].

Theorem 1. *For any irregular decimation mechanism, there always exists a correlation between $\{\tilde{u}_i\}$ and $\{u_i\}$, which is given by*

$$P(\tilde{u}_i = u_i) = 0.5 + \frac{p_1 P_{I_\alpha}}{2^{2n_i+1}} \binom{2n_i}{n_i}. \tag{2}$$

Theorem 1, proved in Appendix A, depicts the basic weakness in any irregular decimation mechanism. It is worth noting that in (2), the only a priori knowledge is the distribution of $\{s_i\}$, other elements such as P_{I_α} and n_i are determined by the choice of α. Besides, from (2), we have $0.5 < P(\tilde{u}_i = u_i) \leq 0.5 + \frac{p_1}{2}$ for any $i \geq 0$, where the upper bound is achieved when $i = 0$. This fact means that the best correlation we can get is determined by the concrete irregular decimation algorithm itself. On the other hand, Theorem 1 shows that $P(\tilde{u}_i = u_i)$ decreases with i increasing. This is also the bottleneck of previous methods in [12, 16, 26]. To get the best correlation $\max_\alpha P(\tilde{u}_i = u_i)$, we can pre-compute the optimal values of α for each i. Table 2 lists the *average* correlations, $\sum_{j=0}^{i-1} P(\tilde{u}_j = u_j)/i$, we get for some i using the optimal values of α corresponding to the two decimation methods in LILI and DECIM. We made experiments to verify the theoretical values in Table 2. In LILI-II, when $i = 10000$, we get the average correlation of 0.518900 for 2^{20} randomly chosen initial states of LFSR c and LFSR d, which is even better than the theoretical estimate.

Table 2. The average biases for some i using the optimum values of α

i	1000	2000	10000	20000	40000	140000
LILI	0.0318732	0.0253801	0.0158841	0.0131732	0.010976	0.00794848
ABSG	0.0273437	0.0217052	0.0135262	0.0112084	0.00933383	0.00675565

Remarks. Theorem 1 is achieved based on the premise that each bit in I_α has the same probability of being correct. In fact, the probability that $u_i = z_r$ is $P(u_i = z_r) = \binom{i}{r}p_1^{r+1}p_0^{i-r}$, which are not the same for different r. Thus, at least in theory, for a given index interval I_α, it is better to use the following measure to make a decision: $\Delta_i = \sum_{z_r \in I_\alpha} P(z_r = u_i)(1 - z_r) - \sum_{z_r \in I_\alpha} P(z_r = u_i)z_r = \sum_{z_r \in I_\alpha} \binom{i}{r}p_1^{r+1}p_0^{i-r}(1-z_r) - \sum_{z_r \in I_\alpha} \binom{i}{r}p_1^{r+1}p_0^{i-r}z_r$. If $\Delta_i > 0$, let $\tilde{u}_i = 0$; otherwise let $\tilde{u}_i = 1$. However, in this case, we have to compute a large number of binomial coefficients for constructing $\{\tilde{u}_i\}$. To our knowledge, the most efficient method for computing binomial coefficient is by the recursion $\binom{i}{r} = \binom{i-1}{r-1} + \binom{i-1}{r}$, which can be computed in about i^2 time. While in Theorem 1, the time for making a decision is at most *linear* in i. Besides, the probability that we made a correct decision, $P(\tilde{u}_i = u_i)$, under the theoretical method is not significantly higher than that in Theorem 1. Our experimental results reveal that the average difference between the two probabilities under the two methods are very small. For example, the gain by the latter method becomes lower than 10^{-4} for $i \geq 2^{16}$. Therefore, we adopt the method in Theorem 1.

Next, we take a closer look at the method in Theorem 1. For each $\tilde{u}_i$, the validity of the majority poll heavily depends on the distribution of 0 and 1 in the index interval I_α. Denote by $n_{i,0}$ and $n_{i,1}$ the number of 0's and 1's in I_α, respectively. Then it is easy to see that the higher the absolute value $|n_{i,0} - n_{i,1}|$, the higher the prediction reliability for $\tilde{u}_i$. This motivates us to make a prediction decision only at those positions where $|n_{i,0} - n_{i,1}|$ is larger than a threshold value θ and otherwise we ignore that position. That is, let

$$\begin{cases} \breve{u}_i = 0, & \text{if } n_{i,0} - n_{i,1} \geq \theta \\ \breve{u}_i = 1, & \text{if } n_{i,1} - n_{i,0} \geq \theta. \end{cases} \tag{3}$$

In this way, we will get a prediction stream $\{\breve{u}_i\}$ at non-consecutive positions, e.g., $(\breve{u}_0, *, *, \cdots, *, \cdots, \breve{u}_i, *, \breve{u}_{i+2}, *, \cdots)$. The following theorem shows that the correlation between $\{\breve{u}_i\}$ and $\{u_i\}$ can be guaranteed.

Theorem 2. *For any irregular decimation mechanism, if we construct $\{\breve{u}_i\}$ following (3), then the correlation between $\{\breve{u}_i\}$ and $\{u_i\}$ for positions i satisfying $|n_{i,1} - n_{i,0}| \geq \theta$ is lower bounded by*

$$P(\breve{u}_i = u_i) \geq 0.5 + P_{I_\alpha} \cdot \frac{\theta}{2(2n_i + 1)}. \tag{4}$$

Theorem 2, proved in Appendix B, shows that we can make $P(\breve{u}_i = u_i)$ greater than or equal to a value determined by both α and θ. Though we can set a large value of θ and a large size of I_α to get a good enough correlation, the probability that we can find such a segment in the keystream is low. There is a tradeoff between $P(\breve{u}_i = u_i)$ and the number of points that having this correlation. This tradeoff is important for the application of Theorem 2.

For simplicity, we denote the prediction stream by $\{\tilde{u}_i\}$ hereafter. In practice, we first set a pointer value i_T and for $i > i_T$, we use (3) to construct $\{\tilde{u}_i\}$; while for $i \leq i_T$, we use Theorem 1 to construct $\{\tilde{u}_i\}$. Let $\lceil x \rceil$ be the smallest integer greater than or equal to x and $\lfloor x \rfloor$ be the biggest integer less than or equal to x. For $i \geq i_T$, let $\mathcal{Z}_i = (z_{t_i}, z_{t_i+1}, \cdots, z_{t_i+2n_i})$ with $t_i = \lfloor ip_1 - \alpha\sqrt{ip_1p_0} \rfloor$. In (3), we actually search for such $\mathcal{Z}_i$ that

$$
W_H(\mathcal{Z}_i) \geq \lceil \frac{2n_i+1+\theta}{2} \rceil \text{ or } W_H(\mathcal{Z}_i) \leq \lfloor \frac{2n_i+1-\theta}{2} \rfloor \tag{5}
$$

holds, where $W_H(\cdot)$ is the hamming weight of the corresponding vector. Let $A_i = \{\mathcal{Z}_i | W_H(\mathcal{Z}_i) \geq \lceil \frac{2n_i+1+\theta}{2} \rceil \text{ or } W_H(\mathcal{Z}_i) \leq \lfloor \frac{2n_i+1-\theta}{2} \rfloor\}$. Then the probability that (5) occurs around the position i is the proportion between the cardinality of $|A_i|$ and all the possible values of $\mathcal{Z}_i$, which is given approximately by

$$
P_i = \frac{2 \cdot \sum_{j=0}^{\lfloor \frac{2n_i+1-\theta}{2} \rfloor} \binom{2n_i+1}{j}}{2^{2n_i+1}} \rightarrow 2 \int_{-\frac{\mu_i}{\sigma_i}}^{\frac{\lfloor \frac{2n_i+1-\theta}{2} \rfloor - \mu_i}{\sigma_i}} \frac{1}{\sqrt{2\pi}} e^{-\frac{x^2}{2}} dx, \tag{6}
$$

where $\mu_i = n_i + 0.5$ and $\sigma_i = \frac{\sqrt{2n_i+1}}{2}$. Since P_j is very close to P_i if j is close to i, for simplicity, we use $P_{i+\lfloor \frac{i}{2} \rfloor}$ as the average value of P_i in the interval $(i, 2i]$. Accordingly, let N_i be the number of points satisfying (3) in $(i, 2i]$ and we choose the same value of α for the points in $(i, 2i]$, then we have $i \cdot P_{i+\lfloor \frac{i}{2} \rfloor} \geq N_i \Rightarrow P_{i+\lfloor \frac{i}{2} \rfloor} \geq \frac{N_i}{i}$. This equation indicates that by properly choosing θ, there are at least N_i positions constructed in $(i, 2i]$ satisfying (3). If $\{z_i\}$ is of length N_z, then by dividing the whole index interval $(i_T, N_z]$ into several segments $(i_T, 2i_T] \cup (2i_T, 4i_T] \cup \cdots \cup (2^{q-1}i_T, 2^q i_T]$ with $q = \lfloor \log_2(N_z - \alpha\sqrt{N_z p_1 p_0}) \rfloor$, a prediction stream $\{\tilde{u}_i\}_{i \geq i_T}$ is constructed segment by segment. In each segment, the average correlation is determined by Theorem 3, proved in Appendix C.

Theorem 3. *In $(i, 2i]$, the average correlation $P(\tilde{u}_i = u_i)|_{(i,2i]}$ can be approximated by $0.5 + P_{I_\alpha} \cdot \frac{\theta}{4}$, where $\theta = \sqrt{2n_i+1} \cdot \beta$ with $P_i = 2\int_{-\infty}^{\beta} \frac{1}{\sqrt{2\pi}} e^{-\frac{x^2}{2}} dx$ specified in (6).*

The validity of Theorem 3 is illustrated in Table 3-6. For the SG, LILI-II, DECIM^{v2} and DECIM-128, we give the theoretical estimates given by Theorem 3 and the corresponding experimental results. For each bias ε in Table 3-6 and the corresponding N_i, we verified the result by randomly assigning the initial state of the cipher 2^{20} times. We see that the simulation results are very close to the theoretical values.

As we know, the correlations given in Table 3-6 are the largest correlations reported for the four stream ciphers for the corresponding keystream length. For

Fig. 2. Constructing $\{\tilde{u}_i\}$ in two directions

Table 3. The theoretical parameters used for constructing $\{\tilde{u}_i\}$ in SG and the corresponding experimental values

interval	α	θ	$P_{i+\lfloor\frac{i}{2}\rfloor}$	ε(Th 3)	ε(found)	N_i(Th 3)	N_i(found)
$(0, 2^{11}]$	1.35	11	2^{-3}	0.0492344	0.059050	2^8	$2^{7.87}$
$(2^{11}, 2^{12}]$	1.35	14	2^{-3}	0.0384059	0.060397	2^8	$2^{7.71}$
$(2^{12}, 2^{13}]$	1.35	20	2^{-4}	0.0384572	0.060687	2^8	$2^{7.60}$
$(2^{13}, 2^{14}]$	1.36	27	2^{-5}	0.0369315	0.055226	2^8	$2^{8.06}$
$(2^{14}, 2^{15}]$	1.345	36	2^{-6}	0.035035	0.054911	2^8	$2^{7.53}$
$(2^{15}, 2^{16}]$	1.36	43	2^{-6}	0.0293113	0.037803	2^9	$2^{8.95}$
$(2^{16}, 2^{17}]$	1.37	51	2^{-6}	0.0245331	0.028811	2^{10}	$2^{9.95}$
$(2^{17}, 2^{18}]$	1.42	61	2^{-6}	0.0205909	0.024350	2^{11}	$2^{11.12}$
$(2^{18}, 2^{19}]$	1.43	73	2^{-6}	0.0172385	0.020377	2^{12}	$2^{12.06}$
$(2^{19}, 2^{20}]$	1.42	87	2^{-6}	0.014522	0.017915	2^{13}	$2^{12.94}$
$(2^{20}, 2^{21}]$	1.42	103	2^{-6}	0.0122084	0.013745	2^{14}	$2^{14.03}$
$(2^{20}, 2^{21}]$	1.42	122	2^{-8}	0.0144604	0.015985	2^{12}	$2^{12.01}$
$(2^{21}, 2^{22}]$	1.39	133	2^{-7}	0.0112604	0.012340	2^{14}	$2^{13.99}$

Table 4. The theoretical parameters used for constructing $\{\tilde{u}_i\}$ in LILI-II and the corresponding experimental values with $\alpha = 1.41$

interval	θ	$P_{i+\lfloor\frac{i}{2}\rfloor}$	ε(Th 3)	ε(found)	N_i(Th 3)	N_i(found)
$(0, 2^{11}]$	11	2^{-3}	0.0514225	0.0541849	2^8	$2^{7.95}$
$(2^{11}, 2^{12}]$	17	2^{-4}	0.0464443	0.0550815	2^7	$2^{7.01}$
$(2^{12}, 2^{13}]$	23	2^{-5}	0.044389	0.0557402	2^7	$2^{7.01}$
$(2^{13}, 2^{14}]$	31	2^{-6}	0.042073	0.045829	2^7	$2^{6.86}$
$(2^{14}, 2^{15}]$	40	2^{-7}	0.038777	0.045625	2^7	$2^{7.14}$
$(2^{15}, 2^{16}]$	51	2^{-8}	0.0349467	0.037282	2^7	$2^{7.49}$
$(2^{16}, 2^{17}]$	65	2^{-9}	0.0314334	0.042365	2^7	$2^{7.19}$
$(2^{17}, 2^{18}]$	82	2^{-10}	0.0281402	0.038911	2^7	$2^{7.12}$

Table 5. The theoretical parameters used for constructing $\{\tilde{u}_i\}$ in DECIM^{v2} and the corresponding experimental values with $\alpha = 1.41$

interval	θ	$P_{i+\lfloor\frac{i}{2}\rfloor}$	ε(Th 3)	ε(found)	N_i(Th 3)	N_i(found)
$(0, 2^{12}]$	12	2^{-3}	0.0413833	0.042008	2^9	$2^{8.78}$
$(2^{12}, 2^{13}]$	16	2^{-3}	0.0320557	0.036443	2^9	$2^{8.89}$
$(2^{13}, 2^{14}]$	19	2^{-3}	0.0268251	0.026698	2^{10}	$2^{10.11}$
$(2^{14}, 2^{15}]$	23	2^{-3}	0.0231502	0.021909	2^{11}	$2^{11.01}$
$(2^{15}, 2^{16}]$	27	2^{-3}	0.0192538	0.037282	2^{12}	$2^{12.05}$
$(2^{16}, 2^{17}]$	32	2^{-3}	0.0161431	0.015987	2^{13}	$2^{12.89}$
$(2^{17}, 2^{18}]$	38	2^{-3}	0.013526	0.013160	2^{14}	$2^{13.90}$
$(2^{18}, 2^{19}]$	45	2^{-3}	0.0113371	0.011347	2^{15}	$2^{15.02}$

the SG, our results are at least 3.6 times larger than those reported in [16]. These large correlations remove the barrier (that the correlation decrease quickly with the keystream length increasing) identified by previous research [16, 26] to a

Table 6. The theoretical parameters used for constructing $\{\tilde{u}_i\}$ in DECIM-128 and the corresponding experimental values with $\alpha = 1.41$

interval	θ	$P_{i+\lfloor \frac{i}{2} \rfloor}$	ε(Th 3)	ε(found)	N_i(Th 3)	N_i(found)
$(2^{11}, 2^{12}]$	14	2^{-3}	0.0392682	0.042436	2^8	$2^{7.69}$
$(2^{12}, 2^{13}]$	16	2^{-3}	0.0320557	0.034141	2^9	$2^{8.89}$
$(2^{13}, 2^{14}]$	27	2^{-5}	0.0381199	0.048058	2^8	$2^{7.93}$
$(2^{14}, 2^{15}]$	35	2^{-6}	0.0352286	0.054168	2^8	$2^{8.12}$
$(2^{15}, 2^{16}]$	46	2^{-6}	0.0328027	0.047282	2^8	$2^{7.76}$
$(2^{16}, 2^{17}]$	55	2^{-7}	0.027746	0.031999	2^9	$2^{9.29}$
$(2^{17}, 2^{18}]$	65	2^{-7}	0.0231366	0.026944	2^{10}	$2^{10.137}$

large extent. Finally, we also need the following theorem to actually construct $\{\tilde{u}_i\}$ in our attack.

Theorem 4. *If we choose an intermediate time point to be 0 and construct a prediction stream $\{\tilde{u}_i\}$ in two directions, as shown in the Figure 2. Then $\{\tilde{u}_i\}$ we get will be of double length, while the average correlation will be the same value as if only half of the keystream (one direction) is employed.*

Theorem 4 is just an observation which can be verified easily. Note that the results in Table 3-6 are only for one direction.

4   Applications

In this section, we use the theoretical results in Section 3 to launch new attacks against the four target ciphers. We need the following notations of the decoding algorithm in [27] involved in our attack.

- L is the length of the involved LFSR.
- k $(k < L)$ is the number of initial state bits to be determined first.
- t is the weight of the parity-checks.
- n is the number of the coefficient patterns appearing in all the parity-checks.

4.1   The Shrinking Generator

For the SG with Krawczyk's parameters, we use the correlations specified in Table 3 to mount a correlation attack as follows. Assume we know the feedback polynomial of the data LFSR. We have $(2^{7.87} + 2^{7.71} + 2^{7.60} + 2^{8.06} + 2^{7.53} + 2^{8.95} + 2^{9.95} + 2^{11.12} + 2^{12.06} + 2^{12.94} + 2^{12.01}) \cdot 2 = 42226$ nonconsecutive bits of $\tilde{u}_i$ associated with u_i by the relation $P(\tilde{u}_i = u_i) = (2^{7.87} \cdot 0.559050 + 2^{7.71} \cdot 0.560397 + 2^{7.60} \cdot 0.560687 + 2^{8.06} \cdot 0.555226 + 2^{7.53} \cdot 0.554911 + 2^{8.95} \cdot 0.537803 + 2^{9.95} \cdot 0.528811 + 2^{11.12} \cdot 0.524350 + 2^{12.06} \cdot 0.520377 + 2^{12.94} \cdot 0.517915 + 2^{12.01} \cdot 0.515985)/(42226) = 0.521744$. In our case, $L = 61$ and we choose the following decoding parameters: $k = 25$, $n = 10$ and $t = 4$. We construct $10 \cdot \frac{\binom{42226}{4}}{2^{61-25}} \approx 2^{24.21}$ parity-checks with a time complexity of $42226^2 = 2^{30.74}$ operations [20, 27]. Thus, according to [27], the time and memory complexities for recovering the internal sate of the data LFSR with a success rate 93% is $(2^{25} \cdot 25 + \frac{\binom{42226}{4}}{2^{61}-2^5} \cdot 29) \cdot 10 \approx 2^{33.1}$ operations and $2^{25} \cdot 32 + 10 \cdot \frac{\binom{42226}{4}}{2^{61}-2^5} \cdot (4 \cdot \log_2 42226 + 61) \approx 2^{31.7}$-bit. From Table 3, the data complexity is $2 \cdot 2^{21} + 2 \cdot 1.42 \cdot \sqrt{2^{21} \cdot 0.25} = 2^{22.1}$ bits. Note that the attack given in [27] has underestimated the pre-computation complexity, i.e., the true complexity is $2^{39.9}$ operations instead of $2^{33.9}$ operations according to [20]. Thus our attack is the best known attack against the SG with Krawczyk's parameters and shows that it is practically insecure. We have implemented the attack in C on a computer running under linux. On average, it takes tens of minutes to recover the state of the data LFSR. This is the first reported experimental result on the SG with Krawczyk's parameters.

4.2   LILI-II

First note that it is equivalent for the keystream generation if we put the irregular clocking after filtering the regularly clocked LFSR d output. Thus, we can combine $\{\tilde{u}_i\}$ with the linear approximations of f_d to get linear approximations on the initial state bits of the LFSR d. First look at the clock control subsystem. We have $c_j = f_c(y_1, y_2) = 2y_1 + y_2 + 1$, where y_1 and y_2 are taken from two stages of LFSR c. If we consider the shrinking-like representation in Section 3 and let $s_0 = 1$, $s_{\sum_{j=0}^{i-1} c_j} = 1$ for $i \geq 1$ and other s_i's equal to 0, then the distribution of $\{s_i\}$ is $P(s_i = 1) = \frac{1}{\sum_{j=1}^{4} \frac{1}{4}j} = 0.4$ and $P(s_i = 0) = 0.6$ from the definition of f_c.

We have computed the Walsh transform of f_d in LILI-II, which is listed in Table 7. The Walsh transform values marked with a star are used in our attack, which give $W = 236 + 552 + 494 + 364 + 100 + 272 + 78 + 384 = 2480$ linear approximations for each bit of $\tilde{u}_i$. We first restore the target state of LFSR d, then the corresponding state of LFSR c can be recovered easily, e.g., using the method from [10]. From Table 7, the average correlation between the regularly clocked output of f_d and its input is $(\frac{3 \cdot 236}{256} + \frac{13 \cdot 100}{1024} + \frac{1 \cdot 552}{128} + \frac{5 \cdot 494}{512} + \frac{11 \cdot 272}{1024} + \frac{7 \cdot 78}{512} + \frac{9 \cdot 364}{1024} + \frac{7 \cdot 384}{1024})/W = 0.50926789$. For $\tilde{u}_i$, we set $i_T = 0$ and choose $\alpha = 1.41$ which corresponds to $P_{I_\alpha} = 0.84146$ for all the positions in our estimates. From Table 4 and 8, the data complexity is $2 \cdot 2^{23} + 2 \cdot 1.41 \cdot \sqrt{2^{23} \cdot 0.4 \cdot 0.6} = 2^{24.1}$ bits. We divide the index segment $(0, 2^{23}]$ into 13 intervals. The corresponding parameters are listed in Table 4 and 8. From Table 4 and 8, there are $2 \cdot 2^7 \cdot 14 = 3584$ bits of $\tilde{u}_i$ constructed and the average correlation between $\{\tilde{u}_i\}$ and $\{u_i\}$ is $0.5 + 2^8 \cdot 0.0514225 + 2^7 \cdot (0.0464443 + 0.042073 + 0.038777 + 0.0349467 + 0.0314334 +$

Table 7. The Walsh spectrum of the function f_d in LILI-II

Value	Number	Value	Number	Value	Number
0	280	$\frac{13}{512}$ *	100	$\frac{11}{512}$ *	272
$\frac{3}{128}$ *	236	$\frac{5}{512}$	256	$\frac{7}{256}$ *	78
$\frac{1}{512}$	28	$\frac{3}{512}$	132	$\frac{7}{512}$ *	384
$\frac{3}{256}$	482	$\frac{1}{64}$ *	552	$\frac{1}{128}$	276
$\frac{1}{256}$	162	$\frac{5}{256}$ *	494	$\frac{9}{512}$ *	364

Table 8. The parameters for constructing $\{\tilde{u}_i\}$ in LILI-II for longer keystream length

interval	θ	ε_2	$P_{i+\lfloor \frac{i}{2} \rfloor}$	N_i
$(2^{18}, 2^{19}]$	103	0.0249915	2^{-11}	2^7
$(2^{19}, 2^{20}]$	129	0.0221166	2^{-12}	2^7
$(2^{20}, 2^{21}]$	160	0.0194221	2^{-13}	2^7
$(2^{21}, 2^{22}]$	199	0.0170798	2^{-14}	2^7
$(2^{22}, 2^{23}]$	246	0.0149264	2^{-15}	2^7

$0.0281402 + 0.0249915 + 0.0221166 + 0.0194221 + 0.0170798 + 0.0149264) / (2^7 \cdot 14) = 0.5333989$. The folded noise in the final linear approximation is $0.5 + 2 \cdot 0.0333989 \cdot 0.00926789 = 0.500619$. The decoding parameters for LILI-II are $t = 6$, $k = 60$ and $n = 16$. We construct $16 \cdot \frac{\binom{2 \cdot 14 \cdot 2^7 \cdot W}{6}}{2^{127-60}} \approx 2^{66.1}$ parity-checks with a time complexity of $(2 \cdot 14 \cdot 2^7 \cdot W)^3 = 2^{69.3}$ operations. Thus, according to [27], the time and memory complexities for recovering the internal sate of LFSR d with a success rate 93.44% is $(2^{60} \cdot 60 + \frac{\binom{2 \cdot 14 \cdot 2^7 \cdot W}{6}}{2^{127-60}} \cdot 64) \cdot 16 \approx 2^{72.32}$ operations and $2^{60} \cdot 32 + 16 \cdot \frac{\binom{2 \cdot 14 \cdot 2^7 \cdot W}{6}}{2^{127-60}} \cdot (6 \cdot \log_2(2 \cdot 14 \cdot 2^7 \cdot W) + 127) \approx 2^{74.1}$-bit. The total time complexity of our attack is $2^{72.32} + 2^{69.3} \approx 2^{72.5}$ operations, which is much faster than the exhaustive search for the 128-bit key. Compared to the distinguishing attack in [11], our attack is a *state recovery* attack with the 2^{79} times smaller data complexity and the 2^{31} times smaller time complexity. We have implemented the attack on a reduced version of LILI-II, (LFSR d is of 40-bit), on the same computer as in the SG case. It takes few minutes to recover the state of the 40-bit LFSR d with $2^{16.1}$-bit keystream.

4.3 DECIM^{v2} and DECIM-128

The attack routine is the same as that in LILI-II. To construct $\{s_i\}$ for the ABSG mechanism, we let the bit pair corresponding to $(\bar{u}, \bar{u})$ be (10) and let the bit string corresponding to $(\bar{u}, u^i, \bar{u})$ be $(010 \cdots 0)$ of length $i + 2$. Note that we have more choices other than $(010 \cdots 0)$ here, e.g., we can let it be $(001 \cdots 0)$. It has no effect on the output keystream. For simplicity, we let $(\bar{u}, u^i, \bar{u})$ be $(010 \cdots 0)$. Since $\frac{1}{\sum_{i=1}^{+\infty} \frac{(1+i)}{2^i}} = \frac{1}{3}$, we have $P(s_i = 1) = \frac{1}{3}$ and $P(s_i = 0) = \frac{2}{3}$ in the ABSG case. In DECIM, the Walsh spectrum of the filter function f has $W = 4096$ points having the value $\frac{1}{64}$, others are all 0.

We also set $i_T = 0$ and $\alpha = 1.41$. First consider the 160-bit reduced version of DECIM^{v2} without buffer. From Table 5 and 9, the data complexity is $2 \cdot 2^{34} + 2 \cdot 1.41 \cdot \sqrt{2^{34} \cdot \frac{1}{3} \cdot \frac{2}{3}} = 2^{35.1}$ bits. We divide the index segment $(0, 2^{34}]$ into 23 intervals. The corresponding parameters are listed in Table 5 and Table 9. From Table 5 and 9, there are $2 \cdot (2^9 + \sum_{i=9}^{21} 2^i + 2^{22} \cdot 8 + 2^{24}) = 54525952$ bits of $\tilde{u}_i$ constructed and the average correlation between $\{\tilde{u}_i\}$ and $\{u_i\}$, shown in Appendix D, is 0.50267488. The folded noise is $0.5 + 2 \cdot 0.0078125 \cdot 0.00267488 = 0.500042$. The decoding parameters are $t = 4$, $k = 66$ and $n = 18$. We construct

Table 9. The parameters for constructing $\{\tilde{u}_i\}$ in DECIM^{v2}

interval	θ	ε_2	$P_{i+\lfloor\frac{i}{2}\rfloor}$	N_i	interval	θ	ε_2	$P_{i+\lfloor\frac{i}{2}\rfloor}$	N_i
$(2^{19},2^{20}]$	53	0.0094566	2^{-3}	2^{16}	$(2^{27},2^{28}]$	296	0.00330107	2^{-5}	2^{22}
$(2^{20},2^{21}]$	63	0.0079407	2^{-3}	2^{17}	$(2^{28},2^{29}]$	470	0.00311483	2^{-6}	2^{22}
$(2^{21},2^{22}]$	75	0.00668815	2^{-3}	2^{18}	$(2^{29},2^{30}]$	517	0.00288294	2^{-7}	2^{22}
$(2^{22},2^{23}]$	89	0.00561395	2^{-3}	2^{19}	$(2^{30},2^{31}]$	667	0.00263001	2^{-8}	2^{22}
$(2^{23},2^{24}]$	106	0.00472731	2^{-3}	2^{20}	$(2^{31},2^{32}]$	851	0.00237274	2^{-9}	2^{22}
$(2^{24},2^{25}]$	126	0.00397451	2^{-3}	2^{21}	$(2^{32},2^{33}]$	1018	0.00212532	2^{-10}	2^{22}
$(2^{25},2^{26}]$	149	0.00332285	2^{-3}	2^{22}	$(2^{33},2^{34}]$	1034	0.00167847	2^{-9}	2^{24}
$(2^{26},2^{27}]$	216	0.00340647	2^{-4}	2^{22}					

Table 10. The parameters for constructing $\{\tilde{u}_i\}$ in DECIM-128

interval	θ	ε_2	$P_{i+\lfloor\frac{i}{2}\rfloor}$	N_i	interval	θ	ε_2	$P_{i+\lfloor\frac{i}{2}\rfloor}$	N_i
$(0,2^{27}]$	324	0.0062582	2^{-9}	2^{18}	$(2^{31},2^{32}]$	665	0.00185414	2^{-6}	2^{25}
$(2^{27},2^{28}]$	158	0.00176206	2^{-2}	2^{25}	$(2^{32},2^{33}]$	869	0.00171327	2^{-7}	2^{25}
$(2^{28},2^{29}]$	251	0.0019793	2^{-3}	2^{25}	$(2^{33},2^{34}]$	1121	0.00156276	2^{-8}	2^{25}
$(2^{29},2^{30}]$	362	0.00201862	2^{-4}	2^{25}	$(2^{34},2^{35}]$	1431	0.00141063	2^{-9}	2^{25}
$(2^{30},2^{31}]$	498	0.00196364	2^{-5}	2^{25}					

$18 \cdot \frac{\left(2 \cdot 2^{12} \cdot (2^9 + \sum_{i=9}^{21} 2^i + 2^{22} \cdot 8 + 2^{24})\right)}{2^{160-66}} \approx 2^{56.22}$ parity-checks with a time complexity of $(2 \cdot 2^{12} \cdot (2^9 + \sum_{i=9}^{21} 2^i + 2^{22} \cdot 8 + 2^{24})^2 = 2^{77.41}$ operations. Thus, according to [27], the time and memory complexities for recovering the internal sate of the reduced length LFSR with a success rate 92% is $(2^{66} \cdot 66 + 2^{77.41} \cdot 70) \cdot 18 \approx 2^{76.3}$ operations and $2^{66} \cdot 32 + 18 \cdot 2^{77.41} \cdot (8 \cdot \log_2(2^9 + \sum_{i=9}^{21} 2^i + 2^{22} \cdot 8 + 2^{26}) + 160) \approx 2^{71.3}$-bit.

For the 256-bit reduced version of DECIM-128 without buffer, we set $i_T = 0$ and $\alpha = 1.41$. The data complexity is $2 \cdot 2^{35} + 2 \cdot 1.41 \cdot \sqrt{2^{35} \cdot \frac{1}{3} \cdot \frac{2}{3}} = 2^{36.1}$ bits. We divide the index segment $(0, 2^{35}]$ into 9 intervals. The corresponding parameters are listed in Table 10. From Table 10, there are $2 \cdot (2^{25} \cdot 8 + 2^{18}) = 537395200$ bits of $\tilde{u}_i$ constructed and the average correlation between $\{\tilde{u}_i\}$ and $\{u_i\}$ is $0.5 + 0.0062582 \cdot 2^{18} + 0.00176206 \cdot 2^{25} + 0.0019793 \cdot 2^{25} + 0.00201862 \cdot 2^{25} + 0.00196364 \cdot 2^{25} + 0.00185414 \cdot 2^{25} + 0.00171327 \cdot 2^{25} + 0.00156276 \cdot 2^{25} + 0.00141063 \cdot 2^{25})/(2^{25} \cdot 8 + 2^{18}) = 0.50178741$. The folded noise in the final linear approximation is $0.5 + 2 \cdot 0.0078125 \cdot 0.00178741 = 0.500028$. The decoding parameters are $t = 6$, $k = 112$ and $n = 18$. We construct $18 \cdot \frac{\left(2 \cdot (2^{25} \cdot 8 + 2^{18})\right)}{2^{256-112}} \approx 2^{92.53}$ parity-checks with a time complexity of $(2 \cdot (2^{25} \cdot 8 + 2^{18}))^3 = 2^{123.1}$ operations. Thus, according to [27], the time and memory complexities for recovering the internal sate of the reduced length LFSR in DECIM-128 with a success rate 73.4% is $(2^{112} \cdot 112 + 2^{92.53} \cdot 118) \cdot 18 \approx 2^{122.98}$ operations and $2^{112} \cdot 32 + 18 \cdot 2^{92.53} \cdot (6 \cdot \log_2 2 \cdot (2^{25} \cdot 8 + 2^{18}) + 256) \approx 2^{117}$ -bit. We can extend the above attacks to the full versions of DECIM by guessing the left 32-bit of the state. The results and comparisons with the time/memory/tradeoffs are given in Appendix E. Our results on DECIM confirms the underlying design idea, but shows an interesting fact that the security of DECIM rely more on the length of the involved LFSR than on the ABSG algorithm. We implemented the attack on a reduced version of DECIM with 40-bit LFSR. It takes several minutes to restore the state of the LFSR with $2^{18.1}$-bit keystream.

5 Conclusions

We presented an improved correlation analysis of arbitrary irregular decimation mechanism and demonstrated the best known attacks on four well-known stream ciphers using irregular decimation. We believe that our correlation analysis can be used to mount efficient attacks against other stream ciphers using irregular decimation, e.g. the alternating step generator and the self-shrinking generator.

Acknowledgements. The author was with State Key Laboratory of Information Security, Institute of Software, Chinese Academy of Sciences, Beijing, 100190, China. This paper is supported by the key programm of the National Natural Science Foundation of China (Grant No. 60833008) and the general programm of the National Natural Science Foundation of China (Grant No. 60603018).

References

1. Babbage, S., De Cannière, C., Canteaut A., et al.: The eSTREAM portfolio, http://www.ecrypt.eu.org/stream/portfolio.pdf
2. Berbain, C., Billet, O., Canteaut, A., Courtois, N., et al.: DECIM^{v2}. In: Robshaw, M.J.B., Billet, O. (eds.) New Stream Cipher Designs. LNCS, vol. 4986, pp. 140–151. Springer, Heidelberg (2008)
3. Berbain, C., Gouget, A., Sibert, H.: Understanding Phase Shifting Equivalent Keys and Exhaustive Search, http://eprint.iacr.org/2008/169.ps.gz
4. Canteaut, A., Trabbia, M.: Improved Fast Correlation Attacks Using Parity-Check Equations of Weight 4 and 5. In: Preneel, B. (ed.) EUROCRYPT 2000. LNCS, vol. 1807, pp. 573–588. Springer, Heidelberg (2000)
5. Chose, P., Joux, A., Mitton, M.: Fast Correlation Attacks: An Algorithmic Point of View. In: Knudsen, L.R. (ed.) EUROCRYPT 2002. LNCS, vol. 2332, pp. 209–221. Springer, Heidelberg (2002)
6. Dawson, E., Clark, A., Golić, J., Fuller, J., et al.: The LILI-128 Keystream Generator. In: Stinson, D.R., Tavares, S. (eds.) SAC 2000. LNCS, vol. 2012, pp. 248–261. Springer, Heidelberg (2001)
7. Clark, A., Dawson, E., Fuller, J., Golić, J., et al.: The LILI-II Keystream Generator. In: Batten, L.M., Seberry, J. (eds.) ACISP 2002. LNCS, vol. 2384, pp. 25–39. Springer, Heidelberg (2002)
8. Coppersmith, D., Krawczyk, H., Mansour, Y.: The Shrinking Generator. In: Stinson, D.R. (ed.) CRYPTO 1993. LNCS, vol. 773, pp. 22–39. Springer, Heidelberg (1994)
9. Courtois, N.T., Meier, W.: Algebraic Attacks on Stream Ciphers with Linear Feedback. In: Biham, E. (ed.) EUROCRYPT 2003. LNCS, vol. 2656, pp. 345–359. Springer, Heidelberg (2003)

10. Courtois, N.T.: Fast Algebraic Attacks on Stream Ciphers with Linear Feedback. In: Boneh, D. (ed.) CRYPTO 2003. LNCS, vol. 2729, pp. 176–194. Springer, Heidelberg (2003)
11. Englund, H., Johansson, T.: A New Distinguisher for Clock Controlled Stream Ciphers. In: Gilbert, H., Handschuh, H. (eds.) FSE 2005. LNCS, vol. 3557, pp. 181–195. Springer, Heidelberg (2005)
12. Ekdahl, P., Johansson, T.: Predicting the Shrinking Generator with Fixed Connections. In: Biham, E. (ed.) EUROCRYPT 2003. LNCS, vol. 2656, pp. 330–344. Springer, Heidelberg (2003)
16. Golić, J.D., Mihaljević, M.j.: A Generalized Correlation Attack on a Class of Stream Ciphers Based on the Levenshtein Distance. Journal of Cryptology 3(3), 201–212 (1991)
14. Golić, J.D.: Embedding and Probabilistic Correlation Attacks on Clocked-Controlled Shift Registers. In: De Santis, A. (ed.) EUROCRYPT 1994. LNCS, vol. 950, pp. 230–243. Springer, Heidelberg (1995)
15. Golić, J.D.: Towards Fast Correlation Attacks on Irregularly Clocked Shift Registers. In: Guillou, L.C., Quisquater, J.-J. (eds.) EUROCRYPT 1995. LNCS, vol. 921, pp. 248–262. Springer, Heidelberg (1995)
16. Golić, J.D.: Correlation Analysis of the Shrinking Generator. In: Kilian, J. (ed.) CRYPTO 2001. LNCS, vol. 2139, pp. 440–457. Springer, Heidelberg (2001)
17. Gouget, A., Sibert, H., Berbain, C., Courtois, N.T., Debraize, B., Mitchell, C.: Analysis of the Bit-Search Generator and Sequence Compression Techniques. In: Gilbert, H., Handschuh, H. (eds.) FSE 2005. LNCS, vol. 3557, pp. 196–214. Springer, Heidelberg (2005)
18. Gouget, A., Sibert, H.: How to Strengthen Pseudo-Random Generators by Using Compression. In: Vaudenay, S. (ed.) EUROCRYPT 2006. LNCS, vol. 4004, pp. 129–146. Springer, Heidelberg (2006)
19. Krawczyk, H.: The Shrinking Generator: Some Practical Considerations. In: Preneel, B. (ed.) FSE 1994. LNCS, vol. 809, pp. 45–46. Springer, Heidelberg (1994)
20. Johansson, T., Jönsson, F.: Fast Correlation Attacks through Reconstruction of Linear Polynomials. In: Bellare, M. (ed.) CRYPTO 2000. LNCS, vol. 1880, pp. 300–315. Springer, Heidelberg (2000)
21. Krause, M.: BDD-Based Cryptanalysis of Keystream Generators. In: Knudsen, L.R. (ed.) EUROCRYPT 2002. LNCS, vol. 2332, pp. 222–237. Springer, Heidelberg (2002)
22. Pasalic, E.: Key Differentiation Attacks on Stream Ciphers, http://eprint.iacr.org/2008/443.pdf
23. Meier, W., Staffelbach, O.: Fast Correlation Attacks on Certain Stream Ciphers. Journal of Cryptology, 159–176 (1989)
24. Molland, H., Helleseth, T.: An Improved Correlation Attack Against Irregular Clocked and Filtered Keystream Generators. In: Franklin, M. (ed.) CRYPTO 2004. LNCS, vol. 3152, pp. 373–389. Springer, Heidelberg (2004)
25. Nakagami, H., Teramura, R., Ohigashi, T., Kuwakado, H.: A Chosen IV Attack Using Phase Shifting Equivalent Keys Against Decim^{v2}, http://eprint.iacr.org/2008/128.pdf
26. Zhang, B., Wu, H., Feng, D., Bao, F.: A Fast Correlation Attack on the Shrinking Generator. In: Menezes, A. (ed.) CT-RSA 2005. LNCS, vol. 3376, pp. 72–86. Springer, Heidelberg (2005)

27. Zhang, B., Feng, D.: An Improved Fast Correlation Attack on Stream Ciphers. In: Avanzi, R., Keliher, L., Sica, F. (eds.) SAC 2008. LNCS, vol. 5381, pp. 214–227. Springer, Heidelberg (2009)

A   Proof of Theorem 1

Proof. Note that an irregular decimation mechanism is characterized by the distribution of $\{s_i\}$ in the shrinking-like representation. From the shrinking-like representation and the majority poll, we have

$$P(\tilde{u}_i = u_i) = \sum_{j=0}^{1} P(s_i = j)P(\tilde{u}_i = u_i|s_i = j) = 0.5p_0 + p_1 P(\tilde{u}_i = u_i|s_i = 1) \tag{7}$$

$$= 0.5p_0 + p_1\{P(\sum_{j=0}^{i-1} s_j \notin I_\alpha|s_i = 1)P(\tilde{u}_i = u_i|\sum_{j=0}^{i-1} s_j \notin I_\alpha, s_i = 1)$$

$$+ P(\sum_{j=0}^{i-1} s_j \in I_\alpha|s_i = 1)P(\tilde{u}_i = u_i|\sum_{j=0}^{i-1} s_j \in I_\alpha, s_i = 1)\}$$

$$= 0.5p_0 + 0.5p_1(1 - P_{I_\alpha}) + p_1 P_{I_\alpha} P(\tilde{u}_i = u_i|\sum_{j=0}^{i-1} s_j \in I_\alpha, s_i = 1),$$

where

$$P(\tilde{u}_i = u_i|\sum_{j=0}^{i-1} s_j \in I_\alpha, s_i = 1) = \sum_{b=0}^{1} P(u_i = b)P(\tilde{u}_i = u_i|u_i = b, \sum_{j=0}^{i-1} s_j \in I_\alpha, s_i = 1) \tag{8}$$

$$= 2 \cdot 0.5 \sum_{j=n_i}^{2n_i} \binom{2n_i}{j}\frac{1}{2^{2n_i}} = \frac{1}{2} + \frac{1}{2^{2n_i+1}}\binom{2n_i}{n_i}.$$

Substituting (8) into (7), we have (2).   □

B   Proof of Theorem 2

Proof. First note that from (3) and $n_{i,0} + n_{i,1} = 2n_i + 1$, we have

$$\max(n_{i,0}, n_{i,1}) \geq \frac{2n_i + 1 + \theta}{2}$$

for positions satisfying $|n_{i,1} - n_{i,0}| \geq \theta$. Then according to the majority poll and the shrinking-like representation, we have

$$P(\breve{u}_i = u_i) = P(\breve{u}_i = u_i, \sum_{j=0}^{i-1} s_j \notin I_\alpha) + P(\breve{u}_i = u_i, \sum_{j=0}^{i-1} s_j \in I_\alpha) \tag{9}$$

$$= P(\sum_{j=0}^{i-1} s_j \notin I_\alpha)P(\breve{u}_i = u_i | \sum_{j=0}^{i-1} s_j \notin I_\alpha) + P(\sum_{j=0}^{i-1} s_j \in I_\alpha)P(\breve{u}_i = u_i | \sum_{j=0}^{i-1} s_j \in I_\alpha)$$

$$= \frac{1}{2} \cdot (1 - P_{I_\alpha}) + P_{I_2} \cdot P(\breve{u}_i = u_i | \sum_{j=0}^{i-1} s_j \in I_\alpha)$$

$$= \frac{1}{2} \cdot (1 - P_{I_\alpha}) + P_{I_2} \cdot \frac{\max(n_{i,0}, n_{i,1})}{2n_i + 1}$$

$$\geq \frac{1}{2} \cdot (1 - P_{I_\alpha}) + P_{I_2} \cdot \frac{\frac{2n_i + 1 + \theta}{2}}{2n_i + 1} = 0.5 + P_{I_\alpha} \cdot \frac{\theta}{2(2n_i + 1)}. \qquad \square$$

C   Proof of Theorem 3

Proof. From Theorem 2, in the interval $(i, 2i]$, we have

$$P(\tilde{u}_i = u_i)|_{(i,2i]} \geq \frac{\sum_{j \in (i,2i]}(0.5 + P_{I_\alpha} \cdot \frac{\theta}{2(2n_j + 1)})}{N_i} \tag{10}$$

$$\geq 0.5 + \frac{1}{N_i} \sum_{j \in (i,2i]} (P_{I_\alpha} \cdot \frac{\theta}{2(2n_j + 1)})$$

$$= 0.5 + P_{I_\alpha} \cdot \frac{\theta}{2} \sum_{j \in (i,2i]} (\frac{1}{N_i(2n_j + 1)}),$$

where $\sum_{j \in (i,2i]}$ only sums over the positions satisfying (3). The precise analysis of $\sum_{j \in (i,2i]}(\frac{1}{N_i(2n_j+1)})$ is complicated in theory. Instead, we use the numerical experiments to determine the value of this term. We made experiments to determine this value in the context of the shrinking generator, LILI-II, DECIM^{v2} and DECIM-128, it turns out that we can take this value as 0.5. This is illustrated by our experimental results, performed 2^{20} times for each ε and N_i, which are given in Table 3-6.

In addition, note that the probability in (6) can be rewritten as

$$2 \int_{-\frac{\mu_i}{\sigma_i}}^{\frac{\lfloor \frac{2n_i + 1 - \theta}{2} \rfloor - \mu_i}{\sigma_i}} \frac{1}{\sqrt{2\pi}} e^{-\frac{x^2}{2}} dx \approx 2 \int_{-\infty}^{\frac{\lfloor \frac{2n_i + 1 - \theta}{2} \rfloor - \mu_i}{\sigma_i}} \frac{1}{\sqrt{2\pi}} e^{-\frac{x^2}{2}} dx \tag{11}$$

for $n_i \geq 50$. This fact determines the choice of β. $\qquad \square$

D   Average Correlation in 160-Bit Reduced Version of DECIM^{v2}

The average correlation between $\{\tilde{u}_i\}$ and $\{u_i\}$ is $0.5 + (0.0413833 \cdot 2^9 + 0.0320557 \cdot 2^9 + 0.0268251 \cdot 2^{10} + 0.0231502 \cdot 2^{11} + 0.0192538 \cdot 2^{12} + 0.0161431 \cdot 2^{13} + 0.013526 \cdot 2^{14} + 0.0113371 \cdot 2^{15} + 0.0094566 \cdot 2^{16} + 0.0079407 \cdot 2^{17} + 0.00668815 \cdot 2^{18} + 0.00561395 \cdot 2^{19} + 0.00472731 \cdot 2^{20} + 0.00397451 \cdot 2^{21} + 0.00332285 \cdot 2^{22} + 0.00340647 \cdot 2^{22} + 0.00330107 \cdot 2^{22} + 0.00311483 \cdot 2^{22} + 0.00288294 \cdot 2^{22} + 0.00263001 \cdot 2^{22} + 0.00237274 \cdot 2^{22} + 0.00212532 \cdot 2^{22} + 0.00167847 \cdot 2^{24}) / (2^9 + \sum_{i=9}^{21} 2^i + 2^{22} \cdot 8 + 2^{24}) = 0.50267488$.

E   Attack Complexity on Full Length Versions of DECIM and Comparisons with TMD

Note that the complexities of the TMD attack are only rough estimates that ignore the logarithmic factors, while the complexities of our attack are much more accurate values. Our attack is at least 2^{35} times faster than the TMD attack in total complexity, while with much smaller data complexity.

Table 11. Attack complexity on full versions of DECIM and comparisons with time/memory/data tradeoff attack

cipher	attack	pre-computation	time	memory	data
DECIM^{v2}	Ours	$2^{77.41}$	$2^{108.3}$	$2^{71.3}$	$2^{35.1}$
	TMD	$O(2^{144})$	$O(2^{96})$	$O(2^{96})$	$O(2^{48})$
DECIM-128	Ours	$2^{123.1}$	$2^{154.98}$	2^{117}	$2^{36.1}$
	TMD	$O(2^{216})$	$O(2^{144})$	$O(2^{144})$	$O(2^{72})$

Author Index